AN IDIOT'S

FUGITIVE ESSAYS ON SCIENCE

EULER at his writing table, old copy of a painting by
EMMANUEL HANDMANN, 1753/1756.

C. TRUESDELL

AN IDIOT'S
FUGITIVE ESSAYS
ON SCIENCE

Methods, Criticism, Training, Circumstances

With 30 Illustrations

Springer-Verlag
New York Berlin Heidelberg Tokyo

C. Truesdell
The Johns Hopkins University
Baltimore, Maryland 21218
U.S.A.

Library of Congress Cataloging in Publication Data
Truesdell, C. (Clifford),
    An idiot's fugitive essays on science.
    Includes index.
    1. Science—History—Addresses, essays, lectures.
    2. Science—Philosophy—Addresses, essays, lectures.
    3. Mechanics—Addresses, essays, lectures.
    4. Scientists—Biography—Addresses, essays, lectures. I. Title.
    Q126.8.T78   1984   500   82-19221

Typeset by J. W. Arrowsmith Ltd., Bristol, England.
Printed and bound by Halliday Lithograph, West Hanover, Massachusetts.
Printed in the United States of America.

9 8 7 6 5 4 3 2 1

ISBN 0-387-90703-3 Springer-Verlag New York Berlin Heidelberg Tokyo
ISBN 3-540-90703-3 Springer-Verlag Berlin Heidelberg New York Tokyo

D. M.

ALICIAE FELDMAN WALKER

N. MDCCCLXXVI OB. MCMLXI

IN CORDE SVO SEP.

POST XXI A. MAERENS

M. PIVS NEPOS

POSVIT

PREFACE

When, after the agreeable fatigues of solicitation, Mrs Millamant set out a long bill of conditions subject to which she might by degrees dwindle into a wife, Mirabell offered in return the condition that he might not thereby be beyond measure enlarged into a husband. With age and experience in research come the twin dangers of dwindling into a philosopher of science while being enlarged into a dotard.

The philosophy of science, I believe, should not be the preserve of senile scientists and of teachers of philosophy who have themselves never so much as understood the contents of a textbook of theoretical physics, let alone done a bit of mathematical research or even enjoyed the confidence of a creating scientist.

On the latter count I run no risk: Any reader will see that I am untrained (though not altogether unread) in classroom philosophy. Of no ignorance of mine do I boast, indeed I regret it, but neither do I find this one ignorance fatal here, for few indeed of the great philosophers to explicate whose works hodiernal professors of philosophy destroy forests of pulp were themselves so broadly and specially trained as are their scholiasts. In attempt to palliate the former count I have chosen to collect works written over the past thirty years, some of them not published before, and I include only a few very recent essays. Thus, for the greater part, such stupidity as my populous college of critics will find in this book must be assigned to failing youthful rather than senile.

Most of my forays that might pass for philosophical were not in their origins so intended. My early research did not fall into any professional category. The only encouragement I received, and it was neither much nor steady, was from older men, all but one now dead: from my teachers, who were BATEMAN, NEMÉNYI, LEFSCHETZ, and BOCHNER, and from HADAMARD, VILLAT, BOULIGAND, SYNGE, HAMEL, PICONE, and FINZI. I doubt that any but NEMÉNYI had gone into what I was doing, and so perhaps what seemed to me encouragement from the rest was only the gentlemanly courtesy of an older generation. I take this occasion to express again my gratitude to NEMÉNYI, who taught me that mechanics was something deep and beautiful, beyond the ken of schools of "applied mathematics" and "applied mechanics" in the 1940s. Those men of my own age who did not simply ignore my work were prone to reject it, not for its contents and occasional errors in mathematics but on general grounds: It did not subserve to their rules of what a scientist ought to do. They found me guilty of felonious thought. I had resort to what I took as higher authority: great mathematicians and physicists of the past. My early historical studies sought mainly inspiration, comfort, and guidance from the classics. For myself, I found those in abundance; my detractors did not recognize the jurisdiction of that tribunal. My early prefaces were apologies for having put pen to paper.

The latter purpose I no longer find worth the effort, but to the former I ever more stoutly hold. That is why I now offer this volume to readers.

"Polemic" is only recently become a word of rebuke. If STOKES was content to describe some of his works as "merely controversial", I am not ashamed to call some of mine "polemic", though I hope they will not be found merely so. Open polemic, which should and sometimes does serve as curative surgery, may be noble; its modern successor, infights and backbiting by grinning gladiators who at best damage each other's reputations and fortunes, at worst damage the science they pretend to promote, can be nothing but base. "Polemic" and "critical" go together. This volume contains, accordingly, several of the some 700 reviews I have written, mainly between 1949 and 1971. I put much effort into them. They will be found out of style, for they are honest estimates drawn from scrupulous perusal and perpension of the work under review. Many of them contain matter I never published elsewhere. The few I have elected to reprint here are among those I think may offer something to the history and philosophy of science. Some of them present my spontaneous first, and perhaps not worst, expression of a thought later, sometimes many years later, at greater pains elaborated, substantiated, and delimited.

No reader should expect to find here a general survey upon history, method, or philosophy of science. My thought about them, like my mathematical researches, has always been eccentric. Some scattered days of servitude excepted, my circumstances have let me remain a seely child who piles up blocks in neat piles because he likes to see them so, an irresponsible *dilettante*, an ἰδιώτης, who follows RABELAIS' naughty counsel: fay ce que vouldras. I have sought a few pretty pebbles on the shore washed by the great ocean of beauty that mathematical science affords.

Natural beauty I like to see by natural light, even if overcast. I hope the strokes of my nocturnal quill reflect no tungstic or neoned glare, gnashing and pitiless successor of the "cheerful kerosene lamps and ever-shining gas burners" whose "hallmark" SALOMON BOCHNER read upon "many bulky scholarly works" of the nineteenth century. Like BALZAC's, may mine be seen "still very close to age-old candlelight, flickering as ever."

Part of the chronic pain of growing older comes from ever sharper cognisance not only of what you do not know but also of what you shall never, with your reduced strength of body and mind, be able to learn. What was not meant to be connex shall not be faulted for gaps, yet I rue the want of much with which the scrip here opened ought by now have been furnished. The shrivelled years coming on may grant me to gather some of that, but surely not all.

January 1, 1983   TRUESDELL
Il Palazzetto, Baltimore

Ex abundantia cordis os loquitur.

Matt. 12, 34

ACKNOWLEDGMENT

I have been helped by many people in many ways and over many years. Even from the most hostile I have often learnt something. In many instances now my progress toward the "dateless night" has left me unable to recall who told me what I remember having been told. Because any acknowledgment to persons will needs be unfairly incomplete, it were better I should omit everything of that kind. Nevertheless I must mention the help CHI-SING MAN has given me in discussions over the last few years, in respect not only to facts and reasoning but also to the literature of the philosophy and history of science.

For the original composition of many of these essays and for the extensive revision that their republication has required I am indebted, deeply indebted, to the partial support given me these past twenty years and more by the U.S. National Science Foundation through its programs in Applied Mathematics, Solid Mechanics, and History of Science.

NOTE REGARDING THE
REPRINTED WORKS

The reprinting is for the most part faithful to the last published text, but I have not hesitated to emend my errors and infelicities of language. Omissions that affect the meaning are indicated by ellipses; additions or changes other than mere limation, by square brackets. Most such parts of the text, whether quoted or my own, as were written in foreign languages, I have here translated into English.

Essays 13, 32bII (Review), 35, 37, and 39 have not been printed before. Essays 8, 10, 33cV, and 41 are so extensively rewritten and augmented as to be essentially new works.

CONTENTS

PART I. AIMS, PROGRAMS, AND METHODS

PART II. CRITICISM: SELECTED REVIEWS

A. WRITING AND TEXTS FOR LIVING SCIENCE

PART IV. TRAINING

PART V. PHILOSOPHY?

PART VI. DIRGE

PART I

AIMS, PROGRAMS, AND METHODS

1. EXPERIENCE, THEORY, AND EXPERIMENT (1955)

Science today is much like government or big business: Scientists are specialists not only in a single science but even in a single problem, each a senior bureaucrat jealous of interference from the multitude of others whose actions are in turn as isolated as his own, and the multiplied sciences themselves are self-fecundating compartments which reproduce, if at all, by division. Everyone is an expert in something. We are accustomed to speaking of every scientist as a leader in a particular field, but often it is difficult to discern any following. That so many experts turn out so much research that no single person can know it all even in a single field, is often brought forward as proving the progress of science. Rate of working, nevertheless, is the product of force by velocity and is not necessarily increased if velocity approaches infinity while force approaches zero. There are costly efforts to gather and review the totality of the literature in various fields, but it might be fitter to find out who, if anyone, reads the typical paper of today.

Fluid mechanics in its various aspects is divided among several types of engineers, a few physicists, and some mathematicians. Since first I began to study fluids, I have had to lose time listening to wrangles among members of these cults, each defending his own while condemning the others. The engineer has the right and duty of knowing fluids as they are met in life for man's direct harm or use; to him, the physicist sets up situations whose only value is their ease of study for physicists, while the mathematician is lost in abstraction and arid brain games. The physicist drills out the true principles of fluids, above both the mere detail and empiricism of the engineer and the purism of the mathematician, who for rigor is ever ready to gloss over essential physical aspects. The mathematician has the assurance of correctness and finality; for him the results of physicist and engineer are alike suspect, mere conjecture and ever subject to possible revision. These views are not without their truth in the cellular science of our day. Permit me to reset them in words less apt to reassure the engineers, physicists, and mathematicians in their respective complacencies. The engineer, blinded by the daily need for design or test of this or that device, will not pause to learn enough of the concepts of modern physics or the methods of modern mathematics to find out whether they can be applied to his problems. The physicist, blinded by the oversimplification and the raw guessing now in vogue, despises alike the phenomena which occur in natural, day-to-day situations and the logical standards of precise reasoning. The mathematician, blinded by a century of ever more abstract pure mathematics, has lost the skill and wish to read nature's book. Whether put as praise or put as blame, the foregoing argument, which each of us must endure at every meeting, is barren. I should like to lay before the community of those who study fluids a motion that we hear no more of it. For my part, I promise to try to speak to you not of one of the professions within fluid mechanics today, but of fluid mechanics itself.

Knowledge of fluids is gained through experience, theory, and experiment. Of these, the first and last are often confused. Experience is sometimes dismissed as the uncomprehending rules of thumb of mere artisans, while experiment is exalted as the foundation of science. The empiricism of some physicists of the last century has been embraced by many philosophers of science and educators of our day, particularly those associated with psychology and the biological and social sciences. Students in some of these doctrines are given instruction in the "scientific method", which is said to consist of controlled experiments and their statistical evaluation, while theory, if mentioned at all, is subsequent curve-fitting. Experience, the straight impress of nature that observant and rational man gains through his unaided senses as he daily encounters the world and which, by his nowadays all too slighted faculty of reason, he puts in verbal generality, is dismissed as primitive and below science. Many prefer to mine nature's darkest and deepest entrails, closed except to the dearest experimental apparatus or voluminous statistics, while leaving the smiling face of earth unheeded. That modern science is experimental science, would follow also from any poll of the scientists themselves: In the biological sciences, little else than experiment exists, while in the physical sciences experiments are often so elaborate that multitudes must be employed upon them. Such force of numbers need not be compelling. A poll of professional musicians would reveal that music today is neither hot jazz nor symphony but sugar stirred in soup. While scientists are more apt than musicians to idolize the means by which they must earn their daily leisure, nevertheless many experimenters will admit by their actions if not by frank confession that theory is the objective of science. The frustrating and so far vain struggles of biologists and social scientists to organize their subjects upon a basis of mathematical theory is apparent in every conversation among them, while it is the successful theories of the physical sciences that distinguish them from other human endeavors and sometimes cause their enthusiasts to put forward thinly disguised claims to the sole possession of knowledge.

The hydraulic engineer is favored in being alike in daily encounter with experience, theory, and experiment. His tasks and problems arise in common experience, which he dares not desert for more voluptuous realms deep hidden from human eye and touch. The flow of water he describes, understands, and controls in terms of the concepts of theory: velocity, pressure, and density, themselves mathematical ideas expressed in symbols and employed in equations. Either to check and correct the results of theory or to find answers to specific and detailed questions, he has recourse to experimental measurement. For this audience, therefore, I can refrain from further generalities on morals and philosophy. Instead I wish to present you two stories from the history of fluid mechanics. These concern the development of two fundamental concepts. The first is one you all use every day, the *static pressure* in a fluid in motion. Its origin is a part of the story of BERNOULLI's Theorem, now more than 200 years old. The second concept, *cross-viscosity*, is one with which few of you are yet likely to be familiar; its story begins little more than ten years ago. I will tell you these stories, not in the fashion of those textbook writers who manufacture historical notices so as to bear out their own views of how science ought have developed, but instead as they really did occur. Since one of these stories is from the earliest period of modern hydraulics and the other is still continuing today, and since despite the lapse of 200 years between them the general outline is much the same, there will be no need for me to add comments or to draw a moral.

For the early development of hydraulics, I refer you to the excellent history by ROUSE & INCE, the first installments of which have now appeared. From it we learn that hydraulic machines are of great antiquity and hence that necessarily man's observation of water flow begins with his history. From many centuries of experience we have records of keen observation and reasoning on matters of principle. For example, THEOPHRASTOS realized that water waves transport motion, not mass, which suggested to him that sound is a similar undulation of the invisible air. LEONARDO DA VINCI made a brilliant comparison of the waves on water with the waves which the wind sends travelling across a wheat field. He went beyond these isolated remarks in asserting that in general the motions of water and of air are of the same kind. This assertion was based on experience. LEONARDO traced streamlines in water by watching small objects cast into the flow, and from his pen we have sketches of waterfalls, river surfaces, and vortices as accurate as a photograph and more beautiful. To follow the motions of air, he watched the leaves blown by the wind and even for gentler motions injected smoke into the current. It is now often claimed that LEONARDO founded the experimental sciences, but I believe this statement is entirely misleading. LEONARDO projected many experiments, some of them reasonable and some of them confused, but he has left us no record of ever having obtained any numerical value by measurement. Rather, he was an observer of undisturbed nature. Like none else he seized upon experience, but experiment lay many decades past his time. In fluid mechanics, as I mentioned, he devised methods for making an existing motion visible, but we have from him no numerical values of discharges or pressures. LEONARDO recorded two quantitative statements in hydraulics. One of these is the principle of continuity in its simplest form. The other is the distinction between the rotation of a wheel and the rotation of an irrotational vortex. LEONARDO's words, as always, are vague; if interpreted as strict proportions, his statements are correct, but it is possible that he intended only qualities and inequalities rather than equations. He gives no indication of how he obtained these basic principles, nor does he apply them.

Whether LEONARDO's notebooks through private and unacknowledged use influenced the numerous writers on hydraulics who came after him, or whether these discovered anew the facts ascertained earlier by him, will always remain a question in debate. I have described LEONARDO's work so as to make clear the keen and abundant experience available before DANIEL BERNOULLI's time. In addition to these fundamental observations, we must notice the increasing popularity of hydraulic machines from 1500 onward. Both for gainful work and for show or pleasure, pumps, presses, screws, fountains, wheels, conduits, and reservoirs were produced in greater and greater number.

After the principle of continuity, the next theoretical statement made by hydraulic writers was TORRICELLI's Law of efflux, of little use for the design of a pump or wheel. DESCARTES was the first philosopher to regard all nature as one great machine, governed by common laws. While nearly all of DESCARTES' physics is wrong in detail, his grand attempt is the beginning of theory in the modern sense, and as a corollary it began in particular a search for a theory of hydraulics based on mechanical principles. This insistence on generality caused the physicists and geometers to reject all empirical rules and resulted in that separation of hydraulic theory from hydraulic practice, still apparent today, to which Professor ROUSE referred in his opening remarks. In nature, the effects of friction, roughness, and turbulence can rarely be neglected and may altogether predominate, just as in ballistics neglect of air resistance would lead to wretched

marksmanship. Nonetheless, just as GALILEO's mental abstraction of
the medium in which all earthly bodies exist was a necessary forestep
to the rational mechanics of solids, a similar abstraction of much of
the daily circumstances of water was a necessary preliminary to the
rational mechanics of fluids. This abstraction was made by NEWTON
in his celebrated attempt to prove TORRICELLI's Law. While NEW-
TON's fiction of the "cataract" is no more than a brilliant course of
imagination and hypothesis, it is the first example of hydraulic theory,
and as such, even though entirely faulty, it showed the possibility of
the field and induced many other savants to attempt the problem. All
these later trials were likewise failures. The forces exerted by fluids at
rest were by now well known, but to consider the effect of motion on
these forces seemed hopelessly difficult.

Such was the scene when DANIEL BERNOULLI, a young mathematician of twenty-five and already famous, took up the study of fluids.
Throughout his life DANIEL BERNOULLI performed both experiments and calculations; while when old he became almost entirely an
experimentist, at the period we are discussing he was in the main a
mathematician, working not only in mechanics but also in analysis and
the theory of numbers. For about five years he gave occasional attention to fluids, and during this time he wrote two important papers on
hydraulics before attacking the simultaneous determination of pressure and velocity. The BERNOULLI Theorem itself he discovered
shortly before 17 July 1730, on which date he wrote to GOLDBACH as
follows:

> For my part, I am entirely plunged in water, which furnishes my
> sole occupation, and for some time now I have renounced all that
> is not hydrostatics or hydraulics.... In these past days I have
> made a new discovery which can be of great use for the design of
> conduits for water, but which above all will bring in a new day in
> physiology: It is to have found the statics of running water, which
> no-one before me has considered, so far as I know.... The prob-
> lem is to find the effort of water which is pushed with an arbitrary
> force in an arbitrary tube.

He goes on to explain "one of the simplest cases", the example indicated by Figure 1. For the height SF of the stagnant water over the
hole in the tube of running water he obtains

$$SF = \left(1 - \frac{1}{n^2}\right)a,$$

where a is the height of the reservoir and $1/n$ is the ratio of the area
of the little hole to the area of the tube. This may not look familiar,

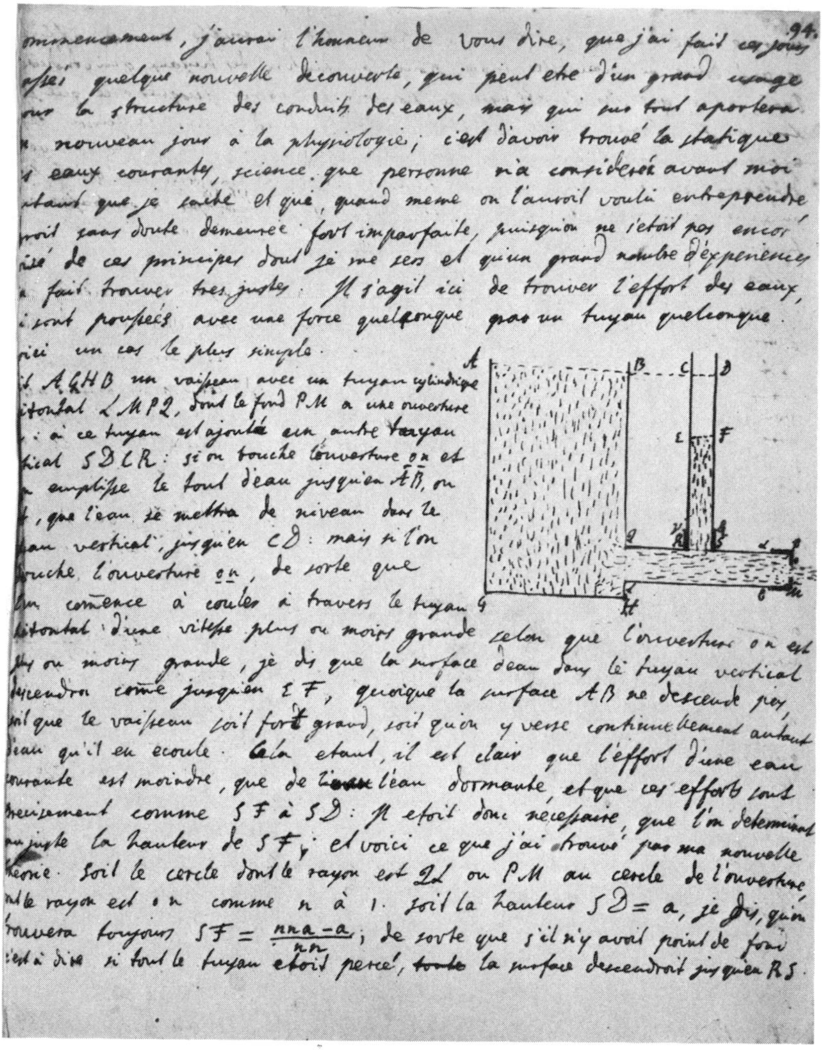

ommencement, j'aurai l'honneur de vous dire, que j'ai fait ces jours
assés quelque nouvelle découverte, qui peut être d'un grand usage
our la structure des conduits des eaux, mais qui sur tout apportera
n nouveau jour à la physiologie; c'est d'avoir trouvé la statique
s eaux courantes, science que personne n'a considérée avant moi
tant que je sache et que, quand même on l'auroit voulu entreprendre,
roit sans doute demeurée fort imparfaite, puisqu'on ne s'étoit pas encor
isé de ces principes dont je me sers et qu'un grand nombre d'expériences
a fait trouver très justes. Il s'agit ici de trouver l'effort des eaux,
i sont poussées avec une force quelconque par un tuyau quelconque.
ici un cas le plus simple.

it AGHB un vaisseau avec un tuyau cylindrique
itontal LMPQ, dont le fond PM a une ouverture
à ce tuyau est ajouté un autre tuyau
ical SDCR: si on bouche l'ouverture on est
n emplisse le tout d'eau jusqu'en AB, on
, que l'eau se mettra de niveau dans le
an vertical, jusqu'en CD: mais si l'on
uche l'ouverture on, de sorte que

n comence à couler à travers le tuyau
itontal d'une vitesse plus ou moins grande selon que l'ouverture on est
us ou moins grande, je dis que la surface d'eau dans le tuyau vertical
cendra comme jusqu'en EF, quoique la surface AB ne descend pas,
il que le vaisseau soit fort grand, soit qu'on y verse continuellement autant
eau qu'il en écoule. Cela étant, il est clair que l'effort d'une eau
urante est moindre, que de l'eau dormante, et que ces efforts sont
ecisement comme SF à SD: Il étoit donc nécessaire, que l'on déterminât
u juste la hauteur de SF; et voici ce que j'ai trouvé par ma nouvelle
éorie. Soit le cercle dont le rayon est Qd ou PM au cercle de l'ouverture
nt le rayon est on comme n à 1. Soit la hauteur SD = a, je dis, qu'on
ouvera toujours SF = $\frac{nna-a}{nn}$; de sorte que s'il n'y avoit point de fond,
est à dire si tout le tuyau étoit percé, toute la surface descendroit jusqu'en RS.

Figure 1. DANIEL BERNOULLI's description of his theorem and experiment, 1730.

but it is in fact the BERNOULLI Theorem for this case, with the press-ure replaced by the height SF. The advantage of this form is that to test it no velocities need be measured, all quantities being geometric. BERNOULLI wrote that an antagonistic senior colleague who belittled his work could not believe the result

> ... until I performed the experiment for him in the presence of other academicians. I made the experiments by means of a very polished iron cylinder which I had caused to be furnished with different covers which had holes of different sizes such as aPMβ; in the middle of the cylinder was welded a little end of tube γRSδ suitable for supporting a glass tube CRSD. All experiments suc-ceeded perfectly.

BERNOULLI's letter makes it clear that his theorem was discovered by theory alone, or, as he put it, *a priori*. Experimental test came afterward. You will note from the diagram that the experiment is devised so as to favor as much as possible the hypotheses under which we derive BERNOULLI's Theorem nowadays.

Some explanation is required before we can recognize the equation written by BERNOULLI as the modern theorem bearing his name. When we turn to the derivation, even in the improved form published in his often cited but never read *Hydrodynamica*, still more explanation is needed. To repeat BERNOULLI's words here would not be helpful. His method is to regard the element EG in Figure 2 as moving down the tube to acdb, where he wishes to find the pressure on the wall. The velocity there is related to the velocity at the small hole o by the principle of continuity. BERNOULLI now imagines the tube down-stream from ab suddenly to break off or dissolve. The element acdb,

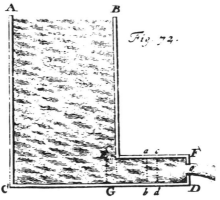

Figure 2. The published diagram for DANIEL BERNOULLI's argument to infer his theorem, 1738.

thus instantly released, suffers an impulsive acceleration. By using the principle of conservation of energy, BERNOULLI calculates this impulse, which in turn he regards as proportional to the pressure on the wall when the tube is not broken off. The argument is intricate; the hypotheses on which it rests are questionable; and the details are confusing.

I have mentioned that DANIEL BERNOULLI set about at once to verify his theorem by experiment. Figure 3 shows the experimental possibilities as presented in the *Hydrodynamica*. It is typical of DANIEL BERNOULLI not only that theory came before experiment but also that when the result was once confirmed by experiment, he regarded his work as finished. It is easy for us today to see that in fact the equation as written by BERNOULLI is not convenient; BERNOULLI's contemporaries saw at once that his derivation was obscure and unconvincing. What is missing from the equation itself and from the proof is the *internal pressure*, not yet invented. It was this same lack that had prevented NEWTON from giving an adequate proof of TORRICELLI's law. In his argument BERNOULLI used four different words, none of them defined, to expresss the forces exerted by the water upon itself and upon the walls of the tube.

[The publication of DANIEL BERNOULLI's book in 1738 brought him doubled fame. It excited his father, the formidable JOHN BERNOULLI, who was then seventy-one years old, ill, and with but a decade of life remaining, to devote his effort thenceforth to the flow of water, and in 1743 he published his own treatise on hydraulics, embellishing it with florid and ugly boasting. By dating it 1732 he provoked a controversy over priority and plagiarism which has lasted until the present day and has tended to dull the glory his treatise deserves because it created hydraulics anew.] JOHN BERNOULLI had

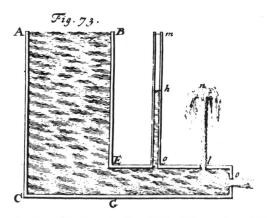

Figure 3. One of BERNOULLI's published illustrations, 1738.

as great a talent for mathematics as any man who ever lived. While less successful in discovering the physical principles of a new field of experience, to derive by suitable new concepts and irreproachable reasoning a result already conjectured, was something natural to his genius. His innovations were profound. First, he separated the kinematic from the dynamic part of the problem. The principle of continuity and the principle of momentum he used consciously as separate basic postulates, as had none before him. Second, he created the idea of hydraulic pressure. In imagination, he isolated a thin slice of water in a tube and introduced a symbol for the force exerted upon it by the fluid on one side. In this way he achieved a differential equation and integrated it to obtain the BERNOULLI Equation for a tube of arbitrary cross-section and position and for flows not necessarily steady. By using the internal pressure he was able to give also a correct derivation from the principle of energy.

The foregoing achievements of old JOHN BERNOULLI were not recognized as due to him until 1955. Perhaps the reason is that he never explained them with any clarity except in a series of letters to EULER. JOHN BERNOULLI lived in a world of challenges, enmities, secret methods, and anagrams. As he wrote to EULER, he derived everything from a certain "principle of the eddy", but even in his letters it is vague, and in the printed treatise he abstained from expressing it from fear lest the "English clowns" accuse him of borrowing the "cataract" of NEWTON. The progress of equations is clear, despite unnecessarily elaborate notations and a mathematical style which was by then obsolete, but mechanical principle is replaced by bombast and boasting.

For EULER, clarity was the hallmark of truth. He saw at once the core of old JOHN BERNOULLI's ideas and disrobed them of vagueness. To him we owe the BERNOULLI theorem in the form and terms today in use. To him we owe also the brilliant imagination of the internal pressure in generality, the pressure field as equipollent to the action of the fluid outside any imaginary closed diaphragm upon that within. This concept, which has been the foundation of all further theory, he achieved ten years after his study of JOHN BERNOULLI's hydraulics. To discuss its formation would carry me afield from hydraulics, but I remark upon it in emphasis of the role of imagination and the importance of quantities which can only be thought of and cannot in themselves be measured. Neither is there time to discuss EULER's papers on hydraulic machines, where his grasp of the concept of internal pressure led him not only to detailed analysis of pumps and turbines but also to criteria for avoiding cavitation. These papers were neglected entirely by the hydraulic engineers of the day, and when EULER died in 1783 even a famous physicist of a younger generation characterized his work on fluids as useless in practice and merely exercises in pure mathematics. EULER did not perform experiments except before he was twenty, and thus he was unable to demonstrate the truth of his discoveries to practical men, who in that day despised calculus as being useless higher mathematics. Nonetheless, EULER was intensely interested in machines, and in 1754 he not only invented the guide wheel for a turbine but even calculated a detailed design and gave a complete hydraulic analysis for the pressure in the rotating machine. EULER wished his turbine to be built, but the engineers at FREDERICK II's court only reflected the ways of the king himself in scoffing at all of higher mathematics. EULER published two papers, one in French and one in Latin, making his invention free to anyone who paused to read, but in fact it was 190 years before his design was tested. In 1944, long after EULER's guide wheel had been rediscovered and adopted in turbine practice, ACKERET found that a model following EULER's plan reached an efficiency of 71%, which may be compared with 78–82% for the best modern turbines of similar capacity and head.

It is easy to praise or blame the actions of long ago, since we are free of responsibility in them. When we find parallel events occurring today, with no lesson learned, we are more ready to find excuses. I turn now to the recent discovery of cross-viscosity.

There are various ways of telling the story, but I prefer to begin with a fact of experience which was a by-product of an experiment. In 1943 MERRINGTON reported that in the course of some measurements of the discharge of rubber solutions or of oils containing metallic soaps he noticed that the fluid column swelled on emerging from the tube (Figure 4). Such an effect obviously cannot follow from any of the classical principles of fluid mechanics. MERRINGTON himself asserted that the fluids in question were visco-elastic and that the swelling was due to their residual elasticity, being in fact recovery from the compression they suffered when forced into the tube. He identified the phenomenon with that observed by BARUS in 1893. BARUS had cut off perfect cylinders of marine glue extruded from a tube and had found that when left free of external load these cylinders continued to deform and in the end converted themselves into cups. Now MERRINGTON's phenomenon occurs in steady flow. The spring of a portion of visco-elastic substance becomes poorer as time passes. Thus if we compare the swelling of steady flows in longer and longer tubes at the same efflux, for a visco-elastic substance this swelling should diminish, since the portion emerging will then have suffered its compression at more and more remote times in the past. If, on the contrary, the swelling is independent of the length of the tube, the phenomenon is not visco-elastic. An experiment should not

Figure 4. MERRINGTON's experiment, 1943.

be difficult, but so far as I know it has never been proposed until today. To return to the story, apparently MERRINGTON did not see any other possible explanation and did not pursue either theory or experiment concerning it.

About the same time several English experimenters noticed a group of new phenomena occurring in rotating fluids. The simplest of these, and the one which goes far to explain the rest, is produced by rotating a vertical rod, the lower end of which projects into a cup of high-polymer solution or oil of the right kind: The fluid climbs up the rod. The results of these experiments were collected and represented schematically by WEISSENBERG in a diagram published in 1947 (Figure 5). He did not suggest any connection with MERRINGTON's phenomenon. He proposed an elastic theory which appears to neglect the usual properties of fluids entirely, and some of the other investigators gave semi-quantitative explanations of a chemical nature. [MERRINGTON's phenomenon is one of several now loosely and with scant historical justice called "the Weissenberg effect".]

Between the dates of these two publications, three theorists began to develop a subject which turned out to be related to these phenomena of experience. The subject is nonlinear viscosity, and the theorists were REINER, RIVLIN, and I. Nonlinear viscosity had been studied

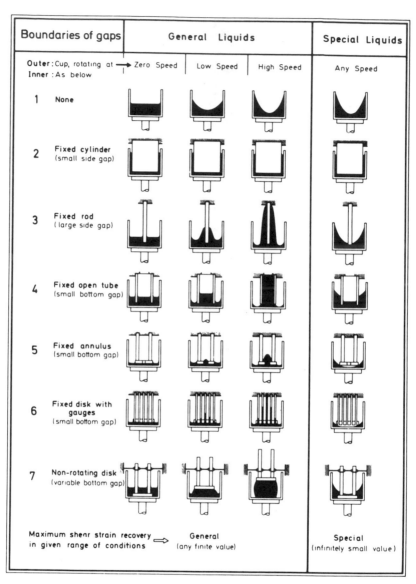

Figure 5. WEISSENBERG's diagrams, 1942.

before, in fact long before, from two different points of view. Some theorists of the last century proposed some general equations as being reasonable, but they did not investigate them sufficiently to get any definite conclusions. More recently a professional group called rheologists had measured departures from linearity in viscometric measurements. The rheologists were accustomed to one-dimensional theories in which a single stress component is taken as a nonlinear function of a single rate of deformation, but none of this literature could reveal a new phenomenon of a kind not included in the classical theory of viscosity at all.

REINER is one of this rheological group and has published much work, both theoretical and experimental, of the type described. In 1945 appeared a paper of his of a different character. In it he attempted to apply to fluids the mathematical methods and mechanical concepts used in the general theory of three-dimensional finite elastic strain, an old though little understood branch of mechanics which had been simplified in the previous decade by the introduction of tensor analysis. REINER had in mind the phenomenon of dilatancy, which REYNOLDS had observed long before in granular materials: "a definite change of bulk, consequent on a definite change of shape." For example, if you shear wet sand in walking upon it, the footprints are dry, since the volume of the sand mass has increased and thus opened greater voids for the water to sink into. REINER proved this phenomenon to be predicted by the general theory of nonlinear viscosity. For his elegant and perspicuous formulation of the general theory itself he acknowledged the assistance of the theoretical physicist RACAH.

RIVLIN was employed by a laboratory investigating the properties of rubber solutions, and it is possible that his work was motivated by the phenomena published by MERRINGTON and WEISSENBERG. In any case, he successfully and simply explained them by a theory of incompressible fluids with nonlinear viscosity which he published in 1947. This theory is a special case of REINER's, but RIVLIN's work is distinguished not only by greater definiteness and clarity but also by explicit and general solutions for the flows in torsional, tubular, and rotating cylindrical viscometers. At first astonishing is the fact that MERRINGTON's swelling, which MERRINGTON himself regarded as recovery from compression, follows from RIVLIN's theory of incompressible fluids. To get the idea behind all these effects, it is easiest to consider simple shearing flow. To produce such a flow, according to the general theory, shearing stress is not enough: Normal pressures on the shear planes must be supplied as well. This phenomenon is called *cross-viscosity* and is a property independent of shear viscosity and bulk viscosity.

Once the idea of the general theory of viscosity and a little experience in tensor analysis or matrix algebra shall have been gained, the explanations are not difficult. Each individual phenomenon can also be explained by physical reasoning, but since these physical arguments came only after the really rather simple mathematics was all worked out, I doubt if they are in fact enlightening, and I will not try to present them.

My own first, crude memorandum on nonlinear viscosity was issued in 1947, when REINER's basic paper was already two years in print and RIVLIN was far ahead of me. I knew of neither their work nor the phenomena of MERRINGTON and WEISSENBERG. Explanation, though no excuse, for my ignorance may be found in the apathy if not hostility of the world of fluid mechanics toward the subject of nonlinear viscosity: All my bosses and colleagues, at that time as now intent upon [what they claimed to be] practical problems and calculations, took no more notice of the work of REINER and RIVLIN than of my own attempts. My formal publication was delayed because a journal of applied mathematics rejected the manuscript on the grounds that no-one was interested in the subject, the paper would be costly to print, and in any case my work was physics rather than applied mathematics. The terms used by the anonymous referee were so harsh that the only logical alternative to suicide was to give up science forever. While my happening to learn the identity of the referee prevented me from resort to either of these extremes, before arranging for publication abroad I spent eighteen months reviewing the fundamentals of mechanics and trying to learn the processes by which the classical theories had been derived by their discoverers. Priority for nonlinear viscosity belongs unquestionably to REINER and RIVLIN, but I speak of my work as well because I have a better knowledge of my own motives and circumstances than of theirs.

For my part, my trouble was that I was employed to study fluids but I could not accept the so-called derivations of the Navier-Stokes equations in textbooks. It seemed as unreasonable to suppose viscous stress a linear function of rate of deformation as to replace every curve by a straight line. I was aware of claims of departures from the Navier-Stokes equations in certain extreme conditions, but I found it more wonderful that the Navier-Stokes equations held at all, and I set out to find the reason. Being naturally both slow and obstinate, I resisted the pressure to calculate or guess useful approximations within accepted theories and to the annoyance of my superiors and the disgust of my senior colleagues insisted on stopping to think. Textbooks hurry the reader on to accept their conclusions as quickly as possible, often replacing a logical gap by asserting that the result is established by experiment, without a reference. Such evasion is not found in original memoirs dealing with matters of principle. The discoverer or first proponent not only has the task of convincing a skeptical public but also often is close to his struggles to convince himself. Moreover, usually there are no appropriate experiments at the time when the theory is first formulated as a plausible model of experience. This statement always surprises believers in the "experimental method". In reply to it they often suggest that if in the history of mechanics theory has usually come before experiment, there must have been many wrong theories proposed. In fact, there were few. Without experience, no explanation definite enough to be considered a mathematical theory is likely to be given; with experience, to expect theorists to propose wholly wrong models suggests rather limited appreciation for the brains of theorists. Examination of the dust pile of mechanics reveals few wrong theories but a host of "approximate" or numerical solutions and experimental measurements concerning details and special cases which have lost their interest, as well as many mathematically erroneous "solutions" within correct theories. Today we are piling up this scrap heap so fast that it is difficult to keep the rare cases of fundamental work out from under.

Going back to the great memoir of STOKES on viscosity, I found none of the dogmatism of the modern texts, but instead an honest hesitancy and search for principle. Taking up STOKES's definition of a fluid, I sought by the aid of tensor analysis to put into mathematical form precisely what STOKES had said in words rather than the mere approximations which were all that the mathematics available in his day could easily handle. My work was influenced also by a study of modern general elasticity and its mathematics, but while REINER had attempted to maintain as close a similarity as possible, to me it was the basic conceptual differences that seemed more important. Some earlier writers had spoken loosely of the Navier-Stokes equations as being valid approximately for "small" rates of deformation, just as classical linear elasticity is valid approximately for small strain. That is plain nonsense. Strain is dimensionless and hence can indeed be small, so that the position of linear elasticity with respect to finite elasticity is clear in this formal sense. Not so for fluids, for rate of deformation has the dimensions of frequency and hence cannot be absolutely small. It can be small with respect to another rate, but what this standard of comparison should be for a fluid, is not obvious. My work on general fluids began at this dilemma, faced it, and resolved it by proposing a theory in which no material parameter of the dimension of time can occur. When I learned of the work of REINER and RIVLIN, I found that they had not considered the role of the time, and I was able in some cases to show differences between fluids that have a natural time-lapse and fluids that do not.

I mentioned the phenomena published by MERRINGTON and WEISSENBERG, which were soon taken up as evidence for the existence of cross-viscosity. After a while it was realized that cross-viscosity was not really new. Everyone knows that you cannot mix paint by rotary stirring, for the paint climbs up the rotor. In the paint industry, other methods of mixing were devised and the phenomenon itself was apparently regarded as chemical. Here the experience lay before us all, but we were blind to its meaning.

Finally came the time for experiment. All old measurements on nonlinear viscosity were made obsolete by the theory, since they measured only small corrections to classical effects and offered no means of detecting even the existence of the new phenomena. The old viscometers have walls supplying lateral pressures of any desired amount with no means of measuring their magnitude. Moreover, it follows fairly generally from the theory that while departures from the classical first-order linear relation is an effect of third order in the rate of deformation, the new phenomena are effects of second order. Thus it is possible that fluids previously believed linear in the range tested are in fact nonlinear. Precise tests in new instruments designed to show the new effects are necessary, first to measure the modulus of cross-viscosity as a function of the rate of deformation, and second to test the consistency of theory with experiment. Such measurements are now coming into print. Some of these appear to confirm the theory of nonlinear viscosity and others do not. In any case we must remember that the classical theory of linear viscosity is a first approximation to several different more general theories, and so not all fluids that obey the classical laws for slow motion can be expected to obey any one particular theory for rapid ones.

The phenomenon of cross-viscosity is typical of nonlinear continuum mechanics. Every more or less plausible theory predicts something of this kind. While a few years ago this phenomenon and others like it seemed[1] outside the domain of mechanics, with the recent development of many new theories we are faced with the opposite difficulty of being unable to take these phenomena as confirming any one theory rather than another. In fact, as I say, these cross effects are typical of mechanics. Everyone knows that if you push a gyroscope, it refuses to move in the direction you push it. This illustrates the general case in mechanics: Only when a body is barely stirred out of its sleep will it answer to your wishes and move approximately as you impel it. A century of unquestioning acceptance of linear theories in mechanics has lulled us into expecting response which is not typical. It

---

[1] Except to the very few persons who knew of certain special results of POYNTING (1909–1913) and later writers concerning shear and torsion of finitely elastic bodies.

appears likely that nonlinear effects will be discovered in increasing number and may eventually have a greater practical importance than is now foreseen.

My two stories are finished, and I have promised to draw no moral. I hope you will not consider my promise broken if I let DANIEL BERNOULLI draw a moral. In the quotation which follows, the word "mathematician" occurs, but it comes from late in BERNOULLI's life when he was absorbed in experiment, and it obviously refers not to a professional group but to a habit of mind. Here is the quotation:

> [T]here is no philosophy that is not founded upon knowledge of the phenomena, but to get any profit from this knowledge it is absolutely necessary to be a mathematician.

Bibliography

J. ACKERET, "Untersuchung einer nach den Euler'schen Vorschlagen (1754) gebauten Wasserturbine", *Schweizerische Bauzeitung* **123** (1944): 9–15.

H. ROUSE & S. INCE, "History of Hydraulics", separately paginated supplements to *La Houille Blanche*, 1954/1955.

C. TRUESDELL, "A new definition of a fluid. I and II", *Journal de Mathématiques Pures et Appliquées* (9) **29** (1950): 215–244; **30** (1951): 111–158.

C. TRUESDELL, Chapter V of "The mechanical foundations of elasticity and fluid dynamics", *Journal of Rational Mechanics and Analysis* **1** (1952): 125–300; **2** (1953): 593–616 [corrected reprint, *Continuum Mechanics. I. The Mechanical Foundations of Elasticity and Fluid Dynamics*, International Review Series, Volume **8**, New York *etc.*, Gordon & Breach, 1966].

C. TRUESDELL, "A program of physical research in classical mechanics", *Zeitschrift für Angewandte Mathematik und Physik* **3** (1952): 79–95 [reprinted with the foregoing].

C. TRUESDELL, "Rational Fluid Mechanics, 1687–1765", pages VII–CXXV of LEONHARDI EULERI *Opera omnia* (II) **12**, 1954.

C. TRUESDELL, "I. Euler's treatise on fluid mechanics (1766); II. The theory of aerial sound, 1687–1788; III. Rational fluid mechanics, 1765–1788", LEONHARDI EULERI *Opera omnia* (II) **13** [1956, pages VII–CXVII].

C. TRUESDELL, "Zur Geschichte der inneren Druck", *Physikalische Blätter* [**12** (1956): 315–326].

Note for the Reprinting

This essay first appeared on pages 3–18 of *Proceedings of the Sixth Hydraulics Conference*, Bulletin 36, State University of Iowa Studies in Engineering, 1956.

Such a foray is rare in my work today and for its date so untypical as to demand explanation. The organizer of the conference had comfortably arranged for a general address by a famous Elder of the Cantabrigian Gospel Mission for Evangelical and Empirical Theology but was disappointed of it at the last minute. In me he found an astonished neighbor who would drive over to Iowa City at low cost. The succeeding speaker on the morning of the lecture was a man who called himself an analytical philosopher. He presented a philosophical theory of the yardstick.

Today, the early history of nonlinear viscosity could be told more completely against a background of deeper understanding. The reader interested in the literature on nonlinear continuum mechanics after 1956 will find an account and bibliography fairly complete through 1964 in *The Non-Linear Field Theories of Mechanics*, Volume **III**/3 of FLÜGGE's *Encyclopedia of Physics*, Berlin *etc.*, Springer-Verlag, 1965. For the later work he will need to consult a dozen monographs on special topics.

Because of SZABÒ's spirited defense of JOHN BERNOULLI (see below, Essay 31), on page 10 I have replaced part of the original text by the passage enclosed in square brackets. Also SZABÒ has quoted eighteenth-century authors on hydraulics to show that what I thought was my historical discovery in 1955 was only a rediscovery. As for "the excellent history of ROUSE & INCE", at the time of the lecture only the first few installments had appeared. My judgment of the later parts may be read further on in this volume, Essay 26.

My note in the *Physikalische Blätter* is completely superseded by "The creation and unfolding of the concept of stress", Essay IV in my *Essays in the History of Mechanics*, New York, Springer-Verlag, 1968.

Likewise, a detailed and documented analysis of LEONARDO's work on mechanics in general and fluid mechanics in particular is given in Essay I of that volume; I have accumulated numerous corrections and additions for it against the possibility of a second edition.

The concluding quotation is from DANIEL BERNOULLI's still unpublished letter of 7 January 1763 to his nephew JOHN III BERNOULLI in Berlin. The original runs as follows:

> [J]e souhaite que le Roi vous ait confié le departement de la Philosophie experimentale. J'ai blanchi dans cette carriere et depuis ma premiere jeunesse j'ai vu a chaque pas, qu'il n'y a d'autre philosophie que celle qui est fondée sur la connoissance des phenomenes: mais il faut etre absolument mathématicien pour tirer parti de cette connoissance.

This occasion reminds me of my gratitude to the late OTTO SPIESS for having loaned me a copy of that letter; I am grateful also to Mme PATRICIA RADELET-DE GRAVE for having located it again for me.

2. THE FIELD VIEWPOINT IN CLASSICAL PHYSICS

Exordium of *The Classical Field Theories* (1960), co-author R. Toupin

Contents

1. Corpuscles and Fields

Today matter is universally regarded as composed of molecules. Though molecules cannot be discerned by human senses, they may be defined precisely as the smallest portions of a material to exhibit certain of its distinguishing properties, and much of the behavior of individual molecules is predicted satisfactorily by known physical laws. Molecules in their turn are regarded as composed of atoms; these, of nuclei and electrons; and nuclei themselves as composed of certain elementary particles. The behavior of the elementary particles has been reduced, so far, but to a partial subservience to theory. Whether they in their turn await analysis into still smaller corpuscles, remains for the future.

Thus in the physics of today, corpuscles are supreme. It might seem mandatory, when we are to deal with extended matter and electricity, that we begin with the laws governing the elementary particles and derive from them, as mere corollaries, the laws governing apparently continuous bodies. Such a program is triply impractical:

A. The laws of the elementary particles are not yet fully established. Even such senior disciplines as quantum mechanics and general relativity remain open to possible basic revision and not yet satisfactorily interconnected.

B. The mathematical difficulties are at present insuperable. (Even on a lower level they remain: As is well known, the "proof" that a quantum-mechanical system may be replaced by a classical system in first approximation is defective.)

C. In such special cases as have actually been treated, the mathematical "approximations" committed in order to get to an answer are so drastic that the results obtained are not fair trials of what the basic laws may imply. When such a result appears in disaccord with experience, we are at a loss whether to assign the blame to the basic laws themselves or to the mathematical process used in the subsequent derivations.

But more than this, such a program even if successful would be illusory:

(a) The future discovery of new entities within the present "elementary" particles would nullify any claim for such results as predictions from "basic" laws of physics. Indeed, *within any corpuscular view the possibility of an infinite regress is logically inevitable.* [See Figure 6.]

(b) The details of the behavior of the corpuscles are extraneous to most mechanical and electromagnetic problems. Materials whose corpuscular structures are quite different may exhibit no perceptible difference of response to stress.

Avoiding illusory complications, we may construct a direct theory of the *continuous field*, indefinitely divisible without losing any of its defining properties. The field may be the seat of motion, matter, force, energy, and electromagnetism. Statements in terms of the field concept are called *phenomenological*, because they represent the immediate phenomena of experience, not attempting to explain them in terms of corpuscles or other inferred [or hypothesized] quantities.

The corpuscular theories and the field theories are mutually contradictory as direct models of nature[1]. The field is indefinitely

---

[1] The formal " derivations " of the field equations from the mass-point equations of mechanics given in many textbooks are illusory, such a derivation being impossible without added assumptions which are rendered superfluous by a direct approach to the continuum. The difficulty can be avoided by a formulation of the fundamental equations as Stieltjes integrals (*cf.* § 201); in essence, this was done by EULER in §§ 20–22 of his "Découverte d'un nouveau principe de mécanique", *Mémoires de l'Académie Royale des Sciences et Belles Lettres* [de Berlin] [6] (1750): 185–217 (1752) = pages 81–108 of LEONHARDI EULERI *Opera omnia* (*II*) 5.

divisible; the corpuscle is not. To mingle the terms and concepts appropriate to these two distinct representations of nature, while unfortunately a common practice, leads to confusion if not to error. For example, to speak of an element of volume in a gas as "a region large enough to contain many molecules but small enough to be used as an element of integration" is not only loose but also needless and bootless.

In a deeper sense, the continuous field and the assembly of corpuscles may be set into entire agreement. Adopting the viewpoint of statistical mechanics, we may consider a classical system of mass-points of any kind whatever and assign a probability to its initial conditions. Extending a notable success by IRVING & KIRKWOOD[2], NOLL[3] has defined certain *phase averages* which he has proved to satisfy exactly the laws of balance for a continuous field. This result, not a limit formula or approximation, is an *exact theorem* on distributions in phase space. Thus those who prefer to regard classical statistical mechanics as fundamental may nevertheless employ the field concept as exact in terms of *expected values*.

While sometimes the phenomenological approach is regarded as only approximate, the result just described shows that in representing matter as continuous rather than discrete we can in fact make no statement that is inconsistent with the statistical view of matter as composed of classical molecules, *so long as we confine attention to the exact and general theory of continuous media*[4].

This treatise presents [an] exact and [fairly] general theory of the continuous field.

---

[2] J. IRVING & J. KIRKWOOD, "The statistical mechanical theory of transport processes. IV. The equations of hydrodynamics", *Journal of Chemical Physics* **18** (1950): 817–829.

[3] W. NOLL, "Die Herleitung der Grundgleichungen der Thermomechanik der Kontinua aus der statischen Mechanik", *Journal of Rational Mechanics and Analysis* **4** (1955): 627–646. In NOLL's paper precise conditions of regularity for the density in phase are stated. The molecules, not restricted in variety, are supposed free of constraints but otherwise subject to arbitrary mutual and extrinsic forces. The expression for the resultant extrinsic force in general does not depend only on the extrinsic forces to which the molecules are subject; otherwise the agreement stated in the text above is unqualified. An extension to quantum-mechanical systems is given by J. IRVING & R. W. ZWANZIG, "The statistical mechanical theory of transport processes. V. Quantum hydrodynamics", *Journal of Chemical Physics* **19** (1951): 1173–1180.

[4] That is, only the *general* equations expressing the balance of mass, momentum, and energy in the continuous field have been derived. There is no indication that any special theory of continuous bodies, such as the theory of perfect fluids, is consistent with statistical mechanics. In fact, a *simple* field theory seems to emerge only in approximation, and from a simple molecular picture an extremely complicated field theory results. Also, the exact agreement does not extend to thermodynamics, which from a statistical standpoint appears to be only an approximate theory.

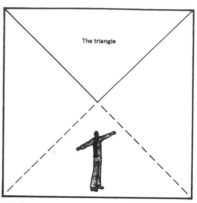

10 000 BC. The inhabitants of the páper square have no conception of the true nature of the universe they inhabit.

1900 AD. Physicists of the square discover a basic subdivision of their universe. They call it the "triangle" and consider it to be the fundamental building block of the universe.

Figure 6 (read across the top and bottom of pages 24 and 25). The progress of "fundamental" physics, conceived by BERKELEY CHEW, published in 1970 by his father, Professor GEOFFREY F. CHEW.

2. CLASSICAL MASS-POINTS AND CLASSICAL FIELDS

From the time of NEWTON until recently, many natural scientists considered the mass-point the fundamental quantity of nature, or at least of mechanics. They believed that matter was composed of many very small particles obeying the laws of classical mechanics, and that, consequently, the behavior of gross matter could be predicted, in principle, to any desired accuracy, from a knowledge of the intermolecular forces. Thus continuum mechanics appears as an approximate or at best secondary theory within classical mechanics. While this

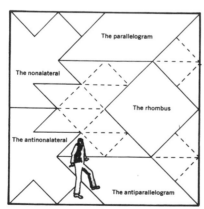

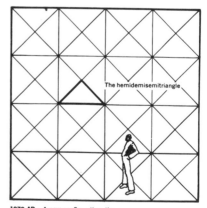

1960 AD. Physicists' conception of their universe is further clouded by new discoveries: the rhombus, the parallelogram, the antiparallelogram, the nonalateral and many others. It is unclear what these discoveries signify.

1970 AD. A new configuration, the "hemidemisemitriangle," is hypothesized, out of which all known configurations of the universe can be constructed. The hemidemisemitriangle is thought to be the fundamental building block of the universe.

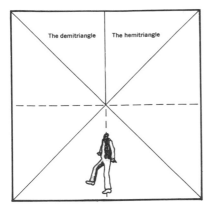

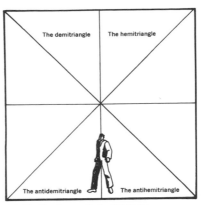

1930 AD. Physicists discover that the triangle can be split. Its parts are termed the "hemitriangle" and the "demitriangle." These are thought to be the fundamental building blocks of the universe.

1950 AD. Mirror images of the hemitriangle and the demitriangle are discovered. These are termed "antihemitriangle" and "antidemitriangle."

tradition clings on in physics teaching today, it defies reality. Aside from the as yet unconquered mathematical difficulties of putting this ideal program into practice, the program itself is out of keeping with modern views on matter. The smallest units of matter are no longer believed to obey the laws of Newtonian mechanics, except approximately and in circumstances rarely occurring in dense matter. Nevertheless, conditions in which the classical laws of momentum and energy fail perceptibly for *tangible portions of matter* are extremely rare if not altogether unknown.

To cite an example, no corpuscular theory based on Newtonian mechanics has produced formulæ for the specific heats of solids which agree with experimental values. Nevertheless, there is not the slightest indication that a solid body when heated and set in motion fails, *as a body*, to obey the classical laws of balance of mass, momentum, and energy. In fact it is almost the rule that *Newtonian mechanics*,

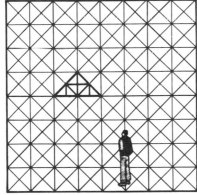

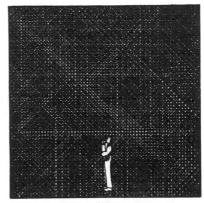

1975 AD. The hemidemisemitriangle is discovered. The following year the hemidemisemitriangle is split.

2000 AD. The inhabitants of this paper square have no conception of the true nature of the universe they inhabit.

while not appropriate to the corpuscles making up a body, agrees with experience when applied to the body as a whole, except for certain phenomena of astronomical scale. Only pædagogical custom has hindered general realization that *as a physical theory, continuum mechanics is better than mass-point mechanics*[5].

Indeed, in physics it is inappropriate to lay down the laws of classical mechanics for small bodies, to which in general they do *not* apply, and thence to derive or state by analogy the corresponding laws for extended bodies, to which they *do* apply. Rather, the process should be reversed: As HAMEL[6] stated, *classical mechanics is the mechanics of extended bodies*. Certain special problems remain, particularly problems in celestial mechanics, ballistics, and mechanisms, where the mechanics of mass-points is accurate. HAMEL[6] remarked that problems of this kind are easily and reasonably regarded as special cases within continuum mechanics.

3. EXPERIMENTS AND AXIOMS

It is becoming fashionable to present the foundations of theoretical physics in terms of experiments. Indeed, since physics is intended to predict numerous phenomena of nature from knowledge of a few, the preconception that a given physical discipline should be derivable from the results of certain basic experiments is appealing. Nevertheless, an experimental approach to mechanics and electromagnetism is not practical. The field, infinite in extent and indefinitely divisible, is by its very nature not measurable directly. The "experiments" sometimes used as the starting point for pædagogical treatments of field theories are *a posteriori* verifications at best; always unperformed and often unperformable, too often they are mere hoaxes. Moreover, they belie the true course by which the field theories have developed. *Experience* has been the guide, *thought* has been the creator[7]. Not only

---

[5] *Cf.* pages 79–80 of C. TRUESDELL, "A program of physical research in classical mechanics", *Zeitschrift für Angewandte Mathematik und Physik* **3** (1952); 79–95, [reprinted on pages 187–203, with annotations of 1962 on pages 215–218, of *Continuum Mechanics. I. The Mechanical Foundations of Elasticity and Fluid Dynamics*, edited by C. TRUESDELL, New York *etc.*, Gordon & Breach, 1966].

[6] *Cf.* G. HAMEL, "Über die Grundlagen der Mechanik", *Mathematische Annalen* **66** (1908): 350–397. Note also that from a theory of phase averages over systems governed by quantum mechanics IRVING & ZWANZIG in *op. cit.* Footnote 3 infer the *classical* equations of balance of mass, momentum, and energy.

[7] *Cf.*, *e.g.*, R. DUGAS, *La Mécanique au XVIIᵉ Siècle*, Neuchâtel, Editions du Griffon. 1954, 620 pages [see Essay 21, below], and C. TRUESDELL, "Experience, theory, and experiment", pages 8–18 of *Proceedings of the 6th Hydraulics Conference*, Bulletin 36 (1957), University of Iowa Studies in Engineering [reprinted as the preceding essay in this volume].

does any theory reduce and abstract experience, but also it over-reaches it by extra assumptions made for definiteness. Theory, in its turn, predicts the results of some specific experiments. The body of theory furnishes the concepts and formulæ by means of which experiment can be interpreted as being in accord or disaccord with it. To *overturn* a theory by the results of experiment, we seek the aid of the theory itself; in terms of the theory, from experiment we may find agreement which develops confidence in the theory; but *establish* a theory by experiment we never can. Experiment, indeed, is a *necessary* adjunct to a physical theory; but it is an adjunct, not the master.

While most theoretical physicists seem to act in accord with the above views, they rarely admit to holding them. Therefore we provide a fuller explanation, largely a paraphrase of a work by SOUTHWELL[8].

The "operational" system accepts as basic only quantities suscep-tible of direct measurement and, connecting them, laws which are to be tested by experiment. From these laws, logical inference is to derive a system shown by actual trial to keep contact with physical experience at every stage.

Apart from the practical limitations in checking any theoretical "law", there is a deeper objection against this view of physics, that it rests on a circularity: No experiment can be interpreted without recourse to ideas in themselves part of the theory under examination. Similarly, no quantity can be measured in the absence of a theory explaining the experiment. Consider the measurement of "mass" by weighing or by impact: In the former way the law of falling bodies and in the latter way the law of conservation of momentum, both employing the concept of mass, are used to complete the measure-ment. If we seek to verify NEWTON's "law"[9] that "Every body con-tinues in its state of rest, or of uniform motion straight ahead, unless compelled to change that state by forces impressed upon it," we require a free body, unavailable because all bodies in the laboratory are subject to the earth's attraction. Indeed, we try to neutralize that attraction, as in "ATWOOD's machine": The body is connected by a light string passing over a freely running pulley with a second body of equal weight, and it is found that, started with any initial velocity, the test body retains its velocity almost unchanged. Casting aside the small observed retardation, doubtless arising from friction, we still cannot accept this result as a proof of the "law" in question. The body found

---

[8] R. V. SOUTHWELL, "Mechanics", *Encyclopædia Britannica* 14th ed., 1929. In the 1944 printing, pages 156–168 of Volume **15**.

[9] Lex I of I. NEWTON, *Philosophiae naturalis principia mathematica*, London, 1687. Translation of the 3rd edition by A. MOTTE, *Sir Isaac Newton's Mathematical Principles of Natural Philosophy and his System of the World*, London, 1729. There are many later editions, reprints, and translations.

to move with substantially uniform speed and direction is *not* a free body, and without the principles of mechanics, themselves dependent upon the law we are supposedly establishing by experiment, we cannot justly assert that the forces present do in fact neutralize each other. More elaborate application of the principles of mechanics is required if we are to reason that the inertia of the pulley has no effect on the ideal experiment. Further, to estimate the "experimental error" in the real experiment, we require a hypothesis of friction and an application of the laws of mechanics both for the effect of this friction and for the partially counteracting effect of the inertia of the pulley.

Such difficulties are avoided by the postulational standpoint, according to which physics, as an abstract discipline, may employ any variables and any consistent initial assumptions or "laws" which are convenient. In construction of this mathematical system it is not necessary to maintain contact with experiment at every stage. The system is an *abstract model*, designed to represent some of the observed phenomena of the physical universe, but directly concerned only with ideal bodies. Some few of the properties of these ideal bodies are postulated; the numerous remainder is to be derived mathematically. Whether these derived properties correspond with physical observation is a separate question, to be decided by subsequent comparison with experiment. But the available tests apply only to the system as a whole: We cannot devise an experiment such as to verify any one of its assumptions apart from the rest.

Naturally it is possible to construct an ideal system without relevance to physics. Nevertheless, since experience is the guide, entirely wrong physical theories have been rare. Rather, a well thought theory usually turns out to square with some phenomena but to err for others. Such is the case with the classical field theories. Their failures are well known and have provided the impetus for "modern" physics. Often their successes are forgotten. It is classical physics by which we grasp the world about us: the heavenly motions, the winds and the tides, the terrestrial spin and the subterraneous tremors, prime movers and mechanisms, sound and flying, heat and light[10].

Thus the classical field theories have won indisputable *permanence* in the language by which we speak of nature. Whatever the future revisions of theories on the structure of matter, the place of the classical field theories will remain unchanged. The permanence, along with the difficulties mentioned in § 1, makes necessary a *complete*

---

[10] *Cf.* the foreword of V. BJERKNES, J. BJERKNES, H. SOLBERG, & T. BERGERON, *Physikalische Hydrodynamik*, Berlin, Springer, 1933.

and independent presentation of the foundations of the classical field theories. Being mathematical disciplines, they should be derived from *axioms*.

Indeed, as his sixth problem HILBERT[11] set the construction of a set of axioms, on the model of the axioms of geometry, for "those branches of physics where mathematics now plays a preponderant part; first among them are probability theory and mechanics." Like all of his problems concerning physical applications of mathematics, his proposal for mechanics has received little attention. The possibility that the future may revise the physics of small corpuscles does not reduce the need for axiomatic treatment of the field theories. Physics, like mathematics, may be constructed precisely at several different levels. The interconnection of the different levels, either exactly or by approximation or by addition of new axioms, then furnishes definite mathematical problems[12].

Having reached agreement that we should base the classical field theories on a set of axioms, we must now admit, ruefully, our inability to do so. In our opinion, none of the attempts to form such a system has been successful. Only in very recent years has an adequate set of axioms for pure mechanics, at last, been constructed; it is the work of NOLL[13]. To present his development of the subject here would be premature, because a correspondingly clear and precise formulation of irreversible thermodynamics is not yet available. We regard the fully invariant formalism for electromagnetic theory given in our Chapter F as being essentially an axiomatization of the subject. COLE-

---

[11] D. HILBERT, Problem 6 in "Mathematische Probleme", *Nachrichten der Gesellschaft der Wissenschaften*, Göttingen, 1900, 253–297. Reprinted with additions, *Archiv der Mathematik und Physik* 1 (1901): 44–63, 213–217. Translated with additions, "Sur les problèmes futurs des mathématiques", pages 58–114 of *Comptes Rendus du 2ème Congrès International de Mathématique* (1900), Paris, 1902. Translation, "Mathematical problems", *Bulletin of the American Mathematical Society* (2) 8 (1902): 437–479.

[12] In mathematics the economy of such independent constructions has long been exploited. *E.g.*, to construct the complex numbers, we presume the properties of the real numbers given; for the real numbers, those of the integers; and for the integers, mathematical logic. To approach fluids in terms of nuclear physics, is like treating functions of a complex variable with the apparatus of formal logic.

[13] W. NOLL, *On the Foundations of the Mechanics of Continuous Media*, Carnegie Institute of Technology Report No. 17, Air Force Office of Scientific Research, 1957; "A mathematical theory of the mechanical behavior of continuous media", *Archive for Rational Mechanics and Analysis* 2 (1958/1959): 197–226 (1958); "The foundations of classical mechanics in the light of recent advances in continuum mechanics", pages 266–281 of *The Axiomatic Method, with Special Reference to Geometry and Physics* (1957), Amsterdam, North Holland Co., 1959. [The second and third of these works are reprinted on pages 1–47 of NOLL's *The Foundations of Mechanics and Thermodynamics, Selected Papers*, New York *etc.*, Springer-Verlag, 1974.]

MAN & NOLL have disclosed to us the outline of what appears to be a
satisfactory basis of general thermodynamics in deformable media.
Thus there are grounds for expecting that HILBERT's program will
shortly be actualized.

Despite the lack of complete axiomatic formulation, the *general
equations* governing the classical fields are known and universally
accepted. The present article is devoted to a *formally precise* study of
these general equations. Any future axiomatization, if successful, will
necessarily lead to these same equations [or to still more general ones
of the same kind].

4. MATHEMATICS AND ITS PHYSICAL INTERPRETATION

That a branch of theoretical physics is a mathematical science, by
no means implies its aim or interests to be those of pure mathematics.
Rather, *the problems are set by the subject.* The developments must
illumine the *physical aspects* of the theory, not necessarily in the narrower sense of prediction of numerical results for comparison with
experimental measurement, but rather for the grasp and picture of
the theory in relation to experience. [In the spirit of KELVIN &
TAIT[14],

> . . . we are engaged specially with those questions which best illus-
> trate physical principles—neither seeking, nor avoiding, difficul-
> ties of a purely mathematical kind.]

Some will reproach us with too much abstract and useless formalism. Not forgetting that such deprecation was directed toward WHIT-
TAKER's *Analytical Dynamics* half a century ago, we are confident that
the reader of half a century hence will regard our compromise of the
moment as erring rather toward insufficient use of the mathematical
tools available.

Any mathematical theory of physics must idealize nature. That
much of nature is left unrepresented in any one theory, is obvious;
less so, that theory may err in adding extra features not dictated by
experience. For example, the infinity of space is itself a *purely mathematical concept*[15], and all theories erected within this space must share

---

[14] § 453 of W. THOMSON & P. G. TAIT, *Treatise on Natural Philosophy* Part I,
Cambridge, Cambridge University Press, 1867. To its loss, this passage is swollen and
stiffened in the second edition, 1879.

[15] § 8 of L. EULER, *Mechanica sive Motus Scientia Analytice Exposita* 1, Petropoli,
1736 = LEONHARDI EULERI *Opera omnia* (II) 1.

in the geometrical idealization already implied. Indeed, it is difficult to find any theory that does not contain infinities, and infinities, by definition, are immeasurable. While at one time certain theoretical statements were regarded as "laws" of physics, nowadays many theorists prefer to regard each theory as a *mathematical model*[16] of some aspect of nature.

In a sense, then, every theory is only "approximate" in respect to nature itself. This unavoidable defect in theory is often taken as a patent for "approximate" mathematics in the deductions from it. Indeed, while mathematics is generally understood to proceed by entirely logical processes, were the "derivations" in some of the accepted physical papers of today translated into common reasoning they would fail to meet the logical standards of a competent historian or bibliographer. All too often is heard the plea that since the theory itself is only approximate, the mathematics need be no better. In truth the opposite follows. Granted that the model represents but a part of nature, we are to find what such an ideal picture implies. A result strictly derived serves as a *test of the model*; a false result proves nothing but the failure of the theorist. To call an error by a sweeter name does not emend it. The oversimplification or extension afforded by the model is not error: The model, if well made, shows at least how the universe *might* behave, but logical errors bring us no closer to the reality of *any* universe. *In physical theory, mathematical rigor is of the essence.*

In this treatise we attempt to keep the argument rigorous. Nevertheless, nothing is gained by laboring elementary details. We presume that the reader knows infinitesimal calculus, simple algebra, and tensor analysis; that he can supply for himself, without repetition on our part, conditions sufficient for interchanging differentiations, inversion of functions, expansions in power series, *etc.* Roughly speaking, our proportion of what is said to what is left unsaid is that which is customary in works on classical differential geometry.

5. EXACT THEORIES AND APPROXIMATE THEORIES

While every theory is a model of nature, and thus not "exact" in relation to it, nevertheless there is a government among theories. A theory is tested by experiment, and a range of confidence in it is established. In this sense, a given theory is "good"; if the range of

---

[16] § 1 of H. V. HELMHOLTZ, *Vorlesungen über Theoretische Physik* **2**, *Dynamik continuirlich verbreiteter Massen* (lectures of 1894), Leipzig, 1902.

application is greater than another's, it is the "better" of the two. For example, the theory of the flow of viscous compressible fluids should suffice to predict definite results, fit for experimental test, concerning the propagation, absorption, and dispersion of sound in fluids. That such results have never been obtained, is only from our lack of sufficient mathematics. Instead, a perturbation scheme has been used to infer equations governing "small" motions. The resulting acoustical theory is presumed to yield an "approximation" to the better but intractable theory of fluids.

Any given theory may be laid down as "exact". It is then a definite mathematical problem to discover the relation of its results to those derived from other theories, considered as "approximate" in respect to it. Problems of this kind are important and difficult, indeed in most cases too difficult for the mathematics available today. We do not attempt to study them in this treatise. Neither do we present unjustified linearizations or formal schemes of perturbation, which occupy much of the literature. Our scope is restricted to *exact treatment*.

6. CLOSED SYSTEMS AND ARMATURES

In the nineteenth century sets of physical laws were sought which should include the maximum range of physical phenomena yet remain sufficiently specific to predict definite results in particular cases. The culmination of this trend came in the systems of JAUMANN and LOHR[17], which postulate an all-embracing set of equations governing mechanics, electrodynamics, chemical reactions, diffusion, heat transfer, electromechanical effects, *etc.* Current knowledge of the structure of matter (*cf.* § 1) has destroyed the *raison d'être* of such closed systems as well as rendering them impractical.

Rather, the classical field theories offer us *armatures* upon which particular models of extended matter and electricity may be built. In this spirit, it is *inclusiveness* rather than particular problems that we seek here. For example, it is often claimed that in nature, if we look closely enough, only conservative forces occur; that such effects as friction are gross appearances resulting only from failure to know the underlying conservative process. But natural problems are not

---

[17] G. JAUMANN, "Geschlossenes System physikalischer und chemischer Differentialgesetze", *Sitzungsberichte der k.u.k. Akademie der Wissenschaften*, Wien, (IIa) **120** (1911): 385–530, and "Physik der kontinuierlichen Medien", *Denkschriften der k.u.k. Akademie der Wissenschaften*, Wien, **95** (1918): 461–562; E. LOHR, "Entropieprinzip und geschlossenes Gleichungssystem", *ibid.* **93** (1917): 339–421.

confined to those on the smallest or largest scale. The world about us, as we see it, must be mastered and controlled. Situations incompletely described are the rule, not the exception, and we must formulate good theories for these *limited aspects* of nature. Our object is a *general framework*[18] for such theories. The most general motions, the most general stresses, the most general flows of energy, and the most general electromagnetic fields furnish the subject of this treatise.

7. FIELD EQUATIONS AND CONSTITUTIVE EQUATIONS

Motion, stress, energy, entropy, and *electromagnetism* are the concepts used to build field theories. Certain laws of *conservation* or *balance* are laid down as relating these quantities in all cases. These basic principles, expressed as relations among integrals[19], in regions where the

---

[18] The field viewpoint is excellently apt to secure such generality, while to overcome the complications of corpuscular theories it is usual to make simplifying hypotheses which sharply lessen their scope. *Cf.* § 2 of HELMHOLTZ, *op. cit.* Footnote 16. As LAGRANGE remarked, the field view has precisely the same mathematical advantage over the corpuscular view as the differential theory of curves over polygonal approximations. *Cf.* ¶ 9 of § II of Section II of Part 1 of his *Méchanique Analitique*, Paris, Veuve Desaint, 1788. *Œuvres de* LAGRANGE 11 and 12 are the 5th edition.

[19] The view that all natural laws should be expressed by integrals is generally attributed to the Göttingen lectures of HILBERT. That jump conditions are not to be derived from smooth solutions was clearly understood by STOKES: "...I wish the two subjects to be considered as quite distinct." See page 353 of his "On a difficulty in the theory of sound", *Philosophical Magazine* 23 (1848): 349–356. A drastically condensed version, with an added note, appears on pages 51–55 of Volume 2 of STOKES's *Mathematical and Physical Papers* [in the reprint of 1966, edited by C. TRUESDELL, the full text is restored].

In recent years mathematicians have created various kinds of "generalized solutions", whereby, granted certain purely *analytic* presumptions as to the intended meaning of a problem formulated in terms of differential equations, discontinuous solutions may be inferred from continuous ones. We regard these approaches not only as demanding unnecessary mathematical apparatus but also as concealing the simple and immediate nature of physical laws. Neither in respect to rigor nor in any other regard do they offer advantages over HILBERT's program of stating physical laws in integral form. While the work of ZEMPLÉN and HELLINGER, which reflects HILBERT's influence, rests upon variational principles, the program of postulating *integral conservation laws*, which we follow here, seems first to have been laid down by KOTTLER. *Cf.* G. ZEMPLÉN, "Kriterien für die physikalische Bedeutung der unstetigen Lösungen der hydrodynamischen Bewegungsgleichungen", *Mathematische Annalen* 61 (1906): 437–449; E. HELLINGER, "Die allgemeinen Ansätze der Mechanik der Kontinua", pages 602–694 of Volume 4₄ of *Encyklopädie der mathematischen Wissenschaften*, 1914; F. KOTTLER, "Newton'sches Gesetz und Metrik", *Sitzungsberichte der k.u.k. Akademie der Wissenschaften*, Wien, (IIa) 131 (1922): 1–14, and "Maxwell'sche Gleichungen und Metrik", *ibid.* 119–146; §§ 8 and 81 of E. CARTAN, "Sur les variétés à connexion affine et la théorie de la relativité généralisée", *Annales de l'Ecole Normale Supérieure* (3) 40

variables change sufficiently smoothly are equivalent to differential *field equations*; at surfaces of discontinuity, to *jump conditions*.

The field equations and jump conditions form an underdetermined system, insufficient to yield specific answers unless further equations are supplied. Within the embracing concept of the balanced fields, certain further conditions define *ideal materials*[20]. These defining conditions are called *constitutive equations*. The most familiar constitutive equations, here expressed in words, are:

The distances between particles do not change (Rigid body).

The stress is hydrostatic (Perfect fluid).

The stress may be determined from the stretching alone (Viscous fluid, perfectly plastic body).

The stress may be determined from the strain alone (Perfectly elastic body).

The flux of energy is a linear function of the temperature gradient (Classical linear heat conduction).

The thermodynamic affinities are linear functions of the thermodynamic fluxes ("Irreversible thermodynamics").

The diffusion velocity of a constituent of a binary mixture is proportional to the gradient of its peculiar density (Classical linear mass diffusion).

(1923): 325–412; **41** (1924): 1–25; **42** (1925): 17–88; and § 3 of N. E. KOTCHINE, "Sur la théorie des ondes de choc dans un fluide", *Rendiconti del Circolo Matematico di Palermo* **50** (1926); 305–344. As VAŇ DANTZIG remarked, it is obvious that notions like differentiability can have no empirical basis at all. *Cf.* his "Some possibilities of the future development of the notions of space and time", *Erkenntnis* **7** (1937): 142–146. Since, on the contrary, such simple theories as the dynamics of perfect fluids and linear electromagnetism are known to furnish inadequate models of experience when the fields appearing are required to be everywhere continuous, no physical principle should be stated in differential form.

The older approach, still followed in many textbooks, either sets up more special postulates for discontinuities or employs an unrigorous limit process using differentiable solutions, presumed to exist.

[20] The program of mechanics was laid out by EULER; *cf.* § 19 of *op. cit.* Footnote 1, and his remarks on rigid bodies and ideal fluids in § 12 of his "Sectio prima de statu aequilibrii fluidorum", *Novi commentarii Academiae scientiarum imperialis Petropolitanae* **13** (1768): 305–416 (1769) = LEONHARDI EULERI *Opera omnia* (II) **13**, 1–72. The program of continuum mechanics in particular was proclaimed by CAUCHY, "Recherches sur l'équilibre et le mouvement intérieur des corps solides ou fluides, élastiques ou non élastiques", *Bulletin de la Société Philomatique* 1823, 9–13 = *Œuvres de* CAUCHY (2) **2**: 300–304, but the later emphasis on linear problems in very special theories caused it to be largely forgotten until it was stated anew by R. v. MISES, "Über die bisherigen Ansätze in der klassischen Mechanik der Kontinua", pages 1–9 of Volume **2** of *Proceedings of the International Congress of Applied Mechanics*, Stockholm, 1930.

The electric displacement is proportional to the electric field; the magnetic induction is proportional to the magnetic intensity (Classical linear electromagnetism).

The constitutive equations and field equations together, along with the jump conditions and boundary conditions, should lead to a definite theory, predicting specific answers to particular problems. For some of the special materials listed above, this definiteness has been proved through theorems of existence and uniqueness.

The present treatise is devoted to the *general principles of balance* alone. Thus we deal only with the *field equations and jump conditions*. Our last chapter mentions guiding principles by which rational constitutive equations may be formulated.

8. The Nature and Plan of This Treatise

We present the *common foundation* of the field viewpoint[21]. We aim to provide the reader a full panoply of *tools of research*, wherewith he himself, put into possession not only of the latest discoveries but also of the profound but all too often forgotten achievements of earlier generations, may set to work as a theorist.

This treatise is intended for the specialist, not the beginner. Necessarily it presents the foundations of the field theories, not as they appeared in the last century and linger on in the textbooks, nor as the experts in some other domains may think they ought to be presented[22], but as they are cultivated by the specialists of today.

This treatise is organized as follows:

1. Kinematics, including conservation of mass, in Chapters B and C.

2. Balance of momentum, Chapter D.

3. Balance of energy, including the thermodynamics of irreversible deformations, Chapter E.

4. Balance of charge-current and magnetic flux, Chapter F.

---

[21] The only single work attempting even a major part of our subject is the book of A. BRILL, *Vorlesungen zur Einführung in die Mechanik raumerfüllender Massen*, Leipzig & Berlin, 1909.

[22] *Cf.* KELVIN & TAIT in the preface to *op. cit.* Footnote 14: "... where we may appear to have rashly and needlessly interfered with methods and systems of proof in the present day generally accepted, we take the position of Restorers, not of Innovators."

5. Guiding principles for constitutive equations and examples of them, Chapter G.

We interpret "classical" in the narrower sense, as confined to phenomena in *Euclidean three-dimensional space*, and governed by *Newtonian mechanical principles*. Nevertheless it would be crippling to maintain this restriction in dealing with electromagnetism. The four-dimensional viewpoint adopted in Chapter F necessitates further kinematical developments there and results in some duplication of subject. On the other hand, to have begun with a world-invariant formalism in the earlier chapters would have greatly lessened their direct usefulness to specialists in mechanics.

That over one half of the work is devoted to kinematics, the mathematical description of motion, is not malapropos. As the need for more and more general field theories has grown, the preliminary light which kinematics unencumbered by physical restrictions can provide, always appreciated by virtuosi of mechanics[23], has become a necessity. In presenting here as our Chapters B and C the first general treatise on the kinematics of continua, we believe that we look toward the future course of the field theories.

9. Tradition

We have tried to supply full and correct attributions, not only for historical perspective but also in plain justice. If the name attached to many a proposition is but a small [or old] one, that is all the less reason that its owner should be pilled of what little he wrought by a no greater name of today, whose slight capacities are scarcely increased by wilful or heedless ignorance of what others have done. Nonetheless the multitude of detailed citations should not prevent the great

---

[23] We follow the tradition of EULER, CAUCHY, and KELVIN. *Cf.* the remarks of H. v. HELMHOLTZ, introduction to "Über Integrale der hydrodynamischen Gleichungen, welche den Wirbelbewegungen entsprechen", *Journal für die reine und angewandte Mathematik* **55** (1858): 25–55 = pages 101–134 of Volume **1** of his *Wissenschaftliche Abhandlungen*. Translation by P. G. TAIT, "On integrals of the hydrodynamical equations, which express vortex-motions", *Philosophical Magazine* (4) **33** (1867): 485–512; Н. Е. Жуковский, "Кинематика Жидкого Тела", *Математический Сборник* **8** (1876): 1–79, 163–238 (also published separately, Moscow, 1876) = (with English summary) pages 7–148 of Volume **2** of his *Полное Собрание Сочинений*, Moscow & Leningrad, 1935; A.-J.-C. BARRÉ DE ST. VENANT, "Géométrie cinématique.—Sur celle des déformations des corps soit élastiques, soit plastiques, soit fluides", *Comptes Rendus des Séances Hebdomadaires de l'Académie des Sciences* (Paris) **90** (1880): 53–56, 209; and G. JAUMANN, introduction to *Die Grundlagen der Bewegungslehre von einem modernen Standpunkte aus*, Leipzig, 1905.

names from emerging. Our subject is largely the creation of EULER and CAUCHY. If we present their results in forms often different from the original, in return we have included many of their discoveries that have not previously found a place in expositions. Not only will their names be the most frequently encountered, but also they appear at the crucial theorems and definitions. Next come STOKES, HELM-HOLTZ, KIRCHHOFF, KELVIN, MAXWELL, and HUGONIOT. In the twentieth century, HADAMARD and HILBERT[24] continued and deepened the tradition. That no one later name is frequently cited, does not indicate that the subject is dead. Rather, after a generation of quiescence, in very recent years it has experienced a revival in a form more compact and general, and, we believe, closer to nature.

Note for the Reprinting

The Classical Field Theories is an article in Part 1 of Volume III of FLÜGGE'S *Encyclopedia of Physics*, Berlin *etc.*, Springer-Verlag, 1960. Its §§ 1–9 are here reprinted essentially unaltered; some references to later sections of the article are removed, some infelicities in expression are emended; the few alterations of sense are put within square brackets.

The statement in § 2 that NEWTON considered the mass-point the fundamental quantity of nature is traditional and wrong. While NEWTON often (but by no means always) considered bodies of small size, I can find no evidence that he ever envisioned mass-points. For details see § 8 of Essay 39, below in this volume.

The portion of § 3 printed in small type, which refutes the "operationalist" philosophy rampant in the 1940s and 1950s, may seem unnecessary now, since that philosophy is no longer in vogue among theoretical physicists. Nevertheless, much as pigtails for boys have disappeared from campuses but now frequently adorn bricklayers and plumbers, operationalism seems to have descended to circles of he-man experiment in engineering. Therefore I append here an early and magisterial refutation of it by J. J. THOMSON in a broadcast address printed in *The Listener*, 29 January, 1930, and reprinted on page 265 of Lord RAYLEIGH's *The Life of Sir J. J. Thomson*, Cambridge University Press, 1942:

There is now a school of mathematical Physicists which objects to the introduction of ideas which do not relate to things which can actually be observed and measured. Thus, before very high vacua were obtained, it would not have been legitimate to speak of the mass or position of a molecule, but only of that of a finite volume of gas. The atomic theory of

---

[24] While HILBERT published nothing about our subject, his influence was widespread and continuing. *Cf.*, *e.g.*, ZEMPLÉN, *op. cit.* Footnote 19: HELLINGER, Footnote 6 of *op. cit.* Footnote 19. The organization of HILBERT's lectures at Göttingen in 1906/1907, *Mechanik der Continua*, has influenced ours; one of us has studied the manuscript notes by W. MARSHALL in the library of the University of Illinois and by A. R. CRATHORNE in the library of Purdue University.

chemistry and the Kinetic Theory of Gases would have had to wait until the technique of high vacua had been developed. A similar view was introduced into metaphysics long ago by Bishop Berkeley, who held that it was impossible to maintain that a quality existed unless one knew how to measure its magnitude. I believe that this view now gets no support from metaphysicians. I think it is bad Physics as well as bad Metaphysics. I hold that if the introduction of a quantity promotes clearness of thought, then even if at the moment we have no means of determining it with precision, its introduction is not only legitimate but desirable. The immeasurable of to-day may be the measurable of to-morrow. A striking example of this is that a movement was started by some chemists at the end of last century to give up thinking in terms of the atomic theory on the ground that the mass of an atom could not be measured. By the irony of fate the movement had hardly begun when a method of measuring the mass was discovered. It is dangerous to base a philosophy on the assumption that what I know not can never be knowledge.

The last line may refer obliquely to a jingle about JOWETT which was still current in my days in Oxford:

> I am the Master of Balliol College;
> What I know not, is not knowledge.

A similar jingle at Cambridge concerned WHEWELL, I am told. I often encounter the same assurance in physicists when matters of classical mechanics or thermodynamics are broached.

While I still hold the view expressed in Footnote 19, I should now word it more carefully. We did not mean to dismiss all generalized solutions but only to voice some reserve regarding them at a time when they were widely regarded as the cure of any difficulty that might arise. Even less did we mean to claim that our statements of the integral laws of balance were final expressions of the conceptual content of the laws themselves. Indeed, I am not convinced that there is such a thing as a final expression of any basic principle of science. To emphasize this openness toward future extension, in this reprint I have weakened the statements with which §§ 1 and 3 ended in the original.

The careful reader of *The Classical Field Theories* will have noticed three passages in which we alluded to possible enlargements of its general framework. One is the discussion of "general moments" in §§ 166 and 205, which may have some claim of paternity for "multipolar" theories. Another is the general description of oriented materials in §§ 60 and 61, from which it is a short and easy step to "micropolar" materials. A third, the most important, is the brief presentation of impulses in §§ 198, 206, and 242. Weak formulations play an essential part in the major memoir in which S. S. ANTMAN & J. E. OSBORN prove the impulse-momentum laws to be equivalent to a principle of virtual work: "The principle of virtual work and integral laws of motion", *Archive for Rational Mechanics and Analysis* **69** (1979): 231–261.

Both § 3 of this essay and the entirety of the preceding would have profited from quotation of parts of Chapter III, called "Experience", in the first edition of KELVIN & TAIT's *Treatise on Natural Philosophy*, Oxford, Clarendon Press, 1867:

Observation and experiment
369. By the term Experience, in physical science, we designate according to a suggestion of Herschel's, our means of becoming

acquainted with the material universe and the laws which regulate it. In general the actions which we see ever taking place around us are *complex*, or due to the simultaneous action of many causes. When, as in astronomy, we endeavour to ascertain these causes by simply watching their effects, we observe; when, as in our laboratories, we interfere arbitrarily with the causes or circumstances of a phenomenon, we are said to experiment.

I can only regret that when I wrote or draughted my explanations, I did not instead quote this paragraph in its magisterial clarity and authority. KELVIN & TAIT continue:

Observation

370. For instance, supposing that we are possessed of instrumental means of measuring time and angles, we may trace out by successive observations the relative position of the sun and earth at different instants. . . .

371. In general all the data of Astronomy are determined in this way, and the same may be said of such subjects as Tides and Meteorology. . . .

372. Even in the instance we have chosen above, that of the planetary motions, the observed effects are complex; because unless possibly in the case of a double star, we have no instance of the *undisturbed* action of one heavenly body on another. . . .

Experiment

373. Let us take a case of the other kind—that in which the effects are so complex that we cannot deduce the causes from the observation of combinations arranged in Nature, but must endeavour to form for ourselves other combinations which may enable us to study the effects of each cause separately, or at least with only slight modification from the interference of other causes.

A stone, when dropped, falls to the ground; a brick and a boulder, if dropped from the top of a cliff at the same moment fall side by side, and reach the ground together. But a brick and a slate do not; and while the former falls in a nearly vertical direction, the latter describes a most complex path. A sheet of paper or a fragment of gold leaf presents even greater irregularities than the slate. But by a slight modification of the circumstances, we gain a considerable insight into the nature of the question. The paper and gold leaf, if rolled into balls, fall nearly in a vertical line. Here, then, there are evidently at least two causes at work, one which tends to make all bodies fall, and that vertically; and another which depends on the form and substance of the body, and tends to retard its fall and alter its vertical direction. How can we study the effects of the former on all bodies without sensible complication from the latter? The effects of Wind, etc., at once point out *what* the latter cause is, the air (whose existence we may indeed suppose to have been discovered by such effects); and to study the nature of the action of the former it is necessary to get rid of the complications arising from the presence of air. Hence the necessity for *Experiment*. By means of an apparatus . . . , we remove the greater part of the air from the interior of a vessel, and in *that* we try again our experiments on the fall of bodies; and now a general law, simple in the extreme, though most important in its consequences, is at once apparent. . . .

KELVIN & TAIT go on to sketch an order of command in theories:

382. Where, as in the case of the planetary motions and disturbances, the forces concerned are thoroughly known, the mathematical theory is absolutely true, and requires only analysis to work out its remotest details. It is thus, in general, far ahead of observation, and is competent to predict effects not yet even observed. . . .

383. Another class of mathematical theories, based to a certain extent on experiment, is at present useful, and has even in certain cases pointed to new and important results, which experiment has subsequently verified. Such are the Dynamical Theory of Heat, the Undulatory Theory of Light, etc. etc. In the former, which is based upon the experimental fact that *heat is motion*, many formulæ are at present obscure and uninterpretable, because we do not know *what* is moving or *how* it moves. Results of the theory in which these are not involved, are of course experimentally verified. The same difficulties exist in the Theory of Light. But before this obscurity can be perfectly cleared up, we must know something of the ultimate, or *molecular*, constitution of the bodies, or groups of molecules, at present known to us only in the aggregate.

384. A third class is well represented by the Mathematical Theories of Heat (Conduction), Electricity (Statical), and Magnetism (Permanent). Although we do not know *how* Heat is propagated in bodies, nor *what* Statical Electricity or Permanent Magnetism are—the laws of their forces are as certainly known as that of Gravitation, and can therefore like it be developed to their consequences, by the application of Mathematical Analysis. . . .

Different species of mathematical theories of physics
386. Mathematical theories of physical forces are in general of one of two species. First, those in which the fundamental assumption is far more general than is necessary. Thus the equation of Laplace's Functions . . . contains the mathematical foundation of the theories of Gravitation, Statical Electricity, Permanent Magnetism, Permanent Flux of Heat, Motion of Incompressible Fluids, etc. etc., and has therefore to be accompanied by limiting considerations when applied to any one of these subjects.

Again, there are those which are built upon a few experiments, or simple but inexact hypotheses, only; and which require to be modified in the way of extension rather than limitation. As a notable example of such, we may give the whole subject of Abstract Dynamics, which requires extensive modifications . . . before it can in general be applied to practical purposes.

In 1982 the examples are no longer the same as in 1867; we no longer regard the Newtonian mathematical theory of the planetary motions as "absolutely true", in reference to heat we do now have at least a rough idea of "*what* is moving" and of "*how* it moves", but the classes distinguished by KELVIN & TAIT still exist (though with different members) and still serve to illustrate the roles of experience, observation, experiment, hypothesis, and theory.

3. MODERN CONTINUUM MECHANICS IN ADOLESCENCE (1962)

a. Preface (1962) to the Reprint of *The Mechanical Foundations of Elasticity and Fluid Dynamics* (1952)

1. CIRCUMSTANCES OF "THE MECHANICAL FOUNDATIONS"

That reprinting of this work is now called for, indicates a great change in circumstances against the time when it was written. When, in 1946, I first began to study the foundations of continuum mechanics, within a few months I had set the whole field in order, to my own satisfaction. I quickly wrote and submitted to an international meeting an expository memoir, which was rejected. In view of the quality of papers accepted by the same meeting, I was naive enough to be astonished as well as disappointed, and I sent the manuscript for criticism to a number of experts. Most of these did not deign to acknowledge it or to reply, but two did. Mr. FRIEDRICHS told me I had underestimated the work of earlier authors. Since my information concerning it was drawn from reputable textbooks, I turned, somewhat taken aback, to the sources they cited, and then to the sources cited by those sources, and so on, until within a year I found out how right he was and how little I had seen of the real issues faced by the great natural philosophers one and two centuries ago. During this period I first saw Mr. REINER's papers of 1945 and 1948. On 23 December 1948 Mr. V. MISES asked me to write a general exposition of recent theories of deformable masses for a volume he was editing. I set to work at once and completed the article, severely condensed so as to keep within twice the space allowed, at the appointed time; on 23 May 1949 Mr. V. MISES acknowledged receipt of the final manuscript. I had first gotten in touch with Mr. RIVLIN in January, although I did not master his work in time for it to assert the influence it ought have had upon the structure of the article, my expositions of it being of the nature of insertions and appendices to the original plan.

These details are mentioned now because they explain some of the shortcomings of the article as printed in 1952. It was never possible to alter the structure. First, the publisher, after holding the manuscript for six months, decreed it had to be retypewritten within three weeks. Working day and night, I took this occasion to go over everything in detail and add references and brief descriptions of what I had learned in the interim. The publisher [—a distressed immigrant who in Germany had owned a respectable press for scientific books—] held the new manuscript for eighteen months before informing the editor that it contained too many symbols for any printer to handle; besides, it was too long and contained too many equations, footnotes, and references. In particular, only citation of recent literature could be useful to scientists.

In those days papers on the foundations of continuum mechanics were rejected by journals of mathematics as being applied, by journals of "applied" mathematics as being physics or pure mathematics, by journals of physics as being mathematics, and by all of them as too long, too expensive to print, and of interest to no-one. The anonymous referees succeeded in displaying not only their contempt for the subject but also their pitiable ignorance of it. It was time to establish a new journal, devoted especially to the foundations of mechanics and to related mathematics. The *Journal of Rational Mechanics and Analysis* was organized in 1951.

I withdrew my article from the publisher who had held it for a year and a half, but he refused to return it, having suddenly discovered it could be printed after all. Again working day and night, within a month I reconstructed it from an imperfect copy and old notes, adding parenthetically much new material. It appeared in January and April, 1952, in the first two numbers of the new *Journal*. Extensive corrections and additions were published in 1953.

Thus, despite references running as late as 1952, *The Mechanical Foundations* is essentially a work of 1949. I record the dates because the years 1949/1952 were decisive ones for modern continuum mechanics. *The Mechanical Foundations* resumes the subject as of 1949, supplemented by notes of detail pointing in the way it had by then started to go.

2. Subsequent Plans for a General Exposition

By September, 1953, when the *Corrections and Additions* were published, I had agreed to write, in collaboration with others, a new exposition for FLÜGGE's *Encyclopedia of Physics*. It was to include everything in *The Mechanical Foundations*, supplemented by fuller development of the field equations and their properties in general; emphasis was to be put on principles of invariance and their representations;

but the plan was much the same. As the work went on into 1954, the researches of RIVLIN & ERICKSEN and of NOLL made the underlying division into fluid and elastic phenomena, indicated in the very title *The Mechanical Foundations of Elasticity and Fluid Dynamics*, no longer a natural one. It was decided to split the projected article into two parts, one on the general principles, and one on the constitutive equations. The former part, with the collaboration of Mr. TOUPIN and Mr. ERICKSEN, was completed and published in 1960 in Part 1 of Volume III of the *Handbuch* as *The Classical Field Theories*. In over 600 pages it gives in full the material sketched in Chapters II and III of *The Mechanical Foundations*.

The second article, to be called *The Non-linear Field Theories of Mechanics*, Mr. NOLL and I are presently engaged in writing. The rush of fine new work has twice caused us to change our basic plan and rewrite almost everything. The volume of grand and enlightening discoveries since 1955 has made us set aside the attempt at complete exposition of older things. It seems better now to let *The Mechanical Foundations* stand as final for most of the material included in it, and to regard the new article not only as beginning from a deeper and sounder basis than could have been reached in 1949 but also as leaving behind it certain types of investigation that no longer seem fruitful, no matter what help they may have provided when new.

To this reason for reprinting *The Mechanical Foundations* may be added a second one. Since it was written, several other expositions of the field have appeared, and with the growth of interest in continuum mechanics, more of them stand ready for the press or in it. None of these seems to me to fill the purpose for which *The Mechanical Foundations* was intended. All are of the nature either of specialized monographs or of "elementary" explanations. Those by experts on modern continuum mechanics present mainly summaries of their authors' previous researches; those by outsiders have mined *The Mechanical Foundations* for content and references but do not always succeed in digesting either before explaining them. Some use a restricted kind of co-ordinates, that on the one hand can, and occasionally does, lead the inexpert to error while on the other has not been shown to produce even in expert hands new discoveries of importance. I can point with pride to several younger men who today are making the finest discoveries in continuum mechanics as having been drawn into the field by a desire to correct and improve what they read and found not to their liking in *The Mechanical Foundations*.

Thus, despite the now all too evident shortcomings of *The Mechanical Foundations*, I have decided to reprint it, especially so as to place in the hands of serious beginners a *direct* intermediary to *all* the earlier sources.

3. The Nature of the Reprint

The present volume makes no attempt to give an exposition of continuum mechanics as it stands today. It contains a *corrected reprint*, incorporating the corrections and additions already published in 1953 and 1954, as well as further corrections of typographical errors or slips in calculation or logic. I wish to reiterate here the expression of gratitude, already published in 1953, to Mr. Ericksen for his great help at that time and to add acknowledgment to Mr. P. L. Sheng (1954) and to Mr. A. J. A. Morgan (1962) for detecting further errors, as well as to thank the dozens of correspondents who have written or telephoned in discussion of one or another point. . . .

In Appendix 1 will be found my *Program of Physical Research* (1953), and in Appendix 3 some comments upon it. That paper was designed to attract physicists; I had intended it for an American journal devoted to review articles, which, of course, was willing to accept it only subject to alterations dictated by an anonymous referee who showed himself innocent of experience or taste in the subject. The somewhat badgered tone in which that paper is written may seem inexplicable to younger men starting to work in continuum mechanics today, when a dozen journals in six countries welcome studies of the foundations, when research in the nonlinear theories does not have to hide under some corner of aerodynamics, strength of materials, computing, or "applied" mathematics, and when a dozen major universities give courses in general continuum mechanics.

While few physicists have been drawn into the field, interest has come from a quarter totally unexpected in 1952: chemists and engineers. For them, some of the considerations in the *Program* are irrelevant. Nevertheless I have chosen to reprint this paper because it still serves its intended purpose as a nonmathematical introduction to *The Mechanical Foundations*.

4. Retrospect upon The Mechanical Foundations

The reader will easily form for himself a judgment of the details, and so their shortcomings need not be listed here. Toward more general ends, nevertheless, he who comes fresh to the subject may not so easily see what was wanting. Having just reread *The Mechanical Foundations*, I find it a good summary of what was done up to 1949 but less successful in showing where the field was going. I tried to keep close to such physical phenomena as then seemed already fairly well covered by theory, and I intentionally avoided what then seemed generality for its own sake. That was wrong. My failure to appreciate at its true value the fundamental memoir of OLDROYD[1] is a symptom of my failure to see then that what the field needed was *generality in order to reach simplicity and clarity.* As a corollary, the mathematical equipment called upon was too meager to represent efficiently and easily the phenomena of nature. The physical experience was long at hand, and some good experiments, too, but we lacked the mathematical maturity to use them well. The newer work rests upon recognition of tensors as operators rather than sets of functions, of the algebraic as fundamental to the differential and integral aspects of theories of invariance, of additive set functions as the building blocks of classical mechanics, and of memory functional as the basic concept for constitutive equations. The reader may see this for himself in the three following volumes of reprints.[2]

b. Two Annotations (1962) to the Reprint of
The Mechanical Foundations of Elasticity and Fluid Dynamics

1. APPROXIMATION AND INTUITION

Except possibly in reflex response to stimuli toward writing papers as a social or professional manifestation, special or approximate theories of elasticity seem to be introduced for two reasons: (1) to serve as basis for deriving equations for stability criteria, plates, shells, or rods, and (2) to enable the solution of special problems too difficult to treat more generally. That there is much and good reason to doubt results growing from the former purpose, is shown by the ocean of literature of this kind, all firmly grounded in "physical" considerations, leading to consequent theories at variance with one another and hence opening the floodgates for subsidiary inundations of further papers "explaining" the differences. The latter purpose is a just one but much abused, since the far preponderant part of the literature concerns problems that can be, and in many cases have been, solved by systematic perturbation procedures in the general theory. Judged pragmatically, the MOONEY-RIVLIN theory is the only

---

[1] J. G. OLDROYD, "On the formulation of rheological equations of state", *Proceedings of the Royal Society* (London) **A202** (1950): 523–541.

[2] *Continuum Mechanics.* II. *The Rational Mechanics of Materials*; III. *Foundations of Elasticity Theory*; IV. *Problems of Non-Linear Elasticity*, being Parts II–IV of International Science Review Series, Volume **8**, edited by C. TRUESDELL, New York *etc.*, Gordon & Breach, 1965.

special one that has proved any real usefulness: Various equations that can be derived but not solved for general incompressible materials can be solved explicitly for the MOONEY-RIVLIN theory, *etc*. In most cases, the type of problem known from the linearized theory can be generalized in a straightforward way by the systematic perturbation procedures, *e.g.*, it is pointless to study combined torsion and tension in a special theory, since for general elastic materials it can be treated once and for all, equally explicitly and to any desired degree of approximation, by anyone sufficiently patient. Special theories serve a useful end mainly for problems of bifurcation, instability, non-periodic oscillation, *etc.—e.g.* for the eversion of a sphere or cylinder, or for the radial oscillation of a tube subject to a very great pressure impulse.

"Physical intuition" seems to have gained nothing solid in this field, since those who invoke it, by the disagreement of their results from those of others who invoke it, negate any claim for the plausibility of their guesses and "approximations". Rather, the goddess to whom they pour out libations seems to be *mathematical simplicity*, a fickle mistress for the student of natural philosophy.

I add these remarks because one might have thought that the collection and comparison of the contents of some 200 papers of this kind in §§ 48–54 would have reduced the subject to endemicity, but since 1952 its seminal potence is witnessed by at least double that number of papers, some from flourishing new colonies in faraway lands.

2. COMPARISON OF RESULTS WITH EXPERIMENT

[My stand on this matter in *The Mechanical Foundations*] was ill taken. The theorist, for clarity as well as for economy, should keep his work as *general* as he can while still getting *concrete* results. It is not his duty to fit data from experiments. Rather, he must prove theorems that demonstrate *distinctions* between results of various theories or classes of theories. At best, he should show not only what assumptions *suffice* to get a theory in agreement with the phenomena but also which ones are *necessary* for that agreement. In an ideal program (never carried out so far), the physical phenomena should dictate the theory. In the past, physics has generally been made to bow to the limitations of the theorists, coming to be regarded as "explained" as soon as *any* theory has somehow been stretched over the observed facts.

c. Two Annotations (1962) to *A Program of Physical Research in Classical Mechanics* (1953)

1. THE PHYSICISTS' DISADVANTAGE WHEN FACING COMMON EXPERIENCE

The training of professional physicists today puts them under heavy disadvantage when it comes to understanding physical phenomena, much as did a training in theology some centuries ago. Ignorance commonly vents itself in expressions of contempt. Thus "physics", by definition, is become exclusively the study of the structure of matter, while anyone who considers physical phenomena on a supermolecular scale is kicked aside as not being a "real" physicist. Often "real" physicists let it be known that all gross phenomena easily *could* be described and predicted perfectly well by structural theories; that, aside from the lack of "fundamental" (*i.e.* structural) interest in all things concerning ordinary materials such as water, air, and wood, the blocks to a truly "physical" (*i.e.* structural) treatment are "only mathematical".

It is curious that these same persons are often sympathetic to non-structural theories on the borders of physics. They do not suggest that a mathematical psychologist investigating models for behavior would do better to try to integrate the equations of motion for the elementary particles making up a rat. They do not even consider a traffic engineer stupid for neglecting to make use of physical and chemical principles determining the motion of the automobile when he sets up his stochastic theories for traffic control. The idea that continuum mechanics is somehow less accurate or "physical" for predicting mechanical phenomena in gross bodies than is a structural theory is a pure illusion, now beginning to give way a little to reason. For most of the physical phenomena of ordinary experience, considerations of the structure of matter do not yield a finer or more accurate theory: They do not yield *any theory at all*, any more than nuclear physics, however true, and however useful for studying nuclei in small numbers, gives *any information at all* about the behavior of a rat, or than classical mechanics, however true, and however useful for explaining the motion of a single automobile, gives *any information at all* about traffic flow. To say that a problem can be solved "in principle" when its solution is conjectured to be obtainable by the combined efforts in calculation of all inhabitants of the earth for 10^6 years, is no more a scientific statement than to say it is solved "in principle" because its solution is presumed known to the mind of GOD.

2. What Experiment Can Do

Cf. Experience, Theory, and Experiment [reprinted as Essay 1 in this volume]. Historical evidence that the discoverers of the classical theories [of the mechanics of fluids and solids in the eighteenth century] did not base them directly on experiments may be found in my prefaces to Volumes 10, 11, 12, and 13 of Series II of LEONHARDI EULERI *Opera omnia*.

While the current adulation of numerical data needs to be purged by a dose of common sense, I am dismayed to hear myself quoted as advising theorists to disregard experiments. Docility is grown so common among scientists as to make it second nature for them, if freed from one bad master, to run straightway to a worse one.

More specifically, to my astonishment I find it necessary nowadays to point out that the lack of decisive experimental tests in favor of the Navier-Stokes equations and the scarcity of decisive experimental evidence for linearized elasticity do not show these theories physically wrong. The scientist of today has been taught to choose sides and swear allegiance immediately; doubt plays no part in his thinking. Lack of evidence is, at worst, lack of evidence, not contrary evidence. Moreover, the best that experiment can do is fail to controvert a theory. Theories such as those of linear viscosity and linearized elasticity, for which we have great and long-standing bodies of indirect experimental evidence that fails, for the most part, to controvert them, deserve to be studied with all attention and respect (though not idolatry), especially when, as they do, they afford mathematically simple representations for natural abstracts of certain *aspects* of everyday physical experience.

Note for the Reprinting

All of the passages reprinted in this number are extracted from *Continuum Mechanics. I. The Mechanical Foundations of Elasticity and Fluid Dynamics*, International Science Review Series, Part I of Volume **8**, New York *etc.*, Gordon & Breach, 1966.

4. PURPOSE, METHOD, AND PROGRAM OF NONLINEAR CONTINUUM MECHANICS

Introduction to *The Non-Linear Field Theories of Mechanics* (1965), co-author W. NOLL

CONTENTS

1. PURPOSE OF THE NONLINEAR THEORIES

Matter is commonly found in bodies consisting of materials. Analytical mechanics turned its back upon this fact, creating the centrally useful but abstract concepts of the mass-point and the rigid body, in which matter manifests itself only through its inertia, independent of its constitution; "modern" physics likewise turns its back, since it concerns solely the small particles of matter, declining to face the problem of how a specimen made up of such particles will behave in the typical circumstances in which we meet it. Materials, nonetheless, continue to furnish the masses of matter we see and use from day to day: air, water, earth, flesh, wood, stone, steel, concrete, glass, rubber, ... All are *deformable*. A theory aiming to describe their mechanical behavior must take heed of their deformability and represent the definite principles it obeys.

The rational mechanics of materials was begun by JAMES BERNOULLI, illustrated with brilliant examples by EULER, and lifted to generality by CAUCHY. The work of these mathematicians divided the subject into two parts. First, there are the *general principles*, common to all media. A mathematical structure is necessary for describing deformation and flow. Within this structure, certain physical laws governing the motion of all finite masses are stated. These laws, expressed nowadays as integral equations of balance, or "conservation laws", are equivalent either to *field equations* or to *jump conditions*, depending on whether smooth or discontinuous circumstances are contemplated. Specifically, the axioms of continuum physics assert the balance or conservation of *mass, linear momentum, moment of momentum, energy, electric charge*, and *magnetic flux*. There is a seventh law, a principle of *irreversibility*, expressed in terms of the *entropy*, but the true form of this law, in the generality we keep here, is not yet known. (Note added in proof: Major progress toward finding this law has been made by COLEMAN. . . .). . .

The general physical laws in themselves do not suffice to determine the deformation or motion of a body subject to given loading. Before a determinate problem can be formulated, it is usually necessary to specify the *material* of which the body is made. In the program of continuum mechanics, such specification is stated by *constitutive equations*, which relate the stress tensor and the heat-flux vector to the motion. For example, the classical theory of elasticity rests upon the assumption that the stress tensor at a point depends linearly on the changes of length and mutual angle suffered by elements at that point, reckoned from their configurations in a state where the external and internal forces vanish, while the classical theory of viscosity is based on the assumption that the stress tensor depends linearly on the instantaneous rates of change of length and mutual angle. These statements cannot be universal laws of nature, since they contradict one another. Rather, they are *definitions of ideal materials*. The former expresses in words the constitutive equation that defines a *linearly and infinitesimally elastic material*; the latter, a *linearly viscous fluid*. Each is consistent, at least to within certain restrictions, with the general principles of continuum mechanics, but in no way a consequence of them. There is no reason *a priori* why either should ever be physically valid, but it is an empirical fact, established by more than a century of test and comparison, that each does indeed represent much of the mechanical behavior of many natural substances of the 'most various origin, distribution, touch, color, sound, taste, smell, and molecular constitution. Neither represents all the attributes, or suffices even to predict all the mechanical behavior, of any one natural material. No natural body is perfectly elastic or perfectly fluid, any more than any is perfectly rigid or perfectly incompressible. These trite observations do not lessen the worth of the two particular constitutive equations just mentioned. That worth is twofold: First, each represents in ideal form an *aspect*, and a different one, of the mechanical behavior of nearly all natural materials, and, second, each does predict with considerable though sometimes not sufficient accuracy the observed response of *many different natural materials in certain restricted situations.*

Pedantry and sectarism aside, the aim of theoretical physics is to construct mathematical models such as to enable us, from use of knowledge gathered in a few observations, to predict by logical processes the outcomes in many other circumstances. Any logically sound theory satisfying this condition is a good theory, whether or not it be derived from "ultimate" or "fundamental" truth. It is as ridiculous to deride continuum physics because it is not obtained from nuclear physics as it would be to reproach it with lack of foundation in the Bible. The conceptual success of the classical linear or infinitesimal field theories is perhaps the broadest we know in science: In terms of them we face, "explain", and in varying amount control, our daily environment: winds and tides, earthquakes and sounds, structures and mechanisms, sailing and flying, heat and light.

There remain, nevertheless, simple mechanical phenomena that are clearly outside the ranges of the infinitesimal theory of elasticity and of the linear theory of viscosity. For example, a rod of steel or rubber if twisted severely will lengthen in proportion to the square of the twist, and a paint or polymer in a rotating cup will climb up an axial rod. Moreover, the finite but discrete memory of the elastic material and the infinitesimal memory of the viscous fluid are obviously idealized limiting cases of the various kinds of cumulative memories that natural materials show in fast or slow or repeated loading or unloading, leading to the phenomena of creep, plastic flow, strain hardening, stress relaxation, fatigue, and failure.

2. METHOD AND PROGRAM OF THE NONLINEAR THEORIES

The *nonlinear field theories* also rest upon constitutive equations defining ideal materials, but ideal materials more elaborate and various in their possible responses. Of course the aim is to represent and predict more accurately the behavior of natural materials, and in particular to bring within the range of theory the effects mentioned above, typical in nature but wanting in the classical linear or infinitesimal theories.

Insofar as a constitutive equation, relating the stress tensor to the present and past motion, is laid down as defining an ideal material and is made the starting point for precise mathematical treatment, the methods of the linear and nonlinear theories are the same, both in general terms and in respect to particular solutions yielding predictions to be compared with the results of experiment in certain definite tests, but in other ways they differ.

α) *Physical range.* When bodies of two different natural materials are brought out of the range in which their responses are close to linearly elastic or linearly viscous, there is no reason to expect their mechanical behaviors to persist in being similar. Rubber, glass, and steel are all linearly elastic in small strain, but their several responses to large strain or to repeated strain differ from one another. It is easy to see mathematically that infinitely many nonlinear constitutive equations, differing not only in quantity but also in quality, may have a common linear first approximation. Thus, both from theory and from physical experience, there is no reason to expect any one nonlinear theory to apply properly to so large a variety of natural substances as do the classical linear or infinitesimal theories. Rather, each nonlinear theory is designed *to model more completely the behavior of a narrower class of natural materials.*

β) *Mathematical generality.* Because of the physical diversity just mentioned, it becomes wasteful to deal with special nonlinear theories unless unavoidably necessary. To the extent that several theories may be treated simultaneously, they surely ought to be. The *maximum mathematical generality* consistent with concrete, definite physical interpretation is sought. The place held by material constants in the classical theories is taken over by material functions or functionals. It often turns out that simplicity follows also when a situation is stripped of the incidentals due to specialization. For example, the general theory of waves in elastic materials is not less definite but is physically easier to understand as well as mathematically easier to derive than is the second-order approximation to it, or any theory resulting from quadratic stress-strain relations.

γ) *Experiential basis.* While laymen and philosophers of science often believe, contend, or at least hope, that physical theories are directly inferred from experiments, anyone who has faced the problem of discovering a good constitutive equation or anyone who has sought and found the historical origins of the successful field theories knows how childish is such a prejudice. The task of the theorist is to bring order into the chaos of the phenomena of nature, to invent a language by which a class of these phenomena can be described efficiently and simply. Here is the place for "intuition", and here the old preconception, common among natural philosophers, that nature is simple and elegant, has led to many great successes. Of course physical theory must be based on experience, but experiment comes after, rather than before, theory. Without theoretical concepts one would neither know what experiments to perform nor be able to interpret their outcomes.

δ) *Mathematical method.* The structure of space and time appropriate to classical mechanics requires that certain *principles of invariance* be laid down. Alongside principles of invariance must be set up *principles of determinism*, asserting which phenomena are to be interconnected, and to what extent. In more popular but somewhat misleading terms, "causes" are to be related to "effects". Principles of these two kinds form the basis for the construction of constitutive equations. General properties of materials such as isotropy and fluidity are related to certain properties of invariance of the defining constitutive equations.

ε) *Product.* After suitably invariant principles of determinism are established, we are in a position to specialize intelligently if need be, but in some cases no further assumptions are wanted to get *definite solutions* corresponding to physically important circumstances. In addition to such solutions, surely necessary for connecting theory with experience and experiment, we often seek also *general theorems* giving a picture of the kind of physical response that is represented and serving also to interconnect various theories.

The physical phenomena these theories attempt to describe, while in part newly discovered, are mainly familiar. The reader who thinks that one has only to do experiments in order to know how materials behave and what is the correct theory to describe them would do well to consult a paper by BARUS[1], published in 1888. Most of the effects BARUS considered had been known for fifty to a hundred years, and he showed himself familiar with an already abundant growth of mathematical theories. That he interpreted his own sequences of experiments as confirming MAXWELL's theory of visco-elasticity, has not put an end either to further experiments reaching different conclusions or to the creation of other theories, even for the restricted circumstances he considered. If the basic problem were essentially experimental, surely 200 years of experiment could be expected to have brought better understanding of the mechanics of materials than in fact is had today.

This example and many others have caused us to write the following treatise with an intent different from that customary in works on plasticity, rheology, strength of materials, *etc.* We do not attempt to fit theory to data, or to apply the results of experiment so as to confirm one theory and controvert another. Rather, just as the geometrical

---

[1] C. BARUS, "Maxwell's theory of the viscosity of solids and its physical verification", *Philosophical Magazine* (5) **26** (1888): 183–217.

figure, the rigid body, and the perfect fluid afford simple, natural, and immediate *mathematical models* for some aspects of everyday experience, models whose relevance or application to each particular physical situation must be determined by the user, we strive to find a rational ingress to more complex mechanical phenomena by setting up clear and plausible theories of material behavior, embodying various aspects of long experience with natural materials.

3. Structure Theories and Continuum Theories

Widespread is the misconception that those who formulate continuum theories believe matter "really is" continuous, denying the existence of molecules. That is not so. Continuum physics presumes *nothing* regarding the structure of matter. It confines itself to relations among gross[2] phenomena, neglecting the structure of the material on a smaller scale. Whether the continuum approach is justified, in any particular case, is a matter, not for the philosophy or methodology of science but for *experimental test*. In order to test a theory intelligently, one must first find out what it predicts. Few of the current critics of continuum mechanics have taken that much trouble[3].

Continuum physics stands in no contradiction with structural theories, since the equations expressing its general principles may be identified with equations of exactly the same form in sufficiently general statistical mechanics[4]. If this identification is just, the variables that are basic in continuum mechanics may be regarded as averages or expected values of molecular actions.

---

[2] The word "macro*scopic*" is often used but is misleading because the scale of the phenomena has nothing to do with whether or not they can be seen ($\sigma\kappa o\pi\epsilon\hat{\iota}\nu$). "Molar", the old antithesis to "molecular", is also a fit term to the extent that only massy bodies are considered.

[3] What it is surely to be hoped is the high-water mark of logical confusion and bastard language has been reached in recent studies of the aerodynamics of rarefied gases, where the term "non-continuum flow" often refers to anything asserted to be incompatible with the Navier–Stokes equations. Even the better-informed authors in this field usually decide by *ex cathedra* pronouncement based on particular molecular concepts, rather than by experimental test, when continuum mechanics is to be regarded as applicable and when it is not.

[4] In their paper "Theory of structured continua. I. General considerations of angular momentum and polarization", *Proceedings of the Royal Society* (London) **A275** (1963): 505–527, H. S. Dahler & L. E. Scriven write "Both approaches, continuum and statistical, yield the same macroscopic behaviour, regardless of the nature of the molecules and submolecular particles of which the physical system is composed."

It would be wrong, nevertheless, to infer that quantities occurring in continuum mechanics *must* be interpreted as certain particular averages. Long experience with molecular theories shows that quantities such as stress and heat flux are quite insensitive to molecular structure: Very different, apparently almost contradictory hypotheses of structure and definitions of gross variables based upon them, lead to the same equations for continua[5]. Over half a century ago, when molecular theories were simpler than they are today, POINCARÉ[6] wrote:

> In most questions the analyst assumes, at the beginning of his calculations, either that matter is continuous, or the reverse, that it is formed of atoms. In either case, his results would have been the same. On the atomic supposition he has a little more difficulty in obtaining them—that is all. If, then, experiment confirms his conclusions, will he suppose that he has proved, for example, the real existence of atoms?

While the logical basis of POINCARÉ's statements remains firm, the evidence has changed. The reader of this treatise is not asked to question the "real" existence of atoms or subatomic particles. His attention is directed to phenomena where differences among such particles, as well as the details of their behavior, are unimportant. However, we cannot give him assurance that quantum mechanics or other theories of modern physics yield the same results. Any claim of this kind must await such time as physicists turn back to gross phenomena and demonstrate that their theories do in fact predict

---

[5] The structural theories of NAVIER are no longer considered correct by physicists, but the equations of linear viscosity and linear elasticity he derived from them have been confirmed by experiment and experience for a vast range of substances and circumstances. MAXWELL derived the NAVIER–STOKES equations from his kinetic theory by using, along with certain hypotheses, a definition of stress as being entirely an effect of transfer of molecular momentum, but experience shows the NAVIER–STOKES equations to be valid for many flows of many liquids, in which no-one considers transfer of momentum as the main molecular explanation for stress. In recent work on the general theory of ensembles in phase space, different definitions of stress and heat flux as phase averages lead to identical field equations for them. Examples could be multiplied.

Cf. § 1 of TRUESDELL's "A new definition of a fluid. I. The Stokesian fluid", *Journal de Mathématiques Pures et Appliquées* (9) **29** (1950): 215–244: "History teaches us that the conjectures of natural philosophers, though often positively proclaimed as natural laws, are subject to unforeseen revisions. Molecular hypotheses have come and gone, but the phenomenological equations of D'ALEMBERT, EULER, and CAUCHY remain exact as at the day of their discovery, exempt from fashion."

[6] Chapter 9 of H. POINCARÉ, *La Science et l'Hypothèse*, Paris, Flammarion, 1902.

them, not merely "in principle" but also in terms accessible to calculation and experiment.

The relative positions of statistical theories, engineering experiment, and the rational mechanics of continua were surveyed as follows by v. MISES[7] in 1930:

> To these brief indications let me add two final remarks. The former is suggested by the extensions of mechanics recently found by the physicists. Certainly no-one would guess that the modifications provided by relativity or wave mechanics could be important for the problems I have discussed here, but statistics is a different matter. It is conceivable that physics expressed by differential equations is incapable of providing a passably satisfying representation of the typical phenomena in solid bodies, that no assumptions either extending or combining those made up to now are such as to represent permanent deformations correctly. In the hydrodynamics of turbulent flows it seems indeed to be established that the statistical character of the phenomenon must be taken account of at the very outset of a usable theory. If we ask, nevertheless, whether we may expect any help in our tasks from "statistical mechanics", the prospect is rather dim. Indeed the reverse has been shown, namely, that when we are quite certain we must deal with statistical material, for example in the mechanics of colloids, the best results attainable follow by incorporating assumptions from continuum mechanics.
>
> The second remark returns . . . to the relation between engineering and rational mechanics in illuminating the mechanical behavior of bodies observable in reality. No doubt at all: The requirements of testing materials press for solutions at least tentatively practical, and they seek them in assumptions of the kind I have described above, but with no rational foundation and in continually increasing confusion of concepts and terms. Countless papers connected with new experimental results seek to define the basic concepts, always starting over from the beginning, and introduce procedures for measurements and even new physical units for the properties of materials—there are at least half a dozen plasticity-meters—altogether without establishing any reasonable foundation in theory. The Americans have already proposed to set up a committee of experts to clarify the state of affairs. I think we

---

[7] R. v. MISES, "Über die bisherigen Ansätze in der klassischen Mechanik der Kontinua", pages 80–89 of Volume **2** of *Proceedings of the 3ʳᵈ International Congress of Applied Mechanics*, Stockholm, 1930.

must first look for progress in a different way: by careful consideration of the logical foundations of theory and of the mathematical assumptions made up to now, development of which is *the only way to direct experimental research into orderly and fruitful paths.*

Since 1930 the data on which v. MISES based this summary have been replaced by other, more compelling facts. While intensive and fruitful work has been carried out both in statistical theories of transport processes and in experiment on materials, on a scale overshadowing all past efforts, the reader of this treatise will see that the rational mechanics of continua has grown in even greater measure.

It should not be thought that the results of the continuum approach are necessarily either less or more accurate than those from a structural approach. The two approaches are *different*, and they have different uses.

First, a structural theory implies more information about *a given material*, and hence less information about *a class of materials*. The dependence of viscosity on temperature in a gas, for example, is predicted by the classical kinetic theory of moderately dense gases, while in a continuum theory it is left arbitrary. For each different law of intermolecular force, the result is different, and for more complicated models it is not yet known. A continuum theory, less definite in this regard, may apply more broadly. The added information of the structural theory may be unnecessary and even irrelevant. To take an extreme example, a full structural specification implies, in principle, all physical properties. From the structure it ought to be possible to derive, among all the rest, the smell and color of a body of the material. A specification so minute will obviously carry with it extreme mathematical complexity, irrelevant to mechanical questions regarding finite bodies.

Second, structural specification necessarily presents *all the attributes* of a material simultaneously, while in continuum physics we may easily separate for special study *an aspect* of natural behavior. For example, the classical kinetic theory of monatomic gases implies a special constitutive equation of extremely elaborate type, allowing all sorts of thermomechanical interactions, with definite numerical coefficients depending on the molecular model. For a natural gas really believed to correspond to this theory, these complexities are sometimes appropriate, and the theory is of course a good one. On the other hand, it is a highly special one, offering no possibility of accounting for many simple phenomena daily observed in fluids. For example, it does not allow for a shear viscosity dependent on density as well as temperature, or for a positive bulk viscosity, both of which are easily handled in classical fluid dynamics. Doubtless it is true that

natural fluids for which such viscosities are significant have a complicated molecular structure, but that does not lessen the need for
theories enabling us to predict their response in mechanical situations, perhaps long before their structure is determined.

Third, a continuum theory may obtain by *a more efficient process*
results shown to be true also according to certain molecular theories.
For example, a simple continuum argument suggests the plausibility
of the Mooney theory of rubber, which was later shown to follow also
from a sufficiently accurate and general theory of long-chain
molecules. A more subtle but more important possibility comes from
the general principles of physics. For example, certain requirements
of invariance and laws of conservation may be applied directly to the
continuum, rendering unnecessary the repeated consideration of
consequences of these same principles in divers special molecular
models, so the continuum method may enable us to derive directly,
once and for all, results common to many different structural
theories. In this way we may separate properties that are sensitive to a
particular molecular structure from those that are necessary consequences of more general laws of nature or more general principles
of division of natural phenomena.

Fourth, the information needed to apply a continuum theory in an
experimental context is *accessible to direct measurement*, while that for a
structural theory usually is not. For example, in the classical
infinitesimal theory of isotropic elasticity it is shown that data
measured in simple shear and simple extension are sufficient to determine all the mechanical response of the material. Both the taking of
the data and the test of the assertion are put in terms of the kind of
measurement for which the theory is intended. The nonlinear
theories show this same accessibility, though in more complex form.

In summary, then, continuum physics serves to *correlate the results of
measurements* on materials and to isolate *aspects* of their response. It
neither conflicts with structural theories nor is rendered unnecessary
by them.

The foregoing observations refer to those structural theories in
which the presumed structure is intended to represent the molecules
or smaller particles of natural materials. In regard to the mechanis-
momorphic structures imagined by the rheologists, we can do no
better than quote some remarks of COLEMAN & NOLL[8] in a more

---

[8] See § 1 of B. D. COLEMAN & W. NOLL, "Foundations of linear viscoelasticity",
Reviews of Modern Physics **33** (1961): 239–249. Reprinted in *Foundations of Elasticity
Theory*, edited by C. TRUESDELL, International Science Review Series, New York *etc.*,
Gordon & Breach, 1965, and in W. NOLL, *The Foundations of Mechanics and Thermody-
namics, Selected Papers*, New York *etc.*, Springer-Verlag, 1974.

special context:

> It is often claimed that the theory of infinitesimal viscoelasticity
> can be derived from an assumption that on a microscopic level
> matter can be regarded as composed of "linear viscous elements"
> (also called "dashpots") and "linear elastic elements" (called
> "springs") connected together in intricate "networks". . . .
>
> We feel that the physicist's confidence in the usefulness of
> the theory of infinitesimal viscoelasticity does not stem from a
> belief that the materials to which the theory is applied are really
> composed of microscopic networks of springs and dashpots, but
> comes rather from other considerations. First, there is the obser-
> vation that the theory works for many real materials. But second,
> and perhaps more important . . ., is the fact that the theory looks
> plausible because it seems to be a mathematization of little more
> than certain intuitive prejudices about smoothness in macroscopic
> phenomena.

4. GENERAL LINES OF PAST RESEARCH ON THE FIELD THEORIES OF MECHANICS

While, reflecting the stature of the researchers themselves, the
early researches on the foundations of continuum mechanics did not
show any preference for linear theories, with the rise of science as a
numerous profession in the nineteenth century it was quickly seen
that linearity lends itself to volume of publication. The linear theories
of heat conduction, attraction, elasticity, and viscosity, along with the
linear mathematical techniques that could be applied in them, were
developed so intensively and exclusively that in the minds of many
scientists down to the present day they are synonymous with the
mechanics of continuous media. It would be no great exaggeration
to say that, in the community of physicists, mathematicians, and
engineers, less was known about the true principles of continuum
mechanics in 1945 than in 1895.

Blame for this neglect of more fundamental study may be laid to
two contradictory misconceptions: first, that the classical linear or
infinitesimal theories account for everything known about natural
materials, and, second, that these two theories are merely crude
"empirical" fits to data. The second is still common among physicists,
many of whom believe that only a molecular-statistical theory of the
structure of materials can lead to understanding of their behavior. The prevalence of the former among engineers seems to have
grown rather from a rigid training which deliberately confined itself

to linearly biased experimental tests and deliberately described every phenomenon in nature, no matter how ineptly, in terms of the concepts of the linear theories.

Of course, at all times there have been a few scientists who thought more deeply or at least more broadly in regard to theories of materials. Various doctrines of *plasticity* arose in the latter part of the last century and have been cultivated diffusely in this. These theories have always been closely bound in motive, if often not in outcome, to needs attributed to engineering and have expatiated at once into detailed approximate solutions of boundary-value problems. Their mechanical foundation is insecure to the present day, and they do not furnish representative examples in the program of continuum physics. Similarly, the group of older studies called *rheology* is atypical in its nearly exclusive limitation to one-dimensional response, to a particular cycle of material tests, and to models suggested by networks of springs and dashpots.

While only very few scientists between 1845 and 1945 studied the foundations of continuum mechanics, among them were some of the most distinguished savants of the period: St. Venant, Stokes, Kirchhoff, Kelvin, Boussinesq, Gibbs, Duhem, and Hadamard. Although phenomena of viscosity and plasticity were not altogether neglected, the main effort and main success came in the theory of finite elastic strain. The success, however, was but small. When the brothers Cosserat published their definitive exposition[9] in 1896, its 116 pages contained little more than a derivation of various forms of the general equations. Beyond the laws of wave propagation and the great theorem on elastic stability obtained shortly afterward by Hadamard, [little] concrete progress was made in the finite theory for the following fifty years. Not only did the want of concepts such as to suggest a simple notation lay a burden of page-long formulæ on the dragging steps of writer or reader, but also there was no evidence of a program of research. Linear thinking, leading to easy solutions for whole classes of boundary-value problems, obviously would not do, but nothing was suggested to take its place, except, perhaps, the dismaying prospect of creeping from stage to stage in a perturbation process.

Nevertheless, in that period many papers on the subject were published. When not essentially repetitions of earlier studies, these con-

---

[9] E. Cosserat & F. Cosserat, "Sur la théorie de l'élasticité", *Annales de la Faculté des Sciences de l'Université de Toulouse pour les Sciences Mathématiques et les Sciences Physiques* **10** (1896): 1–116. A portion of the contents, abbreviated by tensor notation, is contained in the [once] widely read paper by F. D. Murnaghan, "Finite deformation of an elastic solid", *American Journal of Mathematics* **59** (1937): 235–260.

cerned special theories or approximations, most of which have later turned out to be unnecessary in the cases when they are justified. Knowledge of the true principles of the general theory seems to have diminished except in Italy, where it was kept alive by the teaching and writing of SIGNORINI. [SIGNORINI's researches were later seen to cast some light upon the failure of uniqueness in solutions of the boundary-value problem of traction.]

A new period was opened by papers of REINER[10] and RIVLIN[11]. The former was the first to suggest any *general approach* or *unifying principle*[12] for nonlinear constitutive equations; the latter was the first to obtain *concrete, exact solutions* to specific problems of physical interest in nonlinear theories where the response is specified in terms of *arbitrary functions* of the deformation. Both considered not only finitely strained elastic materials but also nonlinearly viscous fluids. RIVLIN was the first to see the far-reaching simplification effected in a nonlinear theory by assuming the material to be incompressible.

In 1952 appeared a detailed exposition, *The Mechanical Foundations of Elasticity and Fluid Dynamics*[13], in which both the old and the new trends were summarized. On the one hand, the numerous special or approximate theories were set in place upon a general frame and

---

[10] M. REINER, "A mathematical theory of dilatancy", *American Journal of Mathematics* **67** (1945): 350–362, reprinted in *Rational Mechanics of Materials*, edited by C. TRUESDELL, International Science Review Series, New York *etc.*, Gordon & Breach, 1965, and "Elasticity beyond the elastic limit", *American Journal of Mathematics* **70** (1948): 433–446, reprinted in *Foundations of Elasticity Theory*, edited by C. TRUESDELL, International Science Review Series, New York *etc.*, Gordon & Breach, 1965.

[11] R. S. RIVLIN, "Large elastic deformations of isotropic materials. IV. Further developments of the general theory", *Philosophical Transactions of the Royal Society* (London) **A241** (1948): 379–397, "The hydrodynamics of non-Newtonian fluids. I", *Proceedings of the Royal Society* (London) **A193** (1948): 260–281, reprinted in *Rational Mechanics of Materials*, cited in Footnote 10; "Large elastic deformations of isotropic materials. V. The problem of flexure", *Proceedings of the Royal Society* (London) **A195** (1949): 463–473, reprinted in *Problems of Non-linear Elasticity*, edited by C. TRUESDELL, International Science Review Series, New York *etc.*, Gordon & Breach, 1965; "Large elastic deformations of elastic materials. VI. Further results in the theory of torsion, shear and flexure", *Philosophical Transactions of the Royal Society* (London) **A242** (1949): 173–195, reprinted with the preceding; "The hydrodynamics of non-Newtonian fluids. II", *Proceedings of the Cambridge Philosophical Society* **45** (1949): 88–91, reprinted in *Rational Mechanics of Materials*, cited in Footnote 10; "A note on the torsion of an incompressible highly-elastic cylinder", *Proceedings of the Cambridge Philosophical Society* **45** (1949): 485–487, reprinted in *Problems of Non-linear Elasticity*, cited above.

[12] Considerations of invariance had occurred earlier, notably in the work of POISSON and CAUCHY, but in special contexts.

[13] C. TRUESDELL, "The mechanical foundations of elasticity and fluid dynamics", *Journal of Rational Mechanics and Analysis* **1** (1952): 125–300; **2** (1953): 593–616; **3** (1954): 801; corrected and annotated reprint, *Continuum Mechanics* I, International Science Review Series, Volume 8, New York *etc.*, Gordon & Breach, 1966.

related to each other insofar as possible, especially to make clear the arbitrary and unsupported physical assumptions and the insufficient if not faulty mathematical processes by which they had been inferred. On the other, the concrete and trenchant gains won by the new approaches were presented in full and with emphasis.

A summary[14] of the researches of 1945–1952, referring especially to problems of flow, has stated:

> By 1949 it could be said that all work on the foundations of rheology done before 1945 had been rendered obsolete. The phenomenon of normal stresses had been shown to be of second order, while departures from the classically assumed linear relation between shearing tractions and rates of shearing are of third order in the rates. The old viscometers, designed without a thought of normal stresses, had fixed opaque walls to help the experimenter overlook the most interesting effect in the apparatus or to prevent his measuring the forces supplied so as to negate it. By theory, the phenomenon of normal stresses was straightway seen to be a universal one, to be expected according to all but very special kinds of non-linear theories. Of course a phenomenon so universally to be expected must have been occurring for a long time in nature, and it was quickly recognized that many familiar effects, such as the tendency of paints to agglomerate upon stirring mechanisms, as well as some concealed mysteries of the artificial fiber industry, are examples of it, though a century of linear thinking in physics had blinded theorists to the possibility that simple mechanics, rather than chemistry, would suffice to explain it. . . .
>
> While . . . [this research] gained a number of theoretical predictions of remarkable completeness, these are the least of what it gave us. Next is the fact that, with little exaggeration, *there are no one-dimensional problems*: A situation which is one-dimensional in a linear theory is automatically two-dimensional or three-dimensional in any reasonable non-linear theory. More important is the *independence in theory* which resulted from the realization that any sort of admissible non-linearity would yield the correct general kind of behavior, and that to account for the phenomena, far from being difficult, was all too easy. Of a theory, we learned that both less and more had to be expected. To calculate the creep in a buckled elliptical column with a square hole in it is too much until the response of materials shall be better understood than it is

---

[14] C. TRUESDELL, "Modern theories of materials", *Transactions of the Society of Rheology* **4** (1960): 9–22. See pages 13 and 15.

today; to be satisfied with a normal stress of the right sign and order, with an adjustable coefficient, is too little until the responses of the *same* material in a *variety* of situations are determined and correlated, with no material constants or functions altered in the process. What is needed is a theory of theories.

Since 1952, it cannot be said that the older type of work has ceased; rather, in the common exuberance of modern publication, an easy place is found not only for continued search of avenues known to be sterile but also for frequent rediscovery of special theories included and criticized in *The Mechanical Foundations* and of special cases of solutions presented there in explicit generality. Beyond this, and heedless of it, a small school of young scientists, of backgrounds and trainings as various as mathematics, physics, chemistry, and engineering, has developed the newer approaches. Not only have major results been obtained in the classical general theory of finite elastic strain, to the point that there is now a technology of the subject, but also success beyond any fair expectation has been met in a very general theory of nonlinear viscosity and relaxation. A great range of the mechanical behavior of materials previously considered intractable if not mysterious has been brought within the control of simple, precise, and explicit mathematical theory. Just a little earlier, appropriate experiments had begun on a material which lends itself particularly well to measurements of the effects of large deformation and flow: polyisobutylene. It should not be thought, nonetheless, that the theories apply only to high polymers. The nonlinear effects are typical of mechanics, and there is reason to think they occur in nearly all materials—for example, in air and in metals—but generally their variety is so great that it is difficult to separate one from another. High polymers are distinguished not so much for the existence as for the simplicity of the nonlinear effects they exhibit. The new researches on the general theories, preceded by the classical foundation established in the last century, form the subject of the present treatise.

Of the several kinds of attack to which the new continuum mechanics has been subject, only two deserve notice, because only these have some basis in truth. First, some scientists of the "practical" kind presume that pages full of tensors and arbitrary functions or functionals can never yield results specific enough to apply to the real world. Second, the analyst who has been taught that everything begins with theorems of existence and uniqueness may reject as being only "physics" or "engineering" anything that does not consist solely of convergence proofs and estimates. We hope that critics of the former kind will notice in our text the multitude of exact or approximate solutions of specific problems for elastic materials and for simple fluids as well as certain explicit calculations for more general materials, with results fit for comparison with measurements; while this treatise is mathematical, we have included by way of an existence proof some tables and graphs of data on experiments done expressly in response to the analyses here summarized. We hope that critics of the latter kind will notice page after page of definite theorems and strict proofs and will allow that mathematics is not confined to any rigid pattern; in particular, we hope that this treatise will be admitted in evidence that mathematics enables us to correlate information available on various aspects of a class of physical theories even when that information is too imperfect to lay down a "well set problem" in the style of the common theories of the last century. Finally, we trust that those who regard as essential to modern science the expense and labor of numerical computation on large machines will easily find for themselves a thousand points in our subject where that taste can be gratified at any time.

5. The Nature of This Treatise

In 1955 it was planned to contribute to this Encyclopedia two articles that would in effect bring *The Mechanical Foundations* up to date and complete it by a correspondingly detailed presentation of aspects of the foundations of continuum mechanics omitted from it. The former portion of the project, concerning the general principles of continuum physics, has been finished and printed as *The Classical Field Theories* in Part 1 of Volume III. The latter portion has had to be modified[15].

In the first place, the flow of important publication on the basic principles of nonlinear theories and on experiment in connection with them has increased tenfold: Scarcely a month passes unmarked by a major paper. What follows here has been not only rewritten but also several times re-organized so as to incorporate researches published after we had begun—in some cases, researches growing straight from the difficulties we ourselves encountered in the writing. Second, the special or approximate theories of elasticity or viscosity, to explaining and interrelating which a considerable part of *The*

---

[15] In the mean time a general exposition of the field has been published by A. C. ERINGEN, *Non-linear Theory of Continuous Media*, New York *etc.*, McGraw-Hill, 1962. *Cf.* the [even-handed and easily substantiable] review by A. C. PIPKIN, *Quarterly of Applied Mathematics* **22** (1964): 172–173. [In consequence of the sociological law that makes lightweights rise in a dense medium this review had to be retracted in a later issue of the *Quarterly.*]

Mechanical Foundations was devoted, have lost their value because of the greater efficiency and enlightenment the more general methods have since been shown to offer. Third, the theories usually named "plasticity" remain in essentially the same state as they were in 1952, when they were by intention omitted from *The Mechanical Foundations*[16].

For these reasons, the present treatise is of lesser scope than was originally planned. First, although we have taken pains to include a new and general foundation for the continuum theory of dislocations, we have not felt able to do more in regard to the usual theories of "plasticity" than to refer the reader to the standard treatises, *e.g.* to the article by FREUDENTHAL & GEIRINGER in Volume VI of this *Encyclopedia*. Second, we have omitted most of the special theories of elasticity and viscosity, for them referring the reader to *The Mechanical Foundations*[17].

Work in this field is often criticized for opaque formalism. Some of those not expert in the subject have implied that the specialists attempt to make it seem more difficult than it is. In the original development of any science, the easiest way is often missed, and the lack of a pre-organized common experience and vocabulary, often called "intuition" by those whose concern is pædagogy or professional amity rather than discovery, makes the path of the creator hard to follow. In writing the treatise we present here, earnest and conscious effort has been put out to render the subject simple, easy, and beautiful, which we believe it is, increasingly with the repeated reconsideration of the groundwork and the major results which have appeared in the last decade. On the other hand, we have not followed the lead of some experts in other fields who have lightly entered this with too hasty expositions that by their slips and gaps prosper in making the subject appear to their unwary readers as being simpler and easier (though less beautiful) than in fact the physical behavior of bodies in large and rapid deformation can be.

Instead of completeness, we have attempted to achieve *permanence*. As the main subjects of this treatise we have selected those researches that formulate and solve *once and for all* certain clear, definite, and broad conceptual and mathematical problems of nonlinear continuum mechanics. We not only hope but also believe that the major part of the contents is not controversial or conjectural, representing

---

[16] Recently GREEN & NAGHDI have proposed a rational theory of finitely deformed plastic materials, but they adopt a yield condition as in the older literature. *Cf.* their paper, "A general theory of an elastic-plastic continuum", *Archive for Rational Mechanics and Analysis* **18** (1965): 251–281.

[17] See §§ 48–54, 60, 81–82 of TRUESDELL, *op. cit.* Footnote 13.

instead unquestionable conquests that will become and remain standard in the subject. After the classic researches done before 1902, nearly everything in this treatise was first published, at least in the form here given to it, after 1952. We do not pretend, however, to be exhaustive even for the most recent work or for citation of it, since we have subordinated detail to importance, and, above all, to *clarity and finality*.

Our citations refer either to the original sources or to works containing related developments not given in this treatise. Thus, since scant service would be done any reader by directing him to the numerous textbooks and pædagogic "introductions", we follow the precedent of *The Classical Field Theories*, criticized by one reviewer for preferring very old or very new references.

Properly, our title should have indicated restriction to classical mechanics, for relativistic field theories lie outside our scope. Since the term "classical" suggests to many a domain long mastered—indeed, one reviewer criticized *The Classical Field Theories* for including material he did not already know—that word seems inappropriate in the title of a treatise devoted mainly to very recent work. Specifically, we consider *the mechanical response of materials in three-dimensional Euclidean space*. While often the dimension 3 can be replaced effortlessly by *n*, the main conceptual structure is closely bound to Euclidean geometry. Relativistic generalization has required major changes in views and details which were not yet known when we laid our plans.

This treatise is written, not for the beginner, but for the specialist in mechanics who wishes to gain quickly and efficiently the solid and complete foundation necessary to do theoretical research, either in applications or in further study of the groundwork, in nonlinear continuum mechanics. We use the term *nonlinear* in the sense of material response, not of mathematical analysis[18].

Accordingly, after an introductory chapter fixing notations and listing a number of mathematical theorems for use in the sequel, this treatise is divided into three major parts, as follows.

Chapter C presents a *general approach*, based upon principles of *determinism*, *local action*, and *material frame-indifference*, to the mechanical properties of materials. For the special case of a *simple*

---

[18] The classical theory of viscous incompressible fluids, for example, is governed by nonlinear partial differential equations, but we do not include it here since its defining constitutive equation is a linear one. In fact, since the acceleration is a nonlinear function of velocity and velocity-gradient, all theories of the motion of continua are nonlinear in the spatial description, and so the analytical distinction is empty except in regard to methods of approximation.

material, in which the stress at a particle is determined by the cumulative history of the deformation gradient at that particle, all three fundamental principles may be expressed in a final and explicit mathematical form. Qualities distinguishing one kind of material from another are then defined by invariant properties of the response functionals; the terms "materially uniform", "homogeneous", "solid", "fluid", and "isotropic" are made precise in terms of mathematical systems constructed from the functionals. Finally, it is shown that if the response functional of a simple material is sufficiently smooth in a certain sense, then BOLTZMANN's equations of linear visco-elasticity result as an approximation in motions whose histories are nearly constant. Thus the general theory of simple materials is seen to furnish a properly invariant generalization of classical visco-elasticity to arbitrary states of deformation and flow; likewise it includes not only as special cases but also in suitable senses of approximation the classical theories of finite elasticity and linear viscosity.

In statics, the stress of any simple material reduces to a function of the finite strain. Materials having this property also in time-dependent deformations are said to be *elastic*, and most of Chapter D is devoted to them. Here we present the theory of *finite elastic strain*, not only its principles but also its general theorems and the known exact solutions and methods of approximation, in generality and completeness not heretofore attempted. When, as proposed by GREEN, the work done in elastic deformation is stored as internal energy, making the stresses derivable from a stored-energy function as a potential, the material is called *hyperelastic*. Nearly all earlier studies concerned only this instance. While we develop its distinguishing properties and general theorems, our emphasis lies on the more embracing concept, due to CAUCHY. Generalizations of hyperelasticity to allow for thermal conduction, polarization, and couple stresses are then sketched. The last sections of the chapter concern the partly more general and partly exclusive concept of *hypo-elasticity*, according to which the rate of change of stress is determined by the stretchings, shearings, and spin at a material element, along with the present stress. The behavior of a hypo-elastic material depends essentially upon the initial stress.

Chapter E concerns *fluidity*. It contains mainly an exhaustive survey of what is known about *simple fluids*. These are distinguished from other simple materials by having the maximum possible isotropy group; necessarily they are isotropic. While they are capable of exhibiting complicated effects of stress-relaxation and long-range memory, these are proved to have no influence on certain special kinds of flow, which turn out to include all those customarily used in viscometric tests. For these special flows, the response functional is shown to manifest itself only through three *viscometric functions*.

One of these may be interpreted as a nonlinear shear viscosity; the other two, as differences of normal stresses. The exact solutions of the dynamical equations are developed for these flows, as well as for some others of similar kind. The chapter closes with remarks upon materials embodying various other concepts of fluidity.

By including general effects of rates and of relaxation, we cover a broader range of physical phenomena than did *The Mechanical Foundations*, although we narrow the topic by omitting most special or approximate theories. The main difference is one of depth. In the present treatise the method of inquiry and formulation, less formal than the approaches known in 1952, goes straight to the physics of each situation. We have sought, and we believe we have often succeeded in finding, *simple and clear* mathematical expression for the physical principles or hypotheses.

Note for the Reprinting

The Non-Linear Field Theories of Mechanics is Volume III/3 of FLÜGGE's *Encyclopedia of Physics*, Berlin *etc.*, Springer-Verlag, 1965. In the years since that volume was published a great deal of fine research on the foundations of mechanics in general and on continuum mechanics, regarding both its structure and its applications to natural phenomena, has appeared, and much of it has been based, in part or altogether, on the views and program set forth in the two prefaces reprinted on the foregoing pages of this volume. The monographs those prefaces introduce are encyclopædic; including accounts of much that is of secondary or even tertiary importance, they were not directed to beginners but to students already active in continuum mechanics or at least desiring to begin research on it. The prefaces were designed to help such scholars see its coherence and its noble origins, to help them defend against occasional onslaughts of prejudice and malice by partisans of the professions their status as independent, unfettered individuals who had chosen to follow a great tradition then largely left in desuetude.

On the basis of the same central ideas, improved and compacted by the discoveries of the following fifteen years, I have composed a textbook for beginners: *A First Course in Rational Continuum Mechanics*, Volume 1, *General Concepts*, New York *etc.*, Academic Press, 1977. The systematic program and an informally axiomatic approach are applied in its first chapter to mechanics as a whole, not just the mechanics of continua, thus encouraging the beginner to take advantage from the very start of the simplicity, compactness, and efficient economy that NOLL's formulation provides. In particular the concept of *system of forces* acting on *pairs* of bodies is developed on the basis of NOLL's axioms of 1965; not only do they clarify the central ideas of every kind of mechanics, but also they facilitate tight proofs of mathematical theorems that establish the several principles of action and reaction, one of which is NEWTON's Third Law of Motion. For further remarks on the axioms of forces, see Essay 39, below in this volume.

5. WAR, SOCIALISM, AND QUANTUM MECHANICS

Extract from the preface to *Essays in the History of Mechanics* (1968)

If these lectures find any favor with professional historians of science, I shall be humbly thankful for their toleration of a book not intended for them. However eager to tell us how scientists of the seventeenth century used their inheritance from the sixteenth, these scholars seem to regard as irrelevant anything a scientist today might think about any aspect of science, including his own debt to the past or reaction against it. Such historians remind me of those taxonomers, perhaps of only fabulous existence, who cannot recognize a particular plant unless they see a sprig of it dead, dried, and pasted to a sheet of paper. For me, mathematical science is alive today, alive not only in its freshest leaves but also in its branches and roots that reach down to the past. I know young men who have read the words of GIBBS and KELVIN and STOKES and CAUCHY, even of EULER and NEWTON, neither so as to decorate a paper of their own by an early reference nor to write a history, but in search of understanding and method, revealed by the speech of giants untranslated by pygmies. For such men, such scientists of our own day, these lectures were composed and are here printed.

The revolutions of world war and socialism and quantum mechanics, however right and necessary and fruitful, have clouded the massive solidity, the serene confidence of classical mechanics. Classical mechanics has weathered through, standing fast behind the smoky, putrid mists. These lectures tell how some parts of it were founded. If the reader may form here his own explanation of how classical mechanics gained solidity so massive and confidence so serene that it not only survived but even is now flourishing anew, my aim will be reached. I will give him a hint. While "imagination, fancy, and invention" are the soul of mathematical research, in mathematics there has never yet been a revolution.

6. THE TRADITION OF ELASTICITY

Extract from the preface to *Introduction to Rational Elasticity* (1973), co-author C.-C. WANG

The theory of elasticity was created by the mathematicians JAMES BERNOULLI, EULER, and CAUCHY. For a long time it was a favorite subject of mathematicians and was regularly taught in mathematics departments. In this century both HADAMARD and HILBERT lectured upon it, as had POINCARÉ and many others in the last. Of the mathematicians of that time who are best known for their work in what is now called "pure" mathematics, we may collect a long list naming those who made at least one important addition to elasticity: BELTRAMI, BETTI, BIRKHOFF, CESÀRO, CHRISTOFFEL, CLEBSCH, FREDHOLM, HADAMARD, KORN, LAMÉ, LEVI-CIVITA, LIPSCHITZ, MORERA, VOLTERRA, WEINGARTEN, WEYL. To these we may add some distinguished Italian mathematicians who specialized in elasticity: ALMANSI, CERRUTI, [DONATI], LAURICELLA, PIOLA, SIGNORINI, SOMIGLIANA, TEDONE, as well as those great British mathematicians whose main interest lay in physics: GREEN, KELVIN, STOKES, and MAXWELL.

Now, unfortunately, this has changed. Although some papers by FRIEDRICHS, JOHN, and FICHERA prove that the rule is not without exceptions, generally mathematicians of our time have abandoned elasticity to the tillers of application. In virtue of GRESHAM's Law, the quality of research on the subject has worsened, whereupon knowledge of it has shrivelled and withered.

This book is written in an attempt to show that now elasticity is again a living branch of mathematics.

In the middle 1940s began a revival and renascence in rational continuum mechanics as a whole. Although this springtide was brought in by a few men who were provoked in the main by a desire to understand and represent the inelastic phenomena of everyday life, which had been foolishly neglected or botched by the professionalized custodians of "application" for some decades, conceptual analysis of those phenomena prerequired reascent of the then almost forgotten rational structure of elasticity itself. While the backgrounds and titles of the men—lonely scouts and barons they were, not "leaders"—were various, respect for mathematics united them. Among the products of the new work is a refounding and recasting of classical elasticity, enriched by a flood of splendid theorems theretofore undreamt of. This rebirth began with the thesis of NOLL in 1954 and owes much to his later work as well as to his strong and benign influence on all who are able to understand his thought....

WANG & I hope that this book will be so good an introduction to the real problems of elasticity that mathematicians will put it quite out of date in a few years.

Note for the Reprinting

I do not mean to imply that only men who were primarily mathematicians made great contributions to mathematical elasticity. Certainly everyone knows that second only to the three founders' work stand the contributions of NAVIER, POISSON, KIRCHHOFF, ST. VENANT, HUGONIOT, DUHEM, BOUSSINESQ, MICHELL, and LOVE. It is my impression that LOVE regarded himself as being primarily a mathematician and was so regarded in his day; the others just listed were primarily engineers or physicists or both; and all were both learned and skilful in mathematics. To LOVE we owe one of the most eloquent defenses of natural philosophy as an end in itself:

> The history of the mathematical theory of Elasticity shows clearly that the development of the theory has not been guided exclusively by considerations of its utility for technical Mechanics. Most of the men by whose researches it has been founded and shaped have been more interested in Natural Philosophy than in material progress, in trying to understand the world than in trying to make it more comfortable. From this attitude of mind it may possibly have resulted that the theory has contributed less to the material advance of mankind than it might otherwise have done. Be this as it may, the intellectual gain which has accrued from the work of these men must be estimated very highly.

This passage appears at the end of the Historical Introduction of A. E. H. LOVE's *A Treatise on the Mathematical Theory of Elasticity*, Cambridge: At the University Press, 2nd edition, 1906, and in all subsequent editions.

Here I rephrase a remark in a part of the preface not reprinted above: The modern status and historical origins of the central theory of *linear* elasticity are masterfully presented in the compact and elegant treatise by Mr. GURTIN in Volume **VIa/2** of the *Encyclopedia of Physics*, 1972, to which may be added the complementary treatise on applications by Mr. VILLAGGIO, *Qualitative Methods in Elasticity*, Leyden, Noordhoff, 1977.

The hope expressed in the last sentence has already been partly realized. Not only are several students recently trained in elasticity also expert mathematicians who have contributed in equal measure to analysis and to elasticity; *vice versa*, several outstanding analysts have recently turned their attention to elasticity after acquiring discernment regarding analytical assumptions natural to mechanics. Many but by no means all of the researches of this kind have appeared in recent volumes of the *Archive for Rational Mechanics and Analysis*. A valuable treatise has just appeared: J. E. MARSDEN & T. R. HUGHES, *Mathematical Foundations of Elasticity*, Englewood Cliffs, Prentice-Hall, 1983.

7. STATISTICAL MECHANICS AND CONTINUUM MECHANICS (1973, 1979)

The ancient Greek philosophers speculated whether matter were an assembly of tiny, invisible, and immutable particles, or a continuous expanse. As the quantitative, mathematical science of the West developed, the debate continued but became more and more definite and detailed. The great theorists proposed specific mathematical theories, restricted to certain specific kinds and circumstances of bodies, for example, to "aeriform fluids" subject to moderate pressures.

Until the first decades of this century it seemed possible that one or another theory would turn out to be the final one, the one that would explain everything about matter and thus be universally accepted as "correct", while all competitors would be defeated. Far from being borne out, this hope now seems childish. Our picture of nature has become less naive. While in the nineteenth century more and more aspects of the sensible world were shown to be mere appearances, mere "applications" of a few fundamental "laws" of physics or biology, the recent enormous production of experimental data has undeceived us of our former simplisms. The line between the living and the inanimate has been blurred if not erased. Within the once indivisible atoms has been found an ever growing host of mysterious "elementary particles" whose nature and function are scarcely clearer than those of dryads and familiar spirits.

Of course these discoveries have brought with them different attitudes toward theories of nature. Those who push forward the frontiers of experiment cannot wait for the thoughtful, critical, and hence cautious and slow analysis that mathematics has always demanded. Mathematicians, for their part, cannot afford to waste their time on physical theories of passing interest.

These contrasting standpoints are reconciled by a keener appraisal of the role a theory is to play. A theory is not a gospel to be believed and sworn upon as an article of faith, nor must two different and seemingly contradictory theories battle each other to the death. A theory is a mathematical model for an aspect of nature. One good theory extracts and exaggerates some facets of the truth. Another good theory may idealize other facets. A theory cannot duplicate nature, for if it did so in all respects, it would be isomorphic to nature itself and hence useless, a mere repetition of all the complexity which nature presents to us, that very complexity we frame theories to penetrate and set aside.

If a theory were not simpler than the phenomena it was designed to model, it would serve no purpose. Like a portrait, it can represent only a part of the subject it pictures. This part it exaggerates, if only because it leaves out the rest. Its simplicity is its virtue, provided the aspect it portrays be that which we wish to study. If, on the other hand, our concern is an aspect of nature which a particular theory leaves out of account, then that theory is for us not wrong but simply irrelevant. For example, if we would analyse the stagnation of traffic in the streets, to take into account the behavior of the elementary particles that make up the engine, the body, the tires, and the driver of each automobile, however "fundamental" the physicists like to call those particles, would be useless even if it were not insuperably difficult. The quantum theory of individual particles is not wrong in studies of the deformation of large samples of air; it is simply a model for something else, something irrelevant to matter in gross.

With this sober and critical understanding of what a theory is, we need not see any philosophical conflict between two theories, one of which represents a gas as a plenum, the other as a numerous assembly of punctual masses. According to the physicists, a real gas such as air or hydrogen is neither of these, nothing so simple. Models of either kind represent aspects of real gases; if they represent those properly, they should entail many of the same conclusions, though of course not all.

A mathematical model for an aspect of nature is a mathematical structure, and as such it must be studied. A theory is not a duplicate of the real world but a diagram of what some simpler but in part similar world might be. Here lies the virtue of theory, for the real world, NEWTON taught us to expect, is simple if only we can learn to find its simplicity. We "understand" and "control" the real world in terms of pictures which make it seem simpler than it is.

Should we wish to trace the paths of a few hundred molecules, we might be able to do so by a theory that represents them individually. If, on the other hand, we design to describe the resistance offered to a solid body by a streaming gas made up of many billions of molecules, the fates of those individual molecules are by their mere number if for no other reason inaccessible to the mind of any one man, even were the most keenly merchandised and costly computer to endigit, ingest, and expectorate them in print. However discrete may be nature itself, the mathematics of a very numerous discrete system remains even today beyond anyone's capacity. To analyse the large, we replace it by the infinite, because the properties of the infinite are simpler and easier to manage. The mathematics of large systems is the infinitesimal calculus, the analysis of functions which are defined on infinite sets and whose values range over infinite sets. We need to differentiate and integrate functions. Otherwise we are hamstrung if we wish to deal effectively, precisely, with more than a few dozen objects able to interact with each other. Thus, somehow, we must introduce the continuum.

There are two ways the mathematics of the continuum may be brought to bear on large assemblies of discrete particles. One is to neglect their discreteness and finite cardinality outright, to represent them as being infinitely numerous and smoothly packed in every part, however small, of a certain region of space. This is the way of continuum mechanics. Aspects of matter such as mass, velocity, momentum, and energy are represented by smooth fields, introduced as primitive quantities and delimited by axioms which set forth their mathematical properties and thus allow us to operate with them. It is this way's advantage to rule out, automatically and axiomatically, the "exceptional" or "unusual" cases of molecular behavior, without ever mentioning a molecule. Aside from this limitation, it is extremely general as well as superbly elegant if kept in the right hands. Its disadvantages are two: it excludes, *a fortiori*, any mathematical test of its own range of validity in application to discrete systems, however large, and, second, it affords at most heuristic means to make use of what we may know about the molecular constitution of bodies.

The second way to bring the mathematics of the continuum into the mechanics of the discrete is usually called "statistics". Instead of stating the positions and velocities of all the molecules, we allow the possibility that these may vary for some reason—be it because we lack precise information, be it because we wish only some average in time or in space, be it because we are content to represent the result of averaging over many repetitions. We thus allow the quantities which describe molecular actions to range over a continuum of values— again, something inherently beyond the range of any experiment by man. We can then assign a probability to each quantity and calculate the values expected according to that probability. Though the system itself is finite, by this broader view we subject it to the mathematics of the continuum. The great advantage of this way is that it lets us take direct account of some aspects of the specification of individual molecules. Its disadvantage lies in the resulting mathematical problems, for they, if taken seriously and not merely nibbled or truncated or mused over or "solved" by declaration, are of enormous difficulty.

The models obtained in these two ways neglect much of nature. As the physicists say, they are "approximate". What one kind neglects, the other partly includes. They start from different assumptions, assumptions which can neither confirm nor contradict one another because they are set in different conceptual frames. Nevertheless, their ranges of intended application to nature may be the same, or nearly so. Thus we can sometimes compare their conclusions, at least in part, by interpreting both within one and the same range of experience with natural bodies.

To compare conclusions, we must first have them. The only way to get a conclusion from a mathematical theory is by logic, by mathematical steps. Any conclusion gotten otherwise, as for example by "physical intuition", blind teamwork on huge machines, or other gilded guessing is really not a conclusion from the theory. At best it is itself some other theory, not a consequence of the one we study. At worst, it is wrong. Once we recognize that a theory is a mathematical model, we recognize also that only rigorous mathematical conclusions from a theory can be accepted in tests of the justice of that theory. If some conclusion of ours is not strict, and if we find it does not square with observations about nature, then we do not know whether to impute its failure to an inexactness of the theory or to our own incapacity as mathematicians. This latter, shared by us all in one degree or another, is to be regretted, but that bears no whit upon the theory itself. Even less is our mathematical incapacity just grounds for the measureless boasting that seems now to be the union label worn by those who regard themselves as gifted in applying mathematics. A proved theorem in a physical theory shows how some part of nature might behave. Mathematical botchery proves no more than the failure of the theorist.

Therefore, in physical theory mathematical rigor is of the essence. Being human beings and hence fallible, we may not always achieve this rigor, but we must attempt it. A result partly proved and honestly presented as such, like a tunnel drilled partly through a mountain, may be useful in giving the next man a better place from which to start, or, if fortune frowns, in showing him that to drive further in this direction is futile. A proof mathematically strict except for certain gaps made plain as the sites for future bridges is not at all the same as a "physical" or "intuitive" argument which claims to be a proof but is no more than a drug to whirl us over high mountains and across deep gorges by illusion, illusion which drops us when we awake at just the point where we started.

This book treats of only one theory: MAXWELL's second kinetic theory of a moderately rarefied, simple, monatomic gas. This was the first molecular theory sufficiently definite and detailed to allow, at least in principle, prediction of the kind of effects and phenomena that the plenum theory of aeriform fluids had already described. MAXWELL's theory is now over 100 years old. Kinetic-molecular theories of much greater complication have been proposed and studied, but MAXWELL's theory remains today the only one that is both consistent with the mechanics of gross bodies and also simple enough in mathematical structure to yield the kind of conclusions we call "theorems"; conclusions proved strictly from the assumptions, by logical steps alone.

It is such conclusions, and such only, that this book designs to develop. We do not attempt to found MAXWELL's theory upon any other one; in particular, we do not "derive" it by imposing "approximations" upon a more refined or less specific statistical mechanics of molecular assemblies[1]. Rather, we set forth the primitive quantities, postulates, and definitions of MAXWELL's kinetic theory in the ordinary mathematical way, along with some motivation and some words about the physical circumstances the theory is intended to represent. Once the axioms of the theory shall have been stated, we will not alter or "approximate" them. In particular, we will not mention linearized replacements such as are sometimes claimed valid when this or that quantity is "small", nor will we consider "models" which replace the theory by another one which is similar to it but mathematically easier. These "approximations" and "models" now occupy so large a part of the literature on the kinetic theory that a beginner may easily gain an impression, altogether false, that the kinetic theory consists of nothing else—an instance of the daily more and more familiar principle that a pile of fake diamonds outshines a small gold nugget.

In fact, there is an exact, mathematical kinetic theory. Although MAXWELL framed the basic definitions and axioms over a century ago, and although in the past thirty years much fine work has been done in this field, the body of results obtained remains small. Thus we can present nearly all of it in this one book. Even so, many of the analyses contain important gaps, as we shall soon see.

Our first purpose is to uncover these gaps and to illuminate them as challenges to future research by mathematicians. To examine a gap, you must first have a sound, strong road to at least one side of it.

---

[1] Those not familiar with the physical concepts on which the theory rests may find sufficient background in TRUESDELL's "Early kinetic theories of gases", *Archive for History of Exact Sciences* **15** (1975): 1–66. That history traces the subject from its beginnings up to 1867, the year in which MAXWELL formulated his second and definitive kinetic theory of gases.

To find such roads, we shall let the conceptual and logical structure of the kinetic theory speak for itself.

Clearly the molecular schema [employed in MAXWELL's second kinetic theory], however plausible it may seem, is not consistent with the principles of the NEWTONian mechanics of punctual systems. If the intermolecular forces extend to infinity, then no encounter can be binary, and the motions of any two molecules are influenced, though possibly indeed not much, by the motions of all the rest. If the intermolecular forces have a cut-off, then binary encounters become possible, but nevertheless we are not justified in assuming that all encounters are of this kind. The motions of a set of mass-points subject to specified mutual forces are determined uniquely by the initial conditions and the dynamical equations. Thus we are not at liberty to assume anything about those motions. Whether or not all encounters of an assembly of mass-points be binary, is a matter for mathematical analysis and proof, not for guesswork. Indeed, if the molecules have finite spheres of action, analytical dynamics leads us to expect that ultimately three such spheres will intersect, not merely two, unless we start the system of molecules in some exceptional way. In general, therefore, MAXWELL's *assumptions regarding the molecular motion contradict the laws of analytical dynamics*[2].

If MAXWELL's assumptions do not generally follow from the principles of mechanics, there is no reason to hope to draw them from the theory of probability, either. There is no general theory of stochastic mechanics, and if there were one, we could not expect it to yield the kind of specific, determinate outcome of an encounter we are here assuming as a prerequisite for even stating the form of the collisions operator.

Rather, that operator must be regarded as a *mathematical model* for the likely effect of the outcome of many collisions, not as the result of calculating that effect or anything regarding it. MAXWELL's kinetic theory is a consequence neither of classical mechanics nor of the axioms of probability theory. Though it is motivated by a masterly and suggestive combination of mechanical and stochastic ideas, it is an independent model of a gas. As such it is to be respected and studied mathematically. The proof of the model lies in its product. The two models of a dissipative fluid that have proved their value again and again are the NAVIER–STOKES theory of linearly viscous fluids and MAXWELL's kinetic theory. Both are far better in product than any

---

[2] [*Cf.* P. G. BERGMANN on page 191 of *The Nature of Time*, edited by T. GOLD, Ithaca, Cornell University Press, 1967: "It is quite obvious that the Boltzmann equation, far from being a consequence of the laws of classical mechanics, is inconsistent with them."]

argument used to motivate or infer them might suggest. Each involves a special kind of nonlinearity that seems somehow to reflect much, though by no means all, of the phenomena seen in natural fluids. These two nonlinearities offer perfect challenges to the student of rational mechanics: They are genuine yet concrete, mathematically so difficult as to afford anyone, no matter how expert he be, opportunity for a lifetime of study, yet not so difficult as to blank all rational inquiry. They enrich most of all the understanding of him who can weigh and value them both. In their predictions they partly agree but mostly disagree widely. They model partly similar and partly different aspects of natural gases.

We have attempted to apply to MAXWELL's theory the standards of conceptual analysis, logical hygiene, and mathematical rigor to which the rational thermomechanics of continua developed in the past quarter-century has accustomed us. The perspective gained from that discipline has been our guide at every turn.

Anybody who does a clean job nowadays with concepts, assertions, and proofs will see his work dismissed by some as mere "axiomatics". While criticism of this kind deserves no reply, we invite those inclined to it to cast their eyes upon the body of formal, explicit calculation we print here. In Chapters XVI and XVII they will see a larger list of exact collisions integrals than has ever been published before; if they study the text, they will learn how to calculate as many more of them as they have time and patience for. In Chapters XXIV and XXV they will see more explicit terms for "transport effects" than ever before published, along with formally exact evaluations of transport coefficients never before determined except through unassessable approximations; if they study the text, they will learn how to calculate as many more of these coefficients as their stomachs will bear. That our calculations are general and are exact at least formally; that our procedures are systematic, explicit, and demonstrably effective rather than "*ad hoc* juggling" of the first few terms in some series; that we specify what we desire to approximate before we start to calculate approximations—"axiomatics" these may seem to some, but they do not render the results themselves any less "physical" than if they had been obtained by the exhortatory and devotional manipulations heretofore accepted in this subject.

Note for the Reprinting

The foregoing was first published as the prologue and a part of Section (vi) of Chapter VII of *Fundamentals of Maxwell's Kinetic Theory of a Simple Monatomic Gas, Treated as a Branch of Rational Mechanics*, co-author R. G.

MUNCASTER, New York *etc.*, Academic Press, 1980. Essentially the same text had appeared in my *Mathematical Aspects of the Kinetic Theory of Gases*, Notas de Matemática Física, Volume **3**, Instituto de Matemática, Universidade Federal do Rio de Janeiro, 1 May 1973.

The reader may notice some affinity to views expressed by EINSTEIN in his address on PLANCK's sixtieth birthday, 1918:

> [T]he theoretical physicist's picture of the world ... demands the highest possible standard of rigorous precision in the description of relations, such as only the use of mathematical language can give. In regard to his subject matter, on the other hand, the physicist has to limit himself very severely: he must content himself with describing the most simple events [that] can be brought within the domain of our experience; all events of a more complex order are beyond the power of the human intellect to reconstruct with the subtle accuracy and logical perfection which the theoretical physicist demands. Supreme purity, clarity and certainty at the cost of completeness. But what can be the attraction of getting to know such a tiny section of nature thoroughly, while one leaves everything subtler and more complex shyly and timidly alone?

(Page 21 of the translation in *The World as I See it*, New York, Covici-Friede, 1934.)

8. OUR DEBT TO THE FRENCH TRADITION: "CATASTROPHES" AND OUR SEARCH FOR STRUCTURE TODAY (1978, 1981)

The connection of my subject with the theory of catastrophes is obvious: Who but a Frenchman could have decided that a catastrophe—$\kappa\alpha\tau\alpha\sigma\tau\rho\sigma\phi\acute{\eta}$, overturn, ruin—is a civilized, regular denizen of that best of all possible worlds, mathematics, demesne of reason and order? Of course, without some assumed or required smoothness of motion, catastrophes become so common as to be altogether stochastic—dull, random events that make up no more than a statistic—but who will deny in a Frenchman innate sense of the right degree of smoothness at the right moment?

For a characteristic example of the far-reaching effects of smoothness we may take the classical problem of the vibrating string. Pretty early in the seventeenth century people thought that a good initial shape would be a triangle. For simplicity the apex was put at the middle, even though every musician knew that was the wrong place to pluck or stroke a string. Nobody got far with the problem, but over a century later D'ALEMBERT derived the partial-differential equation that governs small motions, and he also found the two classes of wave functions that satisfy it. He thought also that a function was identifiable with a formula. It is hard to make this idea clear; D'ALEMBERT, a vague and loose mathematician, never could specify it, but certainly he believed that a single function could not have different algebraic expressions in different parts of its domain. Thus for him a single function could not represent two legs of a triangle. He contended, therefore, that the wave equation had very few solutions. In particular, it could not describe the motion of a string released from an initially triangular form. Within the limitation he imposed, he was right. Why anybody should desire to impose it, is another matter. EULER immediately took up the problem and obtained the general solution in the modern sense of the word. To this end he introduced a broader concept of function, what today we call a mapping of a subset of the reals into the reals. One such mapping represents a triangular initial shape, and EULER found the solution to which it gives rise. To do so, he had to generalize the partial-differential equation. He excused from duty not only the endpoints but also a finite number of others, the possible corners of the figure assumed at any one time. At these, the differential equation makes no sense, for there the figure does not possess a slope, let alone a curvature. Had EULER lived in today's climate of cultural suicide, he might have bestowed upon these exceptional points the lurid, journalese adjective "catastrophic", current now in this sense though that is too recent to be listed in the supplement of 1972 to the *Oxford English Dictionary*. Surely it pertains to the apex of the triangle (Figure 7) if it pertains to anything. As its very first act that apex splits into two new catastrophes, one going to the left and the other to the right. After a quarter period the string undergoes a catastrophe of another kind. It loses its corners and looks just like an undisturbed string: a straight line, as smooth as smooth can be. That happens only at one instant; the tranquil shape then cannot mask its taut motion; at that very moment each and every point of the string has achieved its maximum possible speed. Immediately the two catastrophic corners re-appear, burst out, one from each endpoint, and the shape is again of the trapezoidal kind seen at all but four instants in each period. The motion endures forever; the catastrophes appear, coalesce, divide, and disappear with perfect regularity.

So much for D'ALEMBERT and EULER. Who was the Frenchman? You might think it was D'ALEMBERT, but you would be wrong. When both were alive, people said that in mathematics it was EULER, the master of clarity and the apostle of reason, who was the Frenchman of the two. D'ALEMBERT in affirming as universal a property which nowadays we consider appropriate only to analytic functions was holding to LEIBNIZ's *Law of Continuity*. LEIBNIZ had written, "In nature everything goes by degrees and nothing by jumps, and this rule in regard to changes is a part of my law of continuity." EULER, in philosophy eclectic, adhered fundamentally to one tenet of DESCARTES: *Experientia et Ratione*, which I translate "by facts and by reason". Thus when faced with the facts of acoustic vibration EULER was ready to relax an overriding and up to then fruitful principle of mathematical philosophy, provided reason could be brought to bear so as to produce a decent model. I say "relax", not "sacrifice". In EULER's footsteps, we still regard the partial-differential equation or a corresponding integral relation as the defining statement of the problem. By requiring that the solutions be smooth most of the time at most places, we tame and encage the catastrophes. A catastrophe can appear or disappear only at certain times and places, and it leads an ordered, predictable life.

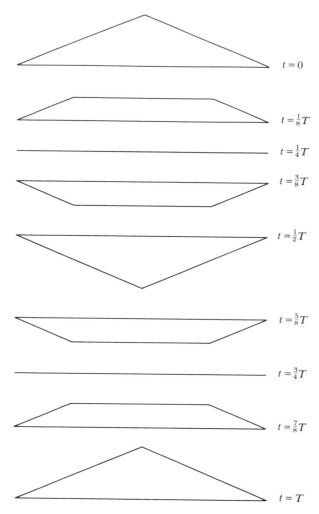

Figure 7. Periodic division, motion, destruction, re-creation, and coalescence of little catastrophes in EULER's solution for vibrations of a taut string whose initial form is an isosceles triangle. T = period, t = time.

It is become traditional for us to formulate a problem smoothly, but we recognize that smoothness may carry the seeds of its own destruction. As the ever mysterious, ever new while old, and ever comical patterns of relations between the sexes suggest—today I should more correctly say relations *among* the sexes—a catastrophe may be the inevitable consequence of smooth beginnings, and a catastrophic start may lead to a smooth solution, though usually for a short time only. If, on the other hand, we were to welcome irregularities as the normal way—currently recommended by those young enough to withstand the fatigues, crises, and shocks of what they call "living a normal life"—the idea of a solution to a problem would disappear. Anything could happen. Physical science has found peace and product in seeking smooth solutions when they exist, and when they do not, in relaxing smoothness just enough to get a mostly smooth solution. LEIBNIZ' universe with no jumps has been replaced by a universe with as few jumps as possible. If D'ALEMBERT's primrose path leads only to *rigor mortis*, EULER's attempt to maximize the primroses while still moving ahead has taught mathematical science, I will not say how to reach heaven, but at least how to predict catastrophes. We could name his idea *the principle of smoothest path*.

We have to be careful, for "smooth" has not been defined. While the example of the vibrating string suggests continuity and differentiability as measures of smoothness, there may be others. For example, smoothness in the context of additive functions should be understood as measurability. In problems leading to bifurcation, like that of a strut subject to axial compression, solutions may have as many derivatives as we like; the straight line is certainly smoothest in the usual sense, but it is not generally the best or true solution; and "smoothest", if we are to retain our principle, must be interpreted here to mean "storing least possible energy".

Applied catastrophe theory today devotes itself to discussing such things as how dogs get angry, how riots occur in prisons, and how wars start. It does so by pointing to unstable points on a cusped surface. The theory is purely static, proposing no laws of motion to describe how one of these phenomena approaches a catastrophe or what happens afterward. The catastrophes considered in that theory are more dramatic than EULER's peaceful, periodic creation, coalescence, and destruction of corners. Apparently a serious catastrophe brings him who suffers it to a point of no return.

Classical mechanics allows those, too. Classical mechanics is a mostly determinate theory of motions. Thus when it predicts something catastrophic, usually it specifies the behavior leading up to it and following afterward. To see that, we need only consider an example from the mechanics of GALILEO, half a century before any equations of motion had been discovered and more than a century before EULER's time. GALILEO gave us a definite theory for the descent of a body falling along a concave polygon (Figure 8). He supposed the body did not jump off but continued in tangential motion. At the corner the velocity vector changes abruptly. GALILEO imposed the requirement that the speed remain constant, thus determining a definite velocity after the body passes the corner. In modern language, the total energy of the body remains constant, although the momentum changes abruptly in accord with the constraint.

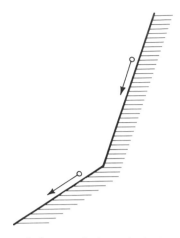

Figure 8. GALILEO's descent of a heavy body along a concave polygon.

Still using GALILEO's ideas, we may consider free motion of a similar kind: A free body strikes an oblique wall, along which it must continue its motion (Figure 9). Now it has two choices: left turn or right turn? Everyone will expect it to persevere in its forward motion as much as it can. Therefore, for the configuration shown in the figure, it will turn to the right. We can express this expectation as a principle of mechanics generalizing the classical statement of the law of inertia: *A free body subjected to frictionless constraints will conserve its total energy while moving in such a way as to alter its forward momentum as little as possible.* This principle provides a unique outcome of the discontinuous motion we are considering. Let us now see what happens when we reverse this motion after the shock has occurred (Figure 10). The body will retrace its former path along the wall. When it reaches the point where it first hit the wall, it will not experience anything to make it change its momentum this time, and so it will not forsake the wall. The motion is locally reversible almost always but not globally so.

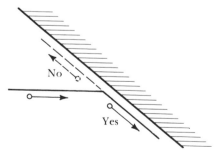

Figure 9. Free body striking a plane, smooth wall.

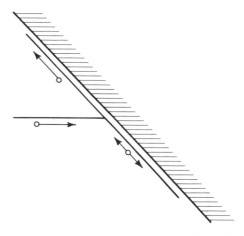

Figure 10. Reversal of the motion shown in Figure 9.

The experiment can be performed with little balls on a grooved table, but it can be made more striking by forming tubes like a branched tree (Figure 11) and letting the balls fall by gravity, one after another. To calculate such a motion, we parallel GALILEO's treatment of the flight of a projectile. Namely, in each shock-free interval we superimpose upon the momentum of the body if free the momentum given by GALILEO's rule for descent of the same body along an inclined plane. No ball will fail to drop finally into the basin. If one of the fallen balls be then shot upward with the reverse of its former terminal velocity, it will ascend the trunk, not the particular set of branches and twigs along which it first descended, and it will drop back into the basin. From the basin there is no final escape. A universe formed from a concatenation of trees like this would allow each soul falling from heaven, heavy with (let us say) her sins, find one and the same resting place, however ramified her travel to it. Should she try to escape, she could do no more than jump straight up and then fall back like the stone of Sisyphos.

We can do still better. As I have remarked elsewhere, my extension of GALILEO's rule of free flight fails to deliver a unique outcome when a punctiform body strikes an acute, solid wedge head on with a velocity whose direction bisects the angle of the wedge. Then determinism fails. We may say that the body in its continuing motion has no reason to prefer one side of the wedge to the other. As the dynamical law commands it to go along one of them (unlike the philosopher's ass equidistant from two bales of hay), some choice must be made; we may regard that choice as an exercise of free will. We may add to the trees like that shown in Figure 11 any number

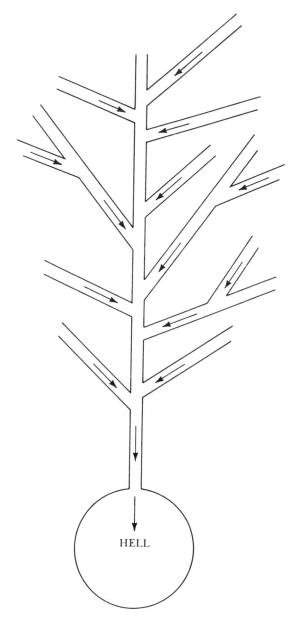

HELL

Figure 11. Deterministic and irreversible fall through twigs and branches of a tubular tree.

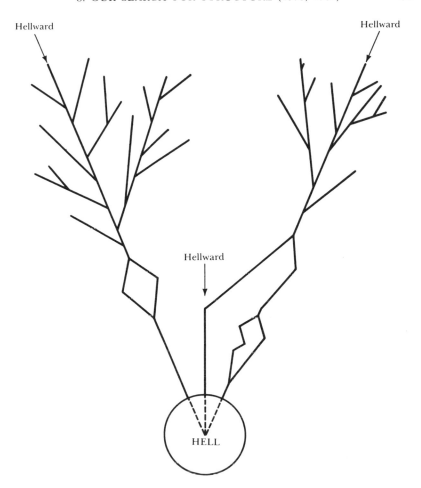

Figure 12. Alternating determinism and free will. The lines are to be conceived as slender tubes within which the heavy soul must fall. She begins her earthly life at some open end. To the irreversible determinism illustrated in Figure 11 are added instances of free choice. The left-hand tree contains one of these; there the final portion of the path is independent of the choice made. The right-hand tree contains two instances of free choice; one of them amounts to a choice between two final roads to hell, while the other makes no difference in the sinner's last moments or the length of her life. Trees with any desired number of distinct paths to the same inevitable end and any number of instances of free choice unaffecting the final outcome are easily conceived.

of instances where the falling soul must apply her free will. Her choice may leave unchanged the direction of her final approach to the fiery furnace and the time it takes her to get there or may alter both, but there she will end without fail. The flabelliform branching may be extended upward and provided with any number of points of indeterminate bifurcation. Those are, as the catastrophists would put it, not generic, but they are not therefore negligible. Since there shall have been, from the creation of Adam until the last trump, but finitely many souls, only finitely many trees are needed to provide each soul with its proper, predestined channel, or several souls may be put initially at the open ends of various twigs on one tree. Philosophers of rosy humor may prefer to appeal to a similar model for the ascent to heaven of souls pre-ordained blest. In analogy with electrified particles they may be supposed to have negative weight (perhaps acquired by sufficient abstinence and other exercise of virtue) and hence by the very gravity that pulls the damned to hell at the earth's center be drawn upward to the celestial sphere. The trees for them should be formed on radial axes, like those for the damned, but with branches pointing downward. Thus the catastrophes compatible with classical mechanics provide a simple model, deterministic overall though not always locally so, of the futility of exercising free will when the Fates have already decided the outcome: In *Flatland* a superior intelligence sees past and future alike because he may call upon use of a third dimension inaccessible to the souls confined to a single plane; the same holds here, too, for both souls and trees may be taken all in one plane, visible in its entirety to a being looking at it from a point not on it. Earthbound philosophers may prefer a three-dimensional model for heaven, hell, and the dismal swamp of life. The wedges providing two asinine choices are then replaced by cones, which allow infinitely many. To see all things at once, the Divinity will need to dispose of a fourth spatial dimension. Here begins a mathematical science of predestination, free will, and the all-knowing, all based upon nothing but my simple and obvious extensions of GALILEO's mechanics. Miracles are even easier to account for here than by quantum mechanics, for no wave functions, Lagrangeans (real or complex), spinors, Cayley–Klein parameters, indeterminacy *etc. etc.* are needed. To the first specialists in this new yet totally classical discipline I leave invention of a model for ultimate salvation after a course seeming to point toward BEELZEBUB's claws (perhaps through soul-weight made variable by free acts of piety) and development of the possibility that free will shall be generic.

Classical mechanics, claim the philosophers, makes all motions reversible, future and past equally predictable. The philosophers call this idea "Laplacian determinism", and they rest their opinion on one sentence at the very beginning of LAPLACE's *Essai philosophique sur les probabilités*:

> An intelligence such as to know at some one instant all the forces
> that give motion to nature and the respective conditions of all

natural beings, could, if only it were vast enough to subject these data to analysis, embrace in a single formula the motions of the biggest bodies of the universe and the slightest atom. Nothing would be useless to such an intelligence, and future and past alike would be open to its vision.

It is not unreasonable to surmise that most philosophers have read no more in the mathematical sciences than this one pronouncement. Many objections could be raised against their interpretation of it, still more to the statement itself. I suspect LAPLACE of teasing. BOCHNER, apparently taking LAPLACE at his word, wrote

> There is nothing more intolerant of discontinuities than this approach . . ..

He went on to remark, in effect, that whatever LAPLACE may have meant or thought, no mathematically literate man of the twentieth century could allow the philosophers' claim:

> And the widespread identification of causality and/or determinism with the "certainty" and "predictability" of a Newtonian mechanical point system was in the twentieth century almost too childish to debate, even if . . . many intellectuals of the "enlightened" West [took this identification for granted].

There are indeed elements of determinism as well as of reversibility in analytical dynamics. These properties are mathematically demonstrable, but *only for short times and for smooth data.* You see from my childish examples that once the restriction to short times and smooth data be relaxed, dynamics even when it is deterministic allows infinitely many possibilities of irreversibility, leaving the mobile beyond all possibility of retracing its path.

Classical mechanics allows catastrophes—shocks as well as weaker disasters, along with strokes of the blindfold goddess that deserve to be called "pieces of good luck".

While inconsequential ramblings such as mine just presented are the privilege of the senescent, I have gone too far. As was said by a scholar of undisputed pre-eminence, just about to enter a final course of study under the direction of MEPHISTOPHELES himself,

> Philosophy is odious and obscure . . ..

LEIBNIZ warned us to have a care of physics, which should be replaced as quickly as possible by pure geometry. If you will allow me to add a single letter to the quotation, we shall find it better said by the older authority just quoted, Dr. FAUSTUS:

> Both Law and Physicke[s] are for petty wits. . ..

Let me return to the French tradition, the only exponent of which I have so far invoked is EULER. Like him, many of its leading exponents never set foot in France. Consider, for example, NEWTON. While the folklore of physics makes NEWTON a disciple of GALILEO, there is no evidence that he had ever read a word of GALILEO's writings on dynamics. The early draughts of the *Principia* do not mention GALILEO at all, and in the book as printed the one reference to the laws of falling bodies, which we today and with scant justice credit to GALILEO, seems to have resulted from the instance of critics who induced the great author, parsimonious as he was whenever it came to crediting anybody else with anything, to insert it. We know he esteemed KEPLER little. Who, then, was his true teacher? DEREK T. WHITESIDE, the editor of NEWTON's *Mathematical Papers* and the only man in our century (so far as I can learn) to have entered deeply into their contents, concludes that it was none other than DESCARTES. So it was for LEIBNIZ, too. Critical as both NEWTON and LEIBNIZ were of the details of DESCARTES' sloppy natural philosophy, they were at one in accepting his view of nature as a great machine, the workings of which could be explained by sufficiently adroit mathematical reasoning applied to mathematical representations of observed facts.

Much of physical science as we know it today was created by the British in the nineteenth century. A good many of these were British of an unusual kind. GREEN had studied alone and in isolation on the top floor of his father's mill in Nottingham; STOKES's early life was spent in an Irish parsonage; KELVIN was the well directed son of a Glasgow professor; and MAXWELL was the heir of a laird in Galloway. Each of them made his peace with Cambridge—in GREEN's case, the peace of death; in STOKES's, surrender to the system and relinquishment of research; in KELVIN's and MAXWELL's, an honorable draw sealed by return to the bonnie braes. One point they had in common with each other and with poor WATERSTON, a Scot whom the system rejected: All were mainly self-trained, and all turned for study and guidance to French masters, mainly LAGRANGE, LAPLACE, POISSON, FOURIER, and CAUCHY. In this way they were able to see the force of imagination disciplined by a sense of order; in this way they could spread their wings and soar above the sodden empiricism that had put the Royal Society in the late eighteenth century into the hands of clockmakers, opticians, speculative chemists, doctors of medicine, and country parsons. Their contemporaries recognized and even admired their foreign ways. "Indeed", Lord RAYLEIGH wrote, "both Green and Stokes may be regarded as followers of the French school of mathematics." Let it not be thought that I wish to put all good men into the French bag. Certainly I should not impute a sense of order and beauty to everyone, nor are the mathematicians many in any period or country who can pass BOILEAU's test:

> Just as our thought is more or less obscure,
> Expression follows it: opaque or pure.
> What's well conceived is clearly, nicely spoken;
> The words to say it come uncalled, unbroken.

The search for beauty and order—for while the two are far from synonymous, as anyone who has looked into a beautiful woman's handbag or bureau drawers will attest, nevertheless the one cannot exist without the other—did not die with STOKES's surrender and MAXWELL's retreat to Dalbeattie. Petty wits raised aloft by the crooked little legs of pygmies standing upon giants' shoulders quickly manufacture a folklore to distort the history of the course of thought into capsules of indoctrination for pygmylets. The smug confusion poured forth from the sewers of aimless precision of experiment dominated by routine obeisance to half-consistent theory half misunderstood that always follows upon and smothers a triumph of creative imagination, drives a thinking man to recourse to higher authority. Such recourse I sought in the 1940s when first I tried to understand the response of materials to severe deformation. My appeal led me at once to the early works of STOKES, and not long thereafter I began to mine the pile of treasure bequeathed us by CAUCHY. For several years there was not a day in which I did not read at least some lines by him. The deep understanding of CAUCHY's work shown by remarks scattered through the papers of Mr. ERICKSEN suggests that he may have had a similar experience. Mr. NOLL drank from a more recent source. A year spent at the old Sorbonne in the brief period of revival and reconstruction, just after the debacle left by the great war of the -isms, the -acies, and the "freedoms" and just before the cancer of Gallic Socialism entered a catastrophic phase, left a mark that can be seen in every line of his writing. What I hope will be recognized as the principal quality of modern continuum mechanics is its clarity. It was designed to shed light, not to dazzle. It is easy. We do not seek the reader's admiration; we hope to instruct him; thus we must find something worth his learning. The aim of rational mechanics is to provide a sound, tight conceptual framework for description of nature as human senses perceive it and to create patterns of systematic inquiry and systematic inference such as to order and interrelate phenomena as thereby conceived. Such was NEWTON's program; such was EULER's; such was CAUCHY's; and such is ours. It presumes taste, not only in authors but also in readers. To a man emancipated from values, a man with no ancestors, that makes no sense. The founders of the Society for Natural Philosophy desired it should remain small, but they saw that to this end they needed no formal limitation, and so they left membership open to one and all.

Today's meeting shows that catastrophes can penetrate even an enclave from the eighteenth century. The prototype, I believe, is a shock wave, just a bit stronger than the acceleration waves familiar since EULER's day. To STOKES we owe the earliest remarks about strong discontinuities, but in his old age he let his friends talk him out of them, and he recanted—recanted with more conviction than the folklore of physics attributes to GALILEO. The first clear treatment of surfaces of discontinuity, I think, is HUGONIOT's. I prefer it to RIEMANN's because it does not begin from the rather limited concept of characteristics of hyperbolic partial-differential equations but instead from a geometric description unrestricted by constitutive relations. We have all heard that a weak wave of condensation must within a finite time grow to infinite strength and so form a shock wave. We have all seen proofs of the former statement, but I doubt if any has been given for the latter. The event, if such it is, provides an example of how a tame, little, reversible catastrophe can grow into a big and irreversible one. More important, I think, is the fact that some motion continues after the catastrophe. It is not sufficient to predict that a catastrophe will occur. If a shock is formed, what are its initial strength and velocity? Do big ones break up into little ones, or do they form even bigger ones? And can a catastrophe simply disappear after a while?

That is not all. We must see also that what is a catastrophe may depend upon who is watching it. One man's regularity is another's ruin. One and the same physical occurrence is modelled as a shock wave in the EULER–HADAMARD theory of perfectly elastic gases, as a smooth but steep layer of transition in the STOKES–KIRCHHOFF theory of viscous gases. Surely it is all one to air and hydrogen themselves which theory we use to describe their motions. The choice is ours. We make or break the catastrophe according to taste, much as we make probable events improbable or *vice versa* by our own assignment of an *a priori* basic probability, for only by such an assignment does the very concept of quantitative consequent probabilities make sense. Our guide is our own estimate of the facts, seen in the light that such mathematical apparatus as we have may cast. In the case of the shock wave in a gas, the more refined model tames the savage, but for some phenomena the opposite holds, and it is refinement that reveals catastrophes that macroscopic vision smoothes over.

What seem to be gross catastrophes need not represent anything outlandish on the molecular scale. The model of a gas as an assembly of ideal spheres presumes all molecular interactions to be "catastrophic"; the model of a gas as an assembly of point molecules subject to mutual forces of infinite range presumes that no molecular action be "catastrophic". In gross what one model delivers differs but in relatively small details from what the other does. Densely packed catastrophes have pretty nearly the same effect as densely packed smooth changes.

References for Quoted or Paraphrased Passages

D'ALEMBERT on the vibrating string:
"Recherches sur la courbe que forme une corde tendue mise en vibration", *Histoire de l'Académie des Sciences de Berlin* [**3**] (1747): 214–219 (1749).

EULER's rebuttal:
"De vibratio chordarum exercitatio", *Nova acta eruditorum 1749*: 512–527 = (transl.) "Sur la vibration des cordes", *Histoire de l'Académie des Sciences de Berlin* [**4**] (1748): 69–85 (1750) = pages 50–77 of LEONHARDI EULERI *Opera omnia* (II) **10**.

EULER on the triangular string:
"De chordis vibrantibus disquisitio ulterior", *Novi Commentarii academiae scientiarium Petropolitanae* **17** (1772): 381–409 (1773) = pages 62–80 of LEONHARDI EULERI *Opera omnia* (II) **11**.

LEIBNIZ on his law of continuity:
"Nouveaux essais sur l'entendement humain", Chapitre VI, § 13, in *Œuvres Philosophiques*, edited by RASPE, 1765, page 267. Also Lettre II à VARIGNON on page 93 of *Leibnizens Mathematische Schriften* **4**, edited by C. I. GERHARDT.

GALILEO on fall along a concave polygon:
Discorsi e Dimostrazioni Matematiche intorno à due nuove scienze Attenenti alla Mecanica & i Movimenti Locali, Leiden, Elsevier, 1638. See Dialogo Terzo, Theor. XXII, Propos. XXXVI.

Flatland:
[EDWIN A. ABBOTT], *Flatland, a Romance of Many Dimensions*, Second and Revised Edition, 1884, Sixth edition, N.Y., Dover Publications, 1952.

LAPLACE on determinism and reversibility in classical mechanics:
Essai philosophique sur les Probabilités, fourth edition [as Introduction to the third edition of *Théorie analytique des Probabilités*], 1820 = pages V–CLIII of *Œuvres Complètes de Laplace*, Volume **VII**, 1886. See pages VI–VII.

BOCHNER on "Laplacian" determinism and its interpretation by "intellectuals": *Einstein Between Centuries*, Rice University Studies **65** (1979), No. 3. See pages 22 and 31.

TRUESDELL on failure of determinism in classical mechanics:
"A simple example of an initial-value problem with any desired number of solutions", *Istituto Lombardo, Accademia di Scienze e Lettere*, Classe de Scienze (A) **102** (1974): 301–304. *Cf.* also pages 19–20 of P. PAINLEVÉ, *Leçons sur la Resistance des Fluides non Visqueux*, Ière Partie, Paris, Gauthier-Villars, 1930.

Dr. FAUSTUS on philosophy and physic[s]:
> The Tragedie of Doctor Faustus in FREDSON BOWERS, editor, The Complete Works of CHRISTOPHER MARLOWE, Cambridge etc., Cambridge University Press, 2nd edition, 1981. See lines 134 and 135.

NEWTON's debt to DESCARTES:
> DEREK T. WHITESIDE, Introduction to Volume **VI** of The Mathematical Papers of Isaac Newton, Cambridge, Cambridge University Press, 1974.

RAYLEIGH on French influence in British mathematical science of the nineteenth century:
> "Obituary Notice of Sir George Gabriel Stokes, Bart. 1819–1903", in Proceedings of the Royal Society, 1903, reprinted as pages ix–xxv of Volume **5** of Mathematical and Physical Papers by the late Sir George Gabriel Stokes . . ., 1905; second edition, edited by C. TRUESDELL, New York & London, Johnson Reprint Corp., 1966.

BOILEAU on clarity:
> NICOLAS BOILEAU-DESPRÉAUX, L'Art Poétique (1674). Chant Premier, lines 150–154.

STOKES on shock waves:
> C. TRUESDELL, preface to the second edition of the Mathematical and Physical Papers of the late Sir George Gabriel Stokes, cited above in regard to RAYLEIGH's estimate of STOKES. The preface is reprinted below as Essay 28 in this volume.

Note for the Reprinting

The foregoing essay derives from the Fifteenth Anniversary Lecture, Society for Natural Philosophy, delivered in the Wren Building, William & Mary College, Williamsburg, Virginia, on April 17, 1978, at the meeting on "Catastrophe Theory". Professor JANE WEBB had suggested the topic.

The authoritative critiques of Applied Catastrophe Theory by SMALE and by SUSSMAN & ZAHLER were then either newly published or just about to be published, and I did not know of them. They are quoted extensively below in Essay 10. They settle, among many issues raised at about the same time in regard to Applied Catastrophe Theory, some gentle and gingerly questions put to the catastrophists who spoke at the meeting.

For reprinting here I have shortened, revised, and augmented the text published in Scientia **117** (1982): 63–77.

9. DRAW FROM THE MODEL AND IMITATE THE ANTIQUE (1979)

Boundary conditions, like field equations, are proposed by theorists who dare represent nature by mathematical hypotheses. "Draw from the model and imitate the antique," said RUBENS. The tradition shows us that only after a theory has been formulated can existence theorems be proved. In framing boundary conditions, just as in framing field equations, the theorist outlines Nature as best he can from what little of herself she lets him see through the fogs with which she modestly covers her sincerity. To do so, he follows the forms and practices that his masters, the great theorists of old, have taught him by example. He demonstrates the properties that solutions must have in order to satisfy his conditions. Like his great forebe-ers he runs the risk that solutions of the kind he analyses may not exist: that all his labor may be spent on describing a few of the countless attributes of the null set. The tradition gives him hope as well as example.

Note for the Reprinting

The sentences reprinted here were first published as the opening of Section (iii) of Chapter XI of *Fundamentals of Maxwell's Kinetic Theory of a Simple Monatomic Gas, treated as a Branch of Rational Mechanics*, New York *etc.*, Academic Press, 1980, co-author R. G. MUNCASTER. While I recalled the above title as a quotation of RUBENS' words, I have been unable to trace it as such. It now seems to me to be an epigraph for RUBENS' masterly essay, *De imitatione statuarum*. Nevertheless, I had not yet read that essay when my memory deceived me, so it is not impossible that the words written were no more than a summary of my decades of uninstructed admiration for the master's works and life. According to the report of 1665 by FRÉART, Sieur de CHANTELOU, BERNINI made a more detailed statement to the same effect in regard to the study of ancient sculpture. Both passages may be read in English in J. R. MARTIN's *Baroque*, New York *etc.*, Harper & Row, 1977.

Here I call attention to RUBENS' statement regarding the decline of art, in which the reader of this book is invited to replace "antique statues" by "classics of mathematical science":

He who has, with discernment, made the proper distinctions [between good and bad antique statues and between statues and real bodies] cannot consider the antique statues too attentively nor study them too carefully; for we of this erroneous age are so far degenerate that we can produce nothing like them: Whether it is that our grovelling genius will not permit us to soar to those heights which the antients attained by their heroick sense and superior parts; or that we are wrapt up in the darkness that overclouded our fathers; or that it is the will of God, because we have neglected to amend our former errors, that we should fall from them into worse; or that the world growing old, our minds grow with it irrecoverably weak; or, in fine, that nature herself furnished the human body, in those early ages, when it was nearer its origin and perfection, with everything that could make it a perfect model; but now being decay'd and corrupted by a succession of so many ages, vices and accidents has lost its efficacy, and only scatters those perfections among many, which it used formerly to bestow upon one. In this manner, the human stature may be proved from many authors to have gradually decreased: For both sacred and profane writers have related many things concerning the age of heroes, giants and *Cyclopes*, in which accounts, if there are many things that are fabulous, there is certainly some truth.

The chief reason why men of our age are different from the antients is sloth and want of exercise . . ..

10. THE ROLE OF MATHEMATICS IN SCIENCE AS EXEMPLIFIED BY THE WORK OF THE BERNOULLIS AND EULER (1979, 1981)

In memoriam magistri dilecti SALOMONIS BOCHNERI

'Αποθανών τις ἐσθ' ὅτ' ἔτυχεν εὐδαιμονίας·
σπανιώτερος δ' ὅστις ἐπεβίωσεν.

CONTENTS

1. SPATIAL FLIGHT: DID MATHEMATICS HAVE ANY PART IN IT?

The flight of man in space is the most astonishing achievement of engineering. Many factors were necessary to it. Everybody knows that

one was numerical calculation on large machines. Another was the
basic equations that the machines were told to solve, thousands and
perhaps millions of times. The equations that govern the motion of
a space capsule were discovered by mathematicians more than
200 years ago. Machines and methods of using them change often.
The basic equations do not. They are permanent.

2. DID NEWTON PROVIDE THE BASIC EQUATIONS OF MOTION?

Insofar as popular accounts mention these equations, they ascribe
them to ISAAC NEWTON who, they tell us, at a single stroke of genius
in 1687 discovered the law of universal gravitation and the axioms
that govern the motions of all bodies. No-one would deny that
NEWTON had forebe-ers whose work he built upon. The folklore
makes NEWTON's achievement seem like a product of the fine arts.
SHAKESPEARE did not create the English language or blank verse;
MICHELANGELO did not invent *Genesis* or fresco painting; but
SHAKESPEARE's plays and MICHELANGELO's ceiling are final and
unsurpassed monuments. There is a difference. We admire and value
Hamlet and *The Creation of Adam*, but we do not apply them to cases,
nor do we correct them or adjust them, nor do we continually refash-
ion them in greater scope. NEWTON's *Principia*, published in 1687, is a
monument of human achievement; it deserves the admiration and
esteem of everyone. Should an engineer study it with a view to using
its contents to determine the motion of a capsule projected into space,
he would be gravelled. Motions there are in abundance, but no general equations. Each motion furnishes a new problem and is treated
by itself. Examples there are, but no algorism: towering concepts and
a magnificent approach, certainly, but no method. Moreover, the only
motions NEWTON succeeded in reducing to mathematics are those of
a single point. He gave us no instance of a motion of as many as three
points precisely determined from his expressed principles.

Not only are the centers of mass of a space capsule, the earth, and
the moon three distinct points, but also each of these bodies comprises
infinitely many points. Today we idealize the earth and the moon,
generally, as rigid bodies, and the space capsule likewise. In travel-
ing forward they can spin, precess, nutate, and even tumble. These
motions, too, as well as the progress of the bodies' centers of mass, are
subject to differential equations, of which no trace, not to mention an
example, can be found in NEWTON's *Principia*.

So much for 1687. If we pick up LAGRANGE's *Méchanique Analitique*, 1788, or POISSON's *Traité de Mécanique*, 1811, we find in them all
the basic equations that describe the motions of systems of points and

rigid bodies, as well as many valuable and enlightening instances worked out and interpreted. These books are textbooks. From them students of middling capacities in the early 1800s were regularly taught mechanics as a branch of mathematics and thereafter were able to apply it in more or less routine ways to problems occurring in nature. The genius of NEWTON was no longer required. Engineers by the thousands could learn mechanics.

Textbooks, however good, are rarely the sources of what they expound. In the present context the two excellent books just mentioned are no exceptions. Before 1788 and after 1687 something had happened to mechanics. What was it?

3. How and When Were the Basic Equations Discovered?

In the century following the appearance of NEWTON's *Principia* mechanics was transformed from an unarticulated set of problems—some solved well, some half solved, some unsolved, a few wrongly solved—into a mathematical science. The men who transformed and ennobled it were JAMES BERNOULLI, JOHN BERNOULLI, DANIEL BERNOULLI, and LEONARD EULER, with lesser contributions by TAYLOR, CLAIRAUT, D'ALEMBERT, and LAGRANGE. They set themselves to solve problems of mechanics that CHRISTIAAN HUYGENS and ISAAC NEWTON had attempted in vain or had failed to attack. Those problems concerned the motions of systems of points, rigid bodies, fluids, flexible or elastic bands and sheets. The story is a long one, parts of which I have tried to tell in several books. Here I can do no more than list some of the stages in the discovery of the differential equations that govern the motions of mass-points and rigid bodies.

1686. JAMES BERNOULLI attempts to relate HUYGENS' theory of a swinging body to the law of the lever. His idea is great in principle but wrong as stated.

1687. NEWTON's *Principia* appears. Its Second Law of Motion asserts that "the change of motion is proportional to the motive force impressed, and it takes place along the right line in which that force is impressed." It does not tell us what is the measure of "motion" or how to determine forces.

1703. Correcting his attempt of 1686, JAMES BERNOULLI sees that for bodies in motion the system of applied forces is equipollent to the system of corresponding reversed accelerations per unit mass. On the basis of this new principle of mechanics he embeds HUYGENS' theorem on swinging bodies, which had been published in 1673 as an isolated statement, into mechanics as a whole.

1713–1750. Differential equations of equilibrium or motion for various collections of bodies are derived by TAYLOR, JOHN BERNOULLI, DANIEL BERNOULLI, EULER, CLAIRAUT, and D'ALEMBERT. The systems include vibrating strings, rigid bodies of specially simple shapes which slide or roll upon each other, compound pendulums, hanging cords and chains, bent elastic bands, compressed struts, elastic bars in vibration, fluids in tubes at rest or in rotation, linked or jointed bars. The methods used include precise instances of NEWTON's Second Law and various other principles proposed as replacements for it or as necessary adjuncts. The most important and fruitful of the adjuncts are two:

(1) The principle of small oscillations (1712 onward): The displacements from equilibrium are as the restoring forces, and all elements of the system oscillate at one and the same frequency.

(2) DANIEL BERNOULLI's principle of superposition (1733–1750): Any small oscillation may be regarded as a set of independent harmonic oscillations, each with its own frequency and amplitude. How to find the parts undergoing those oscillations is left to the adroitness of the student.

1738. In attacking the problem of a ship's oscillation EULER sees that all principles of mechanics so far proposed are insufficient to solve it. He advances a "hypothesis": Any body has three orthogonal axes through its center of mass, about each of which it may oscillate freely in small motion, with arbitrary amplitude and frequency for each.

1746. EULER discovers that a rigid plane sheet cannot generally spin freely about an axis through its center of mass. Permanent rotation about an axis is possible if and only if both products of inertia with respect to that axis are null.

1747–1750. EULER sees that NEWTON's Second Law, properly rendered clear and specific, applies not merely to a body as a whole but also to every part of every body. His statement of it is: the increment of velocity per unit time equals the accelerating power. Today we call this statement the *principle of linear momentum*: The rate of increase of linear momentum of a body equals the resultant force acting upon it. EULER recognizes the role of the mutual actions of parts of a body upon each other: In each case, those forces must be brought into the open as specified unknowns. Otherwise EULER's vast generalization of NEWTON's simple idea would remain jejune. EULER is the first to obtain the differential equations of motion for a general system of mass-points. These are the equations called "Newtonian" today and taught to every student of engineering as the most natural, most easily applicable statement of a basic, necessary, and sometimes sufficient law of mechanics. Later EULER applies his momentum principle to the infinitely many parts of a rigid body. He thus obtains the equations that govern the motion of a top or planet with respect to a fixed point. In so doing, nevertheless, EULER makes an additional assumption which will have to be eliminated in later research. This hidden assumption allows him to claim that the principle of linear momentum is the one and only basic principle of mechanics, which is not so. EULER's analysis puts into his hands the tensor of inertia of a rigid body; he is unable to prove that every rigid body has an axis about which it may spin freely. (In 1755 SEGNER, on the basis of EULER's work, is to prove that every rigid body has at least three mutually orthogonal axes of free rotation.)

1750–1770. EULER and later LAGRANGE exploit the principle of linear momentum by deriving equations of motion for small vibrations of bars, hanging chains, fluids in general flow, flexible membranes, *etc., etc.* EULER shows that the principle of small oscillations is a consequence of the principle of linear momentum in all these systems: For many of them, he proves DANIEL BERNOULLI's principle of superposition, which thus loses its status as an axiom and becomes a demonstrated consequence of the laws of motion.

1771–1775. Noticing that the principle of linear momentum does not suffice to recover the known and accepted theory of bent elastic bars, EULER reverts to an idea of JAMES BERNOULLI. Taking the principle of balance of moments as a basis, he adds to the applied forces reversed accelerations per unit mass. In this way he arrives at the statement we now call the *principle of rotational momentum*. This principle enables him to embed the theory of bent bars within the general scheme of mechanics. EULER recognizes his two great discoveries, the principle of linear momentum and the principle of rotational momentum, as basic *axioms of mechanics*, applicable to every part of every body. The two together suffice to get his equations of 1750 for the motion of a spinning rigid body at once and without the artificial extra assumption he had used to discover them.

4. IN EULER'S RESEARCH WHAT PART WAS PLAYED BY TEAMWORK, COMPUTING, AND EXPERIMENT?

The events I have just skimmed over do not provide an outline of the history of mechanics from 1686 to 1788. That history includes many other researches, some of them different in character. I have tried to recount merely how the basic equations governing interplanetary excursion were first obtained. To do so, I have had to sketch the course of discovery of the basic principles of mechanics, the principles every engineer today is taught to take as his starting point when he chooses to analyse the behavior of a mechanical system. The theory of motions of a rigid body has a double role: Research aimed at discovering it contributed to *discovery of the basic principles of mechanics*, which were known only incompletely before 1775; now it may be regarded as *one of the simplest applications* of these basic principles.

The approach and methods of the BERNOULLIS and EULER may seem strange to the man of science trained in our time. He may ask: How large was EULER's research team, and who paid for it? How much of his success came from numerical calculation? What part did experiment play?

The answer to the first question is easy: EULER had no research team. Only after he was become totally blind did he have assistants who really assisted him. As for calculating numbers, EULER loved it; in some aspects of mechanics, notably in regard to the planetary motions, he did enormous amounts of it; but in the discovery of the basic principles, the principles used today to plan and control the motions of travelers in space, I can find not even one numerical calculation. Experiment? I enter dangerous ground. As ERWIN CHARGAFF puts it, some time ago "natural science came down with a case of galloping experimentitis." H. R. POST in a brilliant inaugural lecture at Chelsea College in 1974 discerned and described "three items of religious worship *inside* present-day science", the third of which is experiment. "[I]n the main the role of experiment", he wrote,

> constitutes a harmless *myth* in the philosophy of scientists. The myth considers experiment to be a generator of theories. In fact the role of experiment . . . is solely to decide between two or more existing theories. . . . Experiment does not generate theories, but rather is suggested by them.

The thoughtful would do well to consider POST's contention; I do not feel able to take a stand. I guess that it is neither altogether untrue nor unexceptionably correct, because experiment for consumption by theorists seems to me something like strong spirits, useful if taken in moderation; POST and CHARGAFF refer not to a stimulating or relaxing amenity but to a disease which may well be compared with alcoholism, jogging, computation, and other dangerous addictions. The experimentist should not take offense at these statements, for they refer, not to experiment in itself, but to *the role of experiment in the creation of mathematical theory*. Everyone agrees, I think, that much of science today gets along very well, at least by its own standards, in being experiment for its own sake. In the century of social democracy purely experimental science is irresistibly attractive because it is almost always team work, and the teams are large; in the industrialized countries experimental science offers the advantage of being costly, and so its value is automatically demonstrated. Thus to deny experiment the rank of *mater Venusque genetrix* of science is now as outrageous as to denigrate home, mother, prayer, and the moral purity of the U.S.A., as heretical as to question the axiom that all races of man are equally endowed with intelligence and genius. Let me assure you that I do not attack the "truths" that to man are dearer than bread, sometimes even than life; I advocate nothing; I merely recount facts, and facts are by their nature limited in scope and restricted in context. I refer only to some facts concerning the development of the theory called "classical mechanics"; in regard to the effect of experiment on theory in other fields, such as the biologies, I have no facts or competence, and so there I remain silent.

Nonetheless we must bear freshly in mind that the words "experiment" and "theory" have lost the meanings they had when, in the seventeenth century, the spring tide of the new humanism, the new natural philosophy brought them in as the touchstones of science. Then upon reading or hearing of an experiment the individual philosopher was not to believe it on faith or to acknowledge it on authority: He was to *try* it himself. Upon receiving a mathematical theorem, he was to perpend the statement and *prove* it for himself; upon receiving the proof of a theorem, he was to *search* it, step by step; he was to accept only such statements and arguments as he found clean. Both "experiment" and "theorem" derive from verbs denoting human acts: to *try* or *test*, and to *gaze upon* or *contemplate*. Science, which means "knowledge", was one of the possessions a man could acquire. He got it and used it himself. The measure of science was a man's experience, thought, and judgment. Though science could envision the unattainably distant and the imperceptibly small, the scale of scientific research was human. Just as explanations in terms of the acts of angels and devils were rejected, there was no such thing as a "black box".

In the late nineteenth century experiment became for the most part too complicated, too difficult, and too costly to be tried by any except an expert who disposed of a laboratory properly fitted out for it. A laboratory for optics did not afford the necessaries for experiments on the strength of solids, and *vice versa*. Experiment came to be done only by professional and specialized experimentists; its public rarely stretched beyond the few experts who could, should they wish, try a particular experiment, and the few witnesses of such trials. Fewer yet were those who would risk on probing someone else's claim the expense, labor, and time they might have used for discovering something of their own, something that they might publish. The desire to publish replaced the desire to know. Theory, likewise, became an activity of scant interest beyond a small group of theorists by profession. These could indeed, and did, range widely over much of scientific theory. They did, indeed, search and recreate for themselves each claimed discovery; thus theory, indeed, preserved the individuality, immediacy, and humanity of earlier science, qualities by then largely lost to experiment. Even so, both experiment and theory were become arcane, accessible only to experts. The interested layman could no more check a proof than try an experiment. A sacerdocy of science stood between mankind at large and knowledge of nature's workings. The glib mouther who can fribble about the latest doctrines of each and every field but knows nothing solid about anything arose to interpret to the ruck the esoteric disclosures of the priests. In the early twentieth century science vastly multiplied its complexity, its corpsmen, and its cost. No longer was it expected that a scientist have any general education, any understanding of arts and humane learning outside his specialty, or any sympathy with them. As successors to the old barons of science, who were (in the main) men both learned and cultured, even noble, arose squads of blindered experts who could drill fast, deep, and straight on but had no need or wish to scan, bridge, or order. In 1918 EINSTEIN remarked,

> For these people any sphere of human activity will do, if it comes to a point; whether they become officers, tradesmen or scientists depends on circumstances.

The ever more diffuse and more enfeebled popular and compulsory "education" produced a huge and dangerous proletariat technically able to read and write a little but stunted in mind. For them, the "science writers" had to translate some part of the flood of science into simplistic snippets in short sentences, employing only a child's vocabulary and syntax, compressed to accommodate a child's span of attention, prerequiring only a child's depth in experience, thinking, and stored knowledge. As CHARGAFF writes in his "Bitter fruits from the tree of knowledge",

> Science is, in many ways, a child of early expansionist capitalism. ... Everything has to be taken on trust; there is no real popularization possible, only a vulgarization that in most instances distorts the discoveries beyond reason.

Traduttore—traditore!

By the middle of the twentieth century physics was already being taught to beginners as a list of declarations and routines to be committed to memory; chemistry likewise; authority, though not associated with anybody's name, again ruled as supreme as had the invoked ARISTOTLE in the Middle Ages, and with a stronger grip. The mathematics of modern theory lay at least four years further on in the curriculum; the experiments, for the most part too costly as well as too dangerous to demonstrate, and even if repeated requiring of him who would understand them at least four more years of specialized study, were come to be presented to beginners largely by animated, gaily colored, schematic diagrams projected on a screen as explanations of what went on within the mysterious installations shown by photographs. Except for its lack of a personal god, ethics, and morality, what was taught to the beginner was no more than a revealed faith labelled "Science".

Though curiosity and participation were lost except to a very few, science was still a human activity for them. The age of the computer now is taking science out of human hands altogether. The computer typically consumes the data of an experiment and presents only the interpretation of them according to some theory, perhaps only a few primitive formulæ, recommended in advance; no human being who might doubt the interpretation can check the data. The computer in producing an application of a theory consumes also the intermediate stages and delivers the answer; no human being could repeat the calculation. Indeed, this quality is boasted of computers by those who promote and sell them. While the enormous growth of experiment as an end in itself blotted out the relation between theory and experiment, at least it left theory undamaged. Science by computer now increasingly hides the basis and structure of theory.

The face of nature, be it searched by experiment, be it limned by theory, is now averted from human experts as well as from human laiety. Instead we have cinematic flow charts which dramatize the mindless battles of rays, electrons, ions, neutrons, quarks, gluons, *etc.*, *ad infinitum.*

In the development of the aspects of mechanics I sketched above, it should be abundantly clear that EULER and the BERNOULLIs had the face of nature before them at all times. It was nature they pondered and scrutinized; to understand nature through mathematical representation was their aim. They took experiment seriously. DANIEL BERNOULLI was himself a great experimenter; JAMES and JOHN BERNOULLI and EULER kept abreast of the results of experiment in many fields, including some that they were never able to tame by mathematics. But, so far as I can learn, the development which led from HUYGENS' and NEWTON's brilliant fragments on mechanics to the systematic, embracing mechanics of EULER never called directly upon experimental data or gave rise to any experimental test. As

Mr. TOUPIN & I wrote long ago[1], "*Experience* has been the guide, *thought* has been the creator." Indeed, when in our time the calculations of trajectories and orbits for astronautics were based upon EULER's differential equations, not one voice was raised to suggest that those equations be first subjected to experimental check lest billions of dollars be wasted in applying what was, after all, just a mathematical theory. Now, of course, we may say that at last, 200 years after their discovery, EULER's equations have been subjected to myriad experimental tests and have emerged safe and sound every time.

Adherents to modernity and "progress" may scorn EULER's approach, may think it as irrelevant to life and science today as to their eyes seem *Macbeth* and *Paradise Lost*. For them I quote from one of the few living men—perhaps the very last—to have made a discovery in theoretical physics that is unanimously esteemed great and permanent: DIRAC. In accepting a prize in 1939 he wrote

> The physicist, in his study of natural phenomena, has two methods of making progress: (1) the method of experiment and observation, and (2) the method of mathematical reasoning. The former is just the collection of selected data; the latter enables one to infer results about experiments that have not been performed. There is no logical reason why the second method should be possible at all, but one has found in practice that it does work and meets with remarkable success. This must be ascribed to some *mathematical quality in Nature*, a quality which the casual observer of Nature would not suspect, but which nevertheless plays an important rôle in Nature's scheme.
>
> One might describe the mathematical quality in Nature by saying that the universe is so constituted that mathematics is a useful tool in its description. However, recent advances in physical science show that this statement of the case is too trivial. . . .
>
> The dominating idea in this application of mathematics to physics is that the equations representing the laws of motion *should be of a simple form*. The whole success of the scheme is due to the fact that equations of simple form do seem to work. The physicist is thus provided with a *principle of simplicity*, which he can use as an instrument of research. . . . The method is much restricted, however, since the principle of simplicity applies only to fundamental laws of motion, not to natural phenomena in general. . . .

---

[1] § 3 of the exordium of the *Classical Field Theories*, Essay 2 in this volume, above.

What makes the theory of relativity so acceptable to physicists in spite of its going against the principle of simplicity is its great *mathematical beauty*. This is a quality which cannot be defined, any more than beauty in art can be defined, but which people who study mathematics usually have no difficulty in appreciating. The theory of relativity introduced mathematical beauty to an unprecedented extent into the description of Nature. ...

We now see that we have to change the principle of simplicity into a *principle of mathematical beauty*. The research worker, in his efforts to express the fundamental laws of Nature in mathematical form, should strive mainly for mathematical beauty. He should still take simplicity into consideration in a subordinate way to beauty. (For example Einstein, in choosing a law of gravitation, took the simplest one compatible with his space-time continuum, and was successful.) It often happens that the requirements of simplicity and of beauty arc the same, but where they clash the latter must take precedence.

DIRAC went on to suggest that the physicist should develop the appropriate pure mathematics first, "at the same time looking for that way in which it appears to lend itself naturally to physical interpretation." EULER, on the contrary, considered physical aspects first and then developed the mathematics fit to pose and solve the problems they offered. Much of his work on pure analysis and geometry arose in this way, and some of his finest discoveries in pure mathematics appear first in his memoirs on hydrodynamics, acoustics, and elasticity.

A still stronger claim in favor of discovery through mathematics had been made by EINSTEIN in an address delivered in 1933:

[A]ny attempt to derive the fundamental concepts and fundamental laws of mechanics from elementary experience is destined to fail.

If, then, it is true that the axiomatic foundation of theoretical physics is not to be inferred from experience but must be freely invented, have we any right to hope that we shall find the correct way? Still more—does this correct way exist at all, save in our imagination? Have we any right to hope that experience will guide us securely ...? I answer with all conviction that ... there is one correct way and that we are capable of finding it. Our experience up to now justifies our faith that nature actualizes the simplest mathematically conceivable ideas. It is my conviction that through purely mathematical construction we can discover those concepts and the necessary connections between them that furnish the key to understanding the phenomena of nature. Experience can probably suggest the mathematical concepts, but they most certainly cannot be deduced from it. Experience, of course, remains the sole criterion of mathematical concepts' usefulness for physics. Nevertheless, the real creative principle lies in mathematics. Thus in a certain sense I regard it true that pure thought can grasp reality, as the ancients dreamed.

We have heard DIRAC and EINSTEIN speaking, not EULER. EINSTEIN himself warned us not to believe statements of that kind:

> If you wish to learn from the theoretical physicist anything about the methods which he uses, I would give you the following piece of advice: Don't listen to his words, examine his achievements. For to the discoverer in that field, the constructions of his imagination appear so necessary and so natural that he is apt to treat them not as the creations of his thoughts but as given realities.

Perhaps EINSTEIN and DIRAC were making the most of their chances to horrify their audiences, which necessarily consisted of ordinary physicists, incapable of taking any advice of the kind being offered. Much of EULER's work in mechanics seems to follow a program such as they wished their hearers to believe that they recommended, particularly in his creation of the field theories of hydrodynamics and one-dimensional elasticity. In those sciences the preceding knowledge was mainly theoretical, some of it very old, and such experimental data as had been collected were scant, unreliable, and often not to the point. EULER employed a different approach to domains where even the simplest theoretical concepts were still to be invented and nearly all knowledge was only and directly experimental. Such fields were in his day the physics of heat, light, electricity, and magnetism. They were then chaotic; reasonable theory even in the simplest situations wanted; new phenomena were discovered rapidly by experiments. Those, we now know from Mr. DAVID SPEISER's researches, EULER eagerly followed. Using mainly his field theory of hydrodynamics as a basis, he sought to represent the appearances by mathematical statements. He it was who first proposed and developed a theory of light as undulation of an æther. His æther was a subtle fluid; it was to be replaced in the nineteenth century by a subtle elastic medium.

In one instance EULER himself designed in mathematical detail an apparatus for experiments, which his son carried out: a lens-shaped glass vessel for measuring the refractive index of liquid confined within it. Instruments of this kind remained in use until less than a century ago.

In considering relations between theory and experiment in EULER's day we must remember that while mathematics was highly developed, experiment was not. The rudiments of chemistry, which enable us to identify pure substances, were yet to be established. Air was the only familiar gas; at common temperatures, water and spirits of wine were the only transparent liquids. Among the fluids tested by EULER's son were French wine and the juices of nutshells and pear leaves. Controllable, identifiable, reproducible variation was achieved only in the nineteenth century. Indeed, and contrary to the folklore of science, the nineteenth century and the early twentieth make the only epoch in which precise, accessible, repeatable experiment on the scale of a human being was available to serve in close comparison with detailed, precise theory. The idea of such a collaboration was invented in the nineteenth century; its practical possibility did not survive the Second World War except in a few pockets. When, as happens to anyone who nowadays receives hospitality in the crenelled turrets and serrate dungeons of professional research, you encounter somebody who boasts achievement in experiment and theory hand-in-hand, you may be pretty sure his egotism exceeds his powers in theory and experiment, either or more likely both.

In 1963 DIRAC, adducing as an example SCHRÖDINGER's having withheld his discovery of the "KLEIN-GORDON equation", warned the theorist expressly against staying too close to experiment:

> If there is not complete agreement between the results of one's work and experiment, one should not allow oneself to be too discouraged, because the discrepancy may well be due to minor features that are not properly taken into account and that will get cleared up with further developments of the theory. . . .

EULER too, sometimes to his loss, took experimental data more seriously than they turned out to deserve.

Two qualities are common to EULER and DIRAC:

(1) Daring imagination in proposing mathematical theories going beyond all known basis in experiment.

(2) Conceptual powers sufficient to achieve beauty and simplicity in their representations of nature.

I do not suggest that other theorists have been or are or will be as successful as EULER was in creating good mathematical theory of physics which was to lack abundant basis in experiment for two centuries. I do suggest that it may be not only naive but stupid to proclaim and in effect to enforce a schema of scientific research that excludes an approach like EULER's.

5. What Methods of Research Did the Bernoullis and Euler Follow?

With one exception, the Bernoullis and Euler have left us no account of their methods. The exception is Daniel Bernoulli. He retained the approach of the great geometers of the preceding century: For each class of mechanical systems he studied, he hypothesized a new governing principle which seemed to include and extend what experience had shown him. His principle of superposition for small motions is an example. He experimented regularly and with brilliance, but he prided himself on constructing the theory first and then using it to suggest the experiments. *Cf.* Essay 1, above. Even so, and great as he was, he does not stand in the first line of discoverers of the mighty mathematical system that is classical mechanics. He rested content, as of necessity had Huygens a century earlier, with loosely linked compartments. As mechanics under the hands of Euler grew into a structure, Daniel Bernoulli turned more and more to other fields, fields not yet sufficiently explored to be ready for system, fields in which clear experiment on an isolated phenomenon was still the best thing that could be got. Such fields then were electricity, magnetism, and physiology, which most of his later work concerns.

To discern the methods used by the two elder Bernoullis and Euler, we must trace the progress of their work, paper by paper, line by line, equation by equation. Different students in so doing will reach different conclusions; I can do no more than state mine.

As I read the works on the foundations of mechanics by Euler, his teacher John Bernoulli, and John Bernoulli's teacher James Bernoulli, I discern a simple pattern.

(1) Always attack a special problem. If possible solve the special problem in a way that leads to a general method. Mr. van der Waerden quotes Hilbert, famous as a creator, organizer, and simplifier of general theories, to the same effect: "You must always start from the very simplest examples!"

(2) With a firm eye on experience, read and digest every earlier attempt at a theory of the phenomenon in question. Perpend with utmost scruple the partial successes and failed attempts of the great masters of the past—for the Bernoullis, these were Huygens, Leibniz, and Newton; for Euler, the same and also the Bernoullis. Partial successes and failed attempts by giants' hands reveal the most plausible concepts and methods that either will not work or may lend themselves to amelioration.

(3) Let a key problem solved be father to a key problem posed. As Paul Valéry was to write in a later century, "The progress of science

can be marked by the number of problems. Each new capacity opens a new list of questions." The new problem finds its place on the structure provided by solution of the old; its solution in its turn will provide further structure.

(4) If two special problems already solved seem cognate, try to unite them in a general scheme. To do so, set aside the differences, and try to build a structure upon the common features, a structure broad enough to admit as alternatives the particularities of each.

(5) Never rest content with an imperfect or incomplete argument. If you cannot complete and perfect it yourself, lay bare its flaws for others to see.

(6) Never abandon a problem you have solved. There are always better ways. Keep searching for them, for they lead to fuller understanding. While broadening, deepen and simplify. A solution using only elementary mathematics is better than one that calls upon advanced apparatus or complicated calculation.

A program less strict but otherwise not much different was written out by DIRAC in 1962:

> I shall attempt to give you some idea of how a theoretical physicist works—how he sets about trying to get a better understanding of the laws of nature.
>
> One can look back over the work that has been done in the past. In doing so one has the underlying hope at the back of one's mind that one may get some hints or learn some lessons that will be of value in dealing with present-day problems. The problems that we had to deal with in the past had fundamentally much in common with the present-day ones, and reviewing the successful methods of the past may give us some help for the present.
>
> One can distinguish between two main procedures for a theoretical physicist. One of them is to work from the experimental basis. For this, one must keep in close touch with the experimental physicists. One reads about all the results they obtain and tries to fit them into a comprehensive and satisfying scheme.
>
> The other procedure is to work from the mathematical basis. One examines and criticizes the existing theory. One tries to pinpoint the faults in it and then tries to remove them. The difficulty here is to remove the faults without destroying the very great successes of the existing theory.
>
> There are these two general procedures, but of course the distinction between them is not hard-and-fast. There are all grades of procedure between the extremes.

Which procedure one follows depends largely on the subject of study. For a subject about which very little is known, where one is breaking quite new ground, one is pretty well forced to follow the procedure based on experiment. In the beginning, for a new subject, one merely collects experimental evidence and classifies it. . . .

With increasing knowledge of a subject, when one has a great deal of support to work from, one can go over more and more towards the mathematical procedure. One then has as one's underlying motivation the striving for mathematical beauty. Theoretical physicists accept the need for mathematical beauty as an act of faith. There is no compelling reason for it, but it has proved a very profitable objective in the past. For example, the main reason why the theory of relativity is so universally accepted is its mathematical beauty.

With the mathematical procedure there are two main methods that one may follow, (i) to remove inconsistencies and (ii) to unite theories that were previously disjoint.

6. What Can Mathematics Bring to Natural Science?

Like Dirac's in our century, the methods of the Bernoullis and Euler were for the most part mathematical, inductive, and synthetic: Conceptual analysis led slowly, very slowly to synthesis of separate solutions into a simple, embracing theory, each step gained through success on some new special problem, the new success having been achieved by pondering old successes and old failures. Mathematics and its realization in nature joined in perfect marriage, each sustaining the other. The level of mathematical rigor was the same in both, rigor in the sense recently defined by André Weil:

Rigor is to the mathematician what morality is to man. It does not consist in proving everything, but in maintaining a sharp distinction between what is assumed and what is proved, and in endeavoring to assume as little as possible at every stage.

Rigor is only one attribute of conceptual analysis. Even more essential is a tireless search for the unifying concept, through which we arrive finally at the simple, the clear, and the beautiful, for only they can reflect the natural and the divine. Newton wrote,

For nature is simple and affects not the pomp of superfluous causes.

EULER in defending his hydrodynamical equations wrote,

> The generality I here take on, far from dazzling our enlighten-
> ment, reveals to us the true laws of Nature in all their brilliance,
> and there we shall find even stronger reasons to admire her
> beauty and her simplicity.

I can paraphrase WEIL's statement about rigor:

> Mathematical discipline is to science what civilization is to man. It
> does not consist in replacing all experiment by reasoning, but in
> making a sharp distinction between what is measured and what is
> derived by reason, and in endeavoring to reduce as much as poss-
> ible the need for measurement at every stage.

More than that:

> Mathematics rightly used brings simplicity, coherence, order, and
> beauty to parts of natural science which formerly seemed com-
> plex, disjoint, unrelated, ugly. Mathematical criticism and syn-
> thesis can convert dogmas and rules of craft—priestly chants and
> trade secrets—into simple understanding which can be taught
> briefly to children.

Finally, EULER and the BERNOULLIs possessed a keen sense of
problem—the problem whose solution is worthwhile in itself and still
more worthwhile for the light it casts upon the science as a whole, the
arrow it shoots into the future. Another name for this sense is taste—
hard to define but easy to recognize. I think that mathematical taste
can be taught, but even such students as have the skill and knowledge
necessary to learn it are sometimes unequal to the discipline or un-
endowed with the imagination sufficient to do so. As POUSSIN put it in
his *Observations on Painting,*

> Style is a particular manner and skill . . . which comes from the
> genius of each individual in his way of applying and using ideals;
> this style, manner, or taste comes from nature and intelligence.

7. HOW ARE MATHEMATICS AND COMPUTATION COMMONLY USED TODAY IN NATURAL SCIENCE?

The approach of EULER and the BERNOULLIs as I have outlined it
is not common today. Mathematics in most of its applications is not
now a messenger of divine order. Today at best a machine tool, all too

often it is a jargon, a lingo used to silence the poor folk who cannot palaver it; at worst it is one of the tricks in a charlatan's bag. Perhaps applied mathematics is not worse off than the rest of science, for which I can do no better than quote CHARGAFF, who refers in particular to molecular biology but whose words would certainly apply equally to physics and chemistry:

> In fact it is the decimal places which make us think we can penetrate deeper into nature, and each further decimal place costs ten times as much as the last. Ever smaller quantities, ever faster reactions, ever tinier structures require ever bigger and dearer apparatus, and the more complicated these become, the more foolish do they make the persons who run them. Finally those persons only skitter around with screwdrivers and try to keep their research gadgets in good humor. Consequently we are arrived at the absurdity where the only depth in natural science of this kind is the engineering of the apparatus. I do not think that such had to be.

Thus this sort of depth in natural science costs a lot of money. A few years back the Western world thought that it could afford all this, and vast sums have flowed into the pockets of construction companies and manufacturers of apparatus for physics, radioactive compounds, etc. To use and even more to watch over all these machines, numerous scientists were needed, and these too were somehow produced and delivered to assure the electronic Molochs a well fed future. Unfortunately, or God be thanked, the general danger of bankruptcy has led to a certain standstill, but when catastrophes can be delayed only by hæmorrhages, I do not think the prognosis favorable. In any case the decimal fever will slow down at least for a while, and we shall be able to sit on the bank of the river of Babel and weep, certainly with a good view of the tower of that name.

Perhaps we could have spent a bit of this time on the question whether this kind of depth—this drunkenness with decimals, this mania for eternal refinement of the tools, to outwit ever more successfully a pseudo-nature discovered for this very purpose— was the only conceivable depth. What took place in science in my time was more extensive than intensive. Without any account of what the sense and goal of pure natural science is, one field after another has been stripped bare until not the tiniest metaphysical leaf remained. The final goal had to be a nature that while fully explicable unfortunately no longer existed. Whether what we have learned from this was really worth the trouble, a later age must decide.

POST provides a similar summary in reference to physics as well as biology:

> The last 40 years in the history of science are remarkable.... Nothing happened. It would be hard to find a comparable period since Galileo during which no change in the foundations of science took place. ...
>
> Accompanying this stagnation in fundamental work there has been a change of style. The kind of look-over-the-shoulder rat race, described in the "Double Helix", is not typical of fundamental research. ... Perhaps the reason why so many scientists cling to external ideologies is that they are themselves victims of alienation in giant research teams. The monthly arrival of a packet of 1000 photographs produced in 10 minutes by a giant machine elsewhere may not create quite that feeling of involvement that individual research gives.

CHARGAFF again:

> Now I will say something naive, something humiliatingly simple: Every human activity worth the name should make him who practises it better, opener, richer in ideas, more luminous. In my circles I almost never saw anything of the sort. Great assiduity, quick trimming to the wind, crabbed amibition, spiteful professional jealousy: These are the qualities of the successful researchers whom I have known well. They were mostly dull, unhappy folk.

8. WHAT IS PLEBISCIENCE?

I do not contest these summaries; I will neither reinforce them nor bewail them. They describe science by, for, and of the demos, in a word, *plebiscience*. Plebiscience demands what POST has called "inevitable research: ... research which is bound to yield **some** results." POST gives as an example "an accelerator with three outlets. Different targets were inserted in each outlet, and one research student hung at each outlet for one year, measuring the scattering intensity in various directions. At the end of the year, he was removed, [he] published his data and [he] was given a Ph.D." We may well post a postscript: "Five years later, the whole matter was programmed on a computer from the first measurement to the composition of the paper for the journal called *Gigahot Flashes of Megamicrophysics*. The research students were thereby rendered unnecessary. Unfortunately the accelerator was by then obsolete. Accordingly, the torch of discovery passed to a multi-national consortium that had a newer and bigger accelerator. A still bigger one is now under construction by an army of robots. Every overdeveloped nation in the world is partner in this supreme effort. Nature's ultimate mysteries, at last, seem to be just around the corner. Success in this new probe, history's biggest, will make the discoveries of LEONARDO DA VINCI, GALILEO, COPERNICUS, NEWTON, and EINSTEIN look like child's play."

Plebiscience first appeared in our own century. ARTHUR GORDON WEBSTER saw it born and warned against it. Speaking in 1914 of "the happy years" he had known when physicists "learned how to do without," he stated

> Sweet are the uses of adversity. In the years since then, many of our American universities have built great laboratories for physics, several of which, costing more than a quarter of a million dollars apiece have seemed to speak the last word of luxury and convenience for the experimenter, but I believe they do not teach the most important lesson, that it is men rather than apparatus or buildings that make progress, and that some of the greatest discoveries have been made with the simplest of apparatus, but by men of genius.

More than half a century later CHARGAFF in an essay called "In praise of smallness" pursued a similar vein:

> ... Talleyrand is quoted as having said that nobody who had not lived before ... the French Revolution ... knew how sweet life could be. I could say something similar, namely, that nobody who had not been at least a graduate student before ... 1942 could know how happy small science can be.

Plebiscience is *big science*. Small science was done by a few great men. Big science calls for many little men. As CHARGAFF sees,

> We are sailing straight into a managerial dictatorship in which the individual scientist can no longer have a voice. ... [S]cience has become thoroughly politicized, a playball of power networks. ...
>
> [T]he trend is all toward the creation of very large scientific conglomerates in which, under the leadership of men with managerial qualifications, the predictable will be discovered in ton lots.
>
> The frightening waste of resources will become evident to anybody who considers how little of value the orgy of goal directed-

ness has actually produced. One could ... argue that our scheme of research support has much more harmed than helped the scientific growth of the individual. ... [I]t is well known that wherever money is abundant charlatans are brought forth by spontaneous generation.

While NEWTON wrote of his having "had entry" into the method of fluxions, as if he had taken a beautiful virgin and made her a beautiful woman, today's heroes of science make "break-throughs" as if to penetrate the lines of a mob of rival gangsters.

Plebiscience, like everything dear to the plebs, is dear for the taxpayer, and in a social democracy value and cost are the same. Nobody grudges a few billions in tax money for some really sharp photographs of Saturn and the news that only a further achievement of big science, of course costing tens of billions, can tell us at last, after centuries of uninformed, indeed savage or theological speculation, whether there is or is not life on Mars. To learn whether there be other suns, each focused in his own hegemony of ellipsing asteroids and planets with their rings and moons, its kinematical monotony relieved now and then by a vagrant comet, will cost us trillions plus a triple "cost overrun". Conversely, as CHARGAFF mentions in his essay called "Triviality in science",

In the eyes of efficiency experts small science must appear as trivial science.

Small science costs too little to be worth anything. Big science, plebiscience is invincible.

9. WHAT IS PROLESCIENCE, AND WHERE WILL IT LEAD US?

Plebiscience is an intermediate stage. The next and last is *prolescience*. In it not only is all research inevitable research, but also only outcomes previously known and accepted are allowed. The function of prolescience will be to confirm and comfort the proletariat in all that it will by then have been ordered to believe. Of course that will be mainly social science.

I know of only one field today which is dominated by prolescience. That is human ethnology. It has a Central Dogma: All races of man are equally endowed with all qualities which might be considered desirable, for example intelligence and aptitude for mathematics. As has been shown by events, any evidence contrary to the Central Dogma is shouted down: Doubt of it may not even be spoken in a seminar at Yale University, let alone broadcast or printed. The only fit object of research in human ethnology is to explain the sources of error in contrary conclusions drawn before the Dogma was proclaimed. It is not strange that the first prolescience was Nazi ethnology. The Central Dogma of current human ethnology is simply a negation of the Nazis' Central Dogma.

That is no joke. CHARGAFF in his "Bitter fruits from the tree of knowledge" reminds us that "The Nazi extermination camps also started as an experiment in eugenics." Now that genetic fiddling rides high, Society by adjusting our brains to be all alike may soon be able to convert our Central Dogma from a belief which it is taboo to question into a self-establishing fact. A Society that clothes its cruelties less elaborately than ours does could effect a similar creation of fact at a more primitive level. To eliminate Discrimination against the blind, it could blind at birth all those born sighted. What might seem a handicap is easily converted into a sanitary protection. A nation of blind men can persuade itself that the sightedness of some, were it allowed to mature, would subject Society to the deadly hazard of Discrimination.

Prolescience will spread. A modest and gentle start may be discerned already in physics. There any theory that does not begin from small particles, Lagrangeans, and complex Hamiltonians is simply ignored by the shamans because, as every good physicist knows, you cannot do good physics in any other way. The invisibly small, the incredibly swift, and the unattainably distant are the true subjects of physics; anyone who deals with what a human being can safely touch is labelled engineer or mathematician, barred forthwith from the guild because his subject is "not fundamental".

10. Is Plebiscience Compulsory Now?

Prolescience is still rare; plebiscience still predominates. Nevertheless, while plebiscience is accepted and lauded and subsidized and Nobelitated, it is not compulsory—not yet. The laws still permit us to approach natural philosophy in the spirit of EULER and the BERNOULLIS. My experience is different from POST's and CHARGAFF's. Assiduity and I are brothers, but my life has been spent tergiversed to computing machines, apparatus, and teams. In the men I have known and do know closely I have never seen quick trimming to the wind, crabbed ambition, or spiteful professional jealousy. They do not attack plebiscience, its trade unions or factions, its leaders or its faithful; they do not figure in its journalism. On the rare occasions when jealousy or frightened ignorance mounts an attack on them, they usually respond only by keeping on with their work, unruffled.

They are not outcast or rebellious or impoverished. They occupy standard posts as hewers and drawers in the common woods and wastes of universities and laboratories. They publish their results in 100 or more journals, all of which print also mainly or partly the products of technicians, teams, computers, and the fancy plumbing and glassware and electronics which the generous taxpayer provides. No arm of plebiscience holds out to tempt them the popular rewards of power, influence, money, and limelight. While not reticent to publish, they have and take the time to let their work mature before they release it, meanwhile circulating it in manuscript to anyone who expresses interest in it. As ANDRÉ WEIL wrote, "For the mathematician there is not even a Nobel prize . . . to deflect him from work fully ripened to a result brilliant but short-lived."

Much as some young people nowadays choose to earn their bread by subsistence farming or by building musical instruments single-handed, a steady trickle of students in several countries joins the little "counterculture" which nourishes mathematical philosophy of nature in the old way. It seeks the four objectives set in CHARGAFF's "In praise of smallness" as means to escape "the dehumanization to which our way of life and therefore our way of doing science condemns us", namely,

> (1) to do away with the deep and fixed grooves in which science has been running lately;
>
> (2) to return to a climate in which real discoveries, that is, unpredictable observations, can again be made;
>
> (3) to liberate the scientist from the atmosphere—half civil servant, half certified public accountant—in which he is now vegetating; and
>
> (4) to bring science back to the laboratory of the individual scientist, which means to permit the real and only seeds of true science to germinate again.

I think it exhibits "some of the spirit of both dignity and high adventure" that scientific research offered in earlier times. I think its "young geniuses" not guilty of the charge levelled in CHARGAFF's "Voices in the labyrinth": "passionately ambitious instead of . . . passionately passionate". I have seen some passionate young geniuses grow up into wise yet still passionate old geniuses and in their turn find among their students passionate young geniuses. According to CHARGAFF,

> Formerly a few scientists lived for science; now many live from it. In this way science has gotten caught in the whole witches' mess of the general disintegration of the western world.

Even now, I think, there are men who live for science, men who strive to bake not witches' mess but giants' bread. What they deliver may be as tedious to the plebs as a lute, but let us not forget that while a cure for cancer may raise the mean age of the population to ninety, a good lute may last to make beautiful music under the hands of men and women yet unborn.

11. WHAT USE ARE CLASSICS OF SCIENCE?

Let me return to interplanetary travel, which I consider only for analysis, expressing neither admiration nor contempt, neither love nor hate of it. We recall that its success, while certainly impossible without enormous use of computers, rested at least for the most part upon knowledge established previously, part of it long previously, in mechanics, mathematics, physics, chemistry, meteorology, and biology. It is truly a triumph of applied science. There was no element of scientific discovery in it. It drew upon classics of science such as the work of the BERNOULLIs and EULER.

There is a difference between classics of science and classics of literature. Rarely if ever does a scientist today read NEWTON and EULER as professors of literature read SHAKESPEARE and HEMINGWAY, seeking to translate into today's pidgin for their students the eternal verities archaically expressed by those ancient masters, or gathering material to use in papers for reverential journals read only by scholiasts of literature, who themselves read only so as to gather material to help them write more papers of the same kind. Nonetheless, some of NEWTON's and EULER's ideas and discoveries make the very ground on which a scientist stands. We cannot think of physics except in their terms, modified in detail and sharpened but still theirs.

12. WHAT KINDS OF MODELS SERVE NATURAL SCIENCE?

The approach of the BERNOULLIs and EULER, followed by later adherents like CAUCHY and MAXWELL, led to classical mechanics and electromagnetism as we know them, to the whole corpus of classical physics, and it is this corpus (*not* quantum mechanics!) in terms of which we face the inanimate circumstances of human life with a good measure of understanding, prediction, and even control. Could that approach do the same for some of the biologies? Today we look upon classical physics as providing us with mathematical models for the behavior of physical objects. We use these models with great caution, for we are deeply aware of their limitations, limitations which reflect our own, the limitations of the minds and imaginations of human beings. The BERNOULLIs had no idea that they were dealing with models; like GALILEO, they thought that nature herself spoke in mathematics. EULER in his middle life began to perceive how much the mathematician replaced nature by his own conceptions. EULER and the BERNOULLIs saw that nature had many aspects they could not neatly estimate or even discern and hence had to leave unrepresented in their theories. In outcome there is no great difference between our attitude toward mathematical theory of nature and theirs, so long as we stay with systematic modelling of particular instances.

POST distinguishes systematic models as follows:

> Traditionally there have been two kinds of models. There is the model in the strict logical sense, an articulation of a particular theory. I shall call this the Deductive Model. For example, models of the ether. There is also the model that has its origin in empirical evidence and some conjectured generalisation. I shall call this the Inductive Model. The history of chemistry provides examples of this. Think of Kekulé's benzene ring.

Such have been the traditional models used in responsible, scrupulous, patient science. Ideally, and sometimes also in practice, a deductive model is finally shown to square with an inductive one. Such reconciliation is sometimes said to "confirm a theory by experiment". An example is provided by EULER's model—model is what we call it today—called an "ideal fluid", which is now known to describe very well the behavior of air and water in many circumstances though to fail in others.

POST continues,

> In recent times, a third type of model has come into use: the "Floating Model". It is neither deductive (either because there is no overriding theory or because such theory is ignored) nor is it inductive, for scientists find interesting mismatches between their model and observation.

For description of a floating model, we may turn to the chapter called "Quantification as Camouflage" in S. ANDRESKI's lovely book, *Social Sciences as Sorcery*:

> Most of the applications of mathematics to the social sciences outside economics are in the nature of ritual invocations which have created their own brand of magician. The recipe for authorship in this line of business is as simple as it is rewarding: just get

hold of a textbook of mathematics, copy the less complicated parts, put in some references to the literature in one or two branches of the social studies without worrying unduly about whether the formulæ which you wrote down have any bearing on the real human actions, and give your product a good-sounding title, which suggests that you have found a key to an exact science of collective behaviour.

Also "The sophisticated mathematical models ... might mislead an unwary reader into believing that he is facing something equivalent to the theories of physics"; ANDRESKI warns us that even in economics, "the branch which has opportunities for measurement unrivalled in the other social sciences, an infatuation with numbers and formulæ can lead to empirical irrelevance and fraudulent postures of expertise."

I prefer to adduce as an example a disaster, already notorious, that mathematicians—of all people, those who ought to know better—have engendered. I refer to "Applied Catastrophe Theory". Some hitherto pure mathematicians, seemingly without experience in classical physics and without the discipline that that experience has always commanded of its votaries, have promulged a "general morphology", which, they claim, with no reference to any structured, organizing theory of physics or chemistry or physiology or economics or society, explains why dogs get angry, how the stock exchange operates, the growth of embryos of chickens, riots in prisons, what happens to spiral nebulæ, the stability of ships, how wars get started, *etc. etc.* The basic idea seems to be that equilibrium of *anything* corresponds "generically" to a point on some surface with a cusp. All you have to do is identify the variables, then read off the answer on the basis of known mathematical classifications of the forms surfaces may have at places near to a cusp. As stated, this claim refers only to changeless circumstances; nevertheless, CHRISTOPHER ZEEMAN, the leader in these applications, and his associates apply it by magic to virtually all "generic" changes in time. Things that are not "generic", they would have us believe, simply do not happen. "Generic" is defined differently from case to case in such a way as to make the answer in each seem plausible, at least to the mathematician who gets it[2].

---

[2] I do not exaggerate. According to SMALE,

> ... the new mathematics associated to Catastrophe Theory really is contained in what is called elementary catastrophe theory (ECT) by Thom and Zeeman. Very briefly, ECT studies a smooth real valued map f as a function (often called a "potential function") of a state x and parameter μ. Here the state lies in some

While sometimes Applied Catastrophe Theory is presented as if it provided inductive models, SUSSMAN & ZAHLER in their analysis of it find that "Catastrophe Theory is an attempt to approach science by trying to impose a preconceived set of [mathematical] structures upon the world . . .." These words could be a paraphrase of ANDRESKI's on the social sciences:

> [Y]ou cannot convert vague and dubious approximations (not to speak of nonsense and half-truths) into a mathematical science simply by transcribing them into the symbolism of mathematics. Not even the most powerful theorems of today's mathematics could be of any help in an attempt to make aristotelian physics into an exact science. The principle that 'nature abhors a vacuum' was a valuable idea at the time—even useful for practical purposes such as making pumps—but no amount of juggling with symbols could convert it into a proposition of mathematical physics. Galileo's mathematics was quite simple, and his achievement did not consist of applying it to the physics which he found, but of creating new concepts capable of yielding genuinely new kinds of information when manipulated with the aid of mathematics.

The mathematicians who have produced the "applications" of Catastrophe Theory are as unscrupulous as social scientists, of whom ANDRESKI writes thus:

> Owing to the continuing widespread ignorance of [mathematics], the utility of mathematical formulæ for the purpose of blinding people with science, eliciting their respect and foisting upon them unwarranted propositions, has hardly diminished. As neither the literary nor the illiterate as well as non-numerate door-to-door sociologists can understand formulæ, while the natural scientists

---

Euclidean space and μ also varies in a (usually low dimensional) Euclidean space. The problem is to find local canonical forms for "generic" f, using a smooth change of coordinates in the variables $x, \mu, y = f(x, \mu)$. In this case, "generic" means for an open dense subset in a suitable function space. A local canonical form is defined in a neighborhood of a pair (x_0, μ_0) at which f is singular. ECT solves this problem when μ is in a low (≤ 5) dimensional space.

Also

... I feel that Catastrophe Theory itself has limited substance, great pretension and that catastrophe theorists have created a false picture in the mathematical community and the public as to [its] power ... to solve problems in the social and natural sciences.

cannot grasp the issues to which these are supposed to apply, and imagine that writings which look like their own must be more scientific than the ones which do not, juggling with mathematical formulæ and words like 'input', 'output', 'entropy' and other importations from the natural sciences, no matter how misplaced, brings kudos for a social scientist. At its best it enables the practitioner to kill two birds with one stone: to avoid having to express his opinion on awkward or dangerous issues, and to score points in the game of academic status-seeking.

Also

> The usage of mumbo-jumbo makes it very difficult for a beginner to find his way; because if he reads or hears famous professors from the most prestigious universities in the world without being able to understand them, then how can he know whether this is due to his lack of intelligence or preparation, or to their vacuity? The readiness to assume that everything that one does not understand must be nonsense cannot fail to condemn one to eternal ignorance; and consequently, the last thing I would wish to do is to give encouragement to lazy dim-wits who gravitate towards the humanistic and social studies as a soft option, and who are always on the lookout for an excuse for not working. So it is tragic that the professorial jargon-mongers have provided such loafers with good grounds for indulging in their proclivities. But how can a serious beginner find his way through the verbal smog and be able to assess the trustworthiness of high ranking academics?

As CHARGAFF remarks,

> [T]he public...takes to champions of every kind, because it erroneously assumes that under every laurel wreath is some head.

In reviewing the writings of ZEEMAN and his collaborators in Applied Catastrophe Theory, SMALE states that they claim almost magical powers; he quotes ZEEMAN as writing that they translate sociological hypotheses into mathematics by use of "deep theorems", then "synthesize the mathematics", and finally "translate the synthesis back into sociological conclusions." Here are ZEEMAN's own words:

> Summarising: we insert seven disconnected elementary local hypotheses into the mathematics, and the mathematics then synthesises them for us and hands us back a global dynamic understanding.

And all this from purely static statements! To make it worse, the mathematics used, as SUSSMAN & ZAHLER remark,

> is very much unlike the kind of mathematics most non-mathematicians are acquainted with. It combines concepts that are completely inaccessible to anyone who is not a professional mathematician—e.g. Whitney topology, genericity, structural stability, unfoldings—with the use of some pictures of amazing simplicity—e.g. the cusp surface. The mystificatory power of the combination is explosive: when confronted with a cusp surface—which anyone can understand—and the claim that, for "deep mathematical reasons" the surface explains why dogs attack, it is hard to resist. Those who demand an explanation will be told something like: "well, consider the natural stratification of the space of smooth families of functions . . .", at which point even the most sophisticated physicist or engineer will prefer to give up.

Writes SMALE,

> Along with this mathematical egocentricity there is a kind of mystification of the subject that is being created by both Thom and Zeeman. Zeeman does this when he speaks of the "deep classification theorem of Thom". . ... Presenting this picture to nonmathematicians and even nontopologists has an intimidating effect. Thom does this by using technical mathematical terms without explanation when addressing nonmathematical audiences, and often writing obscurely. Then Zeeman deepens the mystifying power of Catastrophe Theory by explaining Thom's obscurities with: "When I get stuck at some point in his writing, and happen to ask him, his replies generally reveal a vast new unsuspected goldmine of ideas."

SUSSMAN & ZAHLER describe how Applied Catastrophe Theory attained its repute:

> Write a paper stating an unsupported theory, and this will not cause the theory to be believed. Write a second paper in which you refer to the theory of the first as "well established", and the acceptance level will increase. Let a colleague of yours write a paper referring to your deep work and the level will rise still more. Multiply all this by two hundred, and you obtain something like Catastrophe Theory. By the time the whole thing reaches the average reader, it will be through articles in which the theory is taken for granted. The reader who wishes to pursue the matter

further will be referred to more articles in which the same is done. Few will follow the thread all the way. Those few who do will require such an intellectual effort—since they will have to learn all about maps and diffeomorphisms and singularities and stability— that, when they reach the end and realize that the thread is not tied to any external evidence, but only to itself, it will be hard for them to accept the truth, and to acknowledge that their effort has been in vain.

The next step is to write a potent blurb about yourself on the cover of your book or let your publisher get an admirer to do so. If the claims are extravagant enough to seem sensational, the press will take off from there. The words of CHARGAFF regarding tinkerers with genetics could well be applied here:

> The noise about them was as big as if they had already bred a man who could drink water and piss petroleum.

Not so did EULER and the BERNOULLIS win the great fame that they enjoyed in their own lifetimes. They claimed no more than what their mathematical operations or their experiments unexceptionably delivered, granted the general standards of their times. EULER presented all details of his calculations, which he made as simple as he could, to help anybody with some mathematical education follow them and detect such errors and gaps as, inadvertently or confessedly, there were. In the thousands of pages by the BERNOULLIS and EULER that I have read, there are some errors and some gaps but not a single instance of mystification, not a single finished analysis that requires of its reader anything beyond the will to fix serious attention, the habit of mathematical thinking, and the apparatus then in common use for mathematical research. Praise came not from collaborators and cliques but from outsiders, who were in many instances competitors.

Today no logic, no appeal to experiment, can dompt a floating model promoted by a good press as being an application of that abstruse, arcane thing called mathematics. With Catastrophe Theory the media have displayed their usual ability to sell print by feigning news. The London *Times* in its review of THOM's book stated, "In one sense the only book with which it can be compared is NEWTON's *Principia*." No statement could reveal in brighter glare the boundless gullibility that the press has fostered in the inheritors of BACON's empiricism, which was to place all wits on a level. NEWTON's, the greatest work of all those that sought to discover deductive order and formal structure in a particular and clearly defined aspect of experience with nature, is set level with a loose, placeless model that will float on anything whatever by mere choice of the natural objects with which the mathematical variables are to correlate—an imperial theory, which explains everything "generic". Indeed, our wits today lie at the level of a fellah's before his mullah.

While SUSSMAN & ZAHLER and SMALE have demolished Applied Catastrophe Theory on the basis of at least five criteria for sound science, those do not concern my subject directly. I refer to Applied Catastrophe Theory only as an example of the kind of pseudoscience that can result and has resulted from use of floating models. In a deductive model, because it provides an instance allowed by an over-riding theory concerning a specific class of variables, a theory which it *cannot* contradict, we know at once what we have left out; at the cost of effort we may be able, if need appear, to put some of that back in. A floating model defies discipline. It is like a runner in a race with no rules. If we are displeased with what such a racer does, we can only call for another racer.

13. ARE UNRIGOROUS MATHEMATICS AND NUMERICAL CALCULATION APPLIED TO FLOATING MODELS SAFE?

EULER and the BERNOULLIs used only deductive and empirical models: facts and concepts first, then elaboration by mathematics, and if existing mathematics did not do, they were eager to create new mathematics for the job and able to do so. Floating models, never! Floating models, so-called "applications" of mathematics, are the opposite of mathematical philosophy. Let the mathematical theorist base his model upon what is observed, irrespective of what mathematical disciplines come to hand in his library! Recalling the disaster of Applied Catastrophy Theory, let him heed the advice of SMALE: "Good mathematical models don't start with the mathematics but with a deep study of certain natural phenomena." Let him heed also ANDRESKI's warning against theories "based on threatening people with mathematics: muttering darkly about algebraic matrices and transformations without revealing their exact nature" and mumbo-jumbo with mathematical symbols and terms. Let him learn to apply the patient logical analysis and relentless criticism which alone can convert hypothesis, conjecture, and tentative calculation into a mathematical science! Let him abhor unrigorous mathematics because it greases the path for wrong answers to slip out of right assumptions, for something noxious to man to be by hocus-pocus with the lingo of formulæ and a bore of computed digits whitewashed into something apparently useful. Let the mathematical philosopher mature his work well before releasing it in print; let him learn to express himself directly, to think simply and to use simple words for simple thoughts, to write so as to be read and understood. Finally and above all, let him remain humble before nature and the great thinkers of the past: Let him refrain from positive claims to have explained and predicted!

14. ARE BIOLOGY AND THE PHENOMENA OF SOCIETIES SUSCEPTIBLE OF MATHEMATICAL TREATMENT?

Even so, even granted biological philosophers of a cast scarce today, I dare predict nothing. Mathematical physics rests upon and fosters inanimate models for inanimate things. So far, mathematical models in biology seem to have been essentially physical and hence inanimate models. Mathematical physics abstracts the essence of many phenomena; these phenomena are exhibited by live things and dead things alike; in modelling them, physics does not have to kill what it studies. If we seek by similar methods to model the qualities that living beings lose as soon as they die, have we any reason to expect success?

ANDRESKI finds in social science examples of mathematical models "rooted in a rather simple reification". He writes of "human beings equated with bits of hardware" despite the fact that "one single living cell performs a variety of exceedingly complex homeostatic actions which no computer devised or even envisaged up till now can imitate." SUSSMAN & ZAHLER write, "The idea that anything we talk about can somehow be made into a numerical variable is one of the most pervasive diseases of contemporary science." In Applied Catastrophe Theory—an assembly of floating models or a single great imperial floater for everyman's complaint—they find "a very acute case of the disease of spurious quantification." ANDRESKI again:

> In truth there is no reason whatsoever to presume that amenability to measurement must correspond to importance; and the assumption in question has often led economists to aid and abet the depredations of a soul-destroying and world-polluting commercialism and bureaucratic expansionism, by silencing the defenders of æsthetic and humane values with the trumpets of one-sided statistics.

For an example, he remarks that corruption, which "is immeasurable intrinsically rather than merely owing to insufficient development of the techniques of quantification . . . , would have to disappear in order to become measurable."

15. DOES MATHEMATICS REIFY MAN?

Mathematical philosophy is the willed creation of human beings. Mathematics does not reify: It replaces things by ideas, and it teaches men to think. Far from dehumanizing men, it tempers their minds; only in the aura of order and beauty can it be created. In considering possibilities for extending the scope of mathematics in applications, we must beware of the common prejudice that to quantify requires us to assign numbers as the only measures of magnitudes. Such magnitudes are those that can be read on a scale. It is to them, I believe, that ANDRESKI refers when he writes

> Those who refuse to deal with important and interesting problems simply because the relevant factors cannot be measured, condemn the social sciences to sterility, because we cannot get very far with the study of measurable variables if these depend on, and are closely interwoven with, immensurable factors of whose nature and operation we know nothing.

For example, corruption cannot be disregarded as being beyond the realm of social theory, for only a social system that forbids rewards of any kind for any individual could root out corruption. There are social systems that to us at least seem to be of that kind: the beehive and the anthill, but they have never, so far as I know, been proposed as models of any human society, whether existing or projected, unless it be a great laboratory of physics or chemistry or biology or medicine.

In the main, it is the applied scientists who call for numbers, numbers, numbers. Although the layman conceives mathematics as being the science of numbers, few mathematicians agree, and most of modern mathematics gives to numbers a role at best ancillary or illustrative in the development of concepts. *Mathematics is the science of precise relations.* Mathematicians habitually deal with properties that no measurement could verify or controvert, properties that can only be imagined. I refer to functions defined on infinite sets, continuity, passage to limits, relations of inclusion among infinite sets and in regard to membership in such sets, *etc., etc.* If the hard-headed empiricist replies, "I don't care about any such things, I just get numbers", in most instances the mathematician can easily see that he is deluding himself, for nearly all experimental results in physical science are explained and correlated by use of the operations of differential and integral calculus, in which the infinite is always present, and no finite set of rational numbers suffices for taking even the beginner's first step in that theory. Mathematics might have greater potential use if those who tried to apply it to the biologies and to social phenomena would learn that measurement provides only one of many possible kinds of quantification. The worst advocate for mathematics is the enthusiast who thinks he understands it but does not. For an informed, sober, and concise survey of the pitfalls in mathematical modelling, especially for the biologies, I refer the reader to MAYNARD THOMPSON's "Mathematization in the Sciences".

Beyond that, I do not deny that there may be many relations in nature that cannot be quantified in any sense and hence are inherently unmathematical. If mathematical treatment fails in the biologies, well then, it fails.

Failed mathematics, alas, is fine food for computerized prolescience. Nothing is easier to apply to socio-political phantasmagoria than failed mathematics substantiated by experiments programmed to confirm it.

The bad money drives out the good. It does so because politicians just print it. Our scrupleless demagogues have claimed to repeal GRESHAM's Law by the very means they used to change the dates of WASHINGTON's birthday and Thanksgiving, the Newspeak that lets them call aggression and genocide "defense"; soon they will unite "Independence Day" with the anniversary of some appalling defeat, call the combined holiday "Victory Day", and schedule it for the first Tuesday after Labor Day. FRANKLIN ROOSEVELT decreed gold money bad and paper money good. But gold will out. In the last few years we have learned again that gold is good; not only good, too good to use for money. Paper, the visible excrement of devouring rulers, was to replace gold as the standard of value. It has not done so; it never will. The Baslers always knew that. It was Basel that gave us the BERNOULLIS and EULER.

Note

The references to Basel and to biology derive from the circumstances giving rise to the foregoing essay: a lecture to the Biozentrum, Basel, on 5 December 1979. An essentially faithful text of that lecture has been published in the *Verhandlungen der Naturforschenden Gesellschaft in Basel* **91** (1981): 5–22. That publication includes a passage the interest of which is unlikely to extend beyond Swiss readers and the collectors of bank notes:

> If you look at one of the new 10-franc notes issued this month, you will see EULER's picture on one side, and on the other a diagram of the planets and their moons against a background of what might seem to be a space capsule. Thence you might suspect I had had a hand in designing the bank note, but I did not, and the strange object is EULER's hydraulic turbine. The space capsule could seem more appropriate, for EULER's influence today is greater by far upon interplanetary flight than upon the design of turbines.

The text printed above is augmented by a fuller consideration of "floating models" and shortened by omission of nearly all remarks on numerical calculation and computing machines. In Essay 41, below, the reader will find a fuller treatment of the effects of computers.

I thank Mr. DAFERMOS for providing the comment in elegant Greek put just after the dedication. BOCHNER would have admired it.

Works Referred To

S. ANDRESKI, *Social Sciences as Sorcery*, New York, St. Martin's Press, 1972. The passages quoted may be found on pages 85, 124, 127, 129–131, 141–143.

E. CHARGAFF, "Preface to a grammar of biology", *Science* **172** (1971): 637–642.

E. CHARGAFF, "Bitter fruits from the tree of knowledge: Remarks on the current revulsion from science", *Perspectives in Biology and Medicine* **16** (1973): 466–502. The passages quoted are from §§ III and IX.

E. CHARGAFF, "Variationen über Themen der Naturforschung nach Worten von Pascal und anderen", *Scheidewege* **5** (1975): 365–398. Reprinted in his book: *Unbegreifliches Geheimnis*, Stuttgart, Klett-Cotta, 1980. The passages quoted above in English translation approved by Professor CHARGAFF may be found on pages 366–367, 370, 381, 387 of the original edition, pages 170–172, 175, 189, 196–197 of the reprint.

E. CHARGAFF, "Voices in the labyrinth: Dialogues around the study of nature", *Perspectives in Biology and Medicine* **18** (1975): 251–330. The passage quoted in on page 253.

E. CHARGAFF, "Triviality in science: A brief meditation on fashion", *ibid.* **19** (1976): 324–333. The passage quoted is in § VII.

F. CHARGAFF, "In praise of smallness: How can we return to small science?", *ibid.* **23** (1980) : 370–385. The passages quoted are from §§ IV–VI.

P. A. M. DIRAC, "The relation between mathematics and physics", *Proceedings of the Royal Society of Edinburgh* **59** (1938/9): 122–129 (1939).

P. A. M. DIRAC, "Methods in theoretical physics", pages 21–28 of *From a Life in Physics*, IAEA Bulletin, 1962.

P. A. M. DIRAC, "The evolution of the physicist's picture of nature", *Scientific American* **208** (1963): 43–53.

A. EINSTEIN, "Prinzipien der Forschung", reprinted as the first essay in *Mein Weltbild*, Amsterdam, Querido Verlag, 1933. My quotation is from page 20 of the translation by ALAN HARRIS, *The World as I See It*, New York, Covici-Friede, 1934.

A. EINSTEIN, "Zur Methode der theoretischen Physik", pages 176–187 of *Mein Weltbild*, cited above. There are two English translations: *On the Method of Theoretical Physics*, New York, Oxford University Press, 1933, and pages 30–40 of *The World as I See It*, cited above. Both have helped me make the translations printed in the foregoing text.

H. R. POST, *Against Ideologies* (Inaugural Lecture), London, Chelsea College, 1974. The passages quoted are on pages 7, 9, 12.

N. POUSSIN, quoted on page 419 of J. R. MARTIN's *Baroque*, New York & Hagerstoun, Harper & Row, 1977.

S. SMALE, review of E. C. ZEEMAN's "Catastrophe Theory: Selected Papers", *Bulletin of the American Mathematical Society* **84** (1978): 1360–1368.

H. J. SUSSMANN & R. S. ZAHLER, "Catastrophe theory as applied to the social and biological sciences: a critique", *Synthèse* **13** (1978): 117–216. See §§ 13 and 14.

M. THOMPSON, "Mathematization in the sciences", pages 243–250 of *Mathematics Tomorrow*, edited by L. A. STEEN, New York *etc.*, Springer-Verlag, 1981.

A. G. WEBSTER, address published on pages 60–62 of Volume **3** of *Twenty-fifth Anniversary of Clark University 1889–1914*, Worcester, Massachusetts, Clark University Press, 1914.

A. WEIL, "L'avenir des mathématiques", pages 307–320 of *Les Grands Courants de la Pensée Mathématique*, edited by F. LE LIONNAIS, Paris, Cahiers du Sud, 1947; 2nd edition, Paris, Blanchard, 1962 = pages 359–372 of Volume I of WEIL's *Œuvres Scientifiques*, New York *etc.*, Springer-Verlag, corrected 2nd printing, 1980. See page 360 of the last publication.

A. WEIL, "Mathematical teaching in universities", *American Mathematical Monthly* **61** (1954): 34–36 = pages 118–120 of Volume II of WEIL's *Œuvres*. This article is "the outline of a lecture once given by the author at a joint meeting of the Nancago Mathematical Society and the Poldavian Mathematical Association."

11. CONCEPTUAL ANALYSIS

Address upon receipt of a Birkhoff Prize (1978)

The honor you give me today is double: The first time you invite me to address you indebts me to you also for a splendid award. The occasion is twice doubly dear because it lets me say a little about two men connected with the prize: GEORGE D. BIRKHOFF himself, and a previous recipient, JAMES SERRIN. Both of these men have made an impress upon American and foreign styles in mathematics which I know to be healthful and hope will be lasting.

I met BIRKHOFF just once. It was at the summer meeting of the Society on 12 September 1943, at Rutgers. In those days there were about 2,000 members, of whom some 200 were active mathematicians; at a meeting in the East you could usually encounter about half of those. My teacher, LEFSCHETZ, who was often kind to his students, introduced me to BIRKHOFF and to many other senior men. After dinner there were two speeches. The report[1] published in the *Bulletin* is brief: "Dean R. G. D. Richardson of Brown University spoke of the importance of applied mathematics in the war effort. Professor G. D. Birkhoff of Harvard University urged mathematicians and scientists to maintain a proper balance of values during the emergency." Having survived a summer school of applied mathematics at Brown, I knew what to expect of RICHARDSON; the *Bulletin*'s sentence reproduces his lecture *in toto*. Not so BIRKHOFF's. With commanding dignity and in superb, native English, BIRKHOFF lashed the universities for their subservience to government and warfare. He warned us that by giving junior men heavy teaching loads, as much as eighteen hours, with no assistance, and by admitting unselected and mainly unqualified undergraduates, the universities were destroying what it was their prime duty, above nations and above emergencies, to foster. The many incompetents pressed into instruction were unable to teach, for they did not know, while the competent few were unable to learn because they were left no time.

I wonder what BIRKHOFF would have said, had he lived to see them, about our plebeian universities today, universities which in a

---

[1] *Bulletin of the American Mathematical Society* **43** (1943): 825.

time of peace and ease have forgotten that the task of higher learning is not only to sow, dung, and harvest but also and above all to winnow. I wonder what he would have said about the pollution by social sciences, changing values, team work, computing, sponsored research, involvement in the community, and even soft mathematics, to the point that mathematics of his kind is today decried as being "esoteric". I wonder what he would have said about the iron rule of mediocrity the Government today imposes as the price of the manna called "overhead" it scatters to the voracious and insatiable education-mongeries. I wonder what he would have said about our management by regular corporate administrations which lack, alas, the desire to make a profit and in perfect Parkinsonian policy have but a single real interest alike in students and in employees (some of whom are still called "faculty"): Get money out of them directly or for them through subsidy, in quantity sufficient to sustain exponential growth in number and power of administrators—mad pursuit of bigness, until, one day, every boy and girl will be conscripted to serve a term of unstructured play garnished with social indoctrination called "higher learning", and all universities will be branch offices of one Government bureau.

The centenary of BIRKHOFF's birth is 1984.

After the meeting the next I heard of BIRKHOFF was that he had died. Like BIRKHOFF's speech, the beautiful description of him in the obituary[2] by MARSTON MORSE is fixed so firmly in memory that I can almost quote it by heart:

> ... Birkhoff thought of his contemporaries in Europe, particulary in Germany, as colleagues rather than as teachers. He held Klein lightly, was unenthusiastic over Weierstrass, but gave his full respect to Riemann. Through his papers on non-self-adjoint boundary value problems and asymptotic representations he probably influenced the Hilbert integral-equation school as much or as little as it influenced him. His relations with the members of the French and Italian schools of analysis were close, both personally and scientifically. Levi-Civita and Hadamard were among his best friends. Birkhoff was at the same time internationally minded and pro-American. The sturdy individualism of Dickson, E. H. Moore and Birkhoff was representative of American mathematics "coming of age." The work of these great Americans sometimes lacked external sophistication, but it more than made up for this in penetration, power and originality, and justified Birkhoff's appreciation of his countrymen.

---

[2] *Bulletin of the American Mathematical Society* **52** (1946): 357–391.

MORSE's words should dispel the idea, today often noised, that before the arrival of refugees from HITLER scant mathematical research and advanced teaching were done in this country. Of course, mathematics is above nations and nationalities, but that works both ways; moreover, much of the best teaching is done outside of the classroom and at great distance. MORSE went on to say that "Poincaré was Birkhoff's true teacher."

SERRIN, like BIRKHOFF, emerged from what MORSE called "the dynamic individualism of the Middle West" and also experienced "the environment of tenacious self-sufficiency in New England...." Like BIRKHOFF, SERRIN was trained entirely in this country and largely in the Middle West. I met him when he was a student at the Graduate Institute for Applied Mathematics, Bloomington, Indiana. On 30 September 1950, just after I had arrived there, in my first letter to CHARLOTTE BRUDNO, who was then my assistant and who later graced me by becoming my wife, I wrote: "As for students, at least one in my class is very good .... He has just succeeded in ... giving correct and relatively simple proofs of the theorems of LAVRENTIEFF in free-boundary theory." My next letter to her, two days later, opens with the words "SERRIN is making brilliant discoveries .... Day by day new results appear .... If there is no error, this will be one of the two major achievements of our century in the theory of potential flow of incompressible fluids (the other being LEVI-CIVITA's proof of existence of finite surface waves)." In this same class was JERALD ERICKSEN, also a Westerner by birth and training, a man who developed more slowly but in the end also has made contributions to the rational mechanics of solids and fluids that rank, it seems to me, second to none in our time.

The course was statistical mechanics. That was the first time I taught BIRKHOFF's ergodic theorem. We used KHINCHIN's book, available only in a stupid translation that forced us to create everything halfway afresh. On the final examination SERRIN outlined arguments which indicated his discovery of a formal connection between HILBERT's process and ENSKOG's process in the kinetic theory of gases. His grade was A+. It was the first time I gave that grade, and I cannot recall ever having given it since. Unfortunately SERRIN never worked through his ideas on this central problem, which has since then been clarified by GRAD and MUNCASTER, though still only formally.

When I survey, to the best of my limited ability, the splendid researches of BIRKHOFF and SERRIN, I find several points of similarity. Both deserve MORSE's words of praise for the independent Americans: penetration, power, and originality. Both show a profound sense of problem, the one quality common to great mathematicians of all kinds and all ages. Technique, so prominent a feature in papers on pure analysis today, is there on demand, but it never takes precedence over concept. In neither's work is there a hard line between "pure" mathematics and its applications. About three-quarters of BIRKHOFF's subjects came directly or indirectly from mechanics; in SERRIN's, the proportion seems reversed. BIRKHOFF was able also to face nature and knowledge as such and himself to formulate new theories for them. This part of his work deserves more study than it has received. BIRKHOFF endorsed, as had HELMHOLTZ, the view that a mathematical theory of an aspect of nature is a model, and as such subject not only to mathematical development but also to scrutiny at its foundations. Above all we must not believe uncritically what physicists and engineers tell us. MORSE, perhaps in allusion to the astonishing success of EINSTEIN in the newspapers, wrote "Birkhoff inherited from Poincaré the sentiment that no single mathematical theory of any phenomenon deserves the exclusive attention of physicists, or at least of mathematicians." SERRIN also, in his new work on thermodynamics, faces the phenomena directly, selects the mathematical concepts in which to describe them, and lays down his own axioms to relate them. Theorems, real theorems—rare birds they have been in thermodynamics—come afterward, theorems that are not applications of known techniques but more likely the fathers of new methods in analysis.

Such is the tradition of HILBERT, who in the preface to his famous list of problems for the twentieth century to solve wrote

> While I insist upon rigor in proofs as a requirement for a perfect solution of a problem, I should like, on the other hand, to oppose the opinion that only the concepts of analysis, or even those of arithmetic alone, are susceptible of a fully rigorous treatment. This opinion . . . I consider entirely mistaken. Such a one-sided interpretation of the requirement of rigor would soon lead us to ignore all concepts that derive from geometry, mechanics, and physics, to shut off the flow of new material from the outside world . . .. What an important, vital nerve would be cut, were we to root out geometry and mathematical physics! On the contrary, I think that wherever mathematical ideas come up, whether from the theory of knowledge or in geometry, or from the theories of natural science, the task is set for mathematics to investigate the principles underlying these ideas and establish them upon a simple and complete system of axioms in such a way that in exactness and in application to proof the new ideas shall be no whit inferior to the old arithmetical concepts.

The sixth of HILBERT's problems was "mathematical treatment of the axioms of physics" after the model of geometry . . ., "first of all, probability and mechanics." HILBERT's influence on recent rational mechanics is not so widely known as it should be. While he published scantily in that field, years ago I found in the libraries of Purdue University and the University of Illinois notes taken by two midwesterners in HILBERT's course on continuum mechanics, 1906/1907. After studying those notes, in my own expositions from 1952 onward I followed HILBERT's lead, not in detail but in standpoint and program; basic laws are formulated in terms of integrals, and the science is treated as a branch of pure mathematics, by systematic, rigorous development motivated by its logical and conceptual structure, descending from the general to the particular.

The world "axiom" may confuse, since on the one hand all of mathematics is essentially axiomatic, while on the other hand "axiomatics" is a term often used pejoratively to suggest a fruitless rigorization of what everybody believed already. A mathematician must know that good axioms spring from intrinsic need—details scatter, concept is not clear, or paradox stands unresolved. Formal axioms represent only one response to the call for conceptual analysis before significant new problems can be stated. You cannot solve a problem that has not yet been set. Formal axiomatics make an important part of modern rational mechanics; I refer in particular to WALTER NOLL's solution of the relevant portion of HILBERT's sixth problem. But by far the greater part of modern mechanics is neither more nor less axiomatic than is any other informally stated branch of mathematics. *The essence is conceptual analysis*, analysis not in the sense of the technical term but in the root meaning: logical criticism, dissection, and creative scrutiny. After that comes the poetry of statement and proof, the ornament of illuminating examples.

A mathematical discipline is made by mathematicians. However much they may begin from, digest, clarify, and build upon the ideas of physicists and engineers, it is the mathematicians who make sense out of them.

Now, I fear, many people accept a picture too narrow of mathematical activity. They forget that the great theories which enable us to understand in part the world about us and on which engineers and computers base their applications were created by great mathematicians; they forget also that the problems suggested by new mathematical theories of nature often fail to fall within any of the previously existing fields of "applied mathematics", which fields, all of them, grew out of older theories of nature. They disregard a prophetic warning published in 1924 by that canny and shrewd observer,

RICHARD COURANT:

> ... many analysts are no longer aware that their science and physics ... belong to each other, while ... often physicists no longer understand the problems and methods of mathematicians, indeed, even their sphere of interest and their language. Obviously this trend threatens the whole of science: the danger is that the stream of scientific development may ramify, ooze away, and dry up.

I first read this warning in 1941, when I was a student of BATEMAN; it has been fixed in my memory ever since. I have chosen to mention SERRIN's work as well as BIRKHOFF's because he was trained, not before but long after the danger had appeared, and he proves that mathematics of BIRKHOFF's kind is not impossible even today.

I should like to believe it this kind of mathematics—individual, self-sufficient, neither enslaved to nor isolated from natural science—that the BIRKHOFF Prize is designed to recognize.

If conceptual analysis of physical problems is a part of mathematics, is it pure or is it applied? I think it is both. I believe that here I may count upon the agreement of not only SERRIN, COURANT, BIRKHOFF, HILBERT, and POINCARÉ but also mathematicians of still higher rank: CAUCHY and EULER.

This speech has been longer than it ought have been. There is an excuse. As my first invitation to address the Society has come thirty-six years after I joined it, I have tried to squeeze in everything I had to say, for fear that should a second arrive after the now established lapse of time, it would be not only long after 1984 but also long after the authorities would release me to accept the honor.

Note for the Reprinting

This essay was first published in *The Mathematical Intelligencer* **1** (1978): 99–101. Printer's errors in that text are here corrected. "The Society" mentioned is the American Mathematical Society.

PART II

CRITICISM: SELECTED REVIEWS

A. WRITING AND TEXTS FOR LIVING SCIENCE

12. A COMMENT ON SCIENTIFIC WRITING (1954)

In *Science* for 23 April 1954 there were articles on the frequent wordy emptiness and awkward style in scientific writings. Most of the examples quoted and discussed, although poorly written, were in correct English, but the matter is more serious than that: In the mathematical field, at least, outright errors are not uncommon.

Recently the editors of a journal of a mathematical organization sent me for review a book that furnishes a bad example. It contains a fair proportion of valuable new work. It was written by a group of distinguished scientists. The native language of the majority was German. The result is a chain of slang and stodgy teutonisms, scattered helter skelter among the commas. To qualify this statement I must add that one of the authors born abroad wrote in clear, precise, and correct English, whereas one of the young native Americans wrote miserably. Young scientists are trained by example, and I fear his case is typical. This aspect of the problem, nothing so refined as mere infelicity of expression, was not mentioned in the articles in *Science*, and for this reason the following paragraphs from my unpublished review may be of interest.

> Now to say no more than this about the style might allow misunderstanding. This particular kettle is only somewhat blacker than the pots in its environment. Not to the tolerance of Americans but to their carelessness must be attributed their silent consent with such maculation of their mother tongue. No German or French editor would dare to publish a comparable haggis of blunders and anglicisms. To the triumph of the jargon of comic strips and advertisements has been added the influx of foreign scientists, especially Germans. The splendid additions brought by these foreign scientists to our scientific life make it easy to see how some young Americans, already ill footed in their mother tongue and mistaking in their masters the certainty of knowledge for correct expression of it, have fallen into a ragged

bastardy of language. At the same time there has been an unfortunate pressure on foreign-born scientists to write in English. That their English is better than our German or French, does not make their English correct or clear. That their English is only a little worse than their students' and colleagues', while reflecting little credit to these latter, does not license it. Every editor knows that most manuscripts received contain outright errors which must be corrected silently, while the problem of style is more or less hopeless. I mention all this here because this volume is the worst I have seen: Each manuscript, apparently, is printed in its original purity, making the whole a defining example of *die schönste Lengevitch*.

Moreover, since most of this volume is written by persons whose native languages will not tolerate the mangling to which its free (but nevertheless not inexistent) syntax makes English liable, in this case there was a simple remedy. Had the editors encouraged some of the authors to write in German, some of the articles in this volume would have been expressed in a style commensurate with the value of their contents, and the result would have come nearer to that clarity without which expository works fail of their purpose.

Note for the Reprinting

This note was first published in *Science* **120** (1954): 434. My "unpublished review" was unpublished because I withdrew it after a bigwig of the national society sponsoring the journal that had invited it had declared that he would permit it to be published only if the passage here reprinted were excised. The bigwig was as ephemeral as the volume under review; for more than two decades I have not heard either mentioned; there is no reason to recall from deserved oblivion the name of either.

Times have changed. S. ANDRESKI in the first chapter of his book, *Social Sciences as Sorcery*, New York, St. Martin's Press, 1972, writes as follows:

Not only does the flood of publications reveal an abundance of pompous bluff and a paucity of new ideas, but even the old and valuable insights which we have inherited from our illustrious ancestors are being drowned in a torrent of meaningless verbiage and useless technicalities.

In the second chapter, "The witch doctor's dilemma":

Let me ask the following questions: Which field of activity in America is the least efficient? And which employs the largest number of psychologists and sociologists? The plain answer is, Education. And in which field has the quality of the product been declining most rapidly? And where has the number of psychologists and sociologists been increasing fastest? Again: in education. Or, if instead of comparing it with other sectors within the society, we compare the American educational situation with that of other nations, we get a similar result. ... [Nonetheless] ... there can be no doubt that the American schools are the least efficient in the world, not excluding the poorest countries of Africa or Latin America. I do not think that anywhere else in the world can you find students who have been going to school for at least twelve years but who can read only with difficulty, such as you can meet quite frequently in American universities. What is more, the schools have been getting worse as the number of personnel trained in sociology, psychology and education has been increasing. Perhaps it is all a coincidence. But in no other country can you become a professor at a top university without having first to learn how to write competently. And this doesn't include people of foreign origin or those brought up in a different language, but men and women who know no other language but American English, and yet contravene the rules stated in American grammar books and use words with scant regard to what it says in Webster's dictionary.

These trends are not confined to the U.S.A., and in other countries too a decline in the standards of literary expression has gone in step with the expansion of the social sciences.

My experience as an editor suggests that the young Britons, Germans, and Frenchmen first followed and then surpassed the example of writers turned loose by the Cultural Revolution of the American schools in the 1950s and 1960s.

13. GOLDSTEIN'S *CLASSICAL MECHANICS* (1950)

This book is intended as an advanced text on classical mechanics for the student whose sole desire is to learn quantum mechanics. The author defends teaching mechanics to a physicist despite the fact that "it introduces no new physical concepts . . . nor does it aid him . . . in solving the practical mechanics problems he encounters in the laboratory." He states that "classical mechanics remains an indispensable part of the physicist's education" because it "serves as the springboard for the various branches of modern physics" and "affords the student an opportunity to master many of the mathematical techniques necessary for quantum mechanics . . .." After reading this preamble we are not surprised to learn that GOLDSTEIN finds "the traditional treatment . . . no longer adequate" and that at the same time the basic concepts of mechanics "will not be analyzed critically . . .." His treatment of what is ordinarily called mechanics is little else than formal manipulation; his frequent remarks concerning physics almost invariably fall outside the classical framework and deal with the behavior of small particles. He observes that in some special cases classical principles remain valid even today, and he quotes as illustrations the neutron pile and the V-2 rocket. There are frequent promises of better things to come when the student reaches quantum mechanics. The following subjects are treated: vectorial mechanics of a system of mass-points, LAGRANGE's equations and HAMILTON's principle, the two-body central force problem, the motion of rigid bodies, special relativity, transformation theory, HAMILTON–JACOBI theory, small oscillations, the LAGRANGEan and HAMILTONian formulations for continuous systems.

The book differs from the traditional treatment not only in attitude but also in detail. First, numerous topics are omitted, an example being the entire theory of orbits and their stability. Second, the following unusual topics (many of them to be found as incidental or introductory material in books on quantum mechanics) are included: frequent examples of the motion of particles subject to electromagnetic forces, scattering in a central force field, vibrations of a triatomic molecule. GOLDSTEIN states that "mathematical techniques usually associated with quantum mechanics have been introduced wherever they result in increased elegance and compactness." An example may be found in his treatment of rigid bodies, where in order to introduce linear transformations, matrices, spinors, eigenvalues, dyadics, and tensors in rapid succession he uses sixty-four pages to reach "the so-called Euler equations" by surely the longest derivation in the literature. After all this preparation there are only nineteen pages on the actual motion of a rigid body, containing mostly material found in intermediate texts. GOLDSTEIN makes a point of introducing at length the CAYLEY–KLEIN parameters but does not ever employ them.

The mathematical level of the book may be judged from the fact that GOLDSTEIN avoids elliptic functions, claiming that "such a treatment is not very illuminating". The explanations of some elementary mathematical topics are clear, as for example in the unusually straightforward statement of the formal problem of the calculus of variations and in the discussion of nonholonomic constraints.

At the ends of the chapters are problems, largely concerned with applications to modern physics. "Pedantic museum pieces have been studiously avoided." GOLDSTEIN also fills pages with candid evaluations of his competitors. Of WHITTAKER's *Analytical Dynamics* he says "the development is marred, regrettably, by an apparent dislike of diagrams . . . and of vector notation, and by a fondness for the type of pedantic mechanics problems made famous by the Cambridge Tripos examinations." The *Theorie des Kreisels*, he says ". . . has all the external appearances of the typical stolid and turgid German 'Handbuch'. Appearances are deceiving, however, for it is remarkably readable, despite the handicap of being written in the German language." As far as style is concerned, GOLDSTEIN himself inhabits a glass rather than an ivory tower; infinitives are split and participles suspended as if by rule, and were it not for GOLDSTEIN's repeated disparagement of German we should suspect Teutonic influence in his exclamation points, his isolation of lonely words by commas, his use of "so-called" in the attribution of names, and his hyphenated word-marriages.

In reading this book I have learnt how to solve some problems I failed to find in the classical works which GOLDSTEIN banishes to pedantic museums, but his apologetic attitude toward his subject at every page raised the question, why study classical mechanics at all? The rise of modern particle physics has certainly necessitated a revision of our concepts. Is it not possible to present quantum mechanics as a self-sufficient discipline, without first learning the incorrect principles of classical mechanics only to throw them away again after sharpening our mathematical tools in solving exercises? As prerequisite to thermodynamics we do not set a course in caloric theory, simply because it served as a "springboard" for energetics or because problems of heat conduction illustrate techniques of manipulation. The answer to this question is simple: Like any physical theory, classical mechanics predicts correctly only a certain range of observable phenomena, but this range is so enormous, so far greater than that describable by any other branch of physics, that any person who wishes to understand the world about him must learn classical mechanics *for its own sake.* The writers on quantum mechanics recognize this fact when they invariably derive the classical equations as an approximation. There are eminent physicists who believe that quantum mechanics may require certain ultimate revisions; that Newtonian mechanics must again be somehow included, is unquestioned. GOLDSTEIN expresses his contempt for macroscopic phenomena (*e.g.*, page 15), but *it is only macroscopic phenomena that classical mechanics adequately describes.* Macroscopic phenomena do indeed occur in nature, and their inherent interest to physicists is recognized by the existence of member societies of the Institute and divisions of the Physical Society whose field is wholly or partially classical. The fault in the older treatments of mechanics lies not in their failure to be similar enough to quantum mechanics, but in their being too similar to it—they begin (as does the author) with the Newtonian laws for a mass-point, literally a mathematical point occupying no volume at all, while in fact modern physics has taught us that the classical laws become a poorer and poorer approximation to observed phenomena the smaller is the body. Since classical mechanics yields a correct description of the motions only of rather large bodies, its basic concepts and equations should therefore be put in terms of large bodies, so that the more nearly the physical body approximates the mathematical concept the more accurately the mathematical equation describes its behavior. The *concept* of mass-point may be altogether abandoned. The familiar mass-point *equations*, nonetheless, are satisfied by the centers of mass of large bodies, and the classical mass-point structure thus reappears as an approximation valid when the motions of large bodies relative to their centers of mass are negligible. This truly physical approach to the subject was presented long ago by HAMEL[1] and is summarized in his article on the axioms of mechanics in the (perhaps stolid, German, and turgid) *Handbuch der Physik*[2].

---

[1] G. HAMEL, "Über die Grundlagen der Mechanik", *Mathematische Annalen* **66** (1908): 350–397.

[2] G. HAMEL, "Die Axiome der Physik", GEIGER u. SCHEEL's *Handbuch der Physik*, Volume 5 (1927): 1–42.

I can heartily agree with GOLDSTEIN in wishing a more modern and more physical treatment of mechanics, but I cannot help regretting that his refusal to re-examine the fundamental concepts of his subject prevented him from giving us a confident and substantial book.

Note for the Reprinting

This review of HERBERT GOLDSTEIN's *Classical Mechanics*, Cambridge, Addison Wesley, 1950, was written at the request of the editor of a journal expressly devoted to the teaching of physics, but because of its "unfavorable nature" he refused to publish it. As the book is still widely sold, I print my review now, thinking that perhaps it is not too late to prevent some unwary student from expecting that the contents might live up to the title.

14. MURNAGHAN'S
FINITE DEFORMATION OF AN ELASTIC SOLID (1952)

Apparently intended as a text, this book follows the growing custom of beginning with an introductory chapter containing pure mathematics neither necessary nor sufficient for the applications which follow. The student is deterred from accepting the results without question by frequent interjections of "(why?)", "(prove this)", and other commands, and the pædagogical usefulness of the work is attested by the numerous simple exercises. The pages, whose crowding with symbols and parenthetical expressions in the text suggest that the publisher confused the manuscript with some work on topology or algebra, are a bit frightening.

MURNAGHAN maintains that the theory of finite deformations is most easily presented and understood by the use of matrices. "Do not fall into the error of regarding them as a complicated device invented by mathematicians to make the theory of elasticity harder than it actually is", he warns. Nonetheless, the book did not furnish me recreational reading, and I find MURNAGHAN's former tensorial treatment in the *American Journal of Mathematics*, 1937, not only more complete but also by virtue of its compact explicitness easier to grasp. At one point MURNAGHAN reverts to tensors, although taking care to advise readers to skip this passage, and later in the solution of special problems in curvilinear co-ordinates he employs without derivation equations which would develop naturally if an invariant formulation had been given to start with.

About forty pages are required to derive the classical equations of finite elastic strain. Next MURNAGHAN develops what he calls "the integrated linear theory" of hydrostatic pressure. He obtains this theory by taking the expression for change of volume in the linearized theory, then supposing the LAMÉ constants to be linear functions of pressure, then integrating the result. He shows that by suitable choice of the constants occurring it is possible to get good agreement with some of BRIDGMAN's experiments. He presents the numerical details in full.

To consider MURNAGHAN's theory, let us repeat part of his presentation of the classical proof of the existence of stress-strain relations (pages 55–56). Writing T for the stress matrix, η for the strain matrix, ρ_a and ρ_x for the density before and after deformation, J for the matrix of gradients of deformed with respect to undeformed positions, ψ for the mass density of the energy of deformation, and using a star to denote transposition, he says:

Since [the principle of conservation of energy] must hold for an arbitrary volume ... of our deformable medium, we have (why?)

$$\mathrm{Tr}\,(J^{-1}T(J^*)^{-1}\delta\eta) = \rho_x\,\delta\psi.$$

Since ψ is (by hypothesis) a function (written symmetrically) of η we have (by the very definition of a differential) $\delta\psi = \mathrm{Tr}\,((\partial\psi/\partial\eta)\delta\eta)$ (show this), and so

$$\mathrm{Tr}\,(J^{-1}T(J^*)^{-1}\delta\eta) = \rho_x\,\mathrm{Tr}\left(\frac{\partial\psi}{\partial\eta}\delta\eta\right).$$

Since this relation must hold for an arbitrary (symmetric) matrix $\delta\eta$, we have (why?) ...

$$T = \rho_x J\frac{\partial\psi}{\partial\eta}J^*.$$

The reader notes that if ψ were allowed to depend on T here, MURNAGHAN's formula for $\delta\psi$ would become $\delta\psi = \mathrm{Tr}\,((\partial\psi/\partial\eta)\delta\eta + (\partial\psi/\partial T)\delta T)$, and the proof would fail. That is, the classical finite-strain theory requires that for a given point in the medium at given temperature or entropy, ψ be determined by the strain η *alone*, independently of the existing stress. Thus MURNAGHAN's assumption that the LAMÉ constants are functions of pressure would appear inconsistent with the theory of finite strain, and some readers may prefer to regard his "integrated linear theory" as an isolated semi-empirical statement. The foregoing remarks are not to be confused with the well known theorem, discussed by MURNAGHAN on page 65, which states that if an isotropic elastic body is really subject to initial hydrostatic pressure p_0, but if we choose to neglect that fact and treat the body as if it were unstressed, we can get correct results for small strains if we simply replace λ by $\lambda + p_0$, μ by $\mu - p_0$.

The remainder of the book concerns the consequences of approximating the strain-energy function by a cubic in the strain components. From the general considerations of REINER[1], which

---

[1] M. REINER, "Elasticity beyond the elastic limit", *American Journal of Mathematics* **70** (1948): 433–446.

MURNAGHAN does not mention, it can be shown that to this degree of approximation the classical formulæ relating shear stress to angle of shear, twisting couple to angle of twist, *etc.*, will not be altered, although the characteristic phenomena of nonlinear elasticity will appear in their simplest forms as new stresses not present at all in the linear theory. MURNAGHAN works out the cubic terms for the various types of crystals. In the following discussion of simple tension, simple shear, compression of a circular cylindrical tube and of a spherical shell, and torsion of a circular cylinder, he does not mention the general solutions valid for arbitrary strain energy which have recently appeared in the literature[2]. New, on the contrary, is MURNAGHAN's calculation of the second-order change of dimensions in a state of simple shearing stress, as distinct from a simple shear displacement, and of the similar change of dimensions of a circular cylinder in torsion.

The only historical references in the book tell us that JACOBIans are named after JACOBI, the LAMÉ constants after LAMÉ, besides giving the dates and nationalities of these two persons. Apart from a single reference to some experimental data, the only citations of literature are to MURNAGHAN's other texts. While this shabby practice is become the rule in volumes intended for the pædagogical and undergraduate market, its extension to serious works does not seem altogether commendable. The publishers present this book as an "authoritative exposition". Inclusion of the recent results on finite elastic strain obtained by SIGNORINI, REINER, RIVLIN, and GREEN & SHIELD, which to me seem deep and significant, would not have been unwelcome.

In the preface MURNAGHAN states: "If the mathematical treatment given here serves to stimulate the procurement of experimental knowledge of these phenomena we shall have attained our aim." Abundant and detailed experiments on the very large strain of rubber have been reported by RIVLIN from 1947 onward. In my opinion the results of these experiments fully confirm the predictions of the general theory of elasticity while showing the second-order approximation employed by MURNAGHAN to be insufficient.

Note for the Reprinting

This review of F. D. MURNAGHAN's *Finite Deformation of an Elastic Solid*, New York, Wiley, 1951, was first published in *Bulletin of the American Mathematical Society* **58** (1952): 577–579.

---

[2] R. S. RIVLIN, "Large elastic deformations of isotropic materials. IV. Further developments of the general theory", *Philosophical Transactions of the Royal Society* (London) **A241** (1948): 379–397.

15. NOVOZHILOV'S *FOUNDATIONS OF THE NONLINEAR THEORY OF ELASTICITY* (1953)

Students of mechanics will be grateful to the translators and the publishers for making available the second of the three[1] existing monographs on the general theory of elasticity—the more so, since the Russian original is in this country at least a scarce book.

The translation is in unusually good English (except for "compatability") and the translators have taken unusual care that the exposition of this elaborate subject shall make sense, although they are not always familiar with the terms used in mechanics (*e.g.* on page 58 they use "components of a vortex" for "components of the curl"). Despite its being planographed, and thus repulsive to the eye, the text is readable.

NOVOZHILOV's approach is straightforward, honest, and vigorous. There is little or no nationalism, rhetoric, or pædagogery. NOVOZHILOV gives every evidence of his earnest competence and his respect for a difficult and important group of problems. The book is not scholarly, however; most of the some ninety items in the bibliography are not cited in the text, part of which presents material first published in important papers not listed in the bibliography. Perhaps many results in this book are rediscoveries by the author himself.

This is a serious work, deserving detailed notice. The preface is dated 1947, and the book is on the whole a careful, accurate, and reliable exposition of some of the mechanical aspects of the classical

---

[1] The other two are "Sur la théorie de l'élasticité" by E. COSSERAT & F. COSSERAT, *Annales de la Faculté des Sciences de l'Université de Toulouse pour les Sciences Mathématiques et les Sciences Physiques* **10** (1896): 1–116, and *Finite Deformation of an Elastic Solid* by F. D. MURNAGHAN, New York, 1952. [For a review of the latter, see the preceding essay in this volume. I should have cited here also the three great memoirs of 1904/6 by P. DUHEM, collected in the volume *Recherches sur l'Elasticité*, Paris, Gauthier-Villars, 1906, and the expository works of A. SIGNORINI: "Trasformazioni termoelastiche finite, Memoria 1ᵃ", *Annali di Matematica Pura ed Applicata* (4) **22** (1943): 33–143; "Memoria 2ᵃ", *ibid.* **30** (1949): 1–72. The third and last part was to appear in *ibid.* **39** (1955): 147–201.]

nonlinear theory of elasticity as it stood at that date. It was in 1948 that the numerous publications of RIVLIN, which have enlivened the subject and changed the whole view of it, began to appear[2]. Thus the book, through no fault of its author, cannot be called a definitive exposition of general elasticity, nor even an adequate introduction to it. The term "foundations" in the title is justified by the usually careful treatment of principles: NOVOZHILOV's objective is to set up the governing equations and various approximations to them, without attempt at solutions in special cases.

Nearly two thirds of the text, comprising the first four chapters, are devoted to the basic equations of three-dimensional finite elasticity. In Chapter I the concept of strain is developed with especial care, and various levels of approximation are carefully distinguished (especially §§ 13–15). In speaking of "small elongations and shears", on the contrary, NOVOZHILOV fails to remark that shear is not an invariant concept; what he intends is "small principal extensions".

The treatment of stress in Chapter II is not only rather slipshod but also depressingly elaborate, while to follow through fifteen pages the derivation of stress-strain relations in Chapter III, even though it begs the main question at issue, would require an iron resistance to boredom not easily bred in temperate climes. It is in Chapter III, § 29, that we find the author's most serious oversight. While he reduces the stress-strain relations for isotropic bodies to a material ("Lagrangean") form which is rather simple in appearance, he does not mention the possibility of using spatial ("Eulerian") strain measures. It is this possibility that renders problems of large strain [of isotropic bodies] manageable. It seems unfortunate that JOSEPH FINGER[3], the first to notice this simple but centrally important fact and to obtain the stress-strain relations whose rediscovery facilitated the striking progress in general elasticity since 1947, is unknown in the history of mechanics.

On page 113 NOVOZHILOV states without proof the following invariant form for the "generalized" stress matrix S:

$$S = \Psi_2 \Pi_0 + \Psi_1 \Pi_1 + \Psi_0 \Pi_2,$$

"where Ψ_2, Ψ_1, Ψ_0 are functions of the strain invariants, Π_0 [is] the

---

[2] These are briefly summarized in Chapter IV of my paper, "The mechanical foundations of elasticity and fluid mechanics", *Journal of Rational Mechanics and Analysis* **1** (1952): 125–300; corrections and additions, **2** (1953): 593–616. [A corrected reprint was published in 1966 under the title *Continuum Mechanics I*, New York *etc.*, Gordon & Breach.]

[3] J. FINGER, "Über die allgemeinsten Beziehungen zwischen Deformationen und den zugehörigen Spannungen in aeolotropen und isotropen Substanzen", *Sitzungsberichte der k. u. k. Akademie der Wissenschaften*, Wien, (IIa) **103** (1894): 1073–1100.

unit tensor, Π_1 [is] a tensor whose components are linear combinations of the [material] strain components, and Π_2 [is] a tensor whose components are quadratic combinations of the strain components." Here NOVOZHILOV has found a portion of the basic invariance theorem established simultaneously by REINER[4]: S may be taken as the true stress matrix, Π_1 as any matrix whose proper numbers are analytic functions of the principal extensions and whose principal axes are the principal spatial axes of strain, while Π_2 may be taken as $(\Pi_1)^2$.

In § 31 NOVOZHILOV writes "It follows from the above that for every material a range of small deformations can be established for which HOOKE's law is approximately valid." If you wonder how a mathematical theory could ever establish such a result, you must turn back to § 30, where you find that the strain energy has been assumed analytic because "no negative powers can appear in the series." There is no statement that an assumption has been made, not even a discussion of why fractional powers, for example, might not be appropriate, although a considerable engineering literature devoted to this possibility exists[5]. In fact, the only experimental justification of the assumption of analyticity is the *experimental* validity of HOOKE's Law for many [by no means all!] materials under sufficiently small loads—but this reasoning is the direct opposite of NOVOZHILOV's.

NOVOZHILOV's analysis (§ 32) of HENCKY's theory of plasticity (in this country usually considered with respect to strain-hardening, and often called "the theory of ROŠ, EICHINGER, & SCHMID") was obtained also by C. WEBER[6].

In this book all results are written out at length in rectilinear coordinates. Some sets of formulæ cover most of a page. In the preface we find an explanation: "To make the book as accessible to as wide a circle of readers as possible, the author has attempted to carry out all deductions in the simplest and most intuitive manner, avoiding, in particular, tensor calculus . . . ." In fact, at the top of page 67 the word

---

[4] M. REINER, "Elasticity beyond the elastic limit", *American Journal of Mathematics* **70** (1948): 433–446; [and "A mathematical theory of dilatancy", *ibid.* **67** (1945): 350–362. The theorem is included as a special instance by older theorems on invariants but would be difficult to recognize therein. Priority for its explicit statement is shared by W. PRAGER, "Strain hardening under combined stresses", *Journal of Applied Physics* **16** (1945): 837–840. The early statements and proofs refer to polynomial functions only. The theorem is valid as a purely algebraic statement applicable to all kinds of isotropic mappings of the space of symmetric tensors into itself.]

[5] *Cf. e.g.* R. MEHMKE, "Zum Gesetz der elastischen Dehnungen", *Zeitschrift für Mathematik und Physik* **42** (1897): 327–338.

[6] C. WEBER, "Zur nichtlinearen Elastizitätstheorie", *Zeitschrift für Angewandte Mathematik und Mechanik* **28** (1948): 189–190; **29** (1949): 256.

"tensor" or the idea behind it is avoided with comical precaution: apparently the reader is assumed not to have studied the classical treatises by Voigt and Love. I believe that an intelligent student completely untutored in geometry on reading this page would set himself the problem of formulating and exploring the geometric concept which the two obviously connected results so forcibly separated by the author most plainly suggest. But on page 111 the word "tensor" suddenly appears without explanation and is used several times later. It is somewhat similar with Green's Theorem, which is carefully avoided in the creaking development of the properties of the stress tensor but appears later on page 106 (where, be it noted, it is regarded as so extraordinary as to need two of the five references given in the first 200 pages, the others being to works on orthogonal curvilinear coordinates).

It is not in disrespect to Euler and Cauchy that I say their methods in continuum mechanics are now unnecessarily elaborate; in fact it is in their papers on continuous media that some of the earliest discoveries in the theory of differential invariants occur, and tensor analysis is in part an elaboration of their work. But I think the student who follows in this book the endless pages of dreary resolutions and projections in the Euler–Cauchy style could better spend his time learning tensor analysis, which would enable him to reproduce four fifths of Novozhilov's work in twenty pages, while freeing his attention for the important questions and ideas which are scattered through the remaining one-fifth.

Although Novozhilov founds all his analysis in the fully general theory, his main interest is in the case next in order of generality past the fully linear one, when the extensions are small, but the displacements and rotations may be large. The cause of this restriction, on which he lays great emphasis, is his desire to furnish structural engineers with the basic theories needed for rational solution of their problems of nonlinear elasticity. Since typical structural materials, such as steel, fail to retain their elastic reversibility when subjected to extensions as great as 1%, there are essentially only two such nonlinear problems: (1) elastic stability, which Novozhilov interprets as determining the smallest load at which Kirchhoff's uniqueness theorem breaks down, and (2) bending of "flexible" bodies, such as thin rods, plates, and shells. The last two chapters, the most important in the book though occupying only about eighty pages, are devoted to these two problems.

While Novozhilov's points are well taken, it is instructive to consider a simple analogy. Suppose we are to clarify the problems besetting horological engineers when their pendulums swing in a range beyond that in which the approximation $\sin \theta \approx \theta$ is sufficiently accurate. Doubtless then $\sin \theta \approx \theta - \theta^3/6$ will do for practical problems of this type. Accordingly, if we were to follow the practice which was nearly universal in nonlinear elasticity up to 1948 and is recommended by NOVOZHILOV, we should devote ourselves to the differential equation

$$\ddot{\theta} + k^2(\theta - \theta^3/6) = 0.$$

In so doing, we should lose the simplicity of the linear theory, having to face at once all the complications of nonlinear mechanics—but even if completely successful, all we should have, at great cost, would be a somewhat better approximation. Everyone knows that it is no harder to settle the whole matter rigorously by studying the exact solutions of the exact equation

$$\ddot{\theta} + k^2 \sin \theta = 0.$$

Now a very similar thing has happened in elasticity theory. The work of RIVLIN has shown us that it is feasible and practical to work directly with the exact equations for arbitrarily large strain of a material characterized by an arbitrary strain-energy function. There is not only the scientific satisfaction of solving a general problem for its own sake (*cf.* the last paragraph of the "Historical Introduction" to LOVE's *Treatise on the Mathematical Theory of Elasticity*, 4th edition, Cambridge, 1927), but also the precision of a general analysis leads to simplicity and certainty in the end. While only relatively simple problems can be solved explicitly within the fully general theory, these particular cases are important, and it was the light they cast upon the nonlinear theory that pointed the way to an approximate procedure valid for all problems of prescribed loading[7]. An example of the defectiveness of the approach usual in nonlinear elasticity is furnished by NOVOZHILOV's formulation of the problem of elastic instability within an approximate nonlinear theory. In the absence of a mathematical approximation theorem, we cannot assert with confidence that critical loads obtained from NOVOZHILOV's equations approximate the critical loads which would be obtained from the general theory. But these remarks must not be taken as criticism of NOVOZHILOV's work, which presents in a few pages a relatively simple and cogent development of the problem of elastic instability in the usually received sense.

There is some question also about NOVOZHILOV's distinction between "geometrical" and "physical" nonlinearity (§ 34, and again on page 197). For example, whether or not the rotations are large cannot be determined by "geometric considerations" *a priori*; the rotations

[7] R. S. RIVLIN, "The solution of problems in second order elasticity theory", *Journal of Rational Mechanics and Analysis* 2 (1953): 53–81.

result from loading, and (unless you are using an inverse method) you cannot know in advance whether for given loading of a material defined by a given strain-energy function the nonlinear terms in the strain components will need to be retained or not. True, *after* the problem is solved the question becomes purely geometric, but if we have the exact solution, then it is no longer important whether we can neglect certain terms or not. The question of whether certain approximations are valid *in advance* is avoided by NOVOZHILOV; its treatment would require a new type of approximation theorem for partial-differential equations.

The excellent last chapter is summarized in NOVOZHILOV's conclusion (§ 54):

> Ordinarily, the theory of deformation of flexible bodies (plates, shells, rods) is developed by making certain assumptions which immediately reduce the problem to a two-dimensional one (in the case of plates and shells) or a one-dimensional problem (in the case of rods). However, with such assumptions one necessarily loses sight of the connection between the theory of plates, shells, and rods and the general theory of elasticity. In view of this, many people consider the theory of flexible bodies as a kind of hypothetical superstructure over the general theory of elasticity, as a foreign element in it.
>
> Only in this manner can one probably explain why most contemporary books on the theory of elasticity omit all mention of the problem of deformation of flexible bodies, which is of such practical importance. An attempt was made in Love's book to relate the "hypotheses" of the theory of flexible bodies to the general theory of deformation.
>
> But special work in this direction was carried out by B. G. Galerkin, in whose papers the classical theory of shells and plates truly became a branch of the general theory of elasticity.
>
> The basic idea championed by B. G. Galerkin was that the problems of the bending of plates and shells must always be examined in the context of the general theory. This simple but profound idea was responsible to a large extent for the successful development of the theory of plates and shells in the Soviet Union and turned out to be fruitful not only in the case of thick plates and shells, but also in the case of thin plates and shells.
>
> It is natural to extend this idea to the nonlinear theory of elasticity, since one can expect that many results of this theory may be systematized by starting out from the general equations. The present chapter was an attempt to give a uniform method

for investigating the deformation of flexible bodies on the basis of the general nonlinear theory of deformations. It was our aim to clarify, with the aid of the general equations, those "hypotheses" on which the theory of plates, shells, and rods is ordinarily based, and to examine, from a uniform point of view, all these problems, which are ordinarily treated separately in spite of their common features.

The "basic idea championed by B. G. Galerkin" goes back to CAUCHY and POISSON for the theory of plates, while for slight bending of shells it is NOVOZHILOV himself[8] who has given us the first adequate treatment based on the three-dimensional theory. In the present work he carefully derives from nonlinear three-dimensional elasticity several of the nonlinear theories of rods, plates, shells, taking pains to show that the special hypotheses used are consistent to the degree of approximation considered. The reader not already familiar with this subject, where in the past outright inconsistent assumptions have often been made, may not realize that NOVOZHILOV's treatment deserves the description "simple but profound".

Note for the Reprinting

This review of V. V. NOVOZHILOV's *Foundations of the Nonlinear Theory of Elasticity*, translated from the Russian edition of 1948 by F. BAGEMIHL, H. KOMM, & W. SEIDEL, Rochester, Graylock, 1953, first appeared in *Bulletin of the American Mathematical Society* **59** (1953): 457–473.

---

[8] V. V. NOVOZHILOV, "On an error in a hypothesis of the theory of shells", *Comptes Rendus (Doklady) de l'Académie des Sciences de l'U.R.S.S.* (n.s.) **38** (1943): 160–164. Similar ideas may be found in the thesis of R. BYRNE (1941), "Theory of small deformations of a thin elastic shell", *Seminar Reports in Mathematics, University of California* (Los Angeles) (n.s.) **2** (1944): 103–152; in "The membrane theory of shells of revolution" (based on my Princeton dissertation, accepted 1943), *Transactions of the American Mathematical Society* **58** (1945): 96–166; and in a paper derived from the Toronto thesis (1942) of W.-Z. CHIEN, "Derivation of the equations of equilibrium of an elastic shell from the general theory of elasticity", *Science Reports of the Tsing-Hua University* **A8** (1948): 240–251.
[A treatment based on NOVOZHILOV's is provided in § 213 of TRUESDELL & TOUPIN's "The Classical Field Theories" in Volume **III**/1 of FLÜGGE's *Encyclopedia of Physics*, Berlin *etc.*, Springer, 1960.]

PART II

CRITICISM: SELECTED REVIEWS

B. THE LIGHT OF HISTORY UPON THE PRESENT

16. *CRITICAL PROBLEMS IN THE HISTORY OF SCIENCE* (1961)

This book, handsomely printed and low in price, contains the texts of lectures by R. HALL, G. DE SANTILLANA, A. C. CROMBIE, J. T. CLARK, E. J. DIJKSTERHUIS, D. J. DE S. PRICE, D. STIMSON, H. GUERLAC, C. C. GILLESPIE, L. P. WILLIAMS, T. S. KUHN, I. B. COHEN, J. W. WILSON, J. C. GREENE, C. S. SMITH, and M. BOAS. The periods considered range from classical antiquity to the end of the last century, most of them emphasizing the circumstances of science rather than what a scientist would regard as science itself, though there is some discussion of mechanics, astronomy, chemistry, physics, biology, and metallurgy. As to be expected of unrefereed papers, the quality is uneven, and the book is made still more diffuse by printing nineteen prepared comments, which footnotes indicate to have been possibly more useful to the participants than most are likely to be for the reader, who may be reasonably thought capable of summarizing the papers for himself. In reflection of the concern of historians of science in the U.S.A. to organize, compartment, and promote themselves, we are told how much the symposium cost, where the money came from, and how well the participants thought of each other, with titles, and we are even subjected to two papers on how to teach undergraduates, a matter, like the necessities of the privy, in no way dishonorable but dull for discourse unless provoked by illness or joking.

Within this discouraging matrix are set some fine papers. DIJKSTERHUIS presents a concise, elegantly written, and masterful survey of the origins of mechanics. Since it shows how different are the facts of scientific research from the Victorian fictions still ingrained by pædagogic indoctrination, it deserves to be put into the hands of every teacher or student of physics or of social history, but few will look here for it. The paper called "Contra-Copernicus" by DEREK PRICE is a regular piece of research on the history of science, ably showing that the concrete contribution of COPERNICUS to astronomy, whatever may have been the political-philosophical ends to which his book was later put, was slight. This thesis is developed in more personal terms in

KOESTLER'S *The Sleepwalkers*. Except for these two lectures, the book reflects the tendency of historians of science to forget about mathematics and the dominant influence it has had in the real achievements of the West. For example, the paper of KUHN on the much discussed discovery of the conservation of energy concentrates on those who preached rather than those who did; the whole development of the concept of mechanical work as introduced, named, and put to concrete use in the great papers of EULER and DANIEL BERNOULLI on mechanisms is slipped over by depreciating references to less important writings by those authors, and the one man other than JOULE who really knew what he was doing in proposing the concept of conservation of energy in 1843, namely, J. J. WATERSTON, is not even named.

It is a standard claim of scientists that most historians of science do not have sufficient grasp of science itself to understand the facts rather than the mere circumstances of its history; professional historians, on the other hand, in the words of I. B. COHEN (page 376), are wont to complain of the attempts of "the scientist, whose approach to history often suffers from the consequences of a purely scientific training." Of this latter defect, no examples occur in this book, but of the former, enough. Perhaps the key to the matter may be found in the papers by GILLESPIE and WILLIAMS. These clear, compelling, well documented essays leave the difficult matter of scientific discovery altogether aside and devote themselves to what the politicians and the rabble of the French revolution thought science ought to have been. If the science of the West may simply but justly be distinguished from the sciences of other places and ages, it is in being *mathematical*. It is by the steady, irresistible increase of the power and breadth of mathematical methods applied to the study of nature that we are today, for better or for worse, what we are. That mathematics now influences the daily survival of the most distant and ignorant person, does not mean that he is aware of what is happening to him, much less that he knows any mathematics. Mathematics has always been the preserve of a select few, usually belittled and often punished not only by the masses but also by their greatly more numerous qualitative colleagues. In the two papers on the origins and the effects of the popular attitude toward science in the Revolution we may read the first example of the now-familiar pattern of a "people's" government suppressing "pure" science in favor of the "useful", against the equally familiar background of the stupidity and cowardice of the scientists, especially the typically "arrogant" mathematicians. An attitude toward science much the same as that of some of the Jacobins had been expressed fifty years earlier by FREDERICK II in his management of his Academy, but the moderation of even an absolute monarchy kept the battle bloodless and localized. Nevertheless, within twenty years after

France had beheaded LAVOISIER it stood at the head of Europe in science, led by a school of mathematicians the like of which had never been seen before. Such is the influence of revolutions and politics on the course of thought.

Note for the Reprinting

This review of *Critical Problems in the History of Science,* edited by MARSHALL CLAGETT, Madison, University of Wisconsin Press, 1959, was first published in *Manuscripta* **5** (1961): 101–103. In reprinting it I have restored a passage censored by the editor of that journal.

I am no longer sure that JOULE had any clear idea of the principle of energy in 1843.

17. DUGAS' *HISTOIRE DE LA MÉCANIQUE* (1953)

The view of the history and meaning of mechanics implied by most textbooks is derived from the quite unreliable *Science of Mechanics* by E. MACH[1], which overlooks the deeper conceptual problems, dismissing them as mere "mathematics". There have been few subsequent attempts to treat the subject at large. Among them should be mentioned the [. . .] history of physics by E. HOPPE[2] and DUHEM's original but rather overdrawn *Les Origines de la Statique*[3]. To correct the gross errors of most of the historical statements in current books the best course is to return to the sources. Here E. JOUGUET has done great service in making many short extracts available in his anthology[4].

DUGAS has taken up the monumental task of reading all the sources and forming a new picture of the entire development of mechanics from ARISTOTLE's *Physics* to the quantum mechanics of the 1930s. His work is divided into five books. The first four, each of about 100 pages, consider the "precursors", ending with KEPLER; the formation of classical mechanics, from STEVIN to VARIGNON; the organization and development of principles, from JOHN BERNOULLI to LAGRANGE; and selected, characteristic later work, from LAPLACE to DUHEM. The fifth book, 180 pages long, is about equally divided between relativity and quantum mechanics. The material is arranged in approximately chronological order.

To give a more detailed summary of what the reader will find in this work is not feasible. It will be necessary to restrict attention to DUGAS's method and view of history, besides taking up a few particulars.

---

[1] E. MACH, *Die Mechanik in ihrer Entwicklung historisch-kritisch dargestellt*, 9th edition, Leipzig, Brockhaus, 1933; translation of the 5th edition, Chicago, Open Court, 1942.

[2] E. HOPPE, *Geschichte der Physik*, Braunschweig, Vieweg, 1926.

[3] P. DUHEM, *Les Origines de la Statique*, Paris, Hermann, 2 volumes, 1905/1906.

[4] E. JOUGUET, *Lectures de Mécanique*, Paris, Gauthier-Villars, 2 volumes, 1922.

On pages 623-625 we find:

> To write history is above all to select . . ..
>
> In the course of this book I have quoted the original texts abundantly, not commenting upon them except to clarify them where it seemed to me, rightly or wrongly, to be needed . . .. The essential is to put the reader . . . back into the climate of the century and onto the paths, scattered with pitfalls, that the discoverers *did follow.* That I stress, for even in the eighteenth century Clairaut in his textbooks spoke of the paths that the inventors *could have followed.* This easy way has nothing to do with history. Shall I go so far as to say that I prefer the earliest classics, difficult as they sometimes are to read, just because of the trouble they give us to make contact with a new idea? . . .
>
> Likewise I have dispensed with philosophizing about the principles of mechanics on the margins of history. That field offers a certain interest, but it puts on the stage the critic in place of the actors in positive science. The personality of the historian risks becoming cumbersome, his true role being to select and to appraise. Here and there I indulged myself in a few appraisals, which I trust the reader will forgive me, but *most of the time I have left the reader to make his own judgments* [my italics]. The discussions I have retraced are due for the most part to the actual creators. They have a positive character insofar as they have announced or simply allowed an extension of the principles. The periods in which science confines itself to exhausting the consequences of well defined premisses are periods of latent incomprehension. For want of having to question the basic points of departure, we end up slumbering in deceptive security. . . .
>
> I make no claim to convince those who as a matter of principle regard the history of science as an obsolete cult and think that each new generation should start off in positive science as fast as possible, without a single look backward. . . . Nothing is futile in science, not even contemplation of the past. It is the past that reveals the lesson of our roundabout ways, our scruples, our illusions, and our mistakes. Science has never progressed in that harmonious march we can easily imagine after the fact. Direct acquaintance with old works . . . only enriches the perspectives of the future that opens before us.

To this I can add only that DUGAS has followed his program with scrupulous reserve and entire success. The original analysis is presented faithfully but in condensed form, with frequent quotations. The symbolism is often an intelligent compromise between the original and that now customary. Every line is appropriate, instructive, and a pleasure to read. In numerous cases an evaluation of one master's work by another is quoted. DUGAS' own comments usually explain a difficult passage, add historical background, or point out subsequent usage. His rare critical remarks are penetrating and sometimes urgently necessary, as when he writes "The Renaissance will weaken mediæval mechanics by instigating return to Classical traditions", and when he points out the metaphysical bias of MACH, the declared enemy of metaphysics.

This work is the first single volume from which it is possible to learn what rational mechanics really is. At a time when mechanics is reduced by one group of scientists to a mere subset of the theory of differential equations, by another to a few carelessly stated empirical rules and formal manipulations filling an introductory chapter in a text on quanta, this book presents positive evidence that both these views are false. Needless to say, it describes the entirety of mechanical objects: not only mass-points, but bodies in the general sense, including fluids and solids. Kinetic theory, statistical mechanics, and thermodynamics are excluded. The intersection with MACH's book is nearly null. Experiments are discussed only when they are really relevant [to the bases of theory], as in the case of FOUCAULT's pendulum and gyroscope. But it is particularly instructive to learn from the actual historical data that in the development of relativity and quantum mechanics experiment played the same relatively minor, but of course necessary, part as it did in the history of classical mechanics; that all parts of mechanics have been wrought from experience by intellect, not measurement.

I hazard the conjecture that MACH's empiricism and its successor, operationalism, have run their destructive course. Further, that just as it is a characteristic feature of the modern view of the arts to present them in full consciousness of their history, so also in the mathematical sciences the historical view is to be the view of the next half century. The present book is a herald of this new trend. Destined no doubt to be ignored today, it may well be the handbook of future students of mechanics. (In a related field, reference should be made to the even more scholarly work of J. R. PARTINGTON[5].)

Lest the foregoing appear adulatory, I now point out a few of the numerous omissions and resulting inaccuracies of the work. A better case could be made for ARISTOTLE as a scientist, and even as a

---

[5] J. R. PARTINGTON, *An Advanced Treatise on Physical Chemistry*, London *etc.*, Longmans Green, 3 volumes, 1949/1952.

physicist[6]. In stating that the analysis in D'ALEMBERT's *Essai . . . de la résistance des fluides,* Paris, 1752, is "so long and so full of difficulties I cannot think of summarizing it here", DUGAS misses an important point. In the first place, D'ALEMBERT's analysis refers almost entirely to axially symmetric flows, not to plane ones, contrary to the statement made in historical works and repeated by DUGAS. Second, D'ALEMBERT gives two derivations for the dynamical equation. In the first (§ 45), based essentially on the BERNOULLI equation for streamlines with the added assumption of uniform speed at ∞, he obtains the condition of potential flow. In the second (§ 48), based on "D'ALEMBERT's principle", he obtains a more general result, from which he fallaciously concludes that the flow must be a potential flow. Thus D'ALEMBERT himself gives an example of the unreliability of his "principle" in cases where the answer is not known in advance by other means. In discussing D'ALEMBERT's paradox, DUGAS does not tell us that EULER[7] had discovered and proved and published it more directly seven years earlier. DUGAS's attribution of the "Lagrangean description" to LAGRANGE rather than to EULER is inexcusable, since correct information in this regard is available even in the standard treatises. DUGAS presents as due to LAGRANGE in 1788 the definitive proof that potential flow is only a special case, while in fact the very example LAGRANGE used is taken from a paper of EULER published in 1757. In presenting LAGRANGE's own proof of the velocity-potential theorem, DUGAS neglects to point out that it is false. LAGRANGE's only original contribution to the principles of hydrodynamics was reconciliation of EULER's equations with D'ALEMBERT's principle in variational form, an analysis (1761) DUGAS does not mention. Although he devotes a whole chapter to the principle of least action, nowhere therein does he mention that the trajectories compared must have the same energy; while the analysis of LAGRANGE which he reproduces (page 331) employs this restriction, he does not draw attention to it. There is no material whatever on the general properties of oscillating systems, and nothing on the mathematical theory of bodies moving in resisting media; in particular, there is no discussion of Book II of NEWTON's *Principia,* the most original part of the whole work, though also largely incorrect. Three pages on rigid-body mechanics seem hardly sufficient; and the notion of torque is

---

[6] *Cf. e.g.* I. B. COHEN, "A sense of history in science", *American Journal of Physics* **18** (1950): 343–359.

[7] See Remark 3 on Law I in Chapter 2 of EULER's *Neue Grundsätze der Artillerie . . . ,* Berlin, Haude, 1745 = LEONHARDI EULERI *Opera omnia* (2) **14**, Leipzig & Berlin, Teubner, 1922.

barely mentioned. DUGAS' selection of CAUCHY's paper on finite local rotation of a continuous medium (1841) indicates finesse; his omission of any account of CAUCHY's discovery of the stress tensor and its properties is close to criminal.

In general, DUGAS appears to have followed JOUGUET's selection too closely. But here too the sin committed is of omission. In presenting the actual material of mechanics he ensures that the page as it stands is good without exception.

Note for the Reprinting

This review of R. DUGAS' *Histoire de la Mécanique*, Neuchâtel, Editions du Griffon, 1950, was first published in *Mathematical Reviews* **14** (1953): 341–343.

Perhaps the reader of 1984 will be taken aback by a review so primitive as the foregoing. In 1953 I was altogether naive in presuming that anyone who wrote such a book would have "taken up the monumental task of reading all the sources". I had not yet learned the ways of professional Historians of Science, for most of whom the primrose path of sloth is to let old authors speak for themselves. A modern judgment not passed for whatever reason, inability to comprehend the source quoted being a perfectly good one, is indeed easy to withhold! One might as well praise a eunuch for his continence.

To write history is indeed to choose. My later studies of the history of mechanics, not only more inclusive than DUGAS' but also more specific, confirm and extend my early conjecture that DUGAS' choice followed JOUGUET's too closely. Worse, JOUGUET's reflects the persisting, pernicious influence of LAGRANGE's historical notes. Except for GIBBON, I can name no historian from the late eighteenth century who tried to discover and study all the sources available rather than merely cull and scan them; I can name none who did not infuse his subject with a glow of struggle toward "progress" and relapse into "decline". LAGRANGE noted the latter only by ellipsis. Today the sciences alone permit such a view of history to be even tenable, but LAGRANGE destroyed his own position by his bias toward himself in his experience as a competitor who usually came in second.

Nevertheless, DUGAS's book remains useful. On any topic new to me I consult it before anything else. "The page as it stands" is still good, but many good pages could be replaced by earlier and better ones, and many of the grandest pages of all have nothing but gaps to replace in DUGAS' book.

A "new trend" of the mathematical sciences in the next half century toward "the historical view" has indeed emerged. It is pronounced in researches on rational thermomechanics, which in large measure took up at points where the great thinkers of the last century left off, and which in some cases have looked further back. EULER is not infrequently cited, and so are STOKES and GIBBS and DUHEM—cited, not merely named in the historical penumbra of the textbooks. The "new trend" is less pronounced in other fields of mathematical science, but it is perceptible. The rise of History of Science as a profession has not helped the history of mathematical science directly; indeed, it has impeded comprehension of old mathematics except as a desiccated specimen for use in scholasticism, "case studies", and imaginative anthropology of *homo sciens*; but by its distortions of mathematical truth it has excited in reaction some excellent mathematicians to interest themselves in mathematics of the past. I can only hope that all of those mathematicians and all of their students will in time learn that history, too, has a standard of rigor.

18. JAMMER'S *CONCEPTS OF MASS IN CLASSICAL AND MODERN PHYSICS* (1963)

This booklet, to which a prize has been awarded, succeeds two others, *Concepts of Space* and *Concepts of Force*. The titles suggest that a reader of all three would understand all there is in the science of mechanics. More likely, however, he will end by deciding that mechanics is not understandable. While a physicist writing on the history of physics usually tells us what he thinks the old scientists must have thought, and a historian tells us whom they knew and what books they read, Professor JAMMER tells us mainly what they *said* they did. He tells it to us, moreover, in the jargon of the philosophy classroom today. For example,

> theological reasoning and scholastic cogitation had a decisive impact on concept formation (page 40) . . ..
>
> [R]eviewing . . . the various phases of LEIBNIZ's concept of mass in their formulations corresponding to the different aspects of his philosophical system and its evolution from extensionless atomism to energetic dynamism, it is not difficult to understand that for the philosopher it is a matter of great originality and for the scientist a matter of methodological deficiency (page 80). . . .
> If, however, the "metaxiomatic" requirement of correspondence between primitive notions (at the formal axiomatized level) and observables (at the operational, empirical level) is stipulated—a requirement that naturally has no analogue in the axiomatization of purely mathematical theories—then the concept of mass becomes necessarily a definiendum in the formalized system (page 112).

Certainly every student of mechanics will be glad to use Professor JAMMER's book as a list of scrupulous quotations and citations, in roughly chronological order, and will thank him for finding many good passages previously unnoticed. A history apparently was not intended, since, despite critical remarks here and there, Professor

JAMMER seems to be content with quiet juxtaposition of conflicting opinions. We read what was thought about mass not only by NEWTON and HERTZ but also by ALOIS HÖFLER and CLÉMENTICH DE ENGEL-MEYER. Now if Professor JAMMER had found that HÖFLER and DE ENGELMEYER, although forgotten today, in fact *did* something important in mechanics, everyone should congratulate him on his success as a historian, but when he merely tells us what they *thought*—HÖFLER is quoted as saying, "The tonomonic quantity 'dyne' precedes logically the notion of 'one gram mass' ", and DE ENGELMEYER, "our daily experience prepares us much better for the comprehension of the notion of force than of mass . . ."—then we may well ask, who cares? Indeed, if it had been HUYGENS and EULER who had made the statements just quoted, a historian would do neither their memories nor his readers any service by perpetuating these flat vacuities. Professor JAMMER's democratic content with a name as a name reaches a climax with an appeal to authority on page 110, the authority being a committee of eight "outstanding theorists of mechanics" in Italy in 1907. It is fitting that the eight names are given, but of their deliberations it is recorded only that they failed to reach agreement. Six of the eight names I had never seen before, and so far as I know, none of these experts ever contributed anything to mechanics. Professor JAMMER concludes, "A unanimous agreement on how to introduce the concept of mass in courses on mechanics has not been reached and, in fact, remains a question of some debate today."

1907 is a good date to mention, since the book shows little sign of contact with "classical" mechanics since that date. While I would not presume to criticize the chapters on mass in relativity and quantum mechanics, that on "axiomatized mechanics" gives the unwary reader the idea that mechanics is become the property of pædagogues, philosophers, and logicians. Thus, on page 120,

> In contrast to a purely hypothetico-deductive theory, as for instance axiomatized geometry, where primitive notions (like "point," "straight line," and so forth) can be taken as implicitly defined by the set of axioms of the theory, in mechanics semantic rules or correlations with experience have to be considered and a definiendum, even if defined by an implicit definition, must ultimately be determinable in its quantitative aspects through recourse to operational measurements.

That is simply nonsense. If a physicist says, "I take a sphere of one inch radius weighing one pound," why is only the pound and not the sphere or the inch in need of "operational" definition? Cannot the sort of person who derives comfort from "operational" definitions manufacture them for geometry, too? And when we are told, "MACH did not say what 'mass' really is but rather advanced an implicit definition of the concept relegating the quantitative determination to certain operational procedures," are we really expected to find any meaning here, or is it just a smooth transition to the next chapter in a sociological essay?

Professor JAMMER is scarcely more successful when he becomes concrete, as for example in the passage on mass as a tensor on page 132, which is pure talk since not a word is said to specify the space-time geometry and the group of transformations under which various quantities are asserted to be "contravariant" or "covariant"[1]. Perhaps this talk serves "pour épater le bourgeois"; in any case, it is evidence of the current opinion that mathematics is unnecessary for the historian of science.

In order to write the history of a subject, one must know what it is. To find out what mechanics is, Professor JAMMER seems to have gone to the physicists. One might as well get a definition of anatomy from a surgeon, or of logic from a topologist. Indeed, the physicist, the surgeon, the topologist have learned, after a fashion, mechanics, anatomy, and logic, each in his student days, and each uses some parts of his student's knowledge in his daily practice. This use, however, does not make him a modern expert. Professor JAMMER seems to be unaware that in mechanics, too, there are modern specialists whose ideas do not coincide with his. While his impartiality embraces every philosopher who ever dropped a word about mass, it does not extend to the one modern expert who has advanced a system of axioms[2]. This difference reaches backward, for mechanics *narrowed* after the eighteenth century. Down to, say, 1850, no expert on mechanics would have considered adequate a treatment resting, like Professor JAMMER's, on the tacit assumption that "mass-point" and "body" are one and the same. Professor JAMMER mentions EULER's early work (pages 87–89) but gives not one word to his life-long study of the distinction of mass from inertia, culminating in his great papers on

---

[1] If classical space-time is presumed, there is no need to "require" a metric since the Euclidean metric is already there, and any tensor is covariant or contravariant or mixed, as convenience may dictate, and the whole passage is tautological. If some non-Euclidean space is in question, which is it? And if mass is a tensor, why should it be positive-definite, as the author seems to assume? *Etc., etc.*

[2] W. NOLL, "The foundations of classical mechanics in the light of recent advances in continuum mechanics", pages 266–281 of *The Axiomatic Method, with Special Reference to Geometry and Physics*, Amsterdam, North Holland Co., 1957. Professor JAMMER describes the contents of logicians' papers presented at the same symposium. [NOLL's paper is reprinted in his *Foundations of Mechanics and Thermodynamics*, New York, Springer-Verlag, 1974. *Cf.* also Essay 39, below in this volume.]

the center of mass of a deformable body and the rotary inertia of a rigid body[3]. Also in consequence of Professor JAMMER's oblivion to space-filling bodies, there is not a word about the independence of mass from volume, as developed especially by CAUCHY. *A fortiori*, the reader is left unprepared for modern generalizations of the concept of moment of momentum[4].

This booklet will be useful, but its display of erudition may delude readers into believing it really covers the subject set out in its title.

Note for the Reprinting

This review of MAX JAMMER's *Concepts of Mass in Classical and Modern Physics*, Cambridge, Harvard University Press, 1961, first appeared in *Isis* **54** (1963): 290–291.

---

[3] L. EULER, "Recherches sur la connoissance mécanique des corps", *Mémoires de l'Académie des Sciences de Berlin* [**14**] (1758): 131–153 (1765), and "Du mouvement des corps solides autour d'un axe variable", *ibid.* 154–193. [Both memoirs are reprinted in LEONHARDI EULERI *Opera omnia* (II) **8**.]

[4] *Cf. e.g.* E. HELLINGER, "Die Allgemeinen Ansätze der Mechanik der Kontinua", pages 602–694 of Volume **4/4**, *Encyklopädie der Mathematischen Wissenschaften*, 1914; see especially § 5d. Further generalizations are proposed in a number of papers by J. L. ERICKSEN and R. A. TOUPIN in recent numbers of the *Archive for Rational Mechanics and Analysis*.

19. CLAGETT'S *THE SCIENCE OF MECHANICS IN THE MIDDLE AGES* (1961)

In his *Méchanique Analitique*, published in 1788, LAGRANGE included a history of mechanics which dominates the historical remarks and attributions still current in the teaching of physics. Most of the early references in MACH's romance, *Die Mechanik in ihrer Entwicklung, historisch-kritisch dargestellt* (1883) are those given before by LAGRANGE, and it is from notices and footnotes in MACH's coloring that the history of mechanics is inferred by students today. Let anyone who doubts this try to convince a physicist that the laws of uniformly accelerated motion were well known in the Middle Ages. LAGRANGE said of ARCHIMEDES and GALILEO that "the interval separating these two great geniuses disappears in the history of mechanics." In 1675/1676 NEWTON had written to HOOKE, "If I have seen further it is by standing on ye sholders of Giants." In the century between NEWTON and LAGRANGE, the memory of the giants had shrivelled until even their names were forgotten. NEWTON's statement has been interpreted as an early example of the false modesty which is now required of scientists' locution, but NEWTON was a man loth to say more when less would serve.

Only to one can it be given to discover a whole period of history, and for mechanics in the Middle Ages this one was DUHEM[1]. While to the physicist in the laboratory or classroom the mediæval epoch is still a vanishing interval, DUHEM made it for the historian of science or the student of thought a period of rich and intense analysis and creation. The concepts typical of the western approach, namely: function, inertia, and assignable force, owe their origin to the Middle Ages and are

---

[1] On page xx CLAGETT calls attention to two earlier students of mediæval mechanics, which they regarded as being "merely an offshoot of Greek mechanics and not as an object of independent research": CHARLES THUROT, who in 1868/9 published a history of the principle of ARCHIMEDES, and GIOVANNI VAILATI, who in the late 1800s published essays on the history of statics. The latter recognized the importance of JORDANUS DE NEMORE; the former used manuscript sources as well as printed books.

developments of mediæval ideas. As Professor CLAGETT writes, "So rich were DUHEM's investigations . . . that . . . the succeeding study of medieval mechanics has been largely devoted to an extension or refutation of DUHEM's work."

The task of the aftercomers is necessary but not dramatic. As CLAGETT writes, DUHEM in the exuberant disorder of discovery "made extravagant claims for the modernity of medieval concepts. . .." Also, none could gainsay his inferences, since he used manuscript sources, awkward of access even to those who can profit from them, and he cited "only parts of crucial passages" and "only in French translation". This is not to say DUHEM was wrong or unjust; it only shows where the task of consolidation lies.

First, the essential texts must be published, both in Latin, so that the few who are competent to judge interpretations may do so, and in a modern language, so that members of a wider public may get for themselves some notion of what mediæval science was like. The great bulk of manuscript material in European libraries has to be searched also for sources unknown to DUHEM. Finally, more deliberate study must mature a view of the subject.

All this Professor CLAGETT undertakes in the present volume, which may be regarded as a summary of historical researches by Miss MAIER, Professor KOYRÉ, and himself, as well as an anthology of texts. The mediæval authors represented by liberal selections in English, with the Latin also in cases when there was no modern edition, are JORDANUS DE NEMORE, JOHANNES DE MURIS, ALBERT OF SAXONY, TRIVISANO, GERALD OF BRUSSELS, THOMAS BRADWARDINE, WILLIAM HEYTESBURY, RICHARD SWINESHEAD, JOHN OF HOLLAND, JOHN DUMBLETON, ORESME, GIOVANNI DI CASALI, JACOBUS DE SANCTO MARTINO, BLASIUS OF PARMA, FRANCISCUS DE FERRARIA, FRANCISCUS DE MARCHIA, JOHN BURIDAN, MARSILIUS OF INGHEN. In combination with MOODY & CLAGETT's *Medieval Science of Weights* (Madison, Wisconsin, 1952), the present book makes up the main corpus of source material in handy modern print. As such, it will be the invaluable companion of every student of the history of physics or of mediæval thought.

Professor CLAGETT's general conclusions, expressed in the last chapter, are summarized under twenty propositions, many of them quoted from mediæval authors. Some, like BRADWARDINE's Law, $V \propto \log F/R$, are intermediate steps, erroneous principles later to be rejected. Others, like the Merton definitions of uniform speed, of uniform acceleration, of speed in general, and the theorems concerning uniformly accelerated motion, are permanent discoveries, essential elements of the mechanics later developed by GALILEO. Some, like BURIDAN's statements about falling bodies, are self-contradictory.

Others, like BURIDAN's assertions regarding impetus, introduce concepts intermediate between ancient and modern ones. The last, ORESME's principle of relativity, seems the most modern of all but may be in fact a revival or clarification of an ancient doctrine.

Professor CLAGETT generously and justly wishes us to form our own conclusions from the sources, but he knows how hard these are for the uninitiated to follow. Thus the texts are presented as appendices to the ten descriptive chapters, and each text is followed by a commentary as well. The chapters themselves concern particular groups of problems, such as "The application of two-dimensional geometry to kinematics" and "The free fall of bodies". Copious footnotes in addition give the volume a formidably scholarly appearance, sustained by the scholarly style of the writing itself. It may seem base to carp at the production of a book so carefully printed and above all so cheap, but something about the spacing and arrangement helps to make it a tome difficult to penetrate. There is plenty of blank paper here and there, but almost none in the margin; all kinds of type are more or less the same size and style, and so apparatus criticus merges into text, and on some pages there is more space between paragraphs than between text and notes or between title and text.

What about the physicist who thinks mechanics stopped dead between ARCHIMEDES and GALILEO? Can we put this definitive book into his hands? I fear we should do mediæval science little service. Professor CLAGETT's work cannot be too highly praised as reading for the initiate, but it is hard reading. Coming to it with a predilection for the subject, I read every word, but in small doses, fought out against many a yawn. Despite Professor CLAGETT's immense erudition, it might still be better counsel to the physicist to read DUHEM, exaggerations and inaccuracies and all.

It would of course be unfair to expect of any historian the genius which shines from behind DUHEM's writings. DUHEM was not only the discoverer of mediæval mechanics; he was also a creator himself, and a great one, in rational mechanics, theoretical physics, and physical chemistry. Such a man will sometimes jump to a conclusion that must later be abandoned; he may commit slips in translation, and he will not edit texts. He gives us, however, a depth and a grasp that comes from the habit of creative thought; sometimes, because he knows how scientists think, he comes closer to the creator than does a more painstaking, scrupulous historian. Even in the gross carelessness of MACH we sometimes see flashes of historical insight. In this day when scientists, each in his little, tightly organized field, must conform with the accepted norm of a "real" scientist, and historians of science, likewise organized and eager to promote specialized departments for themselves, are establishing equally tight social norms, it may be futile to try to draw closer to the unity of man, thought, and nature that was the Middle Ages.

But there is something else I miss in Professor CLAGETT's book. It is about the Middle Ages, but a Middle Ages different from any other I have encountered. JORDANUS DE NEMORE wrote when WALTHER VON DER VOGELWEIDE had sung and JOINVILLE was crusading with the Saintly King; San Zeno at Verona had been standing for a century, and, fifty years before, ANTELAMI had finished the *Deposition* at Parma; not long after the Merton kinematical theorems had been proved and JOHN BURIDAN had completed his lectures on dynamics at Paris, GUILLAUME DE MACHAULT wrote his *Coronation Mass*. This life, this color, this drive, is the Middle Ages, but it is not to be found in Professor CLAGETT's book. Does that mean that science in the Middle Ages was really a dull university game, thought divorced from life and art? It would be most unmediæval to think so, and I do not.

Rather, I conjecture that there is much of mediæval science still to be discovered. When many more admirable compilations and close studies such as those of Miss MAIER and Professor CLAGETT shall have been made available, perhaps the picture of scientific thought as a part of mediæval action will come to life.

Note for the Reprinting

This review of M. CLAGETT's *The Science of Mechanics in the Middle Ages*, Madison, University of Wisconsin Press, 1959, was first published in *Speculum* **36** (1961): 119–121.

The statement by LAGRANGE which is quoted in the first paragraph is from his *Méchanique Analitique*, 1788, Seconde Partie, Septième Section (Section X in the second and subsequent editions).

20. STEVIN'S WORKS ON MECHANICS
(1957)

STEVIN is one of those to whom we owe the regrettable custom of publishing in a vernacular rather than a learned language. Therefore his works are mainly known only in SNEL's translation into Latin. Instead of making this acknowledged version available again, the present edition provides a new translation into English which may well be accurate but certainly is not fluent. Comparison with the originals in Dutch, which are reproduced on the facing pages, shows how much the art of printing has regressed since STEVIN's day. Short, careful notes are adjoined.

STEVIN's work in mechanics is limited almost entirely to statics. Not only does it display a thorough understanding of the law of the lever, but also we find an explanation of statically indeterminate cases, as for example suspension by three or more cords in the plane, four or more in space (pages 543–547). As is generally known, STEVIN's work on hydrostatics is superior to PASCAL's in content but not in exposition. Compared with the hypothetical, perhaps even philosophical GALILEO, STEVIN is seen to be a practical scientist. Even if only a small part of the volume concerns experiment, which is always conscientiously separated from theoretical considerations, every page witnesses to a solid footing in experience and experiment, an aspect that in most works of the sixteenth century and its successor we seek mainly in vain. Nevertheless, far from being an empiricist, STEVIN requires and attempts rigorous mathematical proofs starting from explicit hypotheses. More than anyone else of his time, he is the prototype of [the man of science in science's great floraison, 1600–1900, the adherent to DESCARTES'] "experience and reason". It is understandable that he often fails. More remarkable than his abundant success is the keenness of his thought and the clarity of his writing—the latter in notable contrast with GALILEO, whose literary style is certainly more artful.

In a paragraph (page 511) that seems to have escaped general notice, STEVIN writes that in the year 1586 he and JAN DE GROOT "let two spheres of lead, one ten times as large and heavy as the other . . . , fall from a height of thirty feet onto a board or something else that made a noticeable sound." They found that "they hit the board so nearly at the same time that their two blows seemed to make one and the same rap." This experiment is traditionally and wrongly ascribed to GALILEO between 1589 and 1592.

On page 515 we read that "the art of weighing "[*i.e.*, statics] is "a separate, independent branch of mathematics." Next comes an especially clear distinction of geometry, arithmetic, and statics, and a treatment of the relations among them. Here STEVIN shows an insight into the nature and function of theoretical science and especially the connection between rational mechanics and experimental mechanics that many modern scientists seem to lack.

Note for the Reprinting

This review of *The Principal Works of Simon Stevin*, Volume 1, *General Introduction and Mechanics*, edited by E. J. DIJKSTERHUIS, Amsterdam, Swets & Zeitlinger, 1955, is translated from the German original appearing in *Physikalische Blätter* **13** (1957): 578–579.

21. DUGAS' *LA MÉCANIQUE AU XVIIe SIÈCLE* (1956)

Having already given us a general history of mechanics from Greek times until the quantum theories of the interbellum[1], DUGAS now sets himself a task not only more limited but also essentially different. Selecting the former of the two greatest centuries of mechanics, he attempts to show how not only the science of mechanics but also the mechanistic views of science and philosophy developed and influenced one another at that time. Relatively little concerns the solution of specific mechanical problems—in DUGAS' term, "positive" mechanics. The balance between the concepts of mechanics and the relation of those concepts to experience and human thought is such that philosophers will regard the subject as mechanics, scientists as philosophy. In my opinion a measured and critical reading of this book will do good to all parties and perhaps by suggestion reveal some of the sins of today's cellular sciences. In fact, the lines of thought opened here have a certain timeliness. When all about us we see the struggles of natural scientists to achieve definiteness and precision, and even, as they sometimes admit, a mathematical framework for their respective sciences, it is natural to turn to physics as the senior and most mature member of the group. In physics, it is mechanics that is the prototype of a precise, mathematical theory, and it is safe to guess that for every time a nonmechanical principle or formula of physics is used consciously, a mechanical one has been used five times. It was in the seventeenth and eighteenth centuries that this great science reached much of its present clarity, definiteness and extent. How did this happen? On what fundamental experiments does it rest, and how did the theorists use the results of these experiments?

I do not intend to parrot or paraphrase the elegant and sweeping generalities of DE BROGLIE's preface. Such quintessences are cozy comfort for the common reader who is loth to read, but to replace one overstatement by another with a somewhat better choice of personal

---

[1] RENÉ DUGAS, *Histoire de la Mécanique, des origines à nos jours*, Neuchâtel, Editions du Griffon, 1950. [See Essay 17, above.]

names brings us little nearer to the truth. Modern mechanics is the creation of many, including some whose names are unfamiliar, and it is the merit of DUGAS to let them speak for themselves in fairly extensive excerpts. At the same time he adds brief and clear explanations of passages which a reader not accustomed to old science would have trouble following, and also some very short interconnecting statements and summaries of works not quoted. Personalities and anecdotes are almost entirely absent. The book is beautifully written and is so interesting that it is difficult to set aside. It is unfortunate that there is no subject index, especially since the order followed is mainly chronological.

There are nineteen chapters: Antecedents, KEPLER, STEVIN, GALILEO, MERSENNE, GASSENDI, DESCARTES, PASCAL, the CARTESians, HUYGENS, DESCARTES to NEWTON through the English school, NEWTON, CARTESians and the continental reception of NEWTON, LEIBNIZ, GALILEIan dynamics, NEWTONians and CARTESians (three chapters), Conclusion. DUGAS in following his excellent plan of letting the creators speak for themselves stands aside from the several schools of current opinion regarding the origins of mechanics. He is aware of their existence, and his selection tacitly reveals the influence of recent critical thought in his devoting less than thirty pages to GALILEO, less than 100 to NEWTON, but over eighty to DESCARTES and at least fifty more to the CARTESians. This is, I believe, the first book suitable for the general scientific reader to give a real idea of what DESCARTES did for physics. It is not enough to swallow the simple formula: DESCARTES was the first to conceive nature as a single great machine subject to a common set of mechanical laws, and his great attempt to create a theoretical physics, while in every detail a failure, established the task and scope of Western physics. To the best of my knowledge, the foregoing statement is true, but in itself it carries little meaning unless fortified by experience with DESCARTES' writings. Through selected passages from letters and published treatises, DUGAS guides us in following the subtle development of DESCARTES' views. The reader whose background is in positive mechanics will at first lay all to BEECKMAN's credit, the BEECKMAN who was usually right when DESCARTES was wrong. It will not lessen such a reader's ju t admiration for BEECKMAN to learn to understand DESCARTES' impatience with detail, even correct detail, when principles are at stake. At the same time, DESCARTES' manner of the swaggering duelist in scientific controversy will temper our respect and admiration for his intellect with distaste for his manners toward the work and the persons of other scientists, as reflected in the excerpts here.

Another illuminating chapter, sixty pages long, presents the views of LEIBNIZ and helps to explain why the guide lines laid out by this great philosopher, who has joined DESCARTES in the oblivion which positivist historians have put between themselves and the origins of mechanical concepts, dominated much of mathematical physics for half a century.

Coming to the subject from the more recent side, to me the failing of DUGAS' treatment is that, while he takes pains to trace the sources and currents of seventeenth century thought, he gives little indication of where it was going. At the end of the book we find ourselves still in the heroic age of mechanics, furnished with numerous half formulated and half organized principles, along with a few isolated problems well solved, but without the general equations of any single domain of mechanics. Mechanics as we know it today was formulated mainly in the rational eighteenth century and in the first half of its [professionalized] successor. As to what was left for these periods to do, and what views of the seventeenth century were later to be refined or revised, DUGAS gives us no indication. For example, all of DUGAS' treatment of NEWTON concerns the origins of his thought and the aspects of his work which aroused the most immediate contemporary acclaim and opposition. Nearly half of the *Principia*—and to me it seems the more original half—concerns fluids, but of this DUGAS gives us no idea beyond a brief reference to motion in a resisting medium. It was NEWTON's brilliant but faulty analysis of the resistance offered by rare and dense fluids, the figure of a fluid earth, the propagation of surface waves and sound waves, the oscillation of water in a tube, and efflux from a vessel that gave rise to correct solution of all these problems in the eighteenth century. Now this is not mere detail, as some [physicists] would have us believe. For correct solution of these problems, it was necessary to forge the concept of *field*. That some of the giants of whom DUGAS writes made some progress on field problems without the benefit of the field concept, is proof of the astonishing depth and virtuosity of the seventeenth century—but to learn of it, one must look elsewhere than in DUGAS' book.

One might at first find some inconsistency between this neglect not only of NEWTON's fluid mechanics but also of STEVIN's, to which DUGAS gives less than seven pages, and on the other hand the detail regarding PASCAL, whose additions to positive mechanics were trifling. Here we must recall that DUGAS is tracing the history of ideas, and that PASCAL was read easily by persons who could not understand STEVIN or NEWTON. Indeed, DUGAS appears to set value on PASCAL's activity as a publicist, attributing to him "the rejection of authority of any kind", but others may interpret PASCAL's independence as reluctance to acknowledge his sources.

For an example of the strength of DUGAS' view and method, consider his comment on NEWTON's concept of space and time (page 349):

> A positivist would say that Newton is content to affirm the existence of these two absolutes and in his strictly mathematical objective refrain from going further into metaphysics. Instead, it might be thought that Newton simply accepts the metaphysical legacy of his predecessors.

The last sentence is borne out by earlier sections of the book, and DUGAS proceeds to give specimens of NEWTON's metaphysical statements. For an example of weakness, consider the comment on NEWTON's First Law (page 353):

> Even the examples invoked by Newton in support of his First Law: projectiles, ..., tops, planets, and comets—show that here is not the *principle of inertia* in the sense now classic but rather a far more general idea.

What DUGAS fails to tell us is that nowhere in the *Principia* does NEWTON make any use of this more general notion, that neither NEWTON nor any other savant of his century succeeded in formulating the concept of rotary inertia or in solving any typical problem concerning the motion of a rigid body. Referring back to the chapter on HUYGENS, we find a corresponding failure concerning pendulous bodies. DUGAS tells us that HUYGENS' only axiom is the equality of the ascent and the descent of the center of gravity, while "all the rest is geometry, the most elegant and exact that could be." Now that cannot be true. Conservation of energy is sufficient, with certain approximations, to prove the isochrony of small oscillations, but without some idea equivalent to rotary inertia it is impossible to calculate the period. DUGAS spends a page listing the special cases by which HUYGENS groped toward the final solution; he mentions the names of two of HUYGENS' methods but does not tell what they were; and he concludes that he cannot say anything more about it "within the bounds of this study", referring the interested reader to "the learned edition of the Dutch Society of Sciences". This problem, to my mind, marks the beginning of modern mechanics, and I have always wondered what HUYGENS really did, especially since what is usually attributed to him is on the one hand different from anything else from the seventeenth century, on the other, so simple that it would be hard to conjecture any basis for the objections which, as DUGAS tells us, were set against his solution by other great savants of the period. The only help I can derive from DUGAS' pages is his list of references to certain passages in Volume 16 of the works of HUYGENS. DUGAS' frequent mention of the elegant *geometry* of HUYGENS arouses my suspicions: I am reminded of the elegant geometry of ARCHIMEDES, who in his work on hydrostatics drew unwarranted and in fact sometimes untrue dynamical conclusions from purely statical demonstrations.

Perhaps DUGAS' lack of interest in the details of HUYGENS' analysis indicates that he regards it as positive mechanics and hence outside his scope. If so, I find it difficult to agree, for I think that EULER's concept of the rotary inertia of a body is as profound an addition to the principles as is the general concept of mass, which DUGAS regards as one of NEWTON's innovations.

Similarly, DUGAS gives no emphasis to LEIBNIZ's Law of Continuity. On page 486 we find it stated and supported in the midst of a sequence of long quotations from a work written in 1691 but not published until 1860. Earlier there are some oblique statements regarding it. On page 490 is a quotation from a short paper of 1695, in which it is stated prominently but not explained. So far as I know, SPEISER[2] is the only writer who has described properly what this law meant as it was adopted and used by D'ALEMBERT and some other savants of the next century: In modern terms, only *analytic* functions can occur as solutions to physical problems. Hence follow not only a general uniqueness theorem and continuation principle but also a denial of what are now called wave motions in field theories. DUGAS' book does not tell us whether LEIBNIZ himself ever applied his concept of "continuity" to any specific problem and thus gives no definite idea of what the principle, as a statement in mechanics as well as in metaphysics, really meant to him.

At the end of the book DUGAS presents his own brief summary of the century, from which one might try to answer the question with which this review began. Here the French language causes trouble in not offering a ready distinction between *experience* and *experiment*. As DUGAS says, most of the scientists of the seventeenth century experimented passionately, often turning naively to experiment as to an oracle. But what did they conclude from the experiments? In summary, little or nothing, except sometimes the inadequacy of some particular assumed dependence. As an exception, there is PASCAL & PERIER's experiment at Puy de Dôme, which may justify some of the space DUGAS gives to PASCAL. Certainly every scientist of the period based all his effort and all his theory upon experience—experience is the theme of every page of his book. But it is experience balanced

---

[2] A. SPEISER, pages XXII–XXV of LEONHARDI EULERI *Opera omnia* (I) **25**, Bern, 1952.

with reason, the "experientia et ratione" of DESCARTES and HUYGENS. Decisive experiments, I conclude from this book, there were none. "Here will be found not science romanticized but the real romance of science." The scientist, at first impatient with the amount of metaphysics in this book, will grow reconciled and finally thankful as his reading progresses, for only in relation to metaphysics can the true "romance of science" be understood. What DUGAS convincingly and accurately presents us is "rough diamonds mined by great labor . . . in the dark night of metaphysical dramas."

Note for the Reprinting

This review of RENÉ DUGAS, *La Mécanique au XVIIe Siècle* (*des antécedents scolastiques à la pensée classique*), Neuchâtel, Editions du Griffon, 1954, was first printed in *Isis* **47** (1956): 449–452.

My explanation of LEIBNIZ's Law of Continuity here makes it too specific. In LEIBNIZ's works I have found statements he regarded as consequences of his Law but no general enunciation of it.

22. COSTABEL'S *LEIBNIZ AND DYNAMICS*
(1975)

Since the French booklet of which this one is a translation was abundantly reviewed[1] when it appeared in 1960, little further review of the contents is needed. Mathematicians are likely to be puzzled by COSTABEL's use of "force" on pages 22–24 and by LEIBNIZ's whole approach to the problem. Students of mechanics are always surprised to learn that LEIBNIZ in 1692 could dismiss unmentioned NEWTON's *Principia* of 1687 and the entirely different concept of force implied therein, since NEWTONian forces as used later by EULER provide the basis for mechanics as it is taught today. Philosophers and historians seem to care little for REECH's attempt to found mechanics explicitly on the NEWTONian idea of force and to be entirely unaware of the recent work of NOLL, which has given NEWTONian forces a logical and mathematical status strictly level with those of point, line, and mass[2]. Historians of the effective side of mechanics always find it difficult to recognize in LEIBNIZ's work on live forces the LEIBNIZ who published brilliant and concrete analyses of the elastic beam (1684) and the catenary (1690), or the LEIBNIZ whose studies of motion in a resisting medium (1689) demonstrate high competence in plane dynamics. COSTABEL takes seriously LEIBNIZ's claim to have "established a new science", which he called "la Dynamique" (page 65, with confusing and perhaps misprinted quotation marks, also page 104 and elsewhere). Since it would hamstring mechanics to restrict it to cases in which the measure of force is mv^2, and since even in the cases when the sum of *vis viva* and *vis mortua* remains constant, that fact suffices to render specific only the very simplest problems, LEIBNIZ's idea scarcely deserves the rank of a science or a system.

---

[1] PIERRE COSTABEL, *Leibniz et la dynamique*, Paris, Hermann, 1960, reviewed by T. DERENZINI in *Physics* **2** (1960): 267–268, by J. E. HOFMANN in *Archives Internationales d'Histoire des Sciences* **14** (1961): 379–384, by J. W. HERIVEL in *Revue d'Histoire des Sciences* **15** (1962): 81–82, by Y. BELAVAL in *Isis* **53** (1962): 533–535.

[2] W. NOLL, *The Foundations of Mechanics and Thermodynamics*, New York *etc.*, Springer, 1974, especially pages 75–81. The theory of systems of forces was first published in NOLL's lectures included in *Non-Linear Continuum Theories*, Rome, Cremonesi, 1966.

To the extent that a translation of vague metaphysical conceits in KOYRÉ's manner can be accurate, this one is so in all cases I have checked, but too literal to be acceptable English, for example in the sentence on page 104 ending "it was necessary that 'dynamics' take account in particular of that compounding of motions, the logical difficulty of which, users such as Lamy and Varignon, had no suspicion." Also, on page 49, "the text . . . confirms the awareness of LEIBNIZ of the necessity he was under of building a logical structure in order to avoid a battle of words . . .." MADDISON has preserved even the royal "we", disagreeable enough in French but absurdly pompous in English: "Our investigation would have been incomplete if we had not had the good fortune to discover . . ." (page 26), no collaborator being implied.

In a later publication COSTABEL[3] has explained why he regards the mechanics of NEWTON as being a "*Mechanica rationalis*, but . . . not dynamics in the sense of LEIBNIZ." It seems to me to be flawed by the same defect as the book under review: failure to comprehend the nature of a mathematical theory of physical phenomena.

Note for the Reprinting

This review of P. COSTABEL's *Leibniz and Dynamics*, translated by R. E. W. MADDISON, Paris, Hermann, & London, Methuen, & Ithaca (New York), Cornell University Press, 1973, first appeared in *Historia Mathematica* **2** (1975): 360–361. For "the recent work of NOLL" see Essay 39, below in this volume.

---

[3] P. COSTABEL, "Newton's and Leibniz's Dynamics" in *The* Annus Mirabilis *of Sir Isaac Newton, 1666–1966*, Cambridge (Massachusetts), MIT Press, 1970.

23. JOHN BERNOULLI AND L'HÔPITAL
(1958)

What did the BERNOULLIs give to mathematics and mechanics? To answer this question, look in the standard histories. You will find tags such as "brachistochrone" and a few odd details, mostly trivial, associated with the names of JAMES I, JOHN I, and DANIEL, besides some generalities to the effect that they were great men. For their personal relations, everyone has heard vaguely of the quarrel between the great brothers, but even a best-seller featuring gossip about mathematicians gives us beyond that only its author's musings on the BERNOULLI family as material for a study of heredity and environment. No considerable biography of any BERNOULLI has ever been published.

Going to the shelves of any good mathematical library so as to form your own judgment from the sources, you will find the collected works of every Victorian with a name: SMITH, BORCHARDT, CREMONA, FUCHS, HALPHEN, HERMITE, SCHWARZ, TEIXEIRA, STEINER, *etc.*, but unless the library is exceptional, you will find *nothing* by the BERNOULLIs. JOHN I BERNOULLI supervised the publication of his *Opera omnia* in four volumes in 1743, the works of JAMES I were published in two volumes in 1744, while the works of DANIEL never have been collected. Few libraries have the journals in which the BERNOULLIs published: the Leipzig *Acta eruditorum* and the organs of the academies of Paris, Petersburg, and Berlin in the eighteenth century. In fact, in a typical working mathematical library in the U.S.A. there will be found *not a single paper or book* by any BERNOULLI. A library now wishing to obtain the journals and the two sets of *Opera* will do so only at great cost and after years of search.

It is thus plain that mathematicians and historians of science are so little interested in the BERNOULLIs that not only is no modern description available, but even the material from which to draw an analysis is to be found only in the largest libraries. The volume discussed below, despite its importance, was given but cursory notice in the two principal organs of review for mathematics in the U.S.A., organs which not long before had published more or less extensive summaries of works on how Babylonians added fractions and of "new" texts on complex variables.

In view of this evident lack of interest, then, little space should be taken up here. I am compelled, nevertheless, as fully as within my power, to employ the privilege of commenting upon what seems to me the most important event in the history of mathematics in a quarter of a century. This event is the publication of the first volume of the great new edition of the collected works of the mathematicians BERNOULLI.

I. PLAN OF THE EDITION

This edition will collect the letters, diaries, unpublished manuscripts, and printed works of the three great BERNOULLIS, of NICHOLAS I and II, of JOHN II and III, and of JAMES II. Naturally enough it will include also the literary remains of JAMES HERMANN, hardly a compelling figure but one who takes on special interest as a pupil of the deep and enigmatic JAMES I BERNOULLI. The editor has traced over 7500 letters from and to these persons, of which over 2300 involve JOHN I and nearly 3000, JOHN III. This brings us at once to the decision faced by the editor. To publish this entire mass would have set the lesser BERNOULLIS in dominance. The present plan is to publish everything left by the three great BERNOULLIS but only a selection from the five lesser ones and HERMANN. The edition, then, will be dominated by JOHN I, whose existing letters connect him with over 100 correspondents, including every great scientist of the day except HUYGENS.

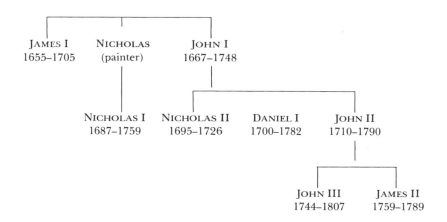

Table (in estimated pages of typescript)

	Previously printed works	Mathematical manuscripts	Courses	Lectures	Letters Unprinted	Letters Previously printed	Biographies	Totals
JAMES I	1300	650	40	30	50	130	200	2400
JOHN I	2100	100	40	280	6480	1230	70	10300
DANIEL	2300	100	100	100	920	280	—	3800
NICHOLAS I	230	250	10	—	1400	90	20	2000
NICHOLAS II	170	—	—	—	70	—	—	240
JOHN II	200	10	—	50	1000	—	40	1300
JAMES II	200	10	—	—	20	—	30	260
JOHN III	200	—	—	40	160	—	1200	1600
JAMES HERMANN	1300	—	80	120	400	200	—	2100
Total	8000	1120	270	620	10500	1930	1560	24000

When it comes to works published before, both papers and books, we find only some 450 from the whole group, making about 14,000 typewritten pages. Again JOHN III is the most prolific, and most of his product is nonmathematical. The three great BERNOULLIS issued about 100 works apiece, 7000 pages in all. Here, as well as in manuscripts, lectures, *etc.*, the editor has decided to print only a small fraction of the output of the lesser figures, taking care to include anything of possible mathematical interest.

Except for the correspondence with LEIBNIZ and EULER, which is left for the collected editions of those scientists, the table opposite is an estimate of the material which this edition is to print. One-half consists of letters, one-third of scientific productions. Three-quarters will be devoted to the preservation of everything concerning the three great BERNOULLIS and HERMANN, and of this about two-thirds for JOHN I; the remainder, a selection from the works of the five lesser savants. This makes a planned edition of 20–25 volumes of 700 printed pages each.

The format and type are smaller than for the great editions of earlier years, but the printing and style are excellent. The cost is low for a work of this magnitude, but the binding is not sturdy enough for a book of permanent value. The present volume contains dozens of facsimiles of the small original line sketches as well as five full page plates with portraits of L'HÔPITAL and PIETER BURMAN and specimens of the handwritings of JOHN I, of L'HÔPITAL, and of his wife.

The edition was started officially in 1935 with the creation of the BERNOULLI Commission under the patronage of the Basler Naturforschende Gesellschaft. The general editor was and is OTTO SPIESS, of whom more below. The financing was originally through small gifts from Swiss industries, and now there is prospect of support from the new Swiss National Fund for Scientific Research. Even if the financial side is in order, the general editor tells us, it is difficult to find competent and willing editors for the individual volumes.

On pages 77–78 SPIESS discusses the relative merits of two possible plans: the *individual*, organized about the persons, and the *collective*, organized about the periods. While he favors the latter as more instructive to the reader, he judges it to be impossible without a more secure financial backing than is assured for the BERNOULLI edition. The individual plan is adopted because it more easily allows the material to be divided among numerous editors whose labor can be extended over many years:

Series 1. Correspondence of JOHN I, in six volumes, followed by one volume apiece for the letters of DANIEL, JOHN II, and NICHOLAS I.

Series 2. Collected works of the two older BERNOULLIS and of NICHOLAS I and NICHOLAS II, followed by two volumes for the letters and works of HERMANN.

Series 3. Works of DANIEL and of JOHN II and his sons, in three volumes.

There is to be a final volume with biography, iconography, and index. The present volume is the first of Series 1. Its contents are such as to serve equally well as a unit for 1690–1705 in an edition organized on the collective plan, which may not be out of the question now that the financial prospect is more favorable than when the work was started.

The editor compares the project with the EULER edition, which, as he reminds us, is but half complete after forty years of steady work. The BERNOULLI edition will contain about two-fifths as much material as the EULER edition.

In respect to critical apparatus, the present volume stands midway between the elaborate commentary of the HUYGENS edition and the bare text of the EULER edition. The correspondences of JOHN I with JAMES I, with L'HÔPITAL, and with miscellaneous persons between 1693 and 1706 are presented separately, each with a special introduction by SPIESS. From a few lines to a half of each page of text is occupied by footnotes. These are most helpful, being sufficient to make the material readable by any patient person who understands calculus and geometry. Sometimes there is a translation of an argument into modern notation; sometimes a missing proof is supplied; the problems discussed in the letters are indexed by numbers and cross-referenced in the notes, which cite all print that relates closely to material in the letters. Missing letters which can be identified are given interpolated numbers, and all known information regarding their content is collected.

OTTO SPIESS has devoted twenty years to the organization of the edition, to the collection of materials for it, and to the editing of this volume. He is now seventy-eight years old. His first product sets a new standard for the editing of mathematical works. Never have I seen an edition so carefully and completely yet unobtrusively and compactly organized, annotated, and cross-indexed. SPIESS's name appears only at the end of his general preface, not on the title page. Not only is such modesty excessive, but it seems to me that for having realized the inception of this great undertaking and for having given to us such important material in so beautiful and useful a form, SPIESS deserves some special tribute from those who cultivate or observe the history of science. I take this occasion to express my heartiest thanks for the generous assistance SPIESS gave to my labor in the midst of his own, when he took the pains to go through enormous files of unpublished material so as to select and send to me copies of everything pertaining to the BERNOULLIS' work on the mechanics of fluids and deformable solids.

II. SPIESS'S INTRODUCTION

The general introduction, eighty-five pages long, explains not only the organization, principles, and plan of the edition but also the history of the manuscripts and of the search for additional material. It is shown that the existing letters make only a small fraction of what was actually written, and reasons for the loss, as well as conjectures regarding its nature, are put forward.

What is especially interesting and instructive, is the fate of the most important existing collection, the great mass of JOHN I's letters. For reasons that will be apparent to the reader of Part IV of this review, it had been JOHN I's plan to publish them in part in his *Opera omnia*, and he sent hundreds of them to CRAMER, the editor. Of these, the 150 letters to and from LEIBNIZ were published in 1745, but the publisher's and editor's means and energy were insufficient for the rest. Eventually the grandsons of JOHN I inherited the manuscripts. In 1790 JOHN III, who revered his family's tradition and showed responsibility toward it but was in reduced circumstances with eleven children to rear, attempted to sell the books and the manuscripts. First he sought a prince or a large library ready to buy the material and publish the letters. Failing this, he tried to induce a press to buy and publish them. It should be added that the prices he asked were most moderate, and that he was ready to adjust them downward in proportion to the amount of material the buyer was willing to publish. All his attempts to secure publication of any part failed. In two lectures, printed in 1803 and 1804 in the *Histoire* of the languishing academy of Berlin, he disclosed that most of the material had been sent to "an illustrious academy of the North", some copies and minor pieces to "the most grand and rich library of a sovereign prince".

Within a few years, the traces of these two great deposits were obliterated. In a sense, the material was never lost. It stayed, safe and untouched, in the libraries of Gotha and the Stockholm Academy. Pages 32–46 of SPIESS's introduction detail the attempts, most futile, of several persons, from the efforts of RUDOLF WOLF in the 1840s down to his own in the 1930s, to locate the missing manuscripts. At various times various persons did indeed know of one or the other collection. Nevertheless, internationally published appeals from those seeking information were fruitless. In particular, ENESTRÖM appears in SPIESS's account as a sort of villain who, it seems, deliberately concealed the existence of the rediscovered Stockholm collection after the Basler Naturforschende Gesellschaft had refused to finance a proposed edition under his direction in 1887. In 1884 ENESTRÖM had founded the *Bibliotheca Mathematica*, which stood under his editorship for thirty years as the principal journal of research in the history of mathematics. Here, over two decades, he spread out small extracts or sections of the BERNOULLI correspondence, mainly with EULER, but apart from a few brief annotations before 1900 he gave no notice of the enormous mass of unpublished material at his disposal. SPIESS, explaining that "he who seeks in the letters of savants of the eighteenth century for new scientific material not in the printed works of the correspondents will generally be disappointed," suggests that ENESTRÖM "had no organ for the particular charm afforded by the personal testimony of great men." Notwithstanding that, it would be misleading to dismiss the contents of the BERNOULLI letters as personal in the narrower sense. Indeed, in my opinion some of the older work on the history of mathematics by confining attention to dry recital of the dates of minutiæ has stifled the interest creative mathematicians of today would naturally feel for the creative processes of their predecessors. Even were it desirable, it would be impossible to explain the tides of thought apart from the pull of the thinkers. In the Enlightenment, personal letters and diaries often disclose the mind and its alternatives more clearly than do the published records, sometimes intended to conceal the origins of results or to avoid antagonizing particular readers.

In any case, the story told by SPIESS shows that during the years when historians of mathematics were most eagerly searching the works of those who in rare examples used ideas of calculus without knowing it, *no one cared* for the splendid flourishing of mathematics at the hands of those whose special pride was the open, explicit cultivation of infinitesimal methods.

III. The Correspondence of James I and John I Bernoulli

The famous open letters from the days of the great quarrel are not reprinted. All that remains are four letters from JOHN to JAMES in 1691, with a partial reconstruction of sixteen more. Here we see JOHN still a student, growing restive and beginning to show his temper and his rebellious independence of his senior brother. The subject is the *velaria*, the profile assumed by a cylindrical sail under certain hypotheses. The correspondence ends just before each brother discovers independently that this curve is a catenary. The letters show rivalry, suspicion, and misunderstanding. The language of JOHN seems rude from the start, but it is possible that even 250 years ago such was permissible between brothers and had to be read with an appropriate tone of voice. A specimen from the third letter follows:

> It is you who wish to defy the world, do you think me blind? If I had the Acta I would show you the place. Would it be possible that I could have by heart for all these years this series giving the quadrature of conic sections, if you had just now invented it? And more, what shows your plagiarism is that you say . . . [while] I say that this is false in general. . . .

The fourth letter, although it contains passages in the same tenor, ends with JOHN's imploring his older brother to recommend him for the professorship at Groningen which in fact he was soon to obtain.

IV. THE CORRESPONDENCE BETWEEN JOHN I BERNOULLI AND L'HÔPITAL

Calculus was made known to the learned world by the brilliant papers of LEIBNIZ and the brothers BERNOULLI from 1690 on. The first textbook of the new science appeared in 1696: The *Analyse des infiniment petits, pour l'intelligence des lignes courbes*, a work of some 200 pages, issued anonymously but known to be by the Marquis de L'HÔPITAL (1661–1704). This book went through several editions and remained a standard for a century. Until the present day it has been considered as a work in large part original, and its famous rule on indeterminate forms is known as L'HÔPITAL's. Beyond this book and a posthumous treatise on conic sections, L'HÔPITAL's works consist in some twenty-five very short notes on special problems. He had become well known through his sixth publication, a four-page note of 1692, giving the solution of DE BEAUNE's problem of the inverse tangent, which had been outstanding for fifty years.

In letters, some of which have been in print for two centuries, JOHN BERNOULLI complained to LEIBNIZ, VARIGNON, and others that most of what was attributed to L'HÔPITAL belonged to himself. He claimed not only the solution of DE BEAUNE's problem and everything else of real interest in L'HÔPITAL's papers but also all except three or four pages of the *Analyse*, which he said was nothing but the first part of the *Course on Differential and Integral Calculus* that he had given or dictated to L'HÔPITAL in Paris. Indeed, it was he who had taught the Marquis the new calculus in 1691, giving him instruction for nearly a year. In his published memoirs BERNOULLI was less positive, though

his claims increased with time after the death of L'Hôpital. Since
Bernoulli was far from reticent in proclaiming his own when
others, even his best friends and closest relatives, were involved, the
moderation and lateness of his accusations against L'Hôpital caused
them to be doubted. Apparently only Leibniz and some Basel
friends believed Bernoulli, and in France his claims were regarded
as ridiculous. In 1742 Bernoulli published Part II, the *Integral Calculus*, of his Course of 1691–1692, but not Part I, whose contents, he
stated, had gone into the well-known *Analyse* of L'Hôpital.

In L'Hôpital's preface is a famous passage in which he expresses
his especial indebtedness to John Bernoulli and asserts that his
book will present the discoveries of various persons without further
acknowledgment. Nevertheless, in the text there are many specific
acknowledgments to half a dozen persons but not one to Bernoulli.

A century ago the correspondence of L'Hôpital with Leibniz and
Huygens was published. Herein may be traced L'Hôpital's own
view, or at least the view he wished his great correspondents to entertain, of his progress in calculus and in writing his treatise. Bernoulli's name is not mentioned. After that, few if any historians
allowed any credit to Bernoulli's accusations.

In 1922 Schafheitlin found in the Public Library of Basel and
published[1] Part I of John Bernoulli's *Course*. It was exactly as its
author had described it, and the work of L'Hôpital was at once
reduced to the exposition, not the content. But the explanation is
more interesting than the fact, and the explanation is to be found only
in the letters of the two principals. The existence and contents of
these letters have been known to a limited circle for some decades, but
the general public will see them for the first time in this volume. Ten
were published in a thesis by O. J. Rebel in 1934, but these are not
the most informative, and in particular the amazing No. 20, from
which I will quote below, is not included. A fair idea of the content of
the collection has been given by Spiess in an earlier publication[2]. The
Course and the letters together *fully substantiate John Bernoulli's
claims* in all but some minor matters.

The fascinating relation between L'Hôpital and John Bernoulli
is traced by Spiess in the special preface, pages 123–157. In this
review I will give only a spare summary, urging the reader to enjoy
for himself Spiess's own words and the following eighty-seven letters.

---

[1] *Verhandlungen der Naturforschenden Gesellschaft in Basel* **34** (1922/1923). German
translation, Ostwalds Klassiker No. 211, Leipzig, 1924.

[2] "Une édition de l'œuvre des mathématiciens Bernoulli", *Archives Internationales
d'Histoire des Sciences* **1** (1947): 356–362.

Figure 13. JOHN BERNOULLI (1667–1748) after an engraving done by G. F. SCHMIDT in 1743 following a painting by J. R. HUBER.

These spread from December 1692, a month after BERNOULLI's return to Basel, to a letter from L'HÔPITAL's widow in 1707.

For BERNOULLI's stay in Paris we must rely on his own autobiography, written just before his death, and on a sequence of unpublished letters detailing his recollections to PIERRE DE MONTMORT in 1718. After the famous meeting of the two savants in the salon of MALEBRANCHE, when BERNOULLI let flash his secret weapon[3], the general formula for the radius of curvature of a curve, L'HÔPITAL immediately engaged BERNOULLI to give him four lessons per week. After six months of that, the scene of instruction shifted to L'HÔPITAL's chateau in the country for three or four more months, and then BERNOULLI returned to Basel.

To be brief, in the following letters we find BERNOULLI giving L'HÔPITAL full information on every current topic of research and full answers to every question. Some of these L'HÔPITAL wrote out and sent to HUYGENS or LEIBNIZ. For every problem of major interest to which L'HÔPITAL has had a claim, a lesson or letter from BERNOULLI stands in the background.

How did that happen? We must remember that in 1691 JOHN BERNOULLI was twenty-four, an unemployed younger son in a modest mercantile family; while a younger brother of a famous mathematician, he had himself published but one important paper. L'HÔPITAL was a marquis of thirty, an established savant; young enough for the ambition of learning and perhaps for learning itself, but old enough for assurance and ease in a worldly society; certain of the income of a marquis, if somewhat improvident in the use of it. While nowadays the difference in social positions seems a trifle, in 1691 it was súrely enough to impress even the ebullient self-confidence of JOHN BERNOULLI when, freed of worldly cares, he was accepted as an equal and intimate friend in the elegant establishment whose presiding deity was a charming and witty Marquise. On the other side, while L'HÔPITAL's originality is annulled and his scientific honesty somewhat tarnished by the relation, not only was his curiosity genuine and extraordinary but also from the moment of meeting it was plain that in the face of his young friend's notorious and ineluctable bluntness, the Marquis would have to put up with a style to which his breeding had hardly accustomed him.

Just what was arranged while BERNOULLI was in France, we do not know. Soon after he returned to Basel, a crisis over DE BEAUNE's

---

[3] That HUYGENS, NEWTON, and LEIBNIZ knew the essence of this formula, did not make it the less secret in Paris, where at this time there was no geometer of the first rank.

problem arose. BERNOULLI had found the solution in the course of his researches on integral calculus and had put it into his *Course* for L'HÔPITAL as Lesson IX. While BERNOULLI was still his guest, L'HÔPITAL sent BERNOULLI's solution to HUYGENS, who naturally inferred that the sender was the author, the more so since in an earlier letter L'HÔPITAL had written that he himself had found a solution. At the same time, L'HÔPITAL published the solution under a pseudonym. A complicated sequence of published and unpublished claims and veiled insults followed. For the plan he had in mind, L'HÔPITAL could not afford to notice even an open affront. After some mutual explanations and a delay of more than half a year, during which BERNOULLI refrained from sending L'HÔPITAL anything of importance, L'HÔPITAL on 17 March 1694 (Letter 20) proposed the greatest anomaly that ever was in science:

> I will be happy to give you a retainer of 300 livres, beginning with the first of January of this year.... I promise shortly to increase this retainer, which I know is very modest, as soon as my affairs are somewhat straightened out.... I am not so unreasonable as to demand in return all of your time, but I will ask you to give me at intervals some hours of your time to work on what I request and also to communicate to me your discoveries, at the same time asking you not to disclose any of them to others. I ask you even not to send here to Mr. Varignon or to others any copies of the writings you have left with me; if they are published, I will not be at all pleased. Answer me regarding all this ....

BERNOULLI's response is lost, but the next letter from L'HÔPITAL indicates that the acceptance was speedy. Thenceforth BERNOULLI was a giant enchained. In letters 33–44 L'HÔPITAL scolds BERNOULLI because, after obediently checking, translating into Latin, and transmitting to Leipzig L'HÔPITAL's solution of a minor problem posed by SAUVEUR, he had been unable to restrain himself from adding a note in which he generalized the problem, identified the resulting curve, and gave for the general case his own analysis consisting in one equation, replacing the twenty-seven used by SAUVEUR to set the special case. L'HÔPITAL reminded BERNOULLI that he was not to publish but to send all his works to him; he promised to *keep them secret*, asserting that he had *no desire to take for himself the honor of these discoveries* (Letter 42). In making his excuses BERNOULLI acknowledged his faults and promised, "You have only to let me know your definite wishes, if I am to publish nothing more in my life, for I will follow them precisely and nothing more by me will be seen." When he wrote those lines in 1695, BERNOULLI was as brilliant a mathematician as any living.

As soon as the *Analyse* appeared, the financial arrangement lapsed.

We should not judge L'HÔPITAL's procedure too harshly. While perhaps financial necessity compelled BERNOULLI to accept the arrangement initially, it continued after he had settled into his professorship at Groningen in 1695. L'HÔPITAL, being a nobleman, was accustomed to pay for the services of others, and what he did would not then have been considered wrong had BERNOULLI been a politician, a lawyer, perhaps even an architect. Certainly it was nothing for L'HÔPITAL to be proud of. Careful examination of the letters in which L'HÔPITAL reported his mathematical progress to LEIBNIZ and HUYGENS shows that with one or two possible exceptions L'HÔPITAL *did not lie* but rather referred to BERNOULLI in a condescending tone without acknowledging any debt to him and in matters of provenance wrote in such a way as to suggest without actually asserting.

Soon JOHN BERNOULLI realized what he had sold away. The financial returns were ephemeral, and even for the few years the agreement was in force L'HÔPITAL did not always pay the full sum due. (It would be unfair to suppose BERNOULLI was his only disappointed creditor.) When grown old BERNOULLI boasted of the princely engagement, magnifying both the recompense and the duration.

In the development of calculus as a tool in geometry and mechanics, nearly every letter from JOHN BERNOULLI to L'HÔPITAL presents an individual achievement. What is most remarkable is the lightning speed of BERNOULLI's conception. His thought and expression in French are no less masterful and far clearer and more direct than in his published works in Latin or his later letters. It would be wasteful to attempt here even to name the problems treated, since these are easily followed by aid of an index at the end of the volume. I can find no better summary of my impression from these letters than the words L'HÔPITAL himself wrote to JOHN BERNOULLI in 1695, when LEIBNIZ, NEWTON, and JAMES BERNOULLI were flourishing: "I am very sure that there is scarcely a geometer in the world who can be compared with you."

Note for the Reprinting

The book reviewed here is *Der Briefwechsel von Johann Bernoulli, herausgegeben von der Naturforschenden Gesellschaft in Basel*, Band **I**, Basel, Birkhäuser Verlag, 1955. The editor was OTTO SPIESS. Though not of an old Basel family, he was saturated with the Basel spirit and mores and traditions, among them "nicht auffallen!" Thus his name does not appear on the title page, and he refused to let me publish his picture with this review, the original title of which was "The New Bernoulli Edition". It was first printed in *Isis* **49** (1958): 54–62.

SPIESS's estimates of length turned out to be far too low.

24. THE WORKS OF JAMES BERNOULLI
(1973)

Some decades ago an edition of the collected works of the BER-
NOULLIS was projected by OTTO SPIESS and after some vicissitudes
given formal support by Swiss organizations. The first volume to
appear was *Der Briefwechsel von Johann Bernoulli*, Band I, in 1955.
This volume was edited by SPIESS himself. I reviewed it in terms as
laudatory as I knew how to write. [See the preceding essay in this
volume.]

Fourteen years later we encounter the second product of this edition, namely, *Die Werke von Jakob Bernoulli*, Band I, published in
1969. The preface is signed by J. O. FLECKENSTEIN and dated 1962.
In a note FLECKENSTEIN states that it was not possible to give this
volume to the press until late in 1966, SPIESS having died early in that
year. The reader who knows the work of JAMES BERNOULLI will have
difficulty in recognizing the author in the items published in this
volume. It contains the writings of his youth, defined by the editor as
1676–1686, the terminal year being that just before the one in which
he took up the professorship of mathematics at Basel.

The first volume contains a long preface by SPIESS recounting
the history of the manuscripts and of the edition. The BERNOULLI-
L'HôPITAL correspondence is provided with an introduction thirty-
five pages long, which recounts the fascinating relations between the
two men and is itself a major memoir in the history of science. The
preface to the volume here under review fills four pages; it seems
to be a general apology for publication of the volume at all, and it
assures the reader that JAMES BERNOULLI's *Jugendschriften* should not
be dismissed as *Jugendsünden*. I fear this pious exhortation will not
convince such readers as regard JAMES BERNOULLI the greatest of
the great family, the creator of probability theory, of the calculus of
variations, of the principle of angular momentum, and of the theory
of the elastica. Nevertheless, the pages are not devoid of interest to the
devotees of the author (I confess to being one of them), since it shows
the miserable swamp of rubbish from which the great man emerged:
CARTESian physics and summaries of lectures on elementary experi-

Figure 14. JAMES BERNOULLI (1655–1705), engraved after the portrait by his brother NICHOLAS in the Alte Aula, Basel. (So far as I can learn, this print by H. PFENNIGER is the only one made before 1800 of a portrait of JAMES BERNOULLI from the life. An earlier engraving by DUPIN in some examples titled "JEAN" and in others "JACQ" represents the former.)

ments, the latter being what the editor, with some mysterious allusion to NEWTON, calls *Philosophia naturalis*. The historical scholar will welcome these last, because they have not been published before; they concern hydrostatics and aerostatics, and so far as I can see are merely standard for the period (1683, 1684, 1686, 1690). Another novelty is the German first draught of BERNOULLI's essay on the orbits of comets, published indeed in 1681, in a volume now very rare. On the title page is a motto which sums the author's life:

Tardè eruuntur, quae tam altè jacent

a motto which cannot be translated into neat English since *altus* means either "high" or "deep" or both. For the second, expanded edition *tardè* was changed to *difficulter*. No word in JAMES BERNOULLI's writing is accidental. We may be sure this change of word reflected a change in BERNOULLI's own estimate of himself.

JAMES BERNOULLI kept a small blank book in which he copied out fair the results of his daily thinking that he regarded as being of possible value. This notebook he entitled *Thoughts, notes, and remarks on theology and philosophy, condensed and collected by me J.B. from the year 1677 onward*. It is preserved in the Basel University Library under the number MSIa3. In it occur passages of marvellous grasp and depth. Moreover, it reveals the gradual conversion of a theologian turned ill-trained teacher of elementary mathematics into a great creator of the modern mathematical sciences. Some of the entries of highest specific interest have been published, and various scholars who have consulted the manuscript have referred to the contents of others. Many students of Western mathematics hoped to see the text of this notebook, properly annotated, appear as a volume of the edition. They will not, for the editors have chosen to publish the articles separately, classified according to subject. The present volume contains some of them, going as far as about 1684. These fall under the heads of astronomy, logic, speculative physics, and experimental physics.

The several parts of the volume are provided with competent prefaces. The plan seems to be the same as SPIESS's, more or less. SPIESS wrote not only his preface but also his running heads and notes in German. In his volume he had the excuse—if excuse it is—that the originals were in several languages and included even a few pages in Dutch, which would offer difficulty to almost anybody but a Dutchman. The text of the present volume, in contrast, is almost entirely in Latin, and so the running heads and notes (except for *Philosophia naturalis*, which the editor apparently regards as insusceptible of

JACOBI
BERNOULLI,
BASILEENSIS,
OPERA.

Tomus Primus.

GENEVÆ
Sumptibus Hæredum CRAMER
& Fratrum PHILIBERT.

M. DCC. XLIV.

Figure 15. Title page of *The Works of JAMES BERNOULLI*, 1744.

translation into German), stick out oddly. What use is the title *Zur Anzahl der logischen Tropoi* to a person who can understand neither the article itself nor BERNOULLI's own title: *An quinque tantum troporum genera?*

In the years 1956–1957 and 1960–1961 I had several conversations with SPIESS about the BERNOULLI edition. He expressed confidence it would remain "in vaterländischen Händen". SPIESS, while he seemed to be a characteristic Basler, was of fairly recent German extraction.

One thing that made the volume produced by SPIESS fascinating was its concentration on the key-problems of the early days of the infinitesimal calculus. This aspect of mathematics was SPIESS's own. He listed the problems as $P_1, P_2, \ldots, P_{94}$ and annotated each item in his volume in those terms. In the present volume there are no specific problems, and so the same method could not be used.

A departure from SPIESS's precedent comes in the format: the paragraphs are not indented. This disease of recent birth, almost without precedent in the history of printing and having no other purpose today than to rub the reader's nose in modernity not only ugly but also ambiguous, has spread now even to scholarly works. This new practice has about the same effect as would replacing every Z by an A and leaving it to the reader to decide which was really meant. While the sections are set apart by leads, the paragraphs are not, and so every time a sentence and a line happen to end together, the reader is left wondering if a new paragraph begins. The answer for the "modern" reader is easy: It does not matter. That is a corollary of a general theorem: What a modern author writes does not matter. This general theorem justifies the army of publishers' clerks, usually holding positions classified as "editors", who by profession lay waste the texts that pass through their hands. The well known corollary of the corollary is that many authors no longer trouble to write a decent text, since they know that "editors" will spoil it anyway. But JAMES BERNOULLI was not a modern author. He did not write for "editors". However PARKINSON's Law may justify the destruction of modern authors' styles by "editors", they should be forbidden to jumble what giants' quills have left us.

Among those giants, JAMES and JOHN BERNOULLI occupy unusually favored positions: Their collected works were published in their own age, published in magnificently printed editions that put modern books to shame, editions edited, as few today are, by men who were mathematicians in their own right and could understand the sense of what they edited. These ancient editions are *Jacobi Bernoulli, Basileenis, Opera*, 2 volumes, Geneva, Cramer and Philibert, 1744 (edited by NICHOLAS I BERNOULLI, with some notes by G. CRAMER), and *Johannis Bernoulli ... Opera omnia*, 4 volumes, Lausanne and Geneva, Bousquet, 1742 (1743), edited by G. CRAMER.

JOHANNIS BERNOULLI,

M. D. MATHESEOS PROFESSORIS,
Regiarum Societatum PARISIENSIS, LONDI-
NENSIS, PETROPOLITANÆ,
BEROLINENSIS, *Socii* &c.

OPERA OMNIA,

TAM ANTEA SPARSIM EDITA,
quam hactenus inedita.

TOMUS PRIMUS,

Quo continentur ea
Quæ ab ANNO 1690 ad ANNUM 1713 prodierunt.

LAUSANNÆ & GENEVÆ,

Sumptibus MARCI-MICHAELIS BOUSQUET & Sociorum.

MDCCXLII.

Cum Privilegio Sacræ Cæsareæ Majestatis , & Sereniss. Poloniæ Regis ,
Elect. Saxon.

Figure 16. Title page of *The Works of JOHN BERNOULLI*, 1743.

In defense of the present *vaterländisch*-limaceous edition it must be said that the annotations to the old *Opera* are reprinted. SPIESS often expressed to me his opinion that there was no urgency to republish anything included in those editions, since one could find them "anywhere". I had been trying for years to buy copies and was grateful to SPIESS for loaning me his while I worked in Basel. Many *ausservater-ländische Universitäten* seemed then, and seem now, not to possess them. Some years later I did acquire them. They were not cheap. Since then, fortunately, the *Opera* of JOHN BERNOULLI have been reprinted by Georg Olms, Hildesheim, 1968; the *Opera* of JAMES BERNOULLI, by *Éditions Culture et Civilisation*, Bruxelles, 1971.

The prospect of waiting an interval of fourteen years between (if we may judge by extrapolation) the successive products of the projected edition of twenty or more volumes of what I in 1958 optimistically called "the new Bernoulli edition" is not encouraging to those whose life expectation is less than 200 years.

I express the urgent hope that some publisher will soon reprint the manuscript of the *Meditationes*.

Note for the Reprinting

This review first appeared in *Isis* **64** (1973): 112–114, under the title "Review of *Die Werke von Jakob Bernoulli*. Band I, Basel, Birkhäuser, 1969". In reprinting it I have made a bibliographical correction.

The prospect is much better than it seemed in 1973. Volume **3** of JAMES BERNOULLI's works and Volume **2** of DANIEL's have appeared; Volume **2** of JOHN BERNOULLI's is expected any time now; and the whole project is progressing rapidly under the able and energetic leadership of Mr. DAVID SPEISER. Best of all, a facsimile edition of the *Meditationes* is planned.

25. DANIEL BERNOULLI'S *HYDRODYNAMICA* (1960)

DANIEL BERNOULLI's name is familiar to every mathematician, but few have any knowledge of his work beyond the vague generalities, inaccurate details, and dubious anecdotes handed down in popular histories. Most mathematical libraries in the U.S.A. do not contain a single work by him. His papers have never been republished in a collected edition; to the best of my knowledge, this printing of BERNOULLI's great classic work, the *Hydrodynamica*, is only the second, the date of the first edition being 1738. The contents of the book are so completely unknown among mathematicians and specialists in fluid mechanics that a detailed review, as if it were a new work, would be appropriate. I have discussed it briefly and have translated a few of the most important paragraphs[1] into English; it is to be hoped that the edition now under weigh in Basel will not be too long delayed.

The volume under review includes a Russian translation of the text and forty-five pages of annotations. Some of these explain the mathematical steps in the notoriously difficult text; others give passages and sketches from hitherto unpublished documents, including the first draught, which BERNOULLI left behind him in Petersburg in 1733. It is a pity the Latin originals are not given along with the Russian translations. Following the text is the brief autobiography of BERNOULLI and a valuable and concrete essay, seventy pages long, on his life and works by V. I. SMIRNOV.

DANIEL BERNOULLI has stood in the shadows of his overbearing father and the dazzling EULER, both of them quicker and more versatile than he. Nonetheless, he had a strong and original mind, independent to the point of blindness to the work of even his nearest colleagues. Certain philosophers of science could point to him and to his hero, HUYGENS, with more justice than to GALILEO and NEWTON as early saints in the control of theory by experiment. That notwithstanding, his work extended into pure mathematics, especially very

---

[1] Part IV of my introduction to LEONHARDI EULERI *Opera omnia* (II)**12**, Lausanne, 1954.

Figure 17. DANIEL BERNOULLI (1700–1782) after a mezzotint of 1744 by J. R. HAID following a portrait by J. R. HUBER.

early and very late in his life; for example, in his last decade he made the capital discovery that a trigonometric series can represent a discontinuous function.

The eighteenth century is the least known period in the history of the mathematical sciences among those for which adequate source material exists. The present volume is a useful contribution to its small historical literature.

Note for the Reprinting

The work reviewed here is *Danielis Bernoulli Hydrodynamica sive de viribus et motibus fluidorum commentarii*, Перевод В. С. Гохмана, Комментарии и Редакция Академика А. И. Некрасова и Профессора К. К. Баумгарта статья Академнка В. И. Смирнова, [Moscow], Ex Officina Academiae Scientiarum FRSS, 1959. The review first appeared in *Mathematical Reviews* **21** (1960): 1169–1170.

To KARL FLIERL we owe a careful translation of the *Hydrodynamica* into German, provided with abundant commentary and scrupulous analysis in which errors of the orignal are corrected: *Des Daniel Bernoulli . . . Hydrodynamik*, München, Deutsches Muscum, 1965. There is also a translation by THOMAS CARMODY & HELMUT KOBUS, *Hydrodynamics by Daniel Bernoulli and Hydraulics by Johann Bernoulli*, with a preface by HUNTER ROUSE, New York, Dover Publications, 1968. Some of the many deficiencies in the translation have been pointed out by A. M. BINNIE & H. J. EASTERLING in their review in the *Journal of Fluid Mechanics* **38** (1969): 855–856.

"The edition [of DANIEL BERNOULLI's hydrodynamical works] now under weigh in Basel" was entrusted to me some years ago. Now it is I who am responsible for the further delay in republishing "the notoriously difficult text".

26. ROUSE & INCE'S *HISTORY OF HYDRAULICS* (1959)

To review this book is difficult in view of its circumstances: (1) it may be regarded as an explanation of hydraulic concepts in their historical origin, written by hydraulic engineers and for hydraulic engineers; (2) it is the first serious attempt, in any language, to cover justly the whole development of hydraulics from the practice of antiquity down to the theory and experiments of living scientists; and (3) it may be regarded as a contribution to the history of science itself. In respect to (1), it is surely excellent; to (2), commendable; to (3), something less.

This review, from its place of publication, must concentrate upon (3) and neglect (1). But in respect to (1), I must say straight out that ROUSE & INCE attempt a necessary work and do it well.

Every page reflects their opinion, finally expressed upon page 244, that "the most profound [recent] change of viewpoint has been the acceptance of the very methods that originally caused the rift between mathematicians and engineers . . . ." Indeed, in addition to vivid and fairly detailed descriptions of principal experiments, there are long passages on hydrodynamic theory formerly taken lightly by many hydraulic engineers.

That their readers may be unprepared to follow the subject, is appreciated by ROUSE & INCE, who present many comments on the general historical scene, the circumstances of the scientists, and developments in related fields such as general mechanics and mathematics. Unfortunately these [attempts to supply general culture] reduce the space given to researches on fluids. Still more unfortunately the comments tend to repeat the stories given by the popular historians of the last century concerning numbers of children, quotable quips, *etc.*, while important things are pushed into dependent and almost parenthetical clauses. ROUSE & INCE sometimes succumb to the practice of embroidery, to which readers of the usual histories are accustomed. On page 92 we read, "During his first sixteen years in Russia [EULER] found Catherine's despotism progressively less attractive, and in 1741 he accepted an invitation to Berlin from Frederick the Great." Apart from the facts that CATHERINE I died the day EULER entered Russia in 1727 and that ROUSE & INCE make it plain they are not referring to CATHERINE II, who brought EULER back in 1766, there is no evidence that governmental "despotism" had any influence on EULER, who, like most scientists, accepted and rejected positions on the basis of the concrete conditions of work and pay they involved. . . .

A reviewer cannot check in detail everything there is in a book, for if he could, he would likely have written it himself. My own experience intersects ROUSE & INCE's material mainly in their Chapters 7, "Seventeenth Century Mathematics and Mechanics", 8, "The Advent of Hydrodynamics", 12, "Classical and Applied Hydrodynamics of the Nineteenth Century", 14, "The Rise of Fluid Mechanics". Criticism of these, which make up less than one-third of the book, may do injustice to authors whose field is hydraulics and who deserve praise for attempting to include hydrodynamics at all. On the other hand, the written words stand, and it would be scarcely right to let their inaccuracy pass unnoted.

To come at once to the trouble, Chapters 7 and 8, at least, seem to be based on secondary sources and thus in large part to propagate the traditional historical errors manufactured in the last century, largely from reading LAGRANGE's *Méchanique Analitique* as a complete and unbiased account of all the rational mechanics that went before it. LAGRANGE made some fruitful effort to ascertain the history of mechanics and included historical sections derived from his reading; when it came to his own century, on the contrary, he was grossly inaccurate. At every opportunity he cites D'ALEMBERT, the patron of his career, going so far as to attribute to D'ALEMBERT things to be found only vaguely or inaccurately limned in D'ALEMBERT's works— "D'ALEMBERT's principle" being one of these. Whenever possible, LAGRANGE avoids citing not only EULER but all of the Basel school, and if he does cite them, it is usually in so vague a way that nothing definite can be inferred. Now that is relevant to the book of ROUSE & INCE, for on page 92, and in the original version also on page 90, we find mention of the *Nouvelle hydraulique* of JOHANN BERNOULLI, and on page 90 its "primary contribution" is described. But there is no such work! It is to JOHANN BERNOULLI's *Hydraulica, nunc primum detecta* that they refer; it was cited by LAGRANGE as the *Nouvelle hydraulique*, and the "primary contribution" he and they find in it would seem rather trivial even if I could verify it, which I cannot. ROUSE & INCE do not tell us that it was in this work that the *internal hydraulic pressure*, a concept used in all subsequent hydraulic researches, first appeared.

An equally important omission is EULER's work on hydraulics, with the sole exception of ten lines on pages 106–107 concerning his reaction turbine. The reader will not learn that from these papers, filtered through the German text books of KAESTNER (1769, 1797) and BRANDES (1805), derives modern theoretical hydraulics, neglecting friction. EULER obtained the first, and correct, hydraulic analyses of cylindrical pumps and centrifugal pumps[1].

There are many questionable details, the more so since a certain vagueness of expression often leaves some doubt of what is really meant, as in the assertion (page 103) that EULER's hydrodynamical equations "differ little" from those used today. Historical inaccuracies resulting from oversight of earlier authors are, of course, unavoidable; e.g., on page 107 the velocity-potential and stream function are attributed to LAGRANGE; correct attributions are to EULER and D'ALEMBERT, respectively. Errors of a more serious kind result from failure to check in detail the sources cited, as when we read (page 107) that in LAGRANGE's work "both D'ALEMBERT's principle of effective force and EULER's basic analysis of fluid acceleration were thoroughly generalized," whereas in fact LAGRANGE contributed nothing whatever to the material on this subject, which he took from much prior papers of EULER. On page 101, CLAIRAUT is both overestimated and underestimated; ROUSE & INCE fail to appreciate his having introduced a general field of force, yet they go too far in saying that "the new science of hydrodynamics" was involved in his work and in attributing to him the celebrated Theorem of MACLAURIN, which they state incorrectly.

Checking back on the relatively few sources cited by ROUSE & INCE, we find only one Latin title, DANIEL BERNOULLI's *Hydrodynamica*, and here they cite also an Italian reference that may give a translation as well as a French work that translates the major passage they present in English. That is not so trivial as it may sound. Essential in the history of hydraulics is DANIEL BERNOULLI's failure to introduce any definite concept of internal pressure in a fluid. For him, "pressio" always means pressure on a wall or in stagnant fluid, and he describes internal effects of such pressure by three other terms, none of which he defines or explains. ROUSE & INCE, apparently following the French source, translate all four words as "pressure" or "overpressure," besides omitting most of the vague original sentence that gives the crux of the argument. Here is the evidence:

DANIEL BERNOULLI, *Hydrodynamica*, 1738, page 258:

> Igitur aqua in tubo tendit ad majorem motum, nisus atuem ejus ab apposito fundo FD impeditur: Ab hoc nisu & renisu com-

[1] J. ACKERET, Introduction to LEONHARDI EULERI *Opera omnia* (II)15, Zürich, 1957.

primitur aqua, quae ipsa compressio coercetur a lateribus tubi, haecque proinde similem pressionem sustinent.

Translation by R. DUGAS, *Histoire de la mécanique*, 1950, page 276:

Donc l'eau dans le tube s'efforce vers un mouvement plus grand, auquel le fond FD fait obstacle. Il en resulte une surpression qui se transmet aux parois.

Translation by ROUSE & INCE, page 97:

Thus the water in the tube tends towards a greater motion, which is obstructed by the end FD. There results an over-pressure which is transmitted to the sides.

My translation in LEONHARDI EULERI *Opera omnia* (2)**12**, Lausanne, 1954, page XXVII:

Therefore the water in the tube tends to a greater motion, but its pressing is hindered by the applied barrier FD. By this pressing and resistance the water is compressed, which compression is itself kept in by the walls of the tube, and hence these too sustain a similar pressure.

The translation by ROUSE & INCE is smoother and easier to understand. My translation attempts to reflect the awkward vagueness of the original, particularly in its use of "compressio" in the root meaning, "a pressing together". Most of all, DANIEL BERNOULLI certainly does not say that any pressure is "transmitted".

Now to try to write a history of science prior to 1800 without consulting Latin sources would be much like trying to learn calculus if one's only language were Eskimo: It is not impossible, but inaccuracies such as we find in the book of ROUSE & INCE are scarcely to be avoided.

Coming to more recent times, we find that in connection with infinitesimal surface waves the name of KELLAND is not mentioned, though most of the simpler results that engineers use are due to his work and are attributed to him in the report of AIRY discussed on pages 199–200. There are, of course, questions of taste. Possibly RIABOUCHINSKY's contribution to hydraulics justifies the authors' giving him as much space as they do to LAPLACE, but when only five lines of the page on LAPLACE (pages 108–109) refer to his work on fluids, and one of these implies a false attribution, while the others are so vague that the reader does not learn what results, if any, LAPLACE achieved, we are left wondering why he is mentioned at all. In fact, in 1779 LAPLACE published the "elementary relationship . . . between the effects of depth and wave length" attributed by the authors on page 200 to AIRY (1845).

Such cases, viewed as individual questions of priority, are of no importance. In totality, nevertheless, they are the infinitesimals which when added together yield historians' traditional underestimation of the eighteenth century in favor of what went before or what came afterward. It is natural that ROUSE & INCE, with the best will to the contrary, inherit part of the traditional view that the geometers of the eighteenth century worked in disregard of experience, experiment, and the applicability of mathematical theories. Notwithstanding that, fuller information[2] was available before their book was printed.

Coming again to the nineteenth century, I am happy to find the work of the great BOUSSINESQ properly appreciated in relation to the later and less complete studies of REYNOLDS. The sketchy discussion of rotational motion and flows with free stream lines (pages 200–201) may indicate that these aspects of fluid mechanics are not yet so well known to hydraulic engineers as I think they will be a few years from now. In the short chapter on the twentieth century ROUSE & INCE manage to name all the important names of today, but for some of them they seem to be unable to specify any particular achievement. Surely MISES' application of the momentum principle to flows in hydraulic machines deserves some mention.

It would be wrong to end in a negative tone the description of this book, conceived and executed in a wholly positive spirit. ROUSE & INCE are at their best in discussions of practical hydraulics, whether in Roman times or in our own. In their book are numerous analyses of experiments, of the development of empirical formulæ, and of ideas growing out of practice with hydraulic machines. The sources here are difficult for a person accustomed to fundamental science to use. If, as it seems, the authors have derived this part of their material from the sources, they have done a service and produced a work of value.

To summarize, the best I can do is state my own sensation on reading the book. The parts not dealing with major figures in fundamental science I found fascinating, and probably I shall refer to them again and again in the future. Nevertheless, before using anything stated by the authors, I shall take care to check the source.

---

[2] C. TRUESDELL, Introductions to LEONHARDI EULERI *Opera omnia* (II)**12** and **13**, Zürich, 1954 and 1956.

Note for the Reprinting

The work reviewed is H. ROUSE & S. INCE's *History of Hydraulics* (reprinted, with corrections, from separately paginated supplements to *La Houille Blanche*, 1954–1956), Iowa City, State University of Iowa for the Iowa Institute of Hydraulic Research, 1957. The foregoing review is reprinted from *Isis* **50** (1959): 69–71.

At Professor ROUSE's request I provided him with the above criticisms and dozens more, but the second edition, published in 1963 by Dover Publications, New York, differs from the first, insofar as I have checked it, only in correcting the two blunders on page 92 and adding two references. In particular, the mistranslation of DANIEL BERNOULLI's description of his theorem remains. It is corrected, finally, by CARMODY & KOBUS in their translation of the whole *Hydrodynamica*, cited at the end of the preceding essay.

The "French source" is reviewed above in Essay 17.

27. HANKINS' *JEAN D'ALEMBERT* (1971)

"This book is intended as a study of the relations between science and philosophy during the Enlightenment as seen through the activities of one of its most prominent spokesmen, Jean d'Alembert." After a brief description of D'ALEMBERT's training, HANKINS details the quarrels with CLAIRAUT, DANIEL BERNOULLI, and EULER which made up most of D'ALEMBERT's life in science. Next comes a description of the project and execution of the *Encyclopédie*, beginning with D'ALEMBERT's limited collaboration, followed by the literary success in the early 1750s which "went to his head" and seduced him into abandon of his real love, geometry, and ending with the rupture between him and DIDEROT. Here HANKINS emphasizes the philosophical differences between the two men and in particular their difference of degree in understanding mathematics. DIDEROT, while respectably trained and competent in the pure mathematics of his day, was at a loss on the deeper issues of mechanics. He thought mathematics had been pushed as far as it could go, and he favored empirical science; nevertheless, his preference for "natural history, anatomy, chemistry, and experimental physics" was not founded upon any real understanding of them but was merely philosophical and did not lead him or those near him to advance these sciences. For D'ALEMBERT, on the contrary, "mathematics was the key to science", though he, too, came to regard mathematics as an exhausted subject, not from any foundation in fact or any preference for other ways of thinking but only from his own inability in his later years to follow the onrush of mathematical discovery by others. D'ALEMBERT showed "a suspicious attitude toward experiment" and satirized the "imagined vague hypotheses" of the experimenters, which "explained anything and its opposite just as well." In celestial mechanics "d'Alembert was working at a disadvantage . . . because he knew almost nothing about observational astronomy." LALANDE in claiming that D'ALEMBERT "had never held a prism in his hand" was only one in a chorus of contemporary scientists who reproached him for the abstractness of his speculations and his ignorance of facts of experiment. Among the mysteries of scientific folklore is the modern reputation of D'ALEMBERT as a man who put experiment first and called for interrelation of theory with it[1], a reputation the opposite of D'ALEMBERT's true character.

The next two chapters present D'ALEMBERT's philosophical position, especially in regard to science and knowledge, and make it plain that his thought on all subjects was dominated by his attachment to methods and examples from rational mechanics. HANKINS then takes up D'ALEMBERT's famous campaign against the concept of force, which he wished to banish from mechanics but in fact could never himself do without. The last chapters present D'ALEMBERT's attempts at formulating general and inclusive laws of motion, his use of virtual velocities, and his part in the *vis viva* controversy.

This book grew out of its author's doctoral dissertation. As a thesis, it is of the highest quality, a work of breadth and depth and erudition extraordinary for a beginner. It is almost free of the typical defects of a thesis: There are few schoolboy blunders[2] and few pontifications in unsupported generalities[3], and on the whole the writing is clear and direct, making an entertaining as well as a learned and reliable book.

---

[1] This folklore is current among hydrodynamicists. It seems to rest on D'ALEMBERT's clearly written prefaces to his otherwise tortuous and obscure mathematical works and on passages where he essentially repeats descriptions of experiments done by others, perhaps intended as finery for his theories; *e.g.* §§ 95–99 of his *Essai d'une Nouvelle Théorie de la Résistance des Fluides*, 1752. There is also the famous sentence concluding the passage in which he presents his rediscovery of a special case of EULER's theorem stating that a perfect incompressible fluid undergoing flow uniform at infinity offers no resistance to a submerged obstacle (the "d'Alembert paradox", *ibid.*, § 70): "Hence one sees how necessary are experiments in the present question."

Also hydrodynamicists sometimes refer to a book which bears D'ALEMBERT's name first on its title page and which contains as its text mainly tables of data from measurements of resistance on ship models towed across a tank: *Nouvelles Expériences sur la Résistance des Fluides*, par MM. D'ALEMBERT, le Marquis de CONDORCET, & l'Abbé BOSSUT . . ., M. l'Abbé BOSSUT, Rapporteur, Paris, Jombert, 1777. The Discours Préliminaire states that since D'ALEMBERT had "solved the problem in question by an analytic method, new and direct, which would leave nothing to be desired if one could integrate rigorously or by convergent series the equations at which he had arrived," he had counselled recourse to experiment. I can find no further mention of D'ALEMBERT in the book, and in the list of theorists on page 130 D'ALEMBERT's name is conspicuously absent, perhaps because it would have been out of place after the opening sentence: "Every theory which is to be applied to practice should be simple in its principles."

In view of D'ALEMBERT's well known thirst for credit, I doubt he had seen any of the text of this book. The experiments seem to have been performed under the supervision of BOSSUT alone, and, as the title page suggests, BOSSUT seems to be the sole author of the text. In 1770 D'ALEMBERT had suffered a second failure of health, and in any case, by his own admission, since the 1760s he had not been capable of the concentration serious scientific work requires.

[2] *E.g.* on page 1 "Turici" is translated as Turin.

[3] *E.g.* on page 7, ". . . Truesdell has overstated his case . . .," with no direct quotation of the "case" and no contrary evidence presented.

HANKINS tells us that he composes "history of science as intellectual history", not "a thoroughgoing biography or a detailed analysis of d'Alembert's scientific writings" but rather an attempt "to place him in the scientific, social, and 'philosophical' communities of the eighteenth century, and to show how one important *philosophe* lived . . . in all three worlds." HANKINS does not fall into the old-fashioned and disagreeable habit of idolizing the author he studies. He is sympathetic enough to write in depth as well as detail about a man whom he neither likes nor much esteems. Few biographers can be so admirably detached:

> d'Alembert was not the intellectual giant he wanted to be and frequently claimed to be; but his role in the eighteenth century was . . . more important than the sum of his specific literary and scientific accomplishments. He stood at the very heart of the Enlightenment. . . . Moreover, d'Alembert was a scientist, the only *philosophe*, except for his protégé Condorcet, who earned his reputation and made his profession in the sciences.

While philosophy is the subject of whole chapters of the book, HANKINS always dwells essentially upon mechanics, its basis in science, and the interpretations and misinterpretations of each other's views and achievements and failures by philosophers and scientists.

The great danger to an author who writes intellectual history of science is that his own intellect may fail to reach the level of his subject. Here HANKINS comes off well so long as he stays with pure mechanics[4], except that he does not make clear to a reader not pre-

---

[4] The one thoroughly bad sentence is on page 49: "In his *Essai* d'Alembert . . . used the concept of fluid pressure to write equations describing the state of the fluid at any point in the 'field'." While this statement is barely true in a weasel lawyer's sense, any but the most expert reader will wrongly conclude from it that the pressure appears in D'ALEMBERT's hydrodynamical equations. Because the main innovation of eighteenth-century mechanics was the concept of internal pressure and other major special cases of the stress principle, and since D'ALEMBERT impeded rather than promoted the creation of this concept, a review must mention this one serious blemish in HANKINS' otherwise mainly satisfactory description of mechanical researches by D'ALEMBERT and others.

Another exceptionally weak statement may be found on page 194. "The notion of mass . . . has no place in mathematics, nor can it be conjured up 'metaphysically'. It is either an arbitrary postulate or it must be defined by reference to the real world." A historian of science should know enough about mathematics to see the difference between "an arbitrary postulate" and an elegant, developed, and immensely important mathematical structure like measure theory. Similar difficulties occur with the concept of force, where HANKINS, along with most other philosophers and historians, seems to think that absence of mathematical structure [whether in fact or only in the mind of the writer] means that no such structure is or was or ever will be desirable or possible.

viously acquainted with rational mechanics that the grand advances, those with the great philosophical implications, all grew inductively from attempts to solve concrete, specific problems for which principles known theretofore were insufficient. When HANKINS comes to astronomy, it is a different matter. I have read three times his account of researches on the motion of the moon's nodes and the controversy over lunar tables among CLAIRAUT, D'ALEMBERT, and EULER, and from HANKINS' text I cannot learn what any of these men really did or wherein their differences lay. The social drama was splendid, but what was the point of mechanics or mathematics that divided the actors? The best HANKINS can do, it seems, is to tell us (page 37) that one used more "observational data" and was interested in "the best lunar tables", while another preferred "refinement of mathematical methods". Since all three were trying to predict the lunar motion by *mathematics* applied to the common principles of mechanics and the still questioned law of gravitation, and since every lunar table *necessarily* is based on certain empirical data about the moon, earth, and sun as a starting point, no real distinction is made by HANKINS' summary, and his claim that it was only "the now familiar problem of the experimental and observational scientist versus the theoretician" is absurd. No experimenter or telescopist, no new law of theoretical physics or newly observed phenomenon of nature was involved. All three principals, apart from an occasional and quickly retracted deviation, were theorists trying to solve a specific *mathematical* problem within *the same, accepted framework* of mechanical theory[5]. An alleged solution of a system of differential equations is right or wrong today on the grounds that made it right or wrong in the eighteenth century, and likewise an approximate solution is better or worse (terms hateful to a social historian) on just the same objective, demonstrable grounds. While proofs of convergence and estimates of error were then wanting, and so there was room for conjecture and controversy from ignorance, today we should be able to judge or at least survey the essentials. Were the differential equations and solutions of them in series by D'ALEMBERT, CLAIRAUT, and EULER the same or different? Were the methods analytic or statistical in whole or part? Did they have greater or lesser domains of convergence or bounds of error?

---

[5] A word of caution should be added. The concepts of small oscillation about relative equilibrium, moments and products of inertia, and instantaneous axis of rotation had become well known through solutions of numerous special problems, mainly by EULER, but the general theory of motion of rigid bodies, due also to EULER, was to develop in the next few years, partly on the basis of the earlier work just mentioned and partly in response to attempts to solve the problem of the moon's apsidal motion.

As I kept puzzling over HANKINS's account of the lunar controversy, I began to wonder why intellectual history could do well on one aspect of mechanics yet fail on another. Then I remembered that while, for pure mechanics, HANKINS was able to draw on and summarize the results of extensive and reliable history of science as science, written by and for today's experts on mechanics, there seems to be as yet no such thing for celestial mechanics and perturbation theory. One wing of his house remains unfounded and hence unbuilt, but that is no reason to try to replace it by a magic-lantern picture.

Note for the Reprinting

The book discussed is THOMAS L. HANKINS' *Jean d'Alembert: Science and the Enlightenment*, Oxford, Clarendon Press, 1970. My review first appeared in *Centaurus* **16** (1971): 56–59.

The question regarding "the differential equations and solutions of them in series" by D'ALEMBERT, CLAIRAUT, and EULER has been answered in the magisterial memoir by CURTIS WILSON, "Perturbations and solar tables from Lacaille to Delambre: the rapprochement of observation and theory", *Archive for History of Exact Sciences* **22** (1980): 53–304.

28. THE *MATHEMATICAL AND PHYSICAL PAPERS* OF G. G. STOKES (1966)

In the recent renascence of continuum mechanics the name of STOKES is often mentioned, and some of his papers have been re-read again and again in the past twenty years. The new interest differs from that which originally greeted these same papers and has little to do with the specific discoveries ascribed to STOKES in the appreciations of his own day, the obituaries, and their epigones, the histories of physical science and the notes in textbooks. Rather, passages in which STOKES wrestled with first principles have been studied, line by line, for the method of inquiry they reveal.

In this regard the most influential paper is that of 1845 on the theories of the internal friction of fluids, beginning on page 75 of Volume 1. STOKES had entered Pembroke College, Cambridge, in 1837. As he recalled[1] in 1901, "In those days boys coming to the University had not in general read so far in mathematics as is the custom at present; and I had not begun the Differential Calculus when I entered College, and had only recently read Analytical Sections. In my second year I began to read with a private tutor, Mr. Hopkins, who was celebrated for the very large number of his pupils who obtained high places in the University examinations for Mathematical Honours." Eight years later STOKES was struggling to over-

---

[1] This recollection and the letters and other personalia quoted below are taken from the *Memoir and Scientific Correspondence of the late Sir George Gabriel Stokes*, ed. J. LARMOR, Cambridge University Press, 1907, 2 volumes.

The correspondence between STOKES and KELVIN was not published. The letters are deposited in the Cambridge University Library.

The present preface does not contribute any new source material regarding STOKES beyond mention that in the Johns Hopkins University Library there are [or were] fifty-two [or more] bound volumes of his collection of offprints on spectroscopy and related subjects. While many of these carry inscriptions by their authors, few if any have annotations by STOKES. [*Note*, 1981. The acute form of bibliophobia called "Library Science" has compounded with ordinary shelf tides to let those volumes suffer attrition, mutilation, and loss.]

All footnotes in this essay except this first one are LARMOR's. [I have put some additions of 1981 within square brackets.]

come the provincial pedantry of a Cambridge education. His great memoir shows on every page evidence of powerful thought and ignorance of the peerless work of CAUCHY on the same subjects, some of it printed while STOKES was a child in the nursery. A remark in the introduction reveals that STOKES saw CAUCHY's *Exercises* after his own work was complete. His tutor, HOPKINS, was capable of publishing later in the same volume results almost entirely repetitive of CAUCHY's, and down into the present century British appreciators of STOKES's analysis of the elasticity of solid bodies quietly ignore the more complete as well as more general work done earlier on the Continent.

Continental response to STOKES was equally slow. Seventeen years after the great memoir on the friction of fluids appeared, ST. VENANT noticed it but did not show any interest in those parts where it did better, rather than worse, than its French predecessors.

Vendôme, (France)
22 January 1862

Sir,

I have read some of your beautiful and learned memoirs several times, and I have always found instruction and great pleasure in them, even though I have scarcely any knowledge of your language.

But regarding one of them I must ask you the favor of some explanation, which I need in order to understand it, because I had it before me for too short a time during my last trip to Paris.

It is the memoir *On the theories of internal friction of fluids in motion, and of the equilibrium and motion of elastic bodies*, read 14 April 1845, and published in the 8th Cambridge volume, 1849.

First, from page 287 up to the supplement on some cases of the motion of fluids, which begins on page 409, *I have found nothing about elastic solid bodies*. And nevertheless it is this memoir that Mr. Maxwell cites in Volume 20 of the *Transactions of the Royal Society*, Edinburgh, 1853, p. 89, where he says that you resort to the general fact of the isochrony of small oscillations so as to equate the pressures to functions of first degree in the displacements*, and that finally you solve the equations in three cases†, 1°, a body pressed equally upon all of its surface, 2°, a stretched shaft, 3°, a twisted cylinder.

---

* Stated in § 15 of the memoir.
† See § 20 of the memoir.

Figure 18. GEORGE GABRIEL STOKES (1819–1903), after a photograph taken in 1892.

I have found nothing of all that in Volume 3 of the Cambridge Transactions, pages 287 to 409.

Should I have looked for it elsewhere?

And then, Sir, I see that you arrive in a simple and ingenious way, without considering as Cauchy does an ellipsoidal surface and its normal vectors, at the equation of third degree

$$\left(e-\frac{du}{dx}\right)\left(e-\frac{dv}{dy}\right)\left(e-\frac{dw}{dz}\right)-\left(e-\frac{du}{dx}\right)\left|\frac{\dfrac{dv}{dz}+\dfrac{dw}{dy}}{2}\right|^2 - \&c.=0,$$

which provides for solids the *three principal dilatations* and for fluids the three principal speeds of extension.

But I admit that I do not see with certainty the principle upon which you rely to reach those conclusions. I see well that if u, v, w are the displacements of the point P in the directions x, y, z (I say *displacements* as if a solid were being considered, because the reasoning should be the same for the *velocities* in a fluid), for the relative displacements of the points P and P' we have

$$\frac{du}{dx}x'+\frac{du}{dy}y'+\frac{du}{dz}z', \qquad \frac{dv}{dx}x'+\cdots, \qquad \frac{dw}{dx}x'+\cdots.$$

I see also that to the relative displacements arising from the absolute displacement u, v, w you add others,

$$\omega'''y'-\omega''z', \qquad \omega'z'-\omega'''x', \qquad \omega''x'-\omega'y',$$

which come from three arbitrary rotations ω', ω'', ω''' about Px, Py, Pz, which gives you for the total displacements

$$U=\frac{du}{dx}x'+\left(\frac{du}{dy}+\omega'''\right)y'+\left(\frac{du}{dz}-\omega''\right)z', \qquad V=\cdots, \qquad W=\cdots,$$

and that you determine these arbitrary rotations in such a way that

$$\frac{dV}{dz'}=\frac{dW}{dy'}, \qquad \frac{dW}{dx'}=\frac{dU}{dz'}, \qquad \frac{dU}{dy'}=\frac{dV}{dx'},$$

which gives you

$$\omega'=\frac{1}{2}\left(\frac{dw}{dy}-\frac{dv}{dz}\right), \qquad \omega''=\frac{1}{2}\left(\frac{du}{dz}-\frac{dw}{dx}\right), \qquad \omega'''=\frac{1}{2}\left(\frac{dv}{dx}-\frac{du}{dy}\right),$$

so ω', ω'', ω''' taken with opposite signs are, as Cauchy showed on p. 231 of Volume 2 of the *Exercices d'analyse et de physique mathématique*, the *mean rotations** of the system about Px, Py, Pz as a result of the displacements u, v, w, and as is easy to show directly.

Hence

$$U = \frac{du}{dx}x' + \frac{1}{2}\left(\frac{du}{dy} + \frac{dv}{dx}\right)y' + \frac{1}{2}\left(\frac{dw}{dx} + \frac{du}{dz}\right)z', \qquad V = \cdots, \qquad W = \cdots;$$

these expressions have the same forms as the numerators of the three algebraic fractions whose denominators are x', y', z', and which consideration of the ellipsoid shows to be equal to the three principal dilatations e (*Exercises de Mathématiques* de Cauchy, 2nd year, 1827, pp. 63 and 68), so indeed

$$\frac{U}{x'} = \frac{V}{y'} = \frac{W}{z'} = e,$$

equations from which it is easy to eliminate x', y', z' and so obtain exactly the equation of 3rd degree in e; and the cosines of the angles of the three principal dilatations with x, y, z are

$$\frac{x'}{\sqrt{x'^2 + y'^2 + z'^2}}, \qquad \frac{y'}{\sqrt{}}, \qquad \frac{z'}{\sqrt{}}.$$

But I do not see, Sir, in your argument any reason to reach this conclusion (doubtless from my failure to have studied it enough). The truth of the conclusion proves the truth of the argument, but I admit I do not understand it. I do not see how, when the conditions

$$\frac{dV}{dz} = \frac{dW}{dy}, \qquad \frac{dW}{dx} = \frac{dU}{dz}, \qquad \frac{dU}{dy} = \frac{dV}{dx},$$

are satisfied, that is to say when the *mean rotations* about Px, Py, Pz

---

* This reference to Cauchy, of date soon after 1840, does not according to St. Venant carry priority in the application of the notion of the differential rotation in fluid motion, as he describes Prof. Stokes' analysis of 1849 (*Leçons de Navier*, p. 733) as "nouvelle et remarquable."

The definition of $(\omega', \omega'', \omega''')$ given by Prof. Stokes in § 14 in terms of the angular momentum in a small portion of the fluid that is spherical at the moment under consideration, and his indication of a proof that this angular momentum remains momentarily constant, carry us in fact to the very threshold of the fundamental theory of vortex motion discovered by Helmholtz in 1858.

[LARMOR somehow fails to see that this same "threshold" had been reached also in CAUCHY's papers here referred to. He fails also to remark, and perhaps he did not know, that HELMHOLTZ's Second and Third Theorems follow elegantly from the formula in CAUCHY's memoir of 1816 that STOKES cites in his § 11.]

are zero, the quotients

$$\frac{U}{x'}, \quad \frac{V}{y'}, \quad \frac{W}{z'}$$

should give* the value of one of the principal dilatations, and x', y', z' should be proportional to the cosines of the angles of its directions with x, y, z.

Please excuse, Sir, the importunity of my question. The interest I have in those matters and the skill with which you treat them make me desire greatly to understand well the considerations you provide.

Please accept, Sir, the expression of my high esteem,

DE ST. VENANT

author of various memoirs (among others, on torsion, *savants étrangers*, Volume XIV, and on flexion, *Liouville's Journal*, 1856).

Mr. Stokes, of the Royal Society, Professor at Cambridge.

P.S. I do not know whether you have read a *Note on the dynamics of fluids* in the Comptes rendus des séances de l'Académie des Sciences, 27 November 1843, Volume 17, page 1240. From a different hypothesis I arrive there at the same equations as you do†.

P.S. Is it possible, Sir, to get from a bookseller your memoir of 1845 on fluids and elastic solids without buying the whole Cambridge volume of 1849?

The notes are by LARMOR, who regarded the letter as showing "how, in advancing subjects, a point of view which is familiar and obvious to one school may differ essentially from the natural course of thought in another. The practical British method of development in mathematical physics, by fusing analysis with direct physical perception or intuition, still occasionally presents similar difficulties to minds

---

* This arises from a misinterpretation of the rather difficultly expressed argument in Prof. Stokes' § 2, in which the principal directions are determined from the fact that for points along each of them the relative displacement is radial.

[STOKES's argument is suggestive but unconvincing. Clearer presentations applied to more general statements due to BELTRAMI and GOSIEWSKI may be found in § 86 of *The Classical Field Theories*, cited in Essay 2 of this volume. The various kinds of time-rates natural to the expression of problems of this kind are specified and explained by C.-C. WANG, "On Gosiewski's theorem", *Archives of Mechanics* **24** (1971): 309–314.]

† Referred to in Prof. Stokes' *Report on Hydrodynamics*, 1846.

trained in a more formal mathematical discipline." I cannot see anything more "practical" or "physical" in STOKES's treatment of the spin; merely different and independent, it can be put in terms of "formal mathematical discipline", but STOKES had been unnecessarily loose on the points in question. [There is nothing national here. From STOKES's time onward Britain has produced mathematicians who felt no need to resort to the "practical" method and whose works easily meet the criteria recognized at the same time on the Continent and across the once colonial seas.]

Indeed, the contrary estimate of RAYLEIGH is the just one: "...both Green and Stokes may be regarded as followers of the French school of mathematicians", and the early work of STOKES, coming just after that of KELLAND, GREEN, and AIRY, and shortly to be taken up by KELVIN, represents the final and total triumph of the LEIBNIZ–BERNOULLI–EULER school of mathematical thought on natural phenomena, even in the stronghold of rigidified NEWTONianism. The mathematics taught in Cambridge in the early nineteenth century was so antiquated that experiment and mathematical theory had turned their backs upon one another. In order to set up a mathematical framework general enough to cover the phenomena of tides and waves and resistance and deformation and heat flow and attraction and magnetism, the young British mathematicians had to turn, finally, straight to what had been until then the enemy camp: the French Academy, where the mantle of the Basel school, inherited from EULER by LAGRANGE, had been passed on to LAPLACE, LEGENDRE, FOURIER, POISSON, and CAUCHY. LARMOR's reference to "the practical British method" transfers backward a distinction which, though real indeed in 1907, did not exist for the young STOKES, more than half a century earlier.

The real purpose of ST. VENANT's letter seems to be carried by the two postscripts, which show that in their personal relations scientists a century ago were not much different from those today.

While it is well known that STOKES corrected a major error of NEWTON in regard to the drag of friction on a cylinder rotating in a fluid, it seems to be less well known that he fully understood the effect of friction on the flow in a tube but from a mistaken faith in irrelevant experiments refused to publish his result. On page 96 of Volume 1 we read, "But having calculated, according to the conditions which I have mentioned, the discharge of long straight circular pipes and rectangular canals, and compared the resulting formulæ with some of the experiments of Bossut and Dubuat I found that the formulæ did not at all agree with experiment." STOKES then decides that the trouble is with the boundary conditions, while in fact it lay in the "complication" indicated long before by NAVIER, the complication today called turbulence. On pages 104–105 STOKES decides that "it may be well to go a certain way towards the solution", and he obtains the parabolic velocity profile but does not evaluate the constant. Therefore, today, the "law of the fourth power" is attributed to HAGEN and POISEUILLE, who had discovered it by experiment not long before and had not seen its basis in theory.

In another case of equally great importance STOKES allowed himself to be intimidated, not by experiment but by colleagues. In the note "On a difficulty in the theory of sound", published in 1848, he had remarked that a particular wave-form would become infinitely steep after a finite time. With the thoughtful and frank daring characteristic of his early work, he had written, "Of course, after the instant at which the expression (A) becomes infinite, some motion or other will go on, and we might wish to know what the nature of that motion was. Perhaps the most natural supposition to make for trial is, that a surface of discontinuity is formed, in passing across which there is an abrupt change of density and velocity. The existence of such a surface will presently be shown to be possible. ..." In the reprint of 1883, on page 54 of Volume 2 STOKES added the footnote: "Not so: see the substituted paragraph at the end", in which he explained that Sir WILLIAM THOMSON, and afterwards independently Lord RAYLEIGH, had pointed out to him that the discontinuous motion involved a violation of the principle of the conservation of energy. Indeed, RAYLEIGH had written as follows:

4 Carlton Gardens, S.W.
June 2/77.

Dear Prof. Stokes,

In consequence of our conversation the other evening I have been looking at your paper "On a difficulty in the theory of Sound", *Phil. Mag.* Nov. 1848. The latter half of the paper appears to me to be liable to an objection, as to which (if you have time to look at the matter) I should be glad to hear your opinion.

$$A$$

$\rightarrow u, \rho$	$\rightarrow u', \rho'$

By impressing a suitable velocity on all the fluid the surface of separation at A may be reduced to rest. When this is done, let the velocities and densities on the two sides be u, ρ, u', ρ'. Then by continuity

$$u\rho = u'\rho'.$$

The momentum leaving a slice including A in unit time $= \rho u \,.\, u'$, momentum entering $= \rho u^2$.

Thus $\qquad\qquad p - p' = a^2 (\rho - \rho') - \rho u (u' - u).$

From these two equations

$$u = a \sqrt{\frac{\rho'}{\rho}}, \qquad u' = a \sqrt{\frac{\rho}{\rho'}}.$$

This, I think, is your argument, and you infer that the motion is possible. But the energy condition imposes on u and u' a different relation, viz.

$$u'^2 - u^2 = 2a^2 \log \frac{\rho}{\rho'},$$

so that energy is lost or gained at the surface of separation A.

It would appear therefore that on the hypotheses made, no discontinuous change is possible.

I have put the matter very shortly, but I dare say what I have said will be intelligible to you.

STOKES, giving up, had answered:

Cambridge,
5th June, 1877.

Dear Lord Rayleigh,

Thank you for pointing out the objection to the queer kind of motion I contemplated in the paper you refer to. Sir W. Thomson pointed the same out to me many years ago, and I should have mentioned it if I had had occasion to write anything bearing on the subject, or if, without that, my paper had attracted attention. It seemed, however, hardly worth while to write a criticism on a passage in a paper which was buried among other scientific antiquities.

P.S. You will observe I wrote somewhat doubtfully about the possibility of the queer motion.

The cancelled pages, which contain the earliest analysis of shock waves and the earliest occurrence of what are called the "Rankine–Hugoniot equations", have been inserted in the present reprint of Volume 2 as an appendix at the end of the paper (page 55A ff.).

The discussion reveals the insufficiency of thermodynamics as it was then (and often still now is) understood. The year 1883 was a fatal one in which to cancel a passage on propagating singular surfaces, for only two years afterward the immortal researches of HUGONIOT began to appear, from which, in time, grew the modern theory of shock waves, created by HADAMARD, DUHEM, and ZEMPLÉN (1899–1905) and then rediscovered or appropriated and elaborated by aerodynamicists. The theory of shock waves, in turn, has served as a guidepost in studies toward a true thermodynamics of continuous media, an object of intense study now.

Another paper of great interest today, "On the conduction of heat in crystals" (Volume 3, 202), contains the earliest nontrivial example of the "reciprocity relations" currently in high favor with cultivators of the linear "thermodynamics of irreversible processes". STOKES here observes that the skew part of the heat conductivity tensor has no energetic significance, and so no measurement of temperature can ever determine that part. STOKES conjectures that it is in fact zero. He observes that thermal symmetry with respect to two planes or under central inversion forces the conductivity to be symmetric, so that "only among crystals which possess a peculiar sort of asymmetry" can we expect to find an experimental test. DUHAMEL's theory, based on "the hypothesis of molecular radiation", had led to a universally symmetric conductivity because it assumed, in effect, that the heat flux at a point was the sum of elementary currents proportional to $\partial u/\partial n$ in each direction n, the temperature being u. This hypothesis STOKES expressly rejects, in conformity with his lifelong suspicion of everything regarding molecules. As a phenomenological reason in favor of the symmetry, STOKES points out that a nonzero skew part makes the character of the lines of heat flow different from those of the temperature gradient. For example, if a source is conceived as a common center for spherical isotherms, the lines of heat flow in a conductor with nonsymmetric conductivity are spiraliform. "This *rotatory* sort of motion of heat, produced by the mere diffusion from the source outwards, certainly seems very strange, and leads us to think, independently of the theory of molecular radiation" that a symmetric thermal conductivity is "the most general possible".

A great figure has polyhedral luster. The faces of his work that shine brightest to one age of men are sometimes turned aslant from another. The short notes I have just mentioned, which currently seem two of STOKES's greatest writings, obviously were esteemed little by his contemporaries and by him. General evaluations of his contribution to science have been written by KELVIN and RAYLEIGH and are reprinted at the beginning of Volume 5. Their bad advice about shock waves aside, no other scientist has had the fortune to be commemorated by two such peers who were also his friends. It would be presumptuous to go over again the ground they have covered, nor can I say that a modern survey would differ essentially from theirs. Rather, the present introduction is intended as a supplement, especially regarding aspects of STOKES's discoveries and method of working that have recently exerted a kind of influence different from what was recognized in histories of physics and mathematics in the last century. I do not claim that this influence is more important than the earlier, but rather that, since it is new and hence an evidence of permanent greatness, and since it is additional, it deserves mention.

In recalling his early days at Cambridge, STOKES wrote "I thought I would try my hand at original research; and, following a suggestion made by Mr. Hopkins while reading for my degree, I took up the subject of Hydrodynamics, then at a rather low ebb in the general reading of the place, notwithstanding that George Green, who had done such admirable work in this and other departments, was resident in the University till he died." The basis upon which STOKES built his great theory of internal friction seems to have been a little reading and much daring and powerful pure thought:

> In reflecting on the principles according to which the motion of a fluid ought to be calculated when account is taken of the tangential force, and consequently the pressure not supposed the same in all directions, I was led to construct the theory explained in the first section of this paper. ... I afterwards found that Poisson had written a memoir on the same subject, and on referring to it I found that he had arrived at the same equations. The method which he employed was however so different from mine that I feel justified in laying the latter before this Society. (The same equations have also been obtained by Navier in the case of an incompressible fluid ..., but his principles differ from mine still more than they do from Poisson's.) The leading principles of my theory will be found in the hypotheses of Art. 1, and in Art. 3.

Imagine the reception some dusty editor, nincompoop professional society, or book manufacturer's clerk would give these lines today if they were submitted by a young college teacher, twenty-four years old and with no more than three short papers to his list of publications! An objective, impersonal, scientific style must be used, not to mention our system of references! The following revision is suggested:

> In the present paper, hereinafter referred to as Ref. 1, a theory of unequal-in-all-directions internal fluid pressure is derived for compressible or incompressible viscous flows. No slip viscosity initial boundary conditions are preferred to slip and stick slip.

Basic operational definitions and laws are given in Sec. 1 and Sec. 3. It is seen that similar relationships have been derived by Poisson (Ref. 2) and Navier (Ref. 3). Pointing out the hypotheticalness of the intermolecular (interatomic) force laws in Ref. 2 and Ref. 3, it is felt by the present author that the physical basis of the theory of Ref. 1 is to hopefully be preferred by the good-physics-knowing theoretician.

The catastrophe that has befallen the language of science in the past hundred years is only the outer dress of the catastrophe to method and thought and taste in natural philosophy. If in STOKES's earliest work the clear light of reason is turned upon experience, by the end of his lifetime theory was struggling to keep its head above the surface of piles of experimental detail, and KELVIN could write "With Stokes, mathematics was the servant and assistant, not the master. His guiding star was natural philosophy." If some recent work has found its inspiration in the period when mathematics was the guiding star for the natural philosopher, examination of STOKES's personal papers and letters shows that KELVIN's total estimate is a just one, and that STOKES moved with his day in allowing to experiment, particularly after the death of MAXWELL in 1879, an ever-increasing tyranny. By the end of the century physical inquiry had become more timid, until often the absence of creative thought seems to try to conceal itself in pages of routine mathematical operations laid out *in extenso* or in the development of computing techniques colored by attaching the names of physical quantities to the mathematical letters. Nearly all the pages reprinted in STOKES's works were first published by 1862, within twenty years of his earliest paper; if much of the contents is mathematical, and some important discoveries in "pure" mathematics are included, in all this period his experimentation and his interest in others' experiments was broader than his publications witness, and in the two thirds of his adult life which lay ahead after his important publication had ended, he experimented with an omnivorous passion. What a scientist gives to the ages, nonetheless, is not the algebraic sum of his endeavors but the best of them, not his philosophy of research but the flowers it bore.

Note for the Reprinting

The foregoing first appeared as my preface to the augmented re-edition of G. G. STOKES's *Mathematical and Physical Papers* (1880–1905), New York & London, Johnson Reprint Corporation, 1966. In reprinting it I have tried to strip off superfluous words and tighten the syntax here and there.

More detailed discussion of STOKES's analysis of the conduction of heat in crystals may be found at the opening of Lecture 7 of my *Rational Thermodynamics, A Course of Lectures on Selected Topics,* New York *etc.*, McGraw-Hill, 1969. To the statement just before (7.7) should be added "This fact was noted by M. LESSEN, 'Note on the symmetrical property of the thermal conductivity tensor', *Quarterly of Applied Mathematics* **14** (1956): 208–209." STOKES remarked also that for materials enjoying various kinds of symmetry the conductivity tensor was necessarily symmetric. In Appendix I to my lecture just cited C.-C. WANG provides an exhaustive analysis of the possibilities for all the kinds of symmetries that are ordinarily considered in applications. Entry 5 in his Table 2 should be corrected to read $\{a\mathbf{1} + b\mathbf{k} \otimes \mathbf{k}\}$.

29. GILLMOR'S *COULOMB* (1973)

The name of CHARLES-AUGUSTIN COULOMB (1736–1806) is known to every student of mechanics and physics: "Coulomb's law" of electrostatics, "Coulomb friction", "Coulomb's criterion for failure" of masonry piers, "Coulomb's equation" for the pressure of earth upon a retaining wall, "Coulomb's theory" of the rupture of arches, "Coulomb's torsional balance" and "law" of torsion. The circumstances of COULOMB's life and work are rarely mentioned and little known. The book under review is the first biography of the great savant. As such it is most welcome, and doubly so because it is the first biography of any great figure of French science (except for LAPLACE and FOURIER, those tiresome and overadvertised genii [and for MONGE]) whose active life began before yet continued through the French Revolution. GILLMOR has searched the records with industry and competence, and the first seventy-nine pages of his book present the facts and circumstances of COULOMB's life. The reader learns what it was to be a *membre adjoint* of the Société des Sciences of Montpellier (1757–1761), student at the engineering school of Mézières (1760–1761), a military engineer stationed in Martinique (1764–1772), a correspondent of the Académie des Sciences (1774–1781) who was on active duty as an engineer at various forts in France, and, finally, an influential academician (1781–1793) and Membre de l'Institut (1795–1806). Everyone interested in the intellectual life of France at this period or in the history of engineering and physics should read this part of GILLMOR's book. For example, on pages 47–48:

> The *ancienne Académie* has been characterized as haughty in its rejection or dismissal of material submitted to it by outsiders. Two comments seem in order here. First, if one examines the many absolutely worthless inventions or plans submitted to the Academy for inspection there is a good case to be made for their rejection. Sometimes these plans consisted of a single sheet of paper bearing a poorly drawn figure. Some indicated a complete lack of both drafting ability and scientific knowledge. Second, at least in the case of Coulomb, the Academy looked at almost the

same number of plans and inventions per year before as after the Revolution. Coulomb's own rate of committee work, both before and after the Revolution, was about fourteen reports a year.

He participated in important work in the committees on hospital reform and on amendment of the system of weights and measures. And in reviewing the design and first operation of the Périer brothers' water pump at the Palais de Chaillot, Coulomb was the first in France to describe publicly the principles of Watt and Boulton's improved steam engine. These are exceptions, however. In examining the actual reports that Coulomb wrote, one must say that few contain much of scientific interest. Most, after all, are engineering opinions on the desirability of developing a particular canal or hydraulic machine or they are "book reviews" of a manuscript submitted for the approbation of the Academy.

More specifically (page 68),

> ... in 1787, Coulomb and Jacques René Tenon were named to travel to England to survey the newest hospital design and operative methods in use there. Coulomb and Tenon spent eight weeks in England, visiting, among others, Sir Joseph Banks and James Watt, and seeing the cities of London, Birmingham, and Plymouth.
>
> [T]his ... episode ... indicates that a number of academicians were deeply interested in various problems of "social engineering" long before the Revolution called for these reforms under the rubric of democracy. The haughtiness of the Academy toward solutions offered by outsiders stemmed not so much from the aristocratic nature of mathematics or the Academy itself, but from hardheaded refusal to accept vague, romantic solutions to problems—be they human or natural.

Most of GILLMOR's pages are filled by efforts to explain COULOMB's permanent contribution to physics and engineering. Although the title of his book contains the word "evolution", GILLMOR seems to feel himself compelled to make his hero responsible for a revolution, in accord with a widespread dogma of professionalized history of science, which may transfer to the domain of science the Liberals' compulsion to exalt revolution for revolution's sake. GILLMOR is hard put to it to find just where this revolution lay. Since a sequence of *géomètres* beginning with HUYGENS and ending with EULER solved problem after problem that today is regarded as an indispensable part of the repertory of every mechanical engineer, GILLMOR must find some way in which COULOMB, late born, could be regarded as a revolutionary. On page 3 he attributes to COULOMB "a rational rather than a traditional empirical engineering". Since COULOMB was indeed, at least in his earlier years, a practising engineer, this distinction may serve: COULOMB did apply his results to practice, if he could, and although his greatest problems transcended practical experience, at least they originated from it. The distinction so made between COULOMB on the one hand, and PARENT, DANIEL BERNOULLI, and EULER on the other, men who were not engineers at all but took a lively and successful interest in some engineering problems, is social rather than intrinsic, and as such may stand.

But GILLMOR is not content with that. He writes (page 175) of "a real break here indicating the emergence of a portion of physics from natural philosophy" and (page 176) of "a generational break"; even more, "His generation may well represent the 'knee' of the curve in the emergence of the empirical physical disciplines." That is pure nonsense, because "the empirical physical disciplines" always must come into being before the experience they concern can be organized into a rational discipline. Rational science is not the beginning but the end, the finished product in a long process of experience, experiment, and reason. By suggesting that empiricism must somehow free itself of rational theory, as BEETHOVEN was once popularly alleged to have "freed" music from the shackles of MONTEVERDI, BACH, and MOZART, GILLMOR flies in the face of all the history of physics. Next we can expect to read that BOHR's model of the atom represented a generational break away from the formalism of HEISENBERG's (subsequent) wave mechanics. When GILLMOR writes (page 158) that "a strict rational mechanical solution to problems in friction and strength of materials did not fully account for the observed phenomena", the reader wonders what on earth "a strict rational mechanical solution" may be. In particular, to what "strict . . . solution" for problems of friction does GILLMOR refer? I have not been able to conjecture what facts of science could be wrested to lend to his assertion any meaning at all.

GILLMOR states his program on pages 137–138:

> Monographs that define a polarity in eighteenth-century physical sciences—rational mechanics versus natural philosophy— give insufficient basis for explaining the later development of the fields of heat, light, electricity, magnetism, and crystallography into the theoretical and experimental physics of the early nineteenth century. These did not emerge merely because analysis was joined to experiment. Perhaps *physique expérimentale* gave the

curiosity, engineering the reality, and rational analysis the harmony that characterize physics.

I believe that these elements distinguish the development of Coulomb's career as well.

His final paragraph is (page 230)

> Perhaps in closing it may be fitting to recall three statements that define Coulomb's approach to his work. First, both in his view of the *Corps du génie* and in public service, Coulomb said that men should be judged on their ability, and that a public service body was a *Corps à talent*. Second, in his engineering work, he called for the use of rational analysis combined with reality in experiment— for the conduct of research in engineering through use of a "mélange du calcul et de la physique." Third, this use of rational analysis and engineering reality, coupled in the pursuit of *physique expérimentale*, led to Coulomb's work in physics and the evaluation of Biot that "it is to Borda and to Coulomb that one owes the renaissance of true physics in France, not a verbose and hypothetical physics, but that ingenious and exact physics which observes and compares all with rigor."

The reader might think, as apparently GILLMOR does, that BIOT here attributes a new direction in physics to BORDA and COULOMB, but that is not so. BIOT gave them credit for *rebirth*, not creation, of "true physics", and only *in France*. It was to the French neo-CARTESians that he referred as being "verbose and hypothetical", possibly with an oblique aspersion upon the arid and complex algebraic formalism of D'ALEMBERT and LAGRANGE, who had dismissed BORDA out of hand. BIOT was a great scholar in the history of physics and especially in the thought of NEWTON, whom he most certainly regarded as a "true" physicist, long before COULOMB and BORDA.

GILLMOR tries to set COULOMB in a category by himself, above the engineers and distinct from the mathematicians (page 8):

> One of the most distinguishing traits of his memoirs, compared to other engineering memoirs of the time, is that Coulomb had the right mathematics for each problem. He stressed the point repeatedly, however, that his mathematical treatment stopped at the edge of reality or practicality and that he left further abstract development to the *géomètres*.

Nevertheless, GILLMOR must admit that the "reality or practicality" he vaunts for COULOMB had to wait a long time for engineers to perceive it (pages 114–115):

> Coulomb's method for arch design was not presented in a manner ready for application by artisans. He provided a method for determining conditions of stability in arches but he gave no definite rules of design. It is probably for this reason that it was largely ignored by civil engineers of his time. As I indicated above, most engineers preferred to use ready-made tables like those compiled by Perronet and Chézy. Coulomb *said* he was designing his method for artists and tradesmen. It is not clear whether he indicated this only to increase the value of the memoir in the eyes of the Academy or whether he actually overrated the abilities of the eighteenth-century engineer. Considering his long experience in Martinique, it is most probable that he realized his memoir would not gain immediate use.

True, indeed, but the same may be said of EULER's abundant works on specific engineering problems, which fill several volumes of his *Opera omnia*; many of these pages employ only very simple mathematics and are explained at great length in simple words, with numerical examples, and also "rules for the artisan" and "ready-made tables", yet they were not read, let alone understood, by the engineers to whom EULER addressed them. The "edge of reality" is just one of those slogans which appeal to advertisers and philosophers: COULOMB's simple but enlightening theories, like all physical theories, describe well only certain cases in nature, and those only approximately—as GILLMOR himself writes on page 100, they are "oversimplified".

According to GILLMOR (page 117)

> [Coulomb's] earth-pressure theory and the "Coulomb Equation" are the fundamental tenets of modern engineering texts in soil mechanics. From 1800 until about 1833, the majority of European bridge builders utilized his theory of design and evaluation of arches. It was his considerations in studying the neutral line in rupture of beams and strength of materials that were to be used in the early nineteenth century. One is not surprised, after all, to know that much of this memoir remained unused for forty years. It required that group of *Polytechniciens*, teachers and students, to appreciate the importance of this work in the context of the new engineering mechanics.

Again, the same may be said of EULER's theory of the buckling of long bars under axial pressure, except that not forty but 140 or more years were required before teachers and students could appreciate the importance of the work in the context of the new engineering mechanics. The same is true also of DANIEL BERNOULLI's theories of the proper frequencies and simple modes of vibrating bodies, which remained the property of mathematicians and then physicists for a century or more before their practical importance was realized. Having "the right mathematics for each problem" was certainly a characteristic also of DANIEL BERNOULLI's and EULER's analyses of machines and machine elements[1]; on the other hand, GILLMOR gives no example to show that COULOMB's work was ever given "further abstract development" by the *géomètres*, and I know of none. These are just the vague, easy generalizations of a journalist of science or a book-speculator's catalogue. COULOMB was in his own way a great innovator, but GILLMOR's special pleading diminishes rather than sustains the uniqueness he desires to attribute to his hero[2]. Again (pages 82–83), after quoting a simplistic summary of EULER's works by BOSSUT (who, though a professional mathematician, was qualified only in the lower ranges of the discipline and did not understand the more advanced parts of rational mechanics as it was then known), GILLMOR writes

It was a fusion of these two types of investigation, the physical and the rational, that Coulomb proposed. This was not a simple process of superposition. It was not merely a mathematization of

---

[1] Had GILLMOR understood the researches of EULER, he would not have attributed the concept of mechanical work ("quantity of action") to COULOMB (pages 24, 78). Estimates for the rate of work, now called the "power", of machines of various kinds run all through EULER's *Scientia Navalis* and many of his memoirs on machines. The concept itself seems to be due to DANIEL BERNOULLI.

[2] On pages 108–109 GILLMOR similarly attributed great originality to LA HIRE (1695): "The first to investigate arch design as a problem of mathematical statics". In fact LA HIRE simply adopted the then well-known theory of HUYGENS, which was based upon a theorem of STEVIN, several times published, and applied in a widely noticed tract by PARDIES (1673). GILLMOR writes on page 108 that "Hooke's elastic approach was favored by mathematicians", but in fact it was the mathematicians HUYGENS, LEIBNIZ, and JAMES BERNOULLI who expressly rejected HOOKE's "law" because it was contradicted by experiment. GILLMOR's statement is doubly wrong because HOOKE did not apply his "law" of elasticity to the arch; rather, he proposed as a model for it the inverted perfectly flexible but *inextensible* chain of discrete links, exactly as did LA HIRE a few years after him.

Although these examples have little to do with GILLMOR's subject and could well have been omitted altogether from his book, they show him succumbing to an all too common temptation of a historian of science: to dismiss as being "mathematics" any bit of science he cannot understand.

physique expérimentale, for experiment itself acquires new definitions and is performed differently and for different reasons. The
mathematics must give real solutions relevant to physics and not
metaphysics, and the experiment must be pertinent.

Pertinent to what? And what is a "real solution"? When, referring to
the evil ways of EULER, GILLMOR writes (page 82) "Physics would not
be regarded as a tennis game at which analysts could test their skill"
and describes the *géomètres* as having "worked in applied mechanics
only as it could serve to show the comprehensive power of analysis",
he resorts to the hoary dogmas of the historians of a century ago,
dogmas which are supported only by blank ignorance of the contents
of the papers of EULER and BERNOULLIS. I am sure nobody would
dismiss EULER's analysis of turbines and his invention of the guide
wheel for them as being "metaphysics" or "a tennis game", although
the former lay 200 years untested, while the latter waited 100 years
until a nineteenth-century engineer invented it anew. Rather, GILL-
MOR grasps at straws to support a distinction that in historical fact
does not exist. So eager is he to exalt COULOMB that he even apolo-
gizes for being sometimes unable to psychoanalyse him: "I cannot
explain why he attacked the problems of electricity first ...." (page
182). No physicist, however great—be he HUYGENS or NEWTON or
EULER—solves or even poses correctly all his own problems, let alone
all problems of science, and that he leaves work for the next generation to do, need not be regarded as the cause of a scientific revolution.

If we go on with GILLMOR's discussion of COULOMB's great memoir
on the statics of solid bodies[3], we reach the part on rupture. Writes
GILLMOR (page 94), "In order to limit his solution to one rupture
plane he assumed that the adherence of the block is infinite
everywhere except along the unknown rupture plane." I can make no

---

[3] GILLMOR fails to mention that comparison of a body's longitudinal and transverse
rupture loads had been introduced by GALILEO and studied again and again subsequently, both in theory and in experiment, for 150 years. That COULOMB
approached this problem afresh, is proof of no originality whatever; rather, it shows
that COULOMB, retaining a splendid and in 1772 already mostly superannuated attitude of seventeenth-century science, did not accept statements made by his predecessors until he himself could confirm them.

On p. 83 GILLMOR overlooks or undervalues the work of AMONTONS and
PARENT on friction, done long before COULOMB was born. Later (pages 87, 120–123)
he revives the work of AMONTONS but dismisses that of PARENT as being only "more
rigorous support", whatever that means. COULOMB's diagrams on page 136 illustrate
an idea which PARENT had proposed and analysed in the case of hemispherical bosses.

sense of that[4], nor of GILLMOR's succeeding description of COULOMB's analysis, and I think GILLMOR misses the real innovation of COULOMB at this point. GILLMOR noted earlier (page 91) that in his analysis of flexure COULOMB had introduced the concept of interior shear stress, as had PARENT before him; now, in his analysis of crushing, COULOMB shows how to use that concept, and he is the first person ever to do so. He *assumes* that failure will occur on a plane of maximum shear stress, and he *proves* that that plane is inclined at 45 degrees to the cross section. This passage is one of the most important in the history of mechanics, because it foreshadows CAUCHY's fundamental theorem and HOPKINS' theorem[5], but GILLMOR misses its contents. Instead (page 83) he praises "the 1773 statics memoir" for "introducing the use of variational calculus in engineering theory" and for "bringing the major civil engineering problems to consider the complexities of nature." For the second phrase it is difficult to guess any sense whatever, not only because we cannot imagine how a problem could consider complexities, but specifically because the interplay of civil engineering (whatever that term might have meant if it had existed in 1773)[6] and the complexities of nature is a subject for journalistic and administrative babble, not for history or science or any combination thereof. More disastrous is GILLMOR's reference to "variational calculus", a doctrine which is used nowhere in any paper of COULOMB and which COULOMB never showed any sign of having learnt. Even those notoriously superficial historians of the 1930s, KARPINSKI, D. E. SMITH and CAJORI, would have been incapable of such a gaffe. COULOMB's "règles des *Maximis* & *Minimis*" are simply the ordinary rules of the calculus of functions of one variable, rules which every competent scientist of the eighteenth century knew by heart and used without comment. It is not clear whether or not GILLMOR understands the difference, for in a footnote to "variational calculus" he writes, "I make no claim whatsoever as to the originality of COULOMB's *mathematical* technique. Taken by itself the maximum-minimum solution here is rather elementary." Yet GILLMOR refers to

---

[4] At the very beginning of his description of the analysis (§ VIII of his great paper in the *Mémoires des Savans Etrangers* for 1773), COULOMB writes, "I continue to suppose the pillar to consist of a homogeneous material, the cohesion of which is δ", and he always treats δ as constant.

[5] *Cf.* § 203 and § App. 46 of "The Classical Field Theories" in FLÜGGE's *Handbuch der Physik*, Volume **III**/1, Berlin *etc.*, Springer-Verlag, 1960. Although this article is not primarily historical, in the second passage cited it specifies COULOMB's contribution to the second of the theorems named above.

[6] It is not in JOHNSON's *Dictionary*, and the *Oxford English Dictionary*'s earliest quotations of it are from the late 1790s. It seems to have been coined by SMEATON.

"variational calculus" again on page 104. I can already see GILLMOR's pat phrase about "variational calculus" rising from page copied from page of the encapsulated historical errors now being petrified in the *Dictionary of Scientific Biography*, that updated successor to the pocket histories of the last century which the catalogues of merchants to book collectors quote *ad nauseam*. Of course the article on COULOMB is by GILLMOR! There he writes that COULOMB "was one of the first to utilize the variational calculus in practical engineering problems", and "he sought to demonstrate the use of variational calculus in formulating methods of approach to fundamental problems in structural mechanics." Such are the facts as reported in a ponderous monument of co-operative modern History of Science, endorsed by all the learned societies of the U.S.A.! GILLMOR's confusion obscures the great originality of the theory COULOMB here proposes. I did explain the matter on page 400 of my book, *The Rational Mechanics of Flexible or Elastic Bodies, 1638–1788*, LEONHARDI EULERI *Opera omnia* (II) 11_2, Zürich, 1960, but I presumed in my readers more comprehension of science than can be expected, it seems, in a historian of it.

COULOMB was by no means the first to apply ordinary calculus (which, along with geometry and algebra, was all the mathematics he gave evidence of knowing) to problems of this kind. It is his virtue to have found important problems he could solve with his limited mathematical tools, problems such creators and specialists in mechanics as DANIEL BERNOULLI and EULER had passed by. That is a great enough achievement for any man.

The notes I collected for this review would enable me to triple its length in the same vein. Since GILLMOR's passages on theoretical mechanics concern pages I had pondered myself, I could unravel his misconceptions and correct his misrepresentations. Electricity and magnetism are outside my range of study, and of what GILLMOR writes concerning COULOMB's work on them, I say only that I can make neither head nor tail. It should not be necessary for the reader of a scientist's biography, in order to understand what the author writes, to be himself an expert on each scientific subject taken up. If the author cannot explain scientific work properly, he should not replace understanding by journalism. Beyond this[7], as far as scientific content is concerned, I can do no more than inform the reader of this

---

[7] In passing I must mention GILLMOR's reference on page 89 in connection with GALILEO's work to "tensile stresses . . . uniformly spread over the cross section", not only gross prolepsis but also injustice to COULOMB, since GALILEO showed no evidence of having any idea of stress at all, while one of COULOMB's greatest merits is his own special yet definite approach to that most important of concepts in mechanics.

review that COULOMB's experiments on mechanics[8] have been recounted and analysed magisterially in a work not yet published when GILLMOR wrote, J. F. BELL's "The Experimental Foundations of Solid Mechanics", *Encyclopedia of Physics* **VIa**/1, Berlin, *etc.*, 1973. BELL finds in COULOMB's work on torsion, where COULOMB had no predecessor, "the origins of an experimental science of mechanics" (page 168). Also (page 173), "The first *experimental* statement that a given solid has a material elastic constant, independent of the specimen and the density of the solid, is of course a major landmark in the history of continuum physics", but GILLMOR lets this landmark pass unmentioned. BELL cites COULOMB on page 1 and page 755 and over sixty intermediate pages. On page 169 he writes

> Other parallels among the features that distinguish excellence for the experimentist may be noted. Apart from the obvious prerequisite of originality, there is perceptivity and taste in the choice of problem, an intuitive feeling for that which may be expected to broaden the understanding rather than terminate the branches; æsthetic simplicity; completeness within the framework of well defined assumptions; and a logical development utilizing different experimental perspectives which lead to new patterns of

---

[8] In regard to preceding experiment GILLMOR is as unreliable as he is for preceding theory. The distinction he makes on page 155, "Unlike many earlier reports of physical experiments, COULOMB's presentation would enable one to repeat the same procedures" is imaginary, for anyone could repeat, and many did repeat them, the experiments of MERSENNE, HUYGENS, SAUVEUR, MUSSCHENBROEK, and DANIEL BERNOULLI. Indeed COULOMB himself repeated some of them: the figure GILLMOR labels on page 88 as "Coulomb's experimental apparatus for tests of tensile rupture in stone" represents no more than a rectilinear version of MARIOTTE's on page 349 of his *Traité du Mouvement des Eaux* (1684) and of MUSSCHENBROEK's in Figure 8 of Plate XVIII of his *Physicae Experimentales . . . Dissertationes* (1729). Only since the Second World War has experiment come to be so loosely described as to be incapable of repetition—a true "generation gap".

GILLMOR passes over in silence the critical if rough experiment of JAMES BERNOULLI on the extension of a gut string, while BELL, who mentions it on about twenty-five of the some 700 pages of a general treatise on experimental solid mechanics, finds it to have provided a *Leitmotif* in experimental and theoretical elasticity for over a century.

On page 97 GILLMOR misrepresents MUSSCHENBROEK's experiments on the rupture of long wooden struts as being based on "the assumption that the column would first *bend*. . . ." MUSSCHENBROEK, who was a splendid experimentist, *observed* the bending, as had others before him. Although GILLMOR's discussion on pages 99–100 is clearer, he still leaves the reader with the impression that COULOMB's work somehow discredits MUSSCHENBROEK's. It does not. There are several different ways in which a solid member may fail in compression. MUSSCHENBROEK and EULER studied one of these, and COULOMB another.

understanding and either definitively separate the physical applicability of prior plausible explanations, or define precisely the main structure which must be included in any new proposed explanation. In the process, the region investigated must be quantitatively and qualitatively established, and associated behavior beyond such boundaries be given at least rudimentary consideration. Coulomb was the first to understand such an approach to experiment in solid mechanics. His work has remained to the present a paragon of the method he inaugurated, and he must be regarded as one of the few outstanding experimentists in this branch of physics.

He substantiates this judgment by ten pages devoted expressly to COULOMB's experiments on the mechanical behavior of solids, and in those ten pages—but the seventy-fifth part of a general treatise not intended primarily as a history of the subject—BELL gives more information about those experiments than does GILLMOR in a chapter of over thirty pages on this aspect of COULOMB's work[9].

---

[9] GILLMOR does not describe the work on torsion accurately. He confuses the use of infinitesimal elements with the true molecules of atomic theories; "the author's conception of COULOMB's torsion theory" on page 160 shows the molecules as lozenges, perhaps in some way suggesting shear. Cf. also the remarks about "macromechanics" and "molecular behavior" and "molecular planes" on page 158, "the simple inertia of the fluid molecules" on page 174. GILLMOR translates COULOMB's "molécules intégrantes" as "integral particles" (page 159), a phrase which carries no meaning to me. In most work of the eighteenth century "molécule" means "a tiny mass", and from COULOMB's use of it I think his "molécule intégrante" was just what now would be called an "element of mass" as used in forming an integral.

On page 152 GILLMOR writes that COULOMB "showed that the equation of motion is equivalent to $Id\Theta/dt^2 = -n\Theta$, where I is the moment of inertia of the body." Apart from the misprint, we ask what body and which of its moments of inertia is intended. An unwary reader might conclude that GILLMOR refers to the moment of inertia of the wire itself rather than to that of the body it suspends, especially since such was stated, erroneously, by A. E. H. LOVE in the preface to his famous treatise, *The Mathematical Theory of Elasticity*, Cambridge, Cambridge University Press, 1895, and later editions; indeed, the moment of inertia of the cross section plays a part in the theory of rods, but only in flexure and not in torsion. Equally, GILLMOR could be understood as claiming that the *general* equation of angular motion of a body rotating about a fixed axis, torque $= Id^2\Theta/dt^2$, where I is the moment of inertia of the body about the axis of rotation, is due to COULOMB, but that equation was the discovery of EULER, some fifty years earlier, a discovery known to COULOMB and to everyone competent in mechanics in the 1770s.

On page 143 GILLMOR says COULOMB "showed that through any moderate angle a deflected needle caused the torsion pendulum to oscillate in simple harmonic motion." In fact COULOMB reported no experiment to discover whether the motion was or was not simple harmonic. He showed the motion to be periodic with period independent of the amplitude, and he then *assumed* it to be simple harmonic. [A historian of mechanics ought to know that "periodic" and "harmonic" are not equivalent qualities of a motion.]

While some recent historians refrain from referring to any ideas of science from a period later than that they are describing, GILLMOR follows the older pattern of making great claims for the lasting influence of his hero, as some of the quotations above confirm. Here, too, he resorts in effect to special pleading by failing to mention COULOMB's errors along with his permanent contribution. Indeed, as GILLMOR writes on pages 157 and 161, COULOMB found that the elastic modulus was unaffected by hardening or annealing, but GILLMOR fails to tell the reader that this is one of COULOMB's few mistakes. Annealing decreases the modulus of extension; BELL writes (page 178; *cf.* also pages 213, 238), "The prestige of COULOMB was such that . . . no one seriously questioned the matter for half a century, until WERTHEIM demonstrated otherwise."

The vague remarks of GILLMOR about the importance of "Coulomb's theory of the role of cohesion" for "others who succeeded him" (page 161), with a footnote listing a number of minor men, mask the enormous influence COULOMB's simple and clear ideas exerted upon theory and experiment in plasticity down to the present time. He mentions only in a footnote (page 155) COULOMB's observation that great tensile force decreases the torsional modulus, yet in 1865 Lord KELVIN described this fact as providing his own "chief interest in solid mechanics" (BELL, page 83), and BELL devotes his entire § 2.20 to "the decrease of moduli with permanent deformation," describing half a century and more of experimentation which arose from this discovery of COULOMB's.

No reader of this review will regard it as an attempt to diminish the importance of COULOMB. It is the reverse: a lament that injustice has been done to the great man by claiming in vague terms that he did what in fact he did not do, while failing to describe the splendid things he did do—an injury to which French scientists, for example, D'ALEMBERT, LAGRANGE, and LAPLACE, have frequently been subjected before now.

It is a tautology to say that before he can apply a mathematical theory or technique, the theorist must first have it. The very greatest theorists were able to create new mathematics in response to the needs of a physical theory, but they were very few indeed[10]: ARCHIMEDES, HUYGENS, NEWTON, JAMES BERNOULLI, EULER, POISSON,

---

[10] On page 221, in reference to COULOMB's "limitations as a mathematician", GILLMOR attributes to BOCHNER "the extremely perceptive observation that in Coulomb's time it was not yet the fashion for *physicists* to 'make up' mathematics to aid their work." GILLMOR's habitual journalese by referring to "the fashion" and "physicists" makes his generalization so vague as to be difficult to refute. Great scientists seldom followed "fashion", and few of the persons regarded as "physicists" in the eighteenth century figure largely in the history of what today is called physics. The matter is complicated

CAUCHY, GREEN, RIEMANN, POINCARÉ, HADAMARD. Most theorists, even great ones, used only mathematics already at hand. Of these, most had a large arsenal of theory at their disposal, but there were some whose tools were few and primitive for their day, yet they could fashion much of great value with them. To this group, which includes GALILEO, CLAUSIUS, BOLTZMANN, and EINSTEIN, belongs COULOMB.

The truth about COULOMB does not make good journalism. He was a fine experimentist, but not the first or the last. His mathematical capacity was limited, but such mathematics as he used was sound and sufficient for his ends. He attacked major problems, and he made major discoveries in several fields of physics. This combination of achievements is rare in the history of science, though neither unique nor unsurpassed by some few predecessors and successors. What COULOMB did should be more than enough to call forth, some day, a biography which does justice to his integrity, his intellect, his devotion to science, and his good taste.

The quotations above reflect the repetitious disorganization of the book under review. There are essentially good passages here and there, such as the summary of "Cartesian" and "Newtonian" mechanics in practice (pages 214–215) and the account of experiments in the observatory (pages 146 ff.), and especially in the narrative at the beginning and the biographical epilogue, but most of the pages are tedious, and many paragraphs are incomprehensible. The text abounds in

---

by the fact that on page 179 of BOCHNER's *The Role of Mathematics in the Rise of Science* (Princeton, 1966), which GILLMOR cites here, there is not a word on this subject.

Nonetheless, on page 188 of BOCHNER's book we read, "A notable exception was Huygens. He was only a physicist. But he was a physicist's physicist . . . ." I join GILLMOR in recognizing that the observations of my revered and admired teacher and friend SALOMON BOCHNER are often "extremely perceptive", but this is not one of those, for here BOCHNER is dead wrong. It was HUYGENS who dismissed JAMES BERNOULLI's solution of the problems of the elastica as "unworthy of geometry" because BERNOULLI took all quadratures as given. Earlier, it was HUYGENS who had delayed publication of his discoveries about the pendulum for at least sixteen years because his proofs did not yet satisfy his criterion of absolute geometrical rigor, although of course he patented his mechanisms and did his best to make money from them. It was HUYGENS who by rigorous though special and almost Hellenic methods had found several major facts which later came to be a part of the differential calculus, such as the formula for the radius of curvature of a general plane curve. It was HUYGENS who, in order to design an ideally isochrone pendulum, discovered and then proved with perfect rigor that the evolute of a cycloid was a cycloid. It was HUYGENS who, almost alone in the West, joined ARCHIMEDES in insisting upon mathematical precision and rigor in the treatment of problems of physics.

Here, as elsewhere in his book, GILLMOR has rested heavily on secondary sources as far as science is concerned, but he has not known how to filter them for truth.

journalese[11], pompous, overstuffed phrases[12], horrid noun piles[13], unattached participles[14], college jargon[15], and outright confusion[16]. Particularly baffling is GILLMOR's mixture of old and new scientific terms without defining or relating them[17]. I confess myself unable to understand how anybody could read as many pages of simple, limpid, elegant French of the eighteenth century as GILLMOR must have done, yet describe their contents in an English not only obscure but even incorrect.

The book is based upon its author's dissertation at Princeton. Not many years ago, no university would have accepted a thesis so verbose, so ill-organized, so self-contradictory as this one, and no university press would have dared publish a book written in such grubby English, unless, perhaps it were by a "social scientist". Those who have allowed this book to be published now have done its author no favor. GILLMOR's first seventy-nine and last nine pages show his talent and industry as a biographer. Had they been issued as a short and personal biography, with the English reasonably pruned and cleaned, they would have served a purpose and gained their author lasting credit, as will the documentary appendices. The remaining 170 pages of text demonstrate again and again that their author had not been

---

[11] "Fiercely proud" (page 5), "poignant letters from the 1760 student" (page 14), "he was posted" (pages 25, 27, 31, 32, 40), "a French 'crash' technical program" (page 19), "a tremendous responsibility... a tremendous experience for Coulomb personally" (page 23), "pertinent" as an absolute (page 83), *etc.*

[12] "He was very desirous of obtaining" to mean "he much wished to get", "ripost" to mean "answer", (page 35), "he was denied access" to mean "he could not find" (page 49), "reparation" to mean "repair" (page 65), reference to himself as "the author" and "I" a few lines apart (page 159), "torsion spectrum" to mean "range of twists" (page 150), "matter on the molecular level" to mean nothing at all (page 151).

[13] "Torsion suspension magnetic compass" (page 49), "electricity and magnetism memoirs" (page 136), "lack of author citations" (page 175), "Coulomb's fluids studies" (page 173), "the 1777 magnetism contest" (page 140), "the 1775 (and 1777) Academy prize committee" (page 142), "the Coulomb statics memoir" (page 115), *et ubique.*

[14] P. 178: "Based on the above two principles, Coulomb then offered...," and "Expressed in modern notation, Coulomb established..." (page 152).

[15] "Entrance exam" (page 7), "assorted texts in math and physics" (page 228).

[16] On page 51, COULOMB's researches were magnetic, and on page 63, filtered water flowed in the Seine. On page 143 COULOMB "examined the parameters", making us wonder if the parameters responded or pleaded the Fifth Amendment.

[17] *E.g.*, on page 143, GILLMOR speaks of "the force of torsion" but at once uses the modern "torque" and tells us that COULOMB called it "momentum de la force de torsion", and then lower down he speaks of the "torsion" as being proportional to the cube of the diameter. Only the expert will know that all four different technical terms on this page mean the same thing.

Most of the terms by which a reader could trace important concepts fail to appear in the index: buckling, collapse, deformation, extension, force, fracture, metal, modulus, pier, pillar, plasticity, pressure, rupture, silk, strain, stress, tension, thread, wire.

subjected to the discipline, necessary before undertaking a long and broad work, of writing short and cogent treatments of limited aspects of a subject. Every line in the book bears witness to GILLMOR's sincerity and devotion to his subject. That he does not understand the science he describes, does not prove he could not have done so, had he had good guidance and the patience to follow it.

Note for the Reprinting

This review of C. STEWART GILLMOR's *Coulomb and the Evolution of Physics and Engineering in Eighteenth-Century France*, Princeton University Press, 1971, first appeared in *Eighteenth-Century Studies* **7** (1973/1974): 213–225.

30. TIMOSHENKO'S *HISTORY OF STRENGTH OF MATERIALS* (1953)

TIMOSHENKO's numerous earlier books are distinguished by their presentations based on original sources and by their abundant and careful attributions. In this volume he publishes in extended form his historical lectures. The main earlier works on this subject are

(1) A.-J.-C. BARRÉ DE ST. VENANT, *Historique abrégé des Recherches sur la résistance et sur l'élasticité des corps solides*, prefaced to NAVIER's Résumé des Leçons ... sur l'Application de la Mécanique ..., Paris, Dunod, 1864.

(2) The abundant notes added by ST. VENANT to the translation *Théorie de l'Élasticité des Corps Solides de* CLEBSCH, Paris, Dunod, 1883; reprinted by Johnson Reprint Corp., 1966.

(3) I. TODHUNTER, *A History of the Theory of Elasticity and of the Strength of Materials from Galileo to Lord Kelvin*, edited and completed by K. PEARSON, Cambridge, Cambridge University Press, 2 volumes, 1886, 1893; reprinted by Dover Publications, 1960. [The contributions by PEARSON, which make up most of the work, are often prejudiced, capricious, and misrepresentative; sometimes they are even wrong.]

TIMOSHENKO has read these works with [some] care, but, besides including [a little] additional material from earlier periods and [a good deal] from later researches down to 1950, he has studied [some of the] sources anew, giving in some cases reasons for dissenting from accepted views. It is evident that this book is the result of great love and understanding for mechanics combined with many years of study and criticism. About 700 authors, many of whom are still living, are mentioned, references (often, unfortunately, incomplete and sometimes inaccurate) being given in each case; a small amount of biographical material, sometimes supplemented by a portrait, is supplied for about seventy. ...

Mathematicians already adept in continuum mechanics will find the book useful and enjoyable. First, it explains in historical setting connections between mathematical theory, engineering calculation or hypothesis, and test or experiment. Second and more important, it explains the guesses and practical expedients which must be used in domains of solid mechanics where adequate mathematical theory has yet to be constructed, whence the mathematical reader will carry away a hearty respect for the difficulty of the problems and the ingenuity with which engineers have faced them. For mathematicians who enjoy not only the solution but also the formulation of mathematical problems, this book with its critical summaries of discussions *pro* and *con* as practice arose will suggest far more challenges than the treatments in textbooks for engineers. It is interesting as a sidelight to learn about the engineering done by such mathematicians as LAMÉ and KLEIN. TIMOSHENKO writes several illuminating discussions of mathematical instruction for engineers in various famous institutions through various periods and of the relation between mathematical knowledge and progress in engineering. The reader is led to infer from numerous examples that, at least in the past, a sound training in advanced mathematics has often led to engineering discoveries even of the most practical type, and that when the training of practical engineers at a place and time has neglected proper foundation in mathematics, the whole practice of design and testing has stood still or degenerated. While the author in his last three chapters carries the work up to the present day, he leaves the reader free to form his own judgment of this period.

TIMOSHENKO's discussion of work before 1823 is somewhat misleading because he often explains it in terms of the concept of stress, which had not yet [been made general and precise]. It is also incomplete. The selection of subjects and authors since 1920 seems somewhat capricious to me. The presentation tends to emphasize the details at the expense of the principles. For example, there is little or no study of the origin of the concepts of shear strain and shear stress. Despite a choppy style and occasional errors in grammar and spelling, the strongly positive character of this book makes it difficult to put down until the last page is reached.

Note for the Reprinting

The volume reviewed is S. P. TIMOSHENKO's *History of Strength of Materials. With a Brief Account of the History of Theory of Elasticity and Theory of Structures*, New York *etc.*, McGraw-Hill, 1953.

The review is reprinted from *Mathematical Reviews* **14** (1953): 1050.

The emendations in square brackets reflect the fact that at the time I wrote this review I had not studied ST. VENANT's *Historique abrégé* with sufficient care. It remains today the fullest and most accurate general history of linear elasticity in the first six decades of the nineteenth century. On earlier work it is maculate. While PEARSON took account of many more papers, especially by British authors, his comprehension of the theory of elasticity and of experiments on the strength of materials was limited and superficial. Valuable additions and corrections for the latter have been supplied by Mr. BELL in his wonderful treatise, *The Experimental Foundations of Solid Mechanics, Encyclopedia of Physics* **VIa**/1, Berlin *etc.*, Springer, 1972.

TIMOSHENKO in the book under review, as also in his other books, relied heavily on the scantness of American engineers' education, which left them unable or at least disinclined to read works in foreign languages. Although his books are almost wholly devoid of originality, they served to acquaint American mechanical and civil engineers with theory and history they were otherwise unlikely to encounter. In this way he helped put them on grounds of equality with their colleagues on the European Continent, enabling them to turn their native ingenuity, intelligence, and industry into scientific channels when and if they were so inclined. The book reviewed above provided such engineers a historical footing in the kind of mechanics that was useful to them. In this function it was unique; its influence on engineers has been and remains good.

When, five or six years after writing the above review, I came to study the development of theory and experiment in the mechanics of solids before the nineteenth century, I found all existing attempts, even ST. VENANT's, so fragmentary as to be almost useless. For that reason I felt compelled to write the history of the early work. A compact summary of my book is provided by my "Outline of the history of flexible or elastic bodies to 1788", *Journal of the Acoustical Society of America* **32** (1960): 1647–1656.

31. SZABÒ'S *GESCHICHTE DER MECHANISCHEN PRINZIPIEN UND IHRER WICHTIGSTEN ANWENDUNGEN* (1979)

For hundreds of years mechanics and physics were almost synonymous; through the middle of the nineteenth century a book of theoretical physics was more than half devoted to mechanics; and even today the disciplines of mechanics in elegant, precise mathematical statement abstract, sum, and correlate the greater part of the physics of ordinary phenomena on the human scale. Mathematical mechanics developed mainly in the seventeenth and eighteenth centuries; mechanical phenomena dominated manufactures and much of everyday life during the following period of industrialization; the philosophies and many of the characteristic attitudes of the West reflect, whether in approbation or contest, mechanics in its universality and mathematical precision; modern life cannot be conceived except intertwined with use and abuse of mechanical principles. Yet general histories of science display such ignorance of mechanics as to give a false picture of science as a whole, and general histories of mechanics in detail or in depth have been few. The first, it seems, is provided by the historical sections LAGRANGE included in his *Méchanique Analitique*, 1788. LAGRANGE's elegant conciseness and dry, impersonal tone have seduced generations of readers to regard his choice of material fair and his description of it accurate. A second stream of historical faith derives from MACH's *Die Mechanik in ihrer Entwicklung*, 1883, which with scant respect for the sources presents mechanics as a collection of coarse rules regarding some simple kinds of bodies. MACH admitted that he was content to sacrifice correct historical detail in his endeavor to promote right thinking about the experimental method. At least until the last few years the physicists have swallowed MACH's prejudices and simplisms lock, stock, and barrel. LAGRANGE's influence has been supreme with students of a mathematical bent. It is reflected in the one fairly comprehensive book we have: R. DUGAS' *Histoire de la Mécanique*, Neuchâtel, 1950, which starts from Hellenic science and ends with the disputes about quantum mechanics in the 1930s. When DUGAS' book appeared, I reviewed it with great favor and some specific criticism (see above, Essay 17 in this volume), but my own study of the sources as it subsequently progressed has made DUGAS' errors and gaps multiply, his decent passages dwindle. DUGAS' scope was too great for any man to master in 1950 or today.

The appearance of a new history of mechanics is a major event in the history of science. SZABÒ's book, less ambitious than DUGAS', covers five aspects:

I. Earliest establishment of the classical mechanics of rigid bodies: NEWTON, EULER, and D'ALEMBERT   41 pp.

II. Controversies and further development of mechanical principles from the seventeenth to the nineteenth centuries   93 pp.

III. History of the mechanics of fluids   172 pp.

IV. History of the linear theory of homogeneous and isotropic elastic materials   108 pp.

V. History of the theory of impact   53 pp.

Apart from some short passages here and there, the earliest author treated is STEVIN; while some references are very recent, most researches SZABÒ describes in detail were done before 1900. The authors most frequently cited are EULER, NEWTON, DANIEL BERNOULLI, GALILEO, LEIBNIZ, and HUYGENS, in that order. COPERNICUS and KEPLER are mentioned several times with great respect, but no account of their work is given. The contents of the book derives from essays SZABÒ has been publishing for several years past, mainly in *Humanismus und Technik,* which he has here revised and woven into consecutive accounts. Since the principles of mechanics grew from analysis of special systems such as mass-points, rigid bodies, elastic materials, and fluids, any treatment of the principles will necessarily have to describe those special systems both as ends in themselves and as contributors to various streams of thought, and so overlap is unavoidable. It is plain that we have here a collection of essays interconnected to form extensive chapters in a general history. We ought not regard the book as a candidate to replace DUGAS', much as DUGAS' needs to be replaced.

SZABÒ brings to his task a background nowadays become most unusual for a historian of science: early training in a *humanistisches Gymnasium* before the war, with the consequent associations and style of a lettered man, absence of which is all too painfully noticeable in most "scholarly" writing today; undergraduate specialization in physics; employment as a mechanical engineer in industry; doctorate in engineering at the Technische Hochschule in Charlottenburg; and thirty years of teaching mechanics to engineers in that institution. During all this time the history of mechanics attracted him, and finally it became his principal interest. In the preface he tells us that he at first accepted the hoary old legends deriving from MACH and the physics teachers, but soon he learned that he must resort to the sources directly, to study the masters, not their students. Like others who have learnt the same lesson, he has declared independence from the traditions of LAGRANGE and MACH, and he has formed a view at variance with theirs. He is not alone in fostering a third picture of the development of mechanics, a picture that reveals the grand principles as achieved by distillation of ideas and methods invented to solve special problems. These special problems grew mainly from the phenomena of everyday life and pondered experience of it, but of course they did not exclude what physicists today call "fundamental" aspects: the invisibly small, the astoundingly swift, and the inaccessibly distant. On the whole SZABÒ shows scant influence of the writings of recent historians of science. An amusing exception occurs when he allows his clarion dislike of philosophers to let him declare war upon a young lady from Wisconsin (pages 84–85), but it is really out of place, for batrachomyomachy of this kind could easily consume a lifetime to no purpose and fill a book bigger than SZABÒ's.

Through perusing the sources on which many passages in DUGAS' history draw I have gradually come to conclude that he did not understand their deeper and sharper aspects. Perhaps he did not try to; perhaps he felt that as a historian he had done his duty when he arrayed extracts, letting the authors speak for themselves. In SZABÒ's book, in contrast, most pages reflect intimacy with mechanics as a whole and knowledge of the sources. We recall that GIBBON laid weight upon the value of his years of experience as an officer of the militia, even though they never brought him into battle or more than a few miles from the comforts of his father's house. It is scarcely possible to review the whole of a book that goes as deeply as SZABÒ's does into so many major aspects of a subtle science. Few if any reviewers will have chosen to probe just the parts of mechanics that SZABÒ has and thus be in a position to write founded criticism. I will confine myself to comments on some passages that to me seem to deserve notice. Another reviewer by selecting different passages might give the reader a different picture of the book. One admiring review that has appeared already shows blank ignorance of the history of science and even follows SZABÒ's habitual misspelling of EARNSHAW's name. It should go without saying that in so short a book as this on so broad a subject, the author's summaries must be not only numerous but also oversimplified. DUGAS cannily avoided this problem by leaving the reader to judge for himself; praiseworthy as such a policy might seem, the excerpts DUGAS provided as a basis from which the reader was to draw his own conclusions were mainly so brief as to warrant no conclusion at all, and those he omitted were often finer than those he chose to quote. SZABÒ, in contrast, tends to write too many sweeping summaries, and many of these are unjust. I remark now on some of these.

The reader who is astonied to see a history of mechanics begin with the theory of rigid bodies must read pages 12–30. There he will recognize in SZABÒ a pupil of HAMEL, whom he quotes as calling the mechanics of points "intellectually unclean" and who in his celebrated *Habilitationsschrift* of 1909 wrote "In what follows I avoid the mechanics of points; what is usually understood by the mechanics of points is nothing more than the theorem on the center of gravity." This attitude, absolutely opposite to the physicists', is tenable; at one time, briefly, I held to it; now, with a firmer grasp upon what a mathematical theory of physics can and should do, I prefer to think of mass as being a LEBESGUE–STIELTJES measure which may be singular at points, lines, and surfaces, and I think this idea renders concrete and rigorous what LAGRANGE in his vague and formal way tried to represent with his famous symbol S (which appears unexplained on SZABÒ's pages 24–26, 30, 40, *etc.*). Nevertheless, pages 12–30 of SZABÒ's book provide not only a delightful introduction to his way of looking at mechanics and its history but also a vigorous and eloquent defense of a general mechanics based upon informal concepts of force and deformable body as primitive and fundamental. Of course, that is the tradition of NEWTON and EULER. There are few men today whose knowledge of mechanics both *as it is now practised* and *in its history* would have sufficed to write these pages.

In SZABÒ's treatment of mathematical hydraulics and the "BERNOULLI equation" (pages 157–192) we find another major departure from older views. The scientific matter here is not new, for JOHN I BERNOULLI's great deserts were explained and justified in modern terms some decades ago. Rather, it is the relations between him and his son DANIEL, and incidentally the relations of both with EULER, that are re-evaluated, on the basis in part of a more careful and thorough study of his *Hydraulica, nunc primum detecta ac demonstrata directe ex fundamentis pure mechanicis, Anno 1732*, first published 1743. No competent reader will fail to see not only that this work presents the hydraulics of frictionless fluids more clearly than DANIEL BERNOULLI had done in his *Hydrodynamica*, 1738, but also that it goes beyond the earlier work in its solution of special problems which illustrate the effects of varying cross-section and unsteady flow. It was long customary nevertheless to attribute to the son not only the fundamental theory for steady flow in a tube of uniform section but also the whole idea of the "BERNOULLI equation", and to regard the father's work on this subject as merely derivative. SZABÒ investigates the matter with care. First, he presents JOHN BERNOULLI's results in more detail than may be found in any other study. Then he traces the origin of the widely accepted idea that the father simply manufactured the date 1732 and in part plagiarized the son's work. SZABÒ is right to condemn what SPIESS wrote in this regard in his biography of EULER, 1929, and repeated in his *Grosse Schweizer*, 1942: that the father took "the best" from the son's work and published it under his own name, "dated back". SPIESS did not understand the least thing about hydraulics and was in no position to decide what was "the best" in DANIEL BERNOULLI's book or to determine what JOHN BERNOULLI's tract contained. SZABÒ finds that the reproach goes back no further than MORITZ CANTOR's general history of mathematics, Volume 2, 1892, whence perhaps SPIESS derived it. Furthermore, SZABÒ finds no trace of anything but respect for JOHN BERNOULLI's independent achievement in the early German expositions of hydraulics: KARSTEN (1770), KÄSTNER (1797), RÜHLMANN (1857, 1880). Above all, of course, stands the immediate judgment of EULER, 18 October 1740, which his old teacher proudly printed at the head of his work. For EULER, close friend of DANIEL BERNOULLI, it was JOHN BERNOULLI who had found the "true and genuine method" for constructing a general theory of hydraulics, and he wrote as much to DANIEL BERNOULLI on 15 September 1740. While in general SZABÒ does not hesitate to point out the shortcomings of earlier authors, with unnecessary tact he passes over in silence my two unjustified general statements in this regard, which appear on lines 6–7 of page XXXII and lines 18–19 of page XXXVII of my account of the whole matter, pages XXXI–XXXVII of L. EULERI *Opera omnia* (II) **12**, 1954. I am not sure that SZABÒ's arguments absolve the old father entirely, but certainly they make me wish I could soften those two statements.

On pages 263–271 SZABÒ makes much of ST. VENANT's priority to STOKES in establishing the NAVIER–STOKES equations from phenomenological reasoning. He blames STOKES for not following the entire international literature. I think SZABÒ is off the mark here. The British of the day, ossified in their NEWTONIAN fluxions and obscure, loose physics, for the most part despised foreign work and especially the monumental papers published in the 1820s and 1830s by the great mathematical physicists of France. STOKES was exceptional; he was a leader in teaching his compatriots to use EULER's mathematics and to study the works of FOURIER, CAUCHY, and POISSON. His successful devotion to the most difficult task in all science, namely, to bring the Cambridge establishment up to date, is to me more important than his having missed a few French papers. Moreover, I do not agree with SZABÒ's opinion of STOKES's derivation. Indeed, it is not complete, but I regard it as one of the early, creeping steps towards reduction of constitutive equations by applying the principle of material frame-indifference, and I regard his long kinematical analysis as adding some beautiful ideas to the corpus which EULER and CAUCHY had provided and which ST. VENANT called upon in his short presentation. ST. VENANT simply *assumed* that the spin had no effect on the stress; STOKES made steps toward *proof* that it could not have any. ST. VENANT just added the hydrostatic term to equations CAUCHY had proposed and motivated as being appropriate to "soft" materials; STOKES took up the entire conceptual burden, calling upon no previously standing theoretical apparatus except what EULERian hydrodynamics provided. Then, too, ST. VENANT did nothing with his general equations, while STOKES in the very same paper and a later one provided classic examples of the effects internal friction could have. Also SZABÒ forgets to tell us that it was STOKES himself, in the same year as DUHEM, who showed that $\lambda + \frac{2}{3}\mu \geqq 0$ and $\mu \geqq 0$. SZABÒ omits the latter relation and forgets the sign of equality in the former, which he describes in such a way as to mislead a reader who does not already know the details. SZABÒ's statement on page 270 about an experimental verification of the NAVIER–STOKES equations is not clear. I miss any mention of the work of STOKES and KIRCHHOFF on the absorption and dispersion of sound due to viscosity and the conduction of heat, especially since in it the coefficient λ makes itself felt in the combination $\lambda + 2\mu$, and KIRCHHOFF did not assume that $\lambda + \frac{2}{3}\mu = 0$, though he remarked that MAXWELL's kinetic theory of moderately rarefied monatomic gases supported that specializing relation.

SZABÒ greatly admires PRANDTL and BECKER. Few now will disagree with him here, but I am not sure the achievements he describes will seem so great fifty years hence. [PRANDTL's chance of survival after the death of his disciples is better than BECKER's.] SZABÒ seems to wish us to think BECKER introduced the shock layer, but it is nothing more nor less than one of the "quasi-ondes" of DUHEM's *Recherches sur l'Hydrodynamique*, 1903. Neither, I am sure, did BECKER's work on the shock layer provide the "clarifying closure" (klärender Abschluss) SZABÒ attributes to it on pages 313–314. He does not tell us that BECKER's conclusions are far from representative because BECKER chose a particular PRANDTL number with no physical basis at all, just because for that number his equations became easy to solve; moreover, BECKER assumed the Stokes relation; general analysis of the shock-layer problem makes its solution depend upon $\lambda + 2\mu$ rather than BECKER's coefficient $\frac{4}{3}\mu$; sufficiently large bulk viscosity has the effect of making the shock layer as thick as may be desired and thus emasculates BECKER's conclusion quoted in fine print on page 313. The reasoning there is illogical anyway, since it tacitly forbids us to interpret continuum fields as representing phase averages from statistical mechanics. SZABÒ finds it astonishing that BECHERT did not know BECKER's work, but he himself does not go on to tell us about the definitive treatment begun by no less a mathematician than WEYL and completed by v. MISES and GILBARG. A brief account of the general theory may be found in SERRIN's article on pages 226–230 of Volume VIII of FLÜGGE's *Encyclopedia of Physics*, 1959. If we accept the NAVIER–STOKES equations, we can determine λ by use of measured shock thicknesses or by measured absorption of ultrasonic waves. Whether or not the Navier–Stokes theory applies to ultrasonic waves and shock waves, is a matter far from clear. Perhaps SZABÒ's remarks may be interpreted as reflecting this state of affairs.

Coming back to an earlier time, I must say that the description of NEWTON's work on sound which SZABÒ provides on pages 282–284 is misleading in its apparent simplicity; it is a modernization which seems to me to betray the original thought while remaining insufficient to convince a modern reader. It is the sort of thing we find in a physicist's textbook. Also I think the explanation of LAPLACE's theory of sound on pages 285–286 is pretty mysterious. Certainly it is incorrect as well. To LAPLACE we owe the idea, expressed with a clarity rare in his work, that the sonorous vibrations are adiabatic. What POISSON did in the paper cited on page 286 was to free the derivation of the complexities in which LAPLACE had hidden it: shells of attracting particles of air and particles of caloric, leading to horrid, unnecessary expansions and approximations. Furthermore, POISSON's law (141) can be read off from LAPLACE's formulæ, and anyway POISSON did not use it to calculate the speed of sound. The reader should be told that LAPLACE's formula (140) does *not* require κ, the ratio of specific heats, to be constant; that the LAPLACE–POISSON law (141) does not hold *unless* κ is constant on the adiabat considered; that while both LAPLACE and POISSON used the caloric theory of heat, they had no need to do so, for all of their results that SZABÒ chooses to present are purely calorimetric.

SZABÒ's description of "the interaction of gas dynamics and thermodynamics" (pages 288–291) is too brief to be helpful. In his equation (148) he introduces the absolute temperature T without telling us what it is or how it differs from the r that he has printed on page 251 in a formally identical equation of EULER. Next he attributes to CARNOT formulæ which not only involve T but also appear nowhere in CARNOT's work and are contradicted by what CARNOT assumed. That was, in SZABÒ's notation, $Q_1 = Q_2$, as SZABÒ himself tells us in different words at the bottom of the page. The whole section on thermodynamics is so grievously counter to history that it should simply be torn from this otherwise largely commendable book. What more positive proof could there be of the sorry mess in which physicists have left thermodynamics than that a man so learned in mechanics as SZABÒ plainly is could have let these pages stand as the best he could make out of the subject? It seems unlikely to me that he would have stated all those positive claims about what CARNOT, MAYER, JOULE, and CLAUSIUS did, had he followed here, too, his scrupulous practice in the rest of the book: to learn from the masters, not their students. Particularly unfortunate in a historical work is use of the entropy to derive the LAPLACE–POISSON relation $p \propto \rho^{\kappa}$ for adiabatic change. That relation antedates all theories of thermodynamics; it was correctly derived in 1823, more than a quarter century before the entropy was introduced (by RANKINE, 1850) and nearly half a century before CLAUSIUS coined the name; it is a consequence of the theory of calorimetry alone and hence enjoys a status superior to theories' relating heat and work. To have reproduced in a historical essay the usual physicists' hocus-pocus with differentials that have signs and can equal or exceed things that are not differentials, has done no service to any student. The reader who is not already expert in the history of thermodynamics naturally gets the idea that CARNOT, MAYER, and CLAUSIUS really made their discoveries and explained them in terms of quasistatic processes and by dividing up differentials (or nondifferentials) into reversible and irreversible parts, which is flatly untrue.

Anybody can find omissions in a book with so broad a scope as this one. It seems to me that the very brief account of COULOMB's work on pages 385–389 could have made the concept of shear stress clearer and ought to have mentioned his experiments on torsion, which provided a major stimulus to the mathematical theory of elasticity, to be created by CAUCHY and applied to torsion by ST. VENANT.

An omission astonishing in a book by a teacher of mechanics to engineers is the concept of work. Work appears unmentioned in the energy theorem expressed by Equation (8) on page 72 in connection with DANIEL BERNOULLI and again on page 174, this time in a passage by JOHN BERNOULLI, and there the word "Arbeit-" appears in parentheses, but it is not listed in the brief and insufficient index. What work means and why we should wish to calculate it, we are not told in either instance. It is my impression that DANIEL BERNOULLI introduced and motivated the definition, and anyone who reads EULER's wonderful papers on machines and ships will encounter it fluently. The history of the concept is still to be traced. Many years ago LIPPMANN remarked that it was SADI CARNOT who took the concept of work, already long and thoroughly familiar to the geometers, and made use of it in physics. LIPPMANN regarded this transfer as the greatest of CARNOT's many achievements; he pronounced it "the beginning of mathematical physics". To this I would add a negative contribution: Thermodynamics is the first mathematical theory of physics to have been created by men who were not mathematicians. It has shown frightful birthmarks ever since.

For me the most interesting chapter of SZABÒ's book is the last, which concerns the mechanics of impact. Impact is particularly important for the foundations of mechanics because it guides us to understanding discontinuities: we encounter instantaneous changes of momentum, not rates of change. Physicists of the nineteenth and twentieth centuries in their textbooks on classical topics, particularly electromagnetism, have tried to explain discontinuous motions as limits of continuous ones. Smooth changes have been second nature to mathematical scientists since the eighteenth century, but as the pioneer studies on impact are earlier than that, it is no wonder that they did not resort to any detour through a sequence of smooth descriptions. Modern studies of the foundations have been almost totally silent on impulse. A welcome change will be seen in the forthcoming[1] paper by ANTMAN & OSBORN in the *Archive for Rational Mechanics and Analysis*. SZABÒ's chapter is easy to read. After reading it I could follow the main lines of thought in a particularly fine paper by EULER which I had not studied before. Next comes an interesting account of a research by POISSON, something always welcome, for POISSON is the most unknown and underestimated of the great figures of mathematical physics. SZABÒ's discussion of MARCI, especially on pages 457–459, shows that more work is to be done there. It is a pity SZABÒ does not take up the theory of HARIOT, published some years ago. While DESCARTES' theory of impact is notoriously bad, I wish SZABÒ had been able to estimate DESCARTES' general contribution (especially pages 54–62) more equably. I do not see that in the contrast between DESCARTES and GALILEO the right lies all on one side. I think about half of DESCARTES' famous critique of the *Two New Sciences* is founded. Although GALILEO protested loudly of the value of mathematics, he was not remarkable in his ability to use it, while DESCARTES compensated somewhat his pronounced weakness in physics and swaggering carelessness in regard to it by the critical power and insight of a great mathematician. More than that, WHITE-SIDE has shown us that it was DESCARTES who was NEWTON's real teacher. If the steely-splendid preface to the *Principia* is not just a bolt

---

[1] ["The principle of virtual work and integral laws of motion", Volume **69** (1979): 231–262.]

from the blue (and where in science are such bolts to be found?), who else than DESCARTES can be its grandsire?

Retrospect upon the factual and interpretive content of SZABÒ's book strengthens a conclusion I reached some years ago in reviewing a conscientious biography of D'ALEMBERT [above, Essay 27 in this volume]: Truthful, competent general history of science can be written only when the writer has at his disposition detailed, thorough, and reliable monographs on the scientific content—timeless experiment and timeless logic—of the scientific works. For many aspects of mechanics such monographs are now available; historians should study them. Among those less well known than they ought to be are some of the prefaces to EULER's *Opera omnia*: I refer in particular to those by ACKERET, FLECKENSTEIN (this being one that SZABÒ consulted with profit and was able to correct in one important regard (see the footnote on page 70)), CARATHÉODORY, and BLANC, which have been joined recently by two excellent essays by HABICHT on the theory of ships, and a fine addition will be D. SPEISER's, now in press, on electricity, magnetism, and heat. SZABÒ's estimate of GALILEO's work might have been more accurate, had he consulted the tract by WINIFRED WISAN, *Archive for History of Exact Sciences* **13** (1974): 103–306. It is a pity that SETTLE's thesis of 1966, Cornell University, still remains unpublished, but it can be consulted nevertheless.

The book is well printed, but the arrangement is confusing [because each new paragraph starts flush left, not indented. This disagreeable practice has even worse consequences than those mentioned in Essay 24, above, in regard to the *Works of JAMES BERNOULLI*, for whenever] a displayed equation ends with a period, the reader must ask, "Does the text run on, or does a paragraph end here?" On page 132 there are six instances; on page 133, five; on page 112, four. . ..

The book is richly illustrated, many of the plates being taken from pieces in the author's collection. I cannot believe that Figure 104 is what it is labelled, a print, for I never before saw a print that reproduced the fabric of canvas. SZABÒ evidently admires EULER enormously, for he gives us no less than five portraits of him! Unfortunately all of them are bad, very bad, except possibly the unexplained Figure 95. Figure 7 represents EULER made into a British country parson by HOLL for KNIGHT's popular portraits, half a century after EULER's death; Figure 35 reproduces what seems to be a crude etching based on the medal by ABRAMSON, who could never have seen EULER; Figure 109 shows a wild zealot from the Committee of Public Safety, seated in a Directorate chair; in Figure 140, a posthumous engraving by HÜBNER, we see some strange person much too shrewd ever to have written *Die Rettung der göttlichen Offenbarung*. Indeed

EULER is not only the most frequently portrayed of mathematicians but also one of the worst portrayed. It seems not to be widely known that four splendid portraits by HANDMANN painted in EULER's Berlin years have come down to us: two in the Kunstmuseum Basel, one (an old copy) in the Deutsches Museum München, one in a private collection (unpublished). The expert, sensitive relief by RACHETTE in the Académie des Sciences, Paris, seems also to be done from the life. These are the least often seen of EULER's portraits.

ARMIN HERMANN's contribution to the gift edition of SZABÒ's book is a revolting performance. It starts out as one of those reviews that contains nothing but what the reviewer would have said, had he written the book. Since what HERMANN chooses to tell us is the general nature of DESCARTES' leadership, which SZABÒ does miss, at least so far it is a positive addition, though unfortunately HERMANN merely pontificates, supplying no evidence to support his bulls. Then, alas, he turns to discourse on how to write history. That is the curse of Historians of Science today. Most of them are so busy arguing with each other and telling us how to write history that they have little time left to write anything historical themselves beyond an occasional note on odds and ends, mainly personalia and society. Be that as it may, why should the fruit of a lifetime of earnest, devoted study by an expert on mechanics be laden with a rider on these vapid generalities? How could the publisher have allowed such an insult to the author?

SZABÒ tells us that his book is written in the first line for students [of mechanics], secondly to provide information which assistants and teachers of mechanics and physics may use in their lectures. He thus wishes to further the growing interest of young people in the origins of what they learn. While he states that he has not written for "scholars active in the history of science", he conjectures that even some of them might be excited by some of his statements and standpoints. I doubt it. His book is not dusty enough, and there is too much science in it. In his main aim, I think SZABÒ is successful. I should like to see his history translated into English and made broadly available to students and teachers of engineering and the applied mechanical sciences. Certainly the few pages on thermodynamics should be cancelled. Beyond that, could SZABÒ meanwhile broaden his scope by fuller treatment of work earlier than NEWTON's and of problems related to systems of mass-points and to celestial mechanics, and could he also soften a bit his exceptionless contempt for all philosophers, even DESCARTES, so much the better.

Note for the Reprinting

The volume reviewed is ISTVÁN SZABÒ's *Geschichte der mechanischen Prinzipien und ihrer wichtigsten Anwendungen*, Basel & Stuttgart, Birkhäuser Verlag, 1976, also issued in a special edition, not for sale, with an accompanying statement by ARMIN HERMANN which occupies ten pages between the author's preface and the table of contents.

My review was first printed in *Centaurus* **23** (1980): 163–175.

The late ISTVÁN SZABÒ had been a respected friend of mine since 1955. Of course I submitted the foregoing text to him for his comments and criticism before I sent it to press. He knew very well the meaning of the last nine lines. Although the book's uniqueness makes it worth translating, everything on thermodynamics and also the attacks upon philosophy in general and DESCARTES in particular should be deleted, and someone with a broader view of mechanics should add sections on great aspects of its development that SZABÒ did not treat. For example, while the importance of the theory of mass-points is crudely and crassly exaggerated by the physicists, the professional historians of science, and the philosophers, their oblivion to or corruption of the nature of classical mechanics does not justify an inverse prejudice that makes a history of the principles ignore punctual systems.

PART III

BIOGRAPHY AND CIRCUMSTANCES

32. GENIUS CONQUERS AND DESPISES THE ESTABLISHMENT: NEWTON

a. Newton's Letters (1960, 1962)

No great mathematician is so difficult to study as NEWTON. For GALILEO, KEPLER, DESCARTES, and HUYGENS, monumental editions have printed, explained, and interconnected every remaining scrap of paper; for the BERNOULLIS and LEIBNIZ, such editions have begun; for LAGRANGE, LAPLACE, and CAUCHY, all the published works have been re-issued in collected editions, in which some letters and personal documents are included; and the great EULER edition of seventy-four part-volumes, also limited in the main to republication, is nearing its end; while every French nineteenth-century figure with any kind of a name (except, unfortunately, NAVIER, POISSON, and ST. VENANT) has been honored by a collection. The only edition of the works of NEWTON was printed nearly 200 years ago; not only far from complete, it is not trustworthy in what it does print. There is not even a critical edition of the *Principia*, let alone a collection of documents and analyses revealing its growth and antecedents. The text of the only easily available English translation is revised without understanding and annotated with puerilities. In such lack of published sources it is no wonder that the histories of mathematics continue to parrot hoary hagiography and Victorian twaddle about the man who is usually made a hero of "the scientific revolution".

There have been several projects of a great collected edition, but all have failed. The initiates always give reasons, and such reasons are adduced by E. N. DA C. ANDRADE in the opening pages of the introduction to the first of the volumes of the Royal Society's edition of NEWTON's letters, but to those unfamiliar with the problems of scholarship in regard to NEWTON they may seem organizational rather than real in view of the difficulties, apparently no more "intractable", that were brilliantly overcome, indeed by great devotion and at some cost, in the editions of HUYGENS and GALILEO. It is sometimes conjectured that the main block to an edition of NEWTON's papers in full is the fear that they would draw a portrait different from that which three centuries of NEWTONian myths have inculcated.

Be that as it may, everyone should welcome the appearance of the first volume of an edition, more modest in scope but certainly completable in short order, of NEWTON's entire correspondence. According to the foreword, three volumes have been prepared for the press by a large committee of experts working for thirteen years under the direction of H. W. TURNBULL; later announcements indicate that there are to be seven volumes in all. It is estimated that about 430 letters by NEWTON and well over 1000 letters addressed to NEWTON or touching upon his work will be printed.

Of the 156 items in the first volume, covering the years 1661–1675, only nineteen have not been printed before, and of these nineteen, only four are by NEWTON. Nevertheless, the contents of the volume make, in effect, a major new fund of source material. First, some of the earlier publications are old and difficult to locate, and so the letters printed therein will not have been seen before except by persons who have searched for them. Second, some of the older publications are inaccurate from carelessness or censorship. Third, and most important, this is the first time that anything like a consecutive and connected view of any part of NEWTON's activity becomes possible, and this material all in one place, properly explained by excellent notes, is of greater use and value than the sum of its formerly scattered parts. The editors have faced the realities of modern learning by appending English translations or paraphrases of all passages written originally in a learned language.

The main correspondents are COLLINS, GREGORY, and OLDENBURG. OLDENBURG, secretary of the Royal Society, transmitted information as he saw fit to various others, including HUYGENS, LEIBNIZ, and HOOKE. Readers will welcome particularly the editor's decision to print letters or passages from contemporary correspondences related to NEWTON's work and to the background of it; these items, making up perhaps a third of the volume, double the value of the rest by making it comprehensible.

The origin of the hostility between NEWTON and HOOKE, sometimes regarded as a principal cause of NEWTON's reluctance to communicate and hence a principal factor in the decline of British mathematics in the century following his death, may be seen here: Nos. 40, 44, 45, 67, 71 (not previously printed), 152, 154. As has been remarked before, OLDENBURG disregarded NEWTON's request (No. 68) to soften any provocative expressions in his replies before transmitting them, and it seems that OLDENBURG kept the matter hot by design. But also HOOKE's evaluation (No. 44) of NEWTON's theory of light was written by order of the Royal Society, and the account of the

Figure 19. ISAAC NEWTON (1642–1727), the Petworth Portrait, 1720, by GODFREY KNELLER, reproduced here by gracious permission of Lord EGREMONT.

matter given by BIRCH implies that the Society, by attaching so great importance to itself and to its ægis, operated in such a way as to make personal warfare almost inevitable among any creative personalities within it.

The main scientific subjects are optics ... and three parts of pure mathematics: numerical solution of equations, infinite series, and quadrature. Most of the mathematics is contained in the correspondence between NEWTON and GREGORY through the intermediacy of COLLINS. TURNBULL writes, "... in contemporary judgement the genius of GREGORY was reckoned to be second only to that of NEWTON, a judgement with which those who have studied the available facts would probably still agree." For example, in No. 24 GREGORY communicates the beginnings of seven series for transcendental functions such as $r \log \tan (\pi/4 + e/2r)$, with the statement, "I have no inclination to publish any thing, safe only to reprint my quadrature of the circle, and to add some little trifles to it." TURNBULL says, "Gregory had now discovered the method of successive differentiation which produces the Taylor expansion...." In reply (No. 26), NEWTON states an elegant rule for the sum of a finite number of terms of $\sum a/(b + nc)$, with bounds of error. The most interesting previously unpublished mathematical piece, No. 90, is also from this correspondence. It contains NEWTON's ingenious substitute for a calculation which later mathematicians would have carried out by a double integral; an equally ingenious method for solving algebraic equations by laying a rule across parallel logarithmic scales; and the construction of conics by movable rulers.

These are but a few of the interesting mathematical problems and methods occurring. The reader will remark the extraordinary intensity of the mathematical spirit at this time, one rarely encountered before or since.

The volume is handsomely printed but not bound stoutly enough for a work of permanence. The editing is all that could be desired to make the volume easy to read and immediately useful to a mathematician. The series will be a standard work, needed by any solid library in the mathematical sciences.

The period covered by the second volume includes that of writing and publishing the *Principia*. As in the preceding volume, not only letters to and from NEWTON but also letters or extracts from other correspondences relevant to it are adjoined, and everything possible is done to help the reader follow the circumstances and arguments.

The scene is well set by the first two letters. In the former NEWTON gives some of his views on alchemy, speaking of the "metallick particles" and "saline particles", which "enter and shake those bodies more fully & by their grossness shake ye dissolved particles more strongly then a subtiler agent would do." The second, from LEIBNIZ, concerns the series for the sine and arcsine.

A good many valuable pieces are printed here for the first time. For example, the correspondence with FLAMSTEED, mainly on comets, is completed and so forms a real source of mathematical astronomy. A side of NEWTON's character which has been somewhat veiled up to the present is now seen in pieces No. 217–221 and 225–226, which present the private aspects of his quarrel with LUCAS, derivative from the earlier public controversy with LINUS over refraction. They show that however desirous he was of avoiding differences, he knew how to pursue one hotly once into it: "Pray trouble your self no further to reconcile me with truth but let us know your own mistakes."

No. 190 is dated 1664 by the editor and is regarded by him as giving "perhaps the earliest statement and proof of the fundamental theorem of the calculus." No. 191, dated 1665 by the editor, gives the binomial theorem for arbitrary rational exponent. No. 192 seems to be connected with the famous "Epistola posterior" sent to OLDENBURG for LEIBNIZ on 24 October 1676. These three pieces, not previously published, are printed along with the better known documents and letters pertaining to the sources of the later controversy over priority in the calculus. It is perhaps through scholarly principle that the notes from the *Commercium epistolicum* of 1742 are reprinted in this connection, yet it is unfortunate that their polemic character is corrected only in the cases arising from demonstrable errors of transcription or interpretation. The editor is not unfair to LEIBNIZ, yet the reprinting of comments from one side only casts a regrettable aura of bias on this part of the volume. Nonetheless, from the documents themselves the reader easily forms the picture of LEIBNIZ, whether earlier or later, standing alone, commanding no squadron of partisans, and dealing openly and elegantly, while NEWTON is put forward as their grudging champion by a school of British who scurry about behind the scenes deciding what is fit to be disclosed and what is not.

The other newly published pieces of main interest to mathematical readers are No. 271, which is an outline of a treatise on mathematics from 1684, and Nos. 312–316, a correspondence with G. CLERKE regarding some difficulties in the *Principia*, then just published. These difficulties concern neither the principles of mechanics nor any general method, only details and terminology.

The volume, like its predecessor, is excellently edited and presented, but unfortunately it wants a preface to orient such readers as are not already at home in the mathematical scene of late seventeenth-century England. Also the editor does not reveal the grounds on which he chose to include certain selected fragments from the great mass of NEWTON's manuscript notes. The reader will be grateful for what he is given, but he may ask why the edition was planned in this unusual way, and when the rest of the non-epistolary manuscripts will be made available.

The third volume calls for similar comments. The mystery of why this collection of letters contains also certain non-epistolary fragments continues. Numbers 347–349 are certainly among the most interesting included: the first, conjectured as being from 1665/1666 and containing NEWTON's correction of passages on falling bodies in GALILEO's *Dialogo*; the second, dated 16 May 1666, on fluxions; the third, conjectured to be from 1672, on the laws of motion. But why are these manuscripts suddenly inserted in a sequence of letters from 1688 to 1694? We are glad to have them anywhere, but might not the editors more profitably have begun a great collected edition of all of NEWTON's works in chronological order, so that early papers such as these, which are the rarest and of course the most important for the history of NEWTON's thought, would have been made generally available all together and at the beginning?

In this volume the annotation and editing leave something to be desired. By a gross slip in translation, NEWTON's name is omitted from the sentence on page 346 describing Figure 2. The notes on the catenary suggest a bias hinted in the review of Volume II. Note (11) on page 348 implies that the identity of the ideal arch with the inverted catenary is a discovery of NEWTON; in fact it was published as an anagram by HOOKE in 1675; HOOKE had demonstrated something about it to the Royal Society in 1671; and it is unbelievable that his result was not known to other Fellows. A committee of experts in thirteen years of work might have been expected to learn more about the catenary than is indicated by notes (11) and (12) on page 167, the quaintness of which is accentuated by the use of French forms for the names of the BERNOULLIS, as was customary in the England of NEWTON's day. [The French, who like the British instinctively reject all foreign pronunciation, replaced the in French easily pronounceable "ou" as in "où" by "oui"; the Basler pronunciation is closer to "o" in the French "olive"; in English it would be the natural and easily pronounceable BERNOLLI, although, alas, the Anglo-Saxons in pronouncing a name so spelt would insert an unwritten K before the N and an unwritten w before the LL.] Of the solutions of LEIBNIZ, JOHN BERNOULLI, and HUYGENS the editors say only, "each of whom communicated a paper on the curve to the *Acta* of June 1691". While the analysis published by GREGORY in 1698 is notoriously wrong, the editors tell us only that "he supplied the proofs by the method of fluxions" and that later he defended his work "against various anonymous objections". They do not mention that the anonymous objectors, who found two cancelling errors, were none other than LEIBNIZ and JAMES BERNOULLI. The editors say also that "according to a memorandum, dated 20 February 1697/1698..., NEWTON informed GREGORY that he had solved the problem of so loading a flexible cord that it should assume the shape of any given curve." They do not mention that this inverse problem is easy and has infinitely many solutions: In fact, we may prescribe the tension arbitrarily as a function of arc-length. In any case, they do not mention that the problem of the general catenary was intensively studied by the two BERNOULLIS and by other Continental geometers, and that by 1697/1698 JAMES BERNOULLI had found the differential equations of equilibrium for a flexible line subject to arbitrary loading. A general history of the catenary problem has been published[1]. Possibly NEWTON had equivalent results; there is every reason to think the problem within his powers, and perhaps documents relative to it exist. But remarks of this kind, from 1687 to the present day, have given rise to suspicions, perhaps unjustified, that the NEWTONians did not and do not understand the work of LEIBNIZ and the BERNOULLIS.

This volume includes some fifty previously unpublished pieces. In addition to the three manuscripts already mentioned, the most interesting mathematically are No. 459, which is a note on the solid of least resistance, and No. 461, which is a memorandum by DAVID GREGORY, lists changes NEWTON intended in 1694 to make in the *Principia* at its next edition. Most of the newly published pieces concern FATIO DE DUILLIER and DAVID GREGORY; they show us that NEWTON really felt strong friendship (an unusual manifestation in NEWTON) for the former, and that in the latter, to whom he confided unusually much, he had chosen a feeble disciple. No. 359, when added to the well known No. 358, according to the editor "places NEWTON in the forefront among biblical scholars of the time"; these two letters occupy 80 pages of print. Nos. 420–426, not new but now for the first time brought together in one place, picture NEWTON's famous "distemper" brought on "by sleeping too often by my fire", as he explained in begging pardon of LOCKE for having accused him of endeavoring "to embroil me wth woemen & by other means". Immediately afterward he wrote his only letter, and a handsome one (No. 427), to LEIBNIZ.

This volume closes with corrigenda to the two previous and with a general index of persons and matters treated in all three.

---

[1] C. TRUESDELL, *The Rational Mechanics of Flexible or Elastic Bodies, 1638–1788*, LEONHARDI EULERI *Opera omnia* (**II**) 11$_2$, Zürich, Orell Füssli Verlag, 1960. See §§ 2–8, 10, 11.

Note for the Reprinting

The three reviews here made one were first published in *Mathematical Reviews* **20** (1960): 1170–1171; **22** (1962): 1594–1595; and **23A** (1962): 692–693. The volumes reviewed are *The Correspondence of Isaac Newton*, edited by H. W. TURNBULL, Volume **I** (1661–1675), Volume **II** (1676–1687), Volume **III** (1688–1694), Cambridge, at the University Press, 1959, 1960, 1961. The edition has since then been completed: Volume **IV** (1694–1709), edited by J. F. SCOTT, 1967; Volumes **V** (1709–1713), **VI** (1713–1718), and **VII** (1718–1727), edited by A. R. HALL & L. TILLING, 1975, 1976, and 1977. The famous and rarely seen Petworth portrait is reproduced in reverse as the frontispiece of Volume **VII**. For the reader who would obviate recourse to a mirror I provide above as Figure 19 an enantiomorph.

The reference to "the scientific revolution" will seem quaint today, now that Historians of Science have discovered that the object of their study is nothing but revolutions, densely packed.

b. Newton's Mathematical Works

I. Introduction to the *Principia* (1973)

In the tribal folklore of physics no saint has a bigger halo than NEWTON's. As it should be with true believers, physicists are loth to come to grips with the sweat and sin, the deviations from the path to Olympus, of their tutelary Herakles. Thus it is no surprise that while any interested person who seeks AUGUSTINE, SHAKESPEARE, or GOETHE need only go to any university library to find a complete, precise, and abundantly annotated edition containing every single word or figure set on paper by those authors and today preserved, until recently the scholar who would study the works of NEWTON was compelled to rely largely on the tiny fraction of his work that had been published, a fraction selected by a mixture of prejudice and caprice. Even now, those to whom NEWTON's native language in science, which was Latin, is inaccessible can consult his masterpiece, the *Principia*, only through a translation notorious for its blunders at critical passages.

This strange contrast results from the natural preference of the masses and the "educated" for great entertainers before great enlighteners. The ill wind of abundant cheap publication, nonetheless, has blown good to NEWTONian scholars by making possible the publication, now in progress, of *The Mathematical Papers of Isaac Newton*, Cambridge University Press, superbly edited by D. T. WHITESIDE, the four volumes going through 1684 now standing in print. In addition there are now four volumes of the *Correspondence of Isaac Newton* through 1709, variously and irregularly edited by a committee, so that the responsibility for random spots of ignorance or national prejudice could be made anonymous. These eight massive volumes, difficult to penetrate, form today the only reliable introduction to NEWTON's thought.

The introduction presently under review is of a different kind. Of course, any introduction is made, not abstractly, but to someone. The quickest inspection of the work shows that I. BERNARD COHEN makes no attempt to speak to any who would follow the brilliantly inventive, perplexing, lacunary, often fallacious and sometimes even contradictory mixture of guesswork and icy mathematics in NEWTON's treatise. Indeed, the content is mentioned only here and there, in passing.

Rather, this introduction, as the appearance of the late ALEX-ANDRE KOYRÉ's name on the flyleaf might suggest, is a contribution to the "new" history of science: the story of the circumstances, rather than the content, of scientific discovery. As such, it is a work of great erudition. The reader is led through the preliminary manuscripts, the writing and publishing of the first text, and all the revisions, both those that did appear in the second and third editions and those that were withheld.

In the task COHEN set himself, he has succeeded, and the present volume will surely stand henceforth as the definitive textual criticism of the *Principia*.

Note for the Reprinting

This review of *Introduction to Newton's "Principia"* by I. B. COHEN, Cambridge, Harvard University Press, 1971, was first published in *Physics Today*, April 1973, page 59.

II. Whiteside's edition of the *Mathematical Works*, especially Volume 6, on the origins of the *Principia*

SUMMARY (1976)

Among the supreme mathematicians NEWTON was for centuries one of the most difficult to know. His icy theorems and proofs seemed to have sprung full grown from his brain onto the page of print. Volume 6 of his *Mathematical Works*, perhaps the most important of the eight projected to make up the edition, contains the successive draughts of the masterpiece published in 1687 as the *Philosophiae Naturalis Principia Mathematica*. From it we may see that while indeed Olympian baroque Latin was NEWTON's native language for science, the pages of the *Principia* were forged by two years of unremitting titan's labor and self-criticism, in isolation.

First is the tract of 1684 on the motion of bodies, written perhaps as a reaction to HOOKE's guesses and unfounded claims about gravitation. It contains NEWTON's earliest treatment of the motions of gravitating bodies; while far from achieving the elegance of the presentation in the *Principia*, it would have gained and retained the rank of a masterpiece of rational mechanics had it been published as soon as it was written, even though the KEPLERian third law is not properly inwoven. Then comes the treatise of 1684–1685, in which NEWTON faces and resolves the difficulty presented by the finite size of the celestial objects, thus launching the theory of the attractions of bodies. Part 1 of the volume ends with augmented versions of what was to become Articles 4–10 of Book I of the *Principia*, provided with proofs. The second part of the volume comprises a sequence of attempts to treat particular problems, some written before and some after the *Principia* was published. The third and last part is a selection from the ameliorations NEWTON wrote in the 1690s, when he planned a second edition that would have been a largely new book.

From start to finish this volume is rational mechanics—the motions of bodies reduced to pure mathematics, mathematics that is faced as mathematics by a prime mathematician. In it there is little of what current physicists call physics. Contrary to the folklore, which makes NEWTON the scion of GALILEO and KEPLER, here we see the real NEWTON at his giant's work: NEWTON the mathematician, building with and upon what he had learned from ARCHIMEDES, APOLLONIOS, HUYGENS, and above all—again contrary to the folklore—from DESCARTES.

Like the preceding volumes in the edition, this one is superbly edited. With what seems to be instant recall and command of all the documents, WHITESIDE reconstructs, explains, and criticizes NEWTON's course of thought. For a specimen of his mastery of fact and literature (both primary and secondary) we may consult footnote 126 on pages 305–308. The reader who wonders, as up to now I wondered, how NEWTON, long after having renounced his philosophical studies, could have with lion's paw smashed BERNOULLI's challenge on the brachistochrone, will find the answer in Footnote 14 on pages 459–461.

Note for the Reprinting

The work under review is *The Mathematical Papers of Isaac Newton*, Volume **VI**, edited by DEREK T. WHITESIDE with the assistance in publication of M. A. HOSKIN and A. PRAG, Cambridge, at the University Press, 1974. The foregoing brief review was first printed in *American Scientist* **64** (1976): 230. The following essay, written 1977, is here published for the first time.

REVIEW (1977)

For the student of mechanics this volume of NEWTON's works will be the most illuminating of all, for it contains and analyses the preliminary draughts for Book I of the masterpiece printed in 1687 under the title *Philosophiae naturalis principia mathematica*:

1. The first tract *De motu Corporum*, Autumn 1684, augmented December 1684?
2. The revised treatise *De motu Corporum*, Winter/early Spring 1684–1685.
3. "Articles" IV–IX of the augmented *De motu Corporum Liber primus*, early Summer 1685/Winter 1685–1686.

These, with appendices concerning related problems and in some cases extracts from the final text for comparison, make up over two-thirds of the volume. The second part comprises a sequence of attempts to treat particular problems, some written before the *Principia* was published and some after. These essays concern mainly the solid of least resistance, which is a subject arid for mechanics but fruitful for the variational calculus, and a heroic application of genius to insufficient principles toward determining the libration of the moon. The third and last part is a selection from the intended corrections of Book I for the second edition which were not incorporated into it, and some essays toward a radical restructuring of Book I which was begun in the early 1690s but was soon abandoned.

Mechanics was scarcely mentioned in the preceding five volumes of the *Works*, covering the years 1664–1684. It is in this volume that the physicists, if they give any notice at all to the publication of NEWTON's work in his own words, will expect to find the hero of their folklore. Indeed, here NEWTON concentrates upon what that folklore regards as the main occupation of his life: "the Kepler problem", with its variants and generalizations. Nevertheless in the NEWTON of this book the NEWTON of the folklore is hidden, indiscernible. There is no sign of his being able "to see almost by Intuition, even without Demonstration". Neither do we find NEWTON in the laboratory or at the telescope. Indeed, astronomical data are by no means disregarded; indeed, they gave rise to all the problems treated here; but in this volume they find no further use, except for one or two cautious remarks. This is a book on rational mechanics, forged and chiselled by a supreme mathematician; it will remain forever closed to those who are not mathematicians; no effort by journalists and sociologers can ever pry it open for the general.

The NEWTON of the folklore is the scion of KEPLER and GALILEO. The documents contain scant evidence of any direct influence upon the real NEWTON by either of these popular heroes of science. KEPLER's work was transmitted to NEWTON much as NEWTON's commonly is to us, namely, through fragmented, distorted tradition rather than by study of the originals, and NEWTON's expressions indicate that he esteemed KEPLER an inaccurate empiricist, no mathematician. GALILEO is not mentioned in the first tract; the acknowledgment to him found in the revision and all later versions may reflect pressure from friends who thought it a good idea NEWTON should cite somebody rather than nobody, which was his own inclination. In earlier researches WHITESIDE and COHEN have concluded that there is no evidence to suggest NEWTON had ever seen GALILEO's *Two New Sciences.* Be that as it may, in the volume under review I discern no trace of GALILEO's methods, point of view, or (beyond a mere acknowledgment) discoveries. On the contrary, here we see the real NEWTON, building with and upon what he had learned from ARCHIMEDES, APOLLONIOS, HUYGENS, and above all—in contradication with the claims of the NEWTONians, those little folk whom for want of better NEWTON in his later years had no choice but regard as disciples—from DESCARTES. From start to finish the motions of bodies are reduced to pure mathematics, mathematics that is faced as mathematics by a colossus among mathematicians. In it there is little of what current physicists call physics, certainly no trace of the irresponsible pontification, the crude bludgeoning, the formal hocus-pocus, the wild extrapolation from scanty experiments, which nowadays witness to "physical intuition".

Among the supreme mathematicians NEWTON was for centuries the most difficult to know. His icy theorems and proofs seemed to have sprung full grown from his brain onto the page of print. This volume shows otherwise. It reveals NEWTON the man, the geometer unsurpassed by any other yet different in degree only, not in kind. Here is the NEWTON who can make mistakes. To me that renders his greatness more accessible but no lesser.

In the hope that somehow I could write something not fatuous upon the titanic struggles, the glorious successes and failures revealed here, I have read and reread the text and notes for over a year but now must confess myself unequal to the task. The best I can do is quote a specimen of NEWTON's mathematical thought and then pass on to discuss the editing and production of the volume.

The original tract *De motu* begins by a thrust to the heart of the problem of two bodies. This directness, covered in the final treatment by pages of preliminaries loosely reminiscent of Greek mathematics, is the mark of a new way to deal with mechanics. Here is the first theorem and its proof, in WHITESIDE's translation:

> *Theorem 1. All orbiting bodies describe, by radii drawn to their centre, areas proportional to the times.*
>
> Let the time be divided into equal parts, and in the first part of the time let the body by its innate force describe the straight line *AB*. It would then in the second part of time, were nothing to impede it, proceed directly[a] to *c*, describing the line *Bc* equal to *AB* so as, when rays *AS*, *BS*, *cS* were drawn to the centre, to make the areas *ASB*, *BSc* equal. However, when the body comes to *B*, let the centripetal force act in one single but mighty impulse and cause the body to deflect from the straight line *Bc* and proceed in the straight line *BC*. Parallel to *BS* draw *cC* meeting *BC* in *C*, and when the second interval of time is finished the body will[b] be found at *C*. Join *SC* and the triangle *SBC* will then, because of the parallels *SB*, *Cc* be equal to the triangle *SBc* and hence also to the triangle *SAB*. By a similar argument, if the centripetal force acts successively at *C, D, E, . . .* , making the body in separate moments of time describe the separate straight lines *CD, DE, EF, . . .* , the triangle *SCD* will be equal to the triangle *SBC*, *SDE* to *SCD*, *SEF* to *SDE* (and so on). In equal times, therefore, equal areas are described. Now let these triangles be infinitely small and infinite in number, such that to each individual moment of time there corresponds an individual triangle (the centripetal force acting now without interruption), and the proposition will be established.

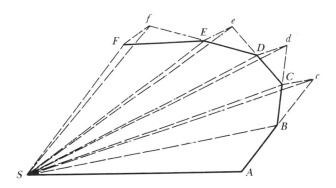

---

[a] Hypoth. 1.
[b] Hypoth. 3.

The statement of this theorem is what we have been taught to call "KEPLER's Law of Areas". The critical reader will see that NEWTON's argument is not a real proof but rather a foothold upon one. The granules of impulse, which he called "forces", can easily be set aside as a device, though it is not clear whether NEWTON so regarded them. The nut in the shell of geometry is the order, "Parallel to *BS* draw *cC*". NEWTON tells us to use Hypothesis 3: "A body is carried in a given time by a combination of forces to the place where it is borne by the separate forces acting successively in equal times." There are defects in this argument. If we try to stay with granules of impulse, we must remark from NEWTON's figure that if the "force" was parallel to *BS* at the beginning of the second instant, it will not be so at the end of that instant. If we regard the granules as only a manner of speaking, it is not easy to extract an interpretation of NEWTON's proof that is sound mathematics. But all this detracts nothing from NEWTON's great idea, namely, that such an assertion *can be proved mathematically* for any "orbiting body", namely a body suffering uniform motion superposed upon an acceleration directed to a fixed center. The proof, as WHITESIDE tells us in a footnote, applies Hypothesis 3 to a body subject to *vis insita* and *vis centripeta*, and NEWTON's tacit replacement of these two *vires* by accelerations shows where his thought is tending.

The discourse descends at once to circular orbits, and as Corollary 5 to his Theorem 2 NEWTON proves what we today expect: "If the squares of the periodic times are as the cubes of the radii, the centripetal forces are reciprocally as the squares of the radii. And conversely so." Theorem 3, the principal theorem of the original tract, concerns an orbit of general form. Here the difficulties met in the proof of Theorem 1 become acute, but it is the applications, set forth in the solution of the ensuing seven problems, that have claimed, for the century and more since this tract was first published, the astonished admiration of every reader trained to use rational mechanics as a living language. Never before nor afterward has anyone cut so swiftly to the core of a great problem of natural philosophy. Had NEWTON published this tract, it would have revealed him instantly as a giant of mechanics. Had he published nothing else, it would have gained him a permanent place in the first rank of mathematical science.

Partly because the gaps and defects are easier to perceive, but more because the work is short and the presentation direct, the original tract gives us a better entry into NEWTON's thought than is offered by the sheer glacier that is the *Principia*. NEWTON seems to prove much from no serious assumptions at all. There are no axioms on forces. Instead we find what purport to be definitions of centripetal and innate forces, both of them using "force" as if everybody knew what that was. Then come

> *Hypothesis 2.* Every body by its innate force alone proceeds uniformly into infinity following a straight line, unless it be impeded by something from without.
> *Hypothesis 3.* A body is carried in a given time by a combination of forces to the place where it is borne by the separate forces acting successively in equal times.
> *Hypothesis 4.* The space which a body, urged by any centripetal force, describes at the beginning of its motion is in the doubled ratio of the time.

Of course NEWTON does not mean what his words say in Hypothesis 3. As is often true of pioneer works, the avowed admissions are neither necessary nor sufficient to the end desired, and only by following the mathematical proofs can we discern the physical assumptions that are really put to use. NEWTON has already decided that the forces acting upon one and the same body are proportional to what we now call accelerations, and of course he knows that accelerations are additive; the proofs are nothing more nor less than calculations of the properties of such motions as result when the radial acceleration is a function of the distance from a fixed point while the azimuthal acceleration is naught. Here is achieved the ideal which LEIBNIZ had expressed in print a few months before NEWTON began to write the tract: "that by these few considerations the whole matter be reduced to pure geometry, which is the unique aim of physics and mechanics."

By now some dozens of reviews of the several volumes of this great edition have appeared. On two points, and two only, the reviewers up to now are unanimous: The production of the Cambridge University Press is splendid, and the price is too high. I voice dissent on both counts.

Indeed, the production of these volumes is probably as good as can be had in a country throttled by labor unions and a world stifled by the universal effective illiteracy brought on by official, fancied universal "education". That is quite another thing from format and typography worthy of NEWTON. It is quite another thing also from the standard set by the edition of the *Œuvres de Christiaan Huygens*, the edition that first comes to mind for comparison in contents and in period. The crowding on pages 74–75 so as to spare half a page of blank space is German in its stinginess, altogether un-English. The format is tight, with running heads too small and too close to the text, with footnotes too crowded. The usefulness of the analytical table of contents, a characteristically British feature that writers of other lands would do well to imitate, the printer has nearly vitiated by setting that table in tiny type, so crowded as to make survey difficult, and with confusing divisions. Without looking for them I noticed in text and footnotes passages so crammed as to suggest a Teutonic compositor, lines improperly justified, a quotation mark misleadingly set as an apostrophe to the following word, a misnumbered reference to a footnote, and enough workups to remind me of American presswork. The halftone illustrations are worthy of the U.S.S.R. The paper is mediocre. The casings of the volumes are not even alike in size, in lettering and placement and scheme of the backstrips, or in color of label. The binding of the copy sent me for review has loops of tangled thread between some of the pages, and the joints have cracked in the course of just one year of careful handling. In a university library, shelf tides will separate the heavy texts from the flimsy casings within a few years, even if nobody takes the books down to read them.

Possibly no cost would have sufficed in Britain today to produce a book meeting the standards the Cambridge University Press maintained for its ordinary books until a few years ago. Certainly $72.50 is too little to make today a book that is worthy of such contents as this one's. A recent history of the kinetic theory of gases, 200 pages longer but in small format and routine as a piece of printing, costs $84. Even ephemeral art books of half the size, mostly pictures, often sell for $100. It is unlikely NEWTON's works will ever be published again, except in the inevitable shoddy reprints. Whatever their price now, were it even $150 a volume, it would be in a few years reduced by inflation to a bargain in memory. Now was the time to do the job right, to make an edition as sound and permanent as the one the Dutch made for HUYGENS, no matter what the cost. A nation that can and does give its most stupid citizen a life of total security while losing billions on a speculation in aircraft might be thought able to afford a decent memorial to its greatest mind. A permanent monument this edition is, nothing to be confused with the muddy pools of "scholarly" minutiæ and misconceptions now poured out by the assembly lines of what one reviewer has called "the Newton industry". Proper dress for it would have cost no more than one year's pay of a few privates in the ever swelling army of bureaucrats. But beauty, however cheap, is too dear for a social democracy.

These words I write, not in smugness because I live in a land for the moment less unfortunate than Britain, but in somber lament upon the extinction of learning and the learned (for the former without the latter is impossible) which goes hand-in-hand with social progress everywhere, extinction planned and carried out by the benign universities and the goose-stepping "intellectual" professions.

Their complacent mediocrities know how to deal with misfits. After eighteen years of entire devotion to his great work WHITESIDE has only lately been granted the scant comforts of a standard academic post. In the same period of time hundreds of co-operative, uniform nobodies gained professorships. Of course they deserve them because they try to teach a few trade skills to the masses (including the masses of historians of science) and serve on university committees. The Royal Society, which sponsors the concurrently appearing, uneven edition of NEWTON's letters, has not even embraced WHITESIDE within its smug numerosity. It could have done so without spending one penny, old or new.

Seen against this background, my criticism of the Cambridge University Press should be deleted. We ought to be grateful for its efforts, sometimes successful, to do a competent job in a world which fears thought and hates a thoughtful man. We are not far from the time when the "scholarly" presses of all lands will issue only paperbacks, offset from the authors' typescripts on paper rejected as substandard by the printers of government propaganda.

I am informed that the volumes of the NEWTON edition are set from handwritten copy. That makes them unique in our time, for they contain not a single blunder introduced by a secretary! We can only marvel that today a compositor can be found who is fluent enough in Latin and mathematical formalism to translate WHITESIDE's handwriting into almost errorless type.

Comparisons with the HUYGENS edition are as timely as a proposal for sound currency. Indeed, the International Monotype Corporation has dubbed this very volume "the best printed book of 1974" for its success in solving the problem—now, it seems, next to intractable—"of containing text, inset figures, translation, and footnotes in the same page foldout"! This twaddle aside, in saddening fact few other presses would have undertaken to publish eight large volumes of seventeenth-century mathematical text and notes by anybody, even by NEWTON and WHITESIDE—volumes certain to run up a big loss. Perhaps no other press would have allowed a fledgling of twenty-seven to function as sole and absolute editor (though not so named), even before he had been sealed with a doctorate and without the usual bondage to a costive committee of senescent "authorities". That proves the old Britain, the Britain in which eccentric individuality was common and courage was a virtue, to be not entirely dead. In the countries to the east, the truly progressive countries, WHITESIDE's independence would not be tolerated in private, let alone endorsed and blazoned in public print. As we are all going down the same road, differing only in our banners and slogans, perhaps the Cambridge

University Press has given us the last example that we are to see of great scholarship recognized in publication.

WHITESIDE's editing has often been praised in general terms, sometimes criticized on principle, but, so far as I know, never taken up in detail. Being one of the other living persons who has edited volumes of the collected works of a supreme mathematician, I hope I may be allowed to remark upon the peculiarities of this edition. As WHITESIDE with eminent fairness again and again points out mistakes in the writings of the great NEWTON—and it is a special virtue of his editing that he does so—a level critic must express a balanced judgment of WHITESIDE's work, too. One reviewer has complained that "... the editor had deliberately, even defiantly, aimed his work at an audience so narrow as to be numbered on the fingers of one hand." I do not estimate combined competence today in mathematics, mechanics, and their histories so low as that, but low it is, and perhaps the fingers and toes together would do for the count by one-to-one correspondence. To an audience no larger, the *Principia* itself was directed. That much of the contents of the *Principia* has found its way, indeed after much transformation, into the common domain of mathematical science, shows that the few guests whom NEWTON invited to his banquet had stomachs to digest the gold and steel he served them. No more than that can be hoped of a great edition like this one. Popularization—and whether that be desirable at all is a question into which I will not here enter—is neither helped nor hindered by an edition for the learned. Learning now is too rare to matter.

The Latin originals and WHITESIDE's translations of them are printed on facing pages. Of course the translations are in sense correct. But their tone is not NEWTON's. Unlike the Continental peoples, the English-speaking make a rigid distinction between literature and the sciences. While GALILEO and D'ALEMBERT and POINCARÉ are recognized as masters of style in their vernaculars by them who speak those today, among the descendants of the Picts, Scots, Angles, Saxons, Jutes, and Normans poetry is now the preserve of a consortium whose members not only do not know what a linear equation is but also despise as unlettered boors those who can solve one. NEWTON, I contend, was as great a writer of English prose as ever we had, even though English was to him pretty nearly a second language rather than his mother tongue. For an example, we may select one of the simpler and shorter sentences from his great letter to BENTLEY on 10 December 1692:

> And to compare & adjust all these things together in so great a variety of bodies argues that cause to be not blind & fortuitous, but very well skilled in Mechanicks & Geometry.

It is unlikely that the man who wrote that sentence would have written also "such that to each interval of time there corresponds..." in his Theorem 1. I should read NEWTON's phrase, "sic, ut singulis temporis momentis singula respondeant triangula," as "so that the several triangles answer to the several moments of time."

Was it necessary or desirable to provide the translations along with the originals? I doubt it. I doubt that they who cannot read the originals be in case to profit from the translations. Much labor for the editor, much expense for the press, and much thickness and weight for the volumes would have been spared, had the edition rested content with NEWTON's choice of language. WHITESIDE's division into volumes witnesses his supreme mastery of the whole corpus of NEWTON's work and is perfect to its end, but the volumes have come out too thick and heavy to handle easily, which without the translations they would not have been.

Introduction and notes are another matter. Those, most certainly, a scholarly edition must provide, and the editor should use the tongue he himself knows best, whether that be his author's or not. English we expect here. It would be too much to expect further that anyone today could write like NEWTON, even after having been submerged for years in NEWTON's prose. Necessarily the editor's notes, be that editor who he may, must be in modern English, and modern English reflects modern life and modern thought: rootless, sprained, cacophonous, verbose yet vague, roundabout and irresponsible, and above all insecure and defensive. NEWTON's sentences consist mainly of monosyllables in simple and direct phrases, majestic in their sonority, firmly yet flexibly interlinked in an easy flow forward, their syntax revealing the various and masterful subtlety of a great poet. Modern English consists of obscure, wordy, and indirect phrases composed of thick polysyllables, strung along higgledy-piggledy as a child would blurt them out, then patched up by hyphens, dashes, and brackets. Even so, some of NEWTON's directness, if not his baronial command, might have been imitated by his editor. That we cannot achieve perfection, is no excuse that we not strive for it. WHITESIDE, while he provides introduction and notes that can have no purpose but to help the student, at the same time puts obstacles in that student's way, especially by two quirks that slow the reading and obstruct the understanding. Every verb, by compulsion it seems, is modified by an adverb, usually a blunting one, and that adverb is invariably placed before the verb rather than after it, where older authors would have put it and where transatlantic ears still expect it. On pages 15–16 we read as part of a long, knock-kneed sentence, "... it was henceforth possible accurately to comprehend—and in principle, exactly to compute—all celestial motions . . .." At first reading I thought "accurately" must modify "possible", while "exactly" modified "in principle". NEWTON used adverbs least of all parts of speech, and nearly always to strengthen, not weaken, the force of the verb. The three adverbs in the passage to BENTLEY quoted above are among the few on the whole page of the text as reprinted; "together" is essential to the sense; "very well" is the only adverbial construction of its kind in the whole letter and comes as a hammer blow, to be read aloud with a rising tone and ictus on three of the four syllables of "very well skilled" in the peroration of a long chain of evidence. WHITESIDE frequently cleaves the verb from its direct object or the parts of a compound verb from each other by a long discursus, all too often interrupted by dashes and brackets, making a trap rather than a path. On page 508 we learn that NEWTON's "drafts and worksheets are . . . able to shed considerable light," but the interstitial words I have indicated by three dots are no less than seventeen, five of which make a parenthesis within the divorce effected upon the unfortunate verbal phrase "are able to shed". In the centrally important Footnote 19 on page 35, where WHITESIDE dissects the proof of Theorem 1, the sentence that begins, "His unanalysed procedure . . ." is scarcely a sentence at all but rather a string of strangely punctuated, syntactically unconnected interpellations.

These minor annoyances set aside, we can only marvel at WHITESIDE's mastery of the sources and their contents, not only the manuscripts and published works of NEWTON but all mathematical science of NEWTON's time and of the periods a little earlier and a little later. Of the typical British bias, extending sometimes even to chauvinism, which experience has taught us to expect and which occasionally we do find in the Royal Society's edition of NEWTON's correspondence, WHITESIDE is altogether free. He is perhaps the first Englishman to set upon the achievements of LEIBNIZ, the BERNOULLIS, and EULER their just values. If there are major historians of science working today, surely one of them is WHITESIDE. Rather than add to the generalities already often expressed, I will mention a few specific points.

First there is the matter of HOOKE's place. Anyone competent in mechanics knows that a theory of gravitation is neither necessary nor sufficient to a general mechanics. The great geometers of the eighteenth century, although they often doubted NEWTONian gravitation, sometimes even to the point of rejecting it either specifically or in terms of action at a distance, took the basic ideas of NEWTON's mechanics as gospel and made of them the mathematical mechanics we call classical. We may rest content to see gravitation as a thing of interest in itself, independent of the basic structure of mechanics. Even here there are misty folk today ready to glory in a misty patron saint who, for them, found physical truth pure and ideal without the tedious preliminary of learning mathematics and the tedious consequence of having to connect by logic what he concluded with what he assumed. WHITESIDE's introduction reminds us that HOOKE, failing to recognize a master when he stumbled across him, had so pestered NEWTON some years before as to make him claim to "bid adew to [Philosophy] eternally, excepting what I do for my private satisfaction." In 1679 HOOKE, nothing daunted, sought to bring him back into the fold of the Royal Society, from which he had tried, in vain, to resign; HOOKE's letter, like all letters from such people, asked him for comments upon a great discovery of his own: "compounding the celestiall motions of the planetts of a direct motion by the tangent & an attractive motion towards the centrall body." Such a tame idea, in itself nothing new for anybody who could understand HUYGENS' mechanics, NEWTON might well have let stand unacknowledged. To our great gain, he did not. Instead, he let himself be drawn into the very sort of vague, conjectural science that was HOOKE's forte, and he guessed wrong. He sketched as the path of a falling body relative to the earth a spiral closing down upon the center of the globe. HOOKE replied that his "theory of circular motion" would make the curve "nothing att all akin to a spirall but rather a kind [of] Elleptueid . . . ." NEWTON accepted the correction and sent HOOKE a figure showing the results of a crude, approximate calculation of the orbit. He was not yet in possession of mathematics sufficient for the job, but, as WHITESIDE writes, "even such . . . unsophisticated reasonings were above Hooke's head . . . ." HOOKE immediately put forward his "supposition that the Attraction always is in duplicate proportion from the Distance to the Center Reciprocall, and consequently that the Velocity will be in a subduplicate proportion to the attraction and consequently as Kepler supposes Reciprocall to the Distance." WHITESIDE remarks that KEPLER's approximate rule, here endorsed by HOOKE, if taken strictly would make the orbit a circle, not a Keplerian ellipse, but HOOKE, impotent in mathematics and hence unable to get anything specific from his grand idea, proclaimed that it "doth very Intelligibly and truly make out all Appearances of the Heavens." The challenge to NEWTON was plain. WHITESIDE in attempting to refute the very early date some historians have suggested for NEWTON's private attempt to answer this challenge gives us an incomprehensible footnote, No. 43 on page 14. The matter was noised about London. HOOKE claimed to WREN and HALLEY that he had calculated "all the Laws of the celestiall motions", but WREN "was little satisfied that he could do it," and he offered HOOKE and HALLEY two months' time "to bring him in a convincing demonstration thereof", for which he would give the author a book worth forty shillings. HOOKE claimed that he had it "but would conceale it for some time that others triing and failing, might know how to value it, when he should make it publick . . . ." HOOKE never produced what he promised, and HALLEY appealed to NEWTON. NEWTON claimed he had done the job some years before but could not find the paper, nor has such a paper ever been found since. Be the paper what it may, NEWTON then composed the tract *De Motu*, and thus the *Principia* began. Footnote 44 reminds us of NEWTON's own account as he wrote it to HALLEY on 27 July 1686:

> Hooke's correcting my Spiral occasioned my finding ye Theorem by wch I afterward examined ye Ellipsis . . . . [Y]et am I not beholden to him for any light into y^t business but only for ye diversion he gave me from my other studies to think on these things for his dogmaticalnes in writing as if he had found ye motion in ye Ellipsis, wch inclined me to try it after I saw by what method it was to be done.

Earlier, on 20 June 1686, NEWTON had written a more succinct summary of the whole "frivolous business":

> Now is this not very fine? Mathematicians that find out, settle & do all the business must content themselves with being nothing but dry calculators & drudges & another that does nothing but pretend & grasp at all things must carry away all the invention as well as of those that were to follow him as of those that went before.

I should like to hope that the whole matter of NEWTON's indebtedness to HOOKE were settled forever by the documents now published and interrelated, but I know such hope futile, for a woolly herd will have a woolly bellwether.

Let no-one expect to read in WHITESIDE's preface and notes the least concession to the physicists' hagiographic approach to their hero. WHITESIDE's findings are distilled from analysis, line by line and equation by equation. Thus (pages 26–28) when he comes to comment upon the last document in this volume, he states that "even Newton's comprehensive mathematical tool-kit . . . was not always adequate to the task he set himself." The matter at issue is the inequalities of the moon. Here "Newton more than met his match. In the *Principia*'s third book . . . he put up a brave public show of deriving . . . periodic and secular inequalities . . ., but the mass of his . . . worksheets . . . tell a different tale of repeated false starts, the myopic pursuit of dead-end trails and a near-total lack of success . . . . [H]e failed utterly to give adequate theoretical justification for his preferred formula . . . ." WHITESIDE goes on to say that a cogent explanation was to be achieved only sixty years later, through the efforts of EULER, D'ALEMBERT, and CLAIRAUT, who discarded NEWTON's approach altogether. Not long ago this simple and just sum of the facts of mechanics would have horrified only the physicists, true believers of the MACH who had preached the gospel of NEWTON the godlike, the greatest and last prophet of classical mechanics, indivisible and final. Today it will shock also the "new" historians.

WHITESIDE's footnotes reflect instant recall and command of all the documents. Footnote 126 on pages 305–308, a little masterpiece, concerns a proposition on the quadrature of ovals which was allowed by no less a critic than JAMES BERNOULLI, was doubted by HUYGENS and LEIBNIZ, was given a new "proof" by WARING, and was accepted through the nineteenth century. WHITESIDE first shows just where NEWTON's error lies; then by an unexceptionable counterexample, one that would have complied even with D'ALEMBERT's prejudices about "equations", he shows that NEWTON's claim is false.

The reader who wonders, as up to now I wondered, how NEWTON long after having renounced his philosophical studies could have with lion's paw smashed JOHN BERNOULLI's challenge on the brachistochrone, will find the answer in Footnote 14 on pages 459–461.

Enough general praise has been published of WHITESIDE's knowledge of seventeenth-century science, of his boundless energy and keen power of analysis, of his critical scholarship, of his devotion to a heavy task peculiarly his, for no-one else would have been equal to it. For comparison, only the HUYGENS edition comes to mind. While I find that edition easier to use, I recall that it was produced by a group of ten or more men working over a period of sixty years. In only eighteen years, working mainly alone, one man has nearly completed the edition of eight volumes of the works of NEWTON: a mass of material equal to over half of HUYGENS' remains (apart from letters), and in many cases even more difficult to penetrate. Only a young man can put out such power, but how can a young man have acquired the necessary knowledge and judgment? That is the miracle of WHITESIDE.

What next? The HUYGENS edition in its lucid completeness nearly explained all of HUYGENS' work in itself and so obviated need for further historical research. I still recall my delight when, twenty years ago, I first consulted that edition and found that a part of the task I had set myself had been done already and done right. Mathematical competence and historical soundness shone from every page. It was like stumbling upon San Ambrogio among the sullen flats of modern Milano. The kinship of HUYGENS' work to others' remained to be established, but there was no question as to what HUYGENS himself had done. I expect the same will be the result of WHITESIDE's edition of NEWTON's mathematical works. Would that we could see now a full stop put to the eruption of studies on NEWTON! Would that the little folk could turn their attentions to HOOKE and BOSCOVICH and the like! No, times have changed. No longer is learning the objective of scholarship. The fact that possible knowledge is boundless is now interpreted as implying that no question, however limited and however particular, is ever settled. By definition, now, there is no learning, because truth is dismissed as an old-fashioned superstition. Instead of learning, there is perpetual "research" on anything and everything. In virtue of PARKINSON's Law, the professional historian must keep on publishing. WHITESIDE's monument to NEWTON, like WREN's masterpiece for ST. PAUL, will soon be hidden by towering concrete hives of new bureaus and new slums.

Note for the Reprinting

It is rare that a publisher follows the advice of a reviewer, still rarer that a publisher goes such advice one better even though the review in which it is given remains unpublished. I refer to the fact that while Volume **VI**, reviewed above, was offered for sale at $72.50, and while I wrote of it that even so much as $150 might be a justified price for such a book if well gotten up, the price of Volume **VII** is $175, and the price of Volume **VIII** is $190. The increases of cost do not reflect any counterpart in quality of composition, presswork, or binding.

For a cogent analysis of the explanations of the inequalities of the moon by EULER, D'ALEMBERT, and CLAIRAUT, see the great memoir by CURTIS A. WILSON, "Perturbations and solar tables from Lacaille to Delambre: the rapprochement of observation and theory", *Archive for History of Exact Sciences* **22** (1980): 53–304.

33. GENIUS TURNS THE ESTABLISHMENT TO PROFIT: EULER

a. Euler's Letters (1960/1977)

The following reviews are reprinted in roughly the order of their appearance. They describe aspects of EULER's life and circumstances as they became known to the general reader.

I. Euler and "friendly and fruitful relations" (1960)

Volume 1 in the series, "Quellen und Studien zur Geschichte Osteuropas herausgegeben von der Historischen Abteilung des Instituts für Slawistik und der Arbeitsgruppe für Geschichte der Slawischen Völker am Institut für Geschichte" contains sixteen essays by East German and Russian professors, teachers, and assistants which were presented at the celebration of the 250th anniversary of EULER's birth held in Berlin in 1957. While EULER is usually thought of as a mathematician, this volume scarcely refers to any aspect of mathematics. The articles dealing with EULER discuss mainly his very considerable activity as an administrator, especially in connection with his life-long service to the Petersburg Academy. There are also articles on philology, bibliography, chemistry, and "cultural relations" having slight connection with EULER.

It would be unfair to take the article by WINTER, who is a member of the Institut für Geschichte der Völker der UdSSR at the Humboldt-Universität, Berlin, as typical, yet there are many passages here and there which fall close to the following lower bound taken from it: "Nevertheless, EULER in 1741 accepted a call from the Academy in Berlin; however paradoxical his reason may sound, it was because in this way he could serve the Enlightenment in Russia even better." P. HOFFMANN's essay gives us the impression that EULER was a kind of cultural secret agent for Russia during his twenty-five years at Berlin.

All the articles consist of dreary lists of facts connected by generalizations about "friendly and fruitful relations" between the peoples. The basis of all, however, is the great collections of unpublished letters and documents in the academies of Berlin and Leningrad. The eighteenth century is a period of particular darkness in the history of science, perhaps because the sources are so voluminous as to be forbidding. Even the scientist of most pragmatic slant might be curious to learn the circumstances in which science was pursued at that time, since the mean production *per man* in the mathematical sciences was then so great. More valuable than any summary would be the full publication of the proceedings of the academies and of the correspondence of EULER. Publication of even the eighty letters which passed between EULER and GERHARD FRIEDRICH MÜLLER in 1754–1756, on which HOFFMANN's article is based, would furnish an excellent introduction to the scientific scene in the Enlightenment.

Note for the Reprinting

The book reviewed here is *Die Deutsch-Russische Begegnung und Leonhard Euler. Beiträge zu den Beziehungen zwischen der deutschen und der russischen Wissenschaft und Kultur im 18. Jahrhundert*, edited by E. WINTER and associates, Berlin, Akademie-Verlag, 1958. The review first appeared in *Isis* **51** (1960): 115.

From the text it is clear that when I wrote it, I did not know of the existence of E. WINTER's earlier volume *Die Registres der Berliner Akademie der Wissenschaften 1746–1766, Dokumente für das Wirken Leonhard Eulers in Berlin*, Berlin, Akademie-Verlag, 1957. In 1962 MIKHAILOV published the 848 references to EULER in the Registers of the Academy at Petersburg, stretching from his arrival in 1727 to his death in 1783. They are included in the volume reviewed below as the first item in Essay 33b.

II. Euler's Correspondence with Müller (1961)

Those who regard the scientific academy of the eighteenth century as a grove of pure devotion to learning will be disillusioned by this volume. EULER and MÜLLER had been colleagues in the Petersburg Academy for about fourteen years in 1741, when EULER went to Berlin. Although their written correspondence begins earlier, while MÜLLER was travelling for a decade as a geographer in BERING's second expedition to Kamchatka, 189 of the 207 letters published in this volume come from the period when EULER was acting as counsellor and virtual editor for the Petersburg Academy while being director of the mathematical class of the Berlin Academy. Their sodden

Figure 20. LEONHARD EULER (1707–1783) after a relief in plaster by J. RACHETTE, 1781, reproduced by permission of the Académie des Sciences, Paris.

pages show us that administrations and administrators have changed little. EULER's abundant energy sufficed to provide an endless sequence of recommendations of mediocrities which would grace the files of a dean today. While EULER himself stood above petty quarrels and intrigues, he could not ignore them, for they made up almost the whole fabric of life in an academy, and the correspondence contains many direct and indirect references to them.

Since MÜLLER, although an industrious author and a geographer of some fame, was no great intellect, there is little that concerns mathematical problems, and the many remarks about natural science are more in the nature of news than of discussion. The most frequent topic is the construction of achromatic lenses, but EULER shows knowledge and interest in every aspect of science. While the editors have not provided an index of subjects, it is replaced by adequately referenced summaries in the interesting and thorough introduction.

It is a truism that everyone who has learned calculus and rigid dynamics since 1750 has done so directly or indirectly from EULER's books. Those who concern themselves with the dissemination of knowledge and those who measure success by the run of the presses may ponder the fact that out of an edition of 500, 406 copies of the *Differential Calculus* remained unsold after six years, and almost as many five years after that. Copies turning up from dealers today are often unopened. In 1760 EULER finished his *Rigid Bodies*; by 1762, only twelve persons had subscribed, though no deposit was requested; when the book appeared in 1765, scarcely any copies had been sold. At this same time the academies, always on the brink of bankruptcy, were pouring great sums into the collective efforts of the nobodies who filled their numerous offices and whose busy little lives have scarcely scratched a groove in the history of science.

Since the correspondence with MÜLLER was official rather than private, only by implication can anything be learned of the miserable treatment EULER received from FREDERICK II after the death of MAUPERTUIS, and of the true reasons causing him to let it be known that he would welcome a chance to return to Russia. Only in one letter, No. 169, dated 27 May/7 June 1763, does EULER give vent to his feelings at what was happening:

> That Mr. d'Alembert has rejected the highly notable and lucrative position in Russia, I should attribute rather to fear lest the matter turn out ill in the end than to philosophy. For despite his unbearable arrogance he has long been able to understand that he is not at all suited for the post. Anyway, his philosophy consists, as Mr. Bernoulli puts it, in an impertinent sufficiency, which makes him try most shamelessly to defend all his mistakes, which come

back to him only too often, and so for many years from vexation he has not wished to touch mathematics. In his hydrodynamic theory he has right cavalierly contradicted most of the theorems of Mr. Bernoulli, which are confirmed by abundant experience, because experience contradicts his own theorems. And he has not been able to overcome his arrogance enough to admit his plain error.

From the informed, his scrimmages with the thorough Mr. Clairaut can bring him nothing but the greatest disgrace. Only here [at Berlin and Potsdam] does he count as a creative mind, a man who comprehends all: But doubtless for the same reason he will not come here, and he is said to have proposed as a substitute for himself in the presidency the Chevalier de Jaucourt. Meanwhile he has upon insistent entreaty decided to make a trip to Cleve, where he was to have arrived yesterday, to decide the entire fate of our academy. That indicates that a bunch of Frenchmen, plainly creative minds, are to be called here.

These paragraphs reflect but do not reveal the extent to which the intellectual history of the eighteenth century remains to be discovered. This volume, well annotated and edited despite a few dubious paragraphs here and there, furnishes a permanent source; much more, unpublished, remains to be seen and used before the science of the Enlightenment can emerge from the conventional fictions repeated in the standard histories.

Note for the Reprinting

The book reviewed here is *Die Berliner und die Petersburger Akademie der Wissenschaften im Briefwechsel Leonhard Eulers.* Volume 1. *Der Briefwechsel L. Eulers mit G. F. Müller, 1735–1767,* edited by A. P. Juškevič, E. Winter, & P. Hoffmann, Berlin, Akademie-Verlag, 1959. The review first appeared in *Isis* **52** (1961): 113–114.

III. Euler's Correspondence with Schumacher (1962)

Nearly all of the second of the volumes of EULER's correspondence regarding the Petersburg Academy consists of letters written by EULER to SCHUMACHER between 1730 and 1757. SCHUMACHER, an Alsatian of no known intellectual capacity, was secretary of the Petersburg Academy and the most powerful individual in it for more than thirty years. While the editors of the volume feel that he was unduly hard on Russian members, the correspondence here published seems to me to witness rather to a lifelong program on his part toward defeat and expulsion of talent wherever it showed itself. In him we see an early specimen of the scientific administrator, now so familiar as to occasion no protest. When EULER objected violently to a paper on imaginaries by KÜHN, a mathematical faker from Danzig, SCHUMACHER found good to publish it nevertheless, since "a few persons" had praised it (Letters 232, 233, 239, 241, 243), and he refused to print EULER's damning summary of it. Then, at a time when nearly every scientist of repute had left, SCHUMACHER could write, like a modern dean, "mediocre savants are of no use to the Academy." SCHUMACHER was also instrumental in the failure of the Academy to pay DANIEL BERNOULLI according to its contract; he delayed payments to EULER, even to reimbursement of out-of-pocket expenses on the Academy's behalf, and it is clear that only the need for EULER's further services can explain his having been, ultimately, paid. While we are inclined to think of those services as producing the unbelievable flood of papers he sent to Petersburg, amounting to some half of the contents of the Academy's *Memoirs* for the entire twenty-five years EULER was in Berlin, the reader of this correspondence will see that SCHUMACHER conceived them differently. Enough of EULER's enormous energy remained, after his own researches and his official duties in Berlin, for him to write SCHUMACHER concentrated information regarding every scientific activity in Europe, endless recommendations and expertises, and advice on every problem, personal or organizational, that faced the Academy, while at the same time he served as *de facto* editor and sole referee for its publications.

EULER's scientific knowledge and interest extended far beyond the subjects of his writings. Apparently he tried, through SCHUMACHER, to steer the Academy's general program along the road of reason. At least I can find no other motive for the trouble he took to write out so much general scientific information for a person who plainly was interested in it only to the extent he could use it for manœuvering personnel. There is no sign of friendship or even friendly interest shown by either correspondent in the other. We must recall that SCHUMACHER was, in effect, EULER's boss as far as concerned the Petersburg money. As in most such relations, the scientist earned ten times over every ruble he received, hoping to serve science as well as Mammon in the process.

The reader of these letters will note EULER's strong and continuing interest in chemistry. Also he might not expect to learn that EULER really cared about practical geography and was proud of the work he had done in younger years, at great cost to his eyesight, on the general map of Russia (Nos. 35, 45, 175).

EULER's personality, heretofore little known because of the objective expression used in his memoirs and purely scientific correspondences, begins to reveal itself through the subsidiary documents the Russians are now publishing. After having been in the Academy some time in a minor position, upon DANIEL BERNOULLI's departure he was proposed for promotion to a professorship. Letters 1–5 concern this appointment. GMELIN, WEITBRECHT, KRAFFT, and MÜLLER, all of whom had been receiving less salary than EULER, were proposed for this same promotion; the new salaries of all five were henceforth to be equal. EULER expressed himself bluntly:

> It seems to me that it is very disgraceful for me that I, who up to now have had more salary than the others, shall now be set equal to them . . .. I think that the number of those who have carried [mathematics] as far as I is pretty small in the whole of Europe, and none of those will come for 1000 rubles.

When he wrote these words, EULER was twenty-four, and he had published seven papers. If the reader thinks this is not the way to talk to a dean, he should remember that we are in the eighteenth century, when stomachs were stronger and not all suits were alike. If EULER thought this kind of talk would get him anywhere, he soon learned the contrary. SCHUMACHER advised the president not to grant him the least concession, since otherwise he would straightway grow impudent. [Such a man will joy to harness a racehorse to a huckster's cart, couple a great charger in a plowyoke with an ox.] EULER had to be content with 400 rubles, like the four little men who were promoted simultaneously. It cannot be said that in later years EULER ever got the upper hand of an administration, but the correspondence with SCHUMACHER shows that he at least lost his illusions and learned that 800 publications are not enough to make one's way in the scientific world.

Another personal matter is cleared by one of the letters here, namely, the ever-repeated story of how the generous D'ALEMBERT charmed EULER and, instead of criticizing him to the hostile king, praised him and recommended that favor be bestowed on him. On 15/26 July 1763 EULER wrote to TEPLOV that he wished he were able to accept immediately the offer to return to Petersburg (a major obstacle having been removed by the death of SCHUMACHER, who would have been the last to offer a lily pad to any big frog).

> Now since Mr. D'ALEMBERT is here, I wished very much that he would have accepted the position [of President of the Berlin Academy], since in that case I should infallibly by this time have gotten my release . . .. But not only has Mr. D'ALEMBERT refused

this position, but also he has done me the ill turn of recommend-
ing me in most emphatic terms to the king .... Thus ... if I
presented my request now, I should be granted some increase,
which I should not dare to refuse, and which would bind me here
forever.

Note for the Reprinting

The volume reviewed is *Die Berliner und die Petersburger Akademie der
Wissenschaften im Briefwechsel Leonhard Eulers*. Volume **2**. *Der Briefwechsel L.
Eulers mit Nartov, Razumovskij, Schumacher, Teplov und der Petersburger
Akademie, 1730–1763*, edited by A. P. JUŠKEVIČ, E. WINTER, P. HOFF-
MANN, & JU. CH. KOPELEVIČ, Berlin, Akademie-Verlag, 1961.
The review is reprinted from *Isis* **53** (1962): 411–413.

IV. Some of Euler's Correspondence with Learned Men
and Administrators (1978)

The third and final volume publishing a selection of EULER's correspondence preserved in Russian libraries concerns mainly bureau-
cratic and personal matters. To what extent the *Opera omnia* will
reprint these letters is not clear. This volume is more interesting than
its predecessors in that it includes correspondence not just with
officials but also with learned men: AEPINUS, HEINSIUS, KRAFFT,
LOMONOSOV, and WETTSTEIN. These correspondences bear witness
to the breadth of EULER's scientific interests and to his friendliness
toward scientists of all kinds, not only those competent in mathematics. He encouraged both AEPINUS (1724–1802) and LOMONOSOV
(1711–1765); the two were enemies, the latter a genius of wider range
than the former but perhaps of lesser depth, and certainly of speculative rather than mathematical bent.

Most letters from EULER's correspondents are merely summarized
in small print, not published. This is rather a pity, since it means
nothing but summaries in the correspondence with AEPINUS (nine
letters); only seven complete letters to HEINSIUS and seventy-seven
summaries; in the correspondence of thirty-nine letters with KRAFFT,
which lies at the level of serious and specific research in mathematics
and astronomy, only five complete texts. The correspondence with
LOMONOSOV, though it concerns mainly chemistry and academic
fights, is by exception reprinted in full. A pretty example from
LOMONOSOV: "Taubert, if he sees a dog in the street bark at me, is
ready to hang on the neck of such a beast and always kiss him under

the tail. And he does so just as long as he needs that barking; then he smashes the dog into the filth and sicks another dog on him." More of the same kind will show the reader that relations among colleagues have not changed much in 200 years, but nowadays few professors know how to express themselves in a lively style.

The correspondence with LOWITZ, a chemist from Göttingen, shows how EULER's work was appreciated by some lesser men and suggests how his discoveries came to be learned and taught as standard fare in some German universities.

The longest correspondence is the fifty-seven letters between EULER and JOHANN CASPAR WETTSTEIN (1746–1759). This WETTSTEIN, born in Basel, was chaplain and librarian to the Prince of Wales. A cousin of his had been one of EULER's teachers. EULER had met him in Petersburg and apparently found him congenial; all but one of the letters are from EULER. The subjects range over all sorts of scientific matters such as expeditions, tables of lunar motions and tides, gravitation, orbits of comets, naval science, the Berlin almanacs, calendars, mortality tables, maps, the culture of mulberries and corn, terrestrial magnetism, theology, optics, telescopes, microscopes, experiments on electricity, and the history of astronomy. They mention also economic matters, the circumstances of persons, and the casualties of war.

While these letters have been available through copies sent in 1843 to the Petersburg Academy, they are little known; such extracts as have been published before are translations into Russian. The originals of most of the letters turned up recently in a medical library in Britain. WETTSTEIN seems not to have known much mathematics but to have been an amiable person, interested in general science, and EULER wrote more freely and widely to him than to anyone else. It is strange that the two Baslers chose to communicate with each other in French. In his letter of 16 July 1746, EULER states that if his election to the Royal Society "could be effected in such a way that nobody could suspect I had sought it, I should be very grateful." The rule was that fellowship in that august body had to be applied for, and EULER was known to be unwilling to make any move in this regard. In the immediately following sentence he thanks WETTSTEIN warmly for his efforts to send some good tobacco. EULER was thought to share the dislike that his teacher, JOHN BERNOULLI, carried as far as contempt for the *scurri anglicani*, only NEWTON excepted. Be that as it may, EULER's election to the Royal Society on 22 January 1747 soon produced an astonishing reaction: In his letter of 5 March 1748 we read

... there is no country where I had rather live than in England
.... I remark that here the taste for belles-lettres gets more and

more the better of that for mathematics, and I have reason to fear that soon I shall become useless. In such a case I should not wish to return to Petersburg, because my family could not hope for any solid position there. Since it is very numerous, I do not see a place suitable for me either in our native land or anywhere else, unless it be in England. But I know also that I should not be fit for any ordinary post, and I should have to be granted some unusual pension, no less than what I have here. If you think I could hope for any such solid position, I beg you to act on my behalf. Perhaps what little reputation I have acquired, joined with your recommendation, could lead some great lords to procure a pension sufficient for me to subsist with my family. But as it is a very delicate affair, I pray you to take every precaution lest anyone here be able to suspect I had disclosed these things to you.

Two years earlier, EULER had written to this same WETTSTEIN,

After leaving the Academy of St. Petersburg I have every reason to be completely happy with my fate. The King gives me the same pension as I had in St. Petersburg, which amounts to 1600 écus, and I am responsible directly to His Majesty. I can do just what I wish, and nobody demands anything of me. The King calls me His Professor, and I think I am the happiest man in the world.

WETTSTEIN's reply to EULER's plea for a post in England, though written only seven weeks later, was delayed. EULER, assuming that the response to his request was negative, meanwhile wrote that he so understood; when the response came, it was indeed negative. England offered freedom but no money: "even though foreigners are allowed to settle in England, that was far from meaning they would be given pensions." In 1765 the English were to soften EULER again by voting that "a summ of money, not exceeding three hundred pounds in the whole, shall be paid to you, as a reward for having furnished Theorems, by the help of which the late Mr. Professor Mayer of Gottingen constructed his Lunar Tables...." As ANDREAS SPEISER was wont to remark with a sly smile, here is the only instance of a government's paying money for theorems. If we are amazed because it was the *British* government that chose to reward a foreign furnisher of theorems, we can recover our balance by reading the rest of the story: The tables were *useful* for navigation, the same Parliament paid the computer £5,000, and the theorems, of merely ephemeral interest, are among the least valuable as well as the least beautiful of the hundreds that EULER discovered.

The tone of EULER's letters to WETTSTEIN is the easiest in all of EULER's correspondence, but they could not be called intimate. They add to the already abundant evidence of EULER's generosity in scientific matters, his freedom from envy or competition (but of course nobody could compete with him), and his consistent, level judgments of others. The correspondence closes in 1759, when WETTSTEIN was dying. Thus it does not enlighten the circumstances of EULER's angry departure from Berlin.

EULER, born free, was to spend all his adult life in lands ruled by despots—despots benevolent or not, according to circumstances or whims, but despots who paid salaries.

As for Berlin, EULER's correspondence with the BERNOULLIS makes it plain he soon regretted he had too quickly left Petersburg[1], especially as his pay was not increased by the change. FREDERICK II did not even fully honor the contract with which he had tempted EULER away; on 21 January 1743 he wrote EULER the first in what was to be a long series of sarcastic insults. FREDERICK's petty squabbles with his subjects were not only unroyal (we cannot imagine LOUIS XIV or even the affable CHARLES II descending so far, except, perhaps, in the bedroom), they were juvenile. Nevertheless, EULER held out for twenty-five years in Berlin, indeed until the treatment he received had become intolerable. The truth was, the literary king hated mathematics.

EULER's letters do not tell us much about his personal affairs. He kept all his correspondents at a distance. His three letters of March 1741 to BREVERN, who was then the president of the Petersburg Academy, show how difficult it was to get permission to resign. We are left wondering whether EULER's expressed reason, namely, that he feared to lose his eyesight entirely, was fact or excuse: "if I cannot get quickly to a milder climate, before my eyes I see nothing else than the total ruin of my health and certain death." On the one hand, twenty-five years later EULER did return, and eagerly, to the very same climate; on the other, he did lose his remaining eyesight soon thereafter; as for certain death, he found that, too, but at the age of seventy-six. Urgent as the tone of the letter is, EULER manages to mention that in a milder climate he could salvage his health as much as possible, "and hence I could continue to serve the Imperial Academy." That is exactly what he was to do, in return for a pension, of course.

---

[1] *Cf.* the remarks of E. WINTER, pages 21–22 of *Die Registres der Berliner Akademie der Wissenschaften 1746–1766, Dokumente für das Wirken Leonhard Eulers in Berlin*, Berlin, Akademie-Verlag, 1957. They may go beyond the evidence on which they are based.

While nearly all the correspondences reflect the self-control that ruled EULER's life, there is a famous exception: EULER's resentful anger when the ratios of his salary to those of four insignificant colleagues were reduced. We know the episode from EULER's letters to SCHUMACHER, published in Volume 2 of this series. [See above, part aIII of this essay.] Here we may confirm and extend our picture of it by reading the letters to BLUMENTROST. The affair, which is as shocking today as it was in 1731, shows that the administrative process was practised very well long before management science became a university discipline. The editors write, "It is a classic expression of the struggle of the genius against the bureaucracy. It ends, as most frequently happens, with an apparent victory of the bureaucrats at first." I am puzzled by the words "apparent" and "at first", for EULER never got a kopec in return for his outcry. Perhaps JUŠKEVIČ & WINTER allude to the fact that had not SCHUMACHER and BLUMENTROST mistreated EULER, nobody today would ever have heard of them.

Note for the Reprinting

The volume reviewed is *Die Berliner and die Petersburger Akademie der Wissenschaften im Briefwechsel Leonhard Eulers.* Volume **3**. *Wissenschaftliche und wissenschaftsorganisatorische Korrespondenzen 1726–1774*, edited by A. P. JUŠKEVIČ & E. WINTER, Berlin, Akademie-Verlag, 1976.
  The review is reprinted from *Isis* **69** (1978): 301–303.

V. The Catalogue of Euler's Letters (1977)

Every mathematician will joyfully tell you again and again that EULER used divergent series with heedless formal abandon; few know that his works contain a precise discussion of the convergence and divergence of the geometric series in CAUCHY's sense and at least one definition of the sum of a divergent series acceptable even today for a complex power series at points on its circle of convergence. With even greater glee, every mathematician will tell you that EULER used a definition of a function that will cover, essentially, only algebraic functions; few will know that it was EULER himself who soon thereafter saw that this definition was inadequate and then proceeded to introduce, almost word for word, the definition now used universally, a definition that was reintroduced in circuitous terms and with an unnecessary limitation by DIRICHLET, nearly a century later[1]. With a finality that no mere fact can shake, every mathematician will tell you that EULER was unrigorous by ABEL's standards. He will not know that so, usually, were GAUSS and CAUCHY and RIEMANN; even less will he know that not only in contrast with his teacher, JOHN BERNOULLI, with his contemporary, D'ALEMBERT, and with his successor, LAGRANGE, but also in comparison even with the great founders of infinitesimal analysis, namely, NEWTON and LEIBNIZ—in contrast or comparison with all these, EULER was a scrupulous, careful mathematician, determined to replace obscurity and mysticism by precision and system wherever he could. For this last, two examples should suffice here. (1) It was EULER who first presented the derivative systematically as the limit of a difference quotient and founded the differential calculus upon a developed, exact calculus of finite differences. (2) It was EULER who showed, once and for all, that differentials of second and higher orders, which populate the pages of analysis through the middle of his century, were unnecessary, to the point that they were banished forthwith and have left scarcely a trace upon mathematics as it is taught today[2].

Mathematicians may do many different things. They may solve outstanding problems which baffled their predecessors; create new concepts and disciplines, providing fields of research for future generations; organize and unify previously scattered theorems and theories; carry through difficult calculations or obviate them so as to reach deep results through new and penetrating lines of thought; render trivial important facts or theories which previously seemed difficult; apply mathematics successfully to problems of natural science; invent and illustrate schemes of numerical computation, or contrive means of rendering such computation unnecessary. Each of these activities provides a standard. Any mathematician today who will take the trouble to learn what EULER really did will be forced to concede that by every one of these standards, EULER was unsurpassed by anyone—unsurpassed even by ARCHIMEDES, NEWTON, LEIBNIZ, GAUSS, or RIEMANN. In breadth of command and in clarity of exposition, no-one came near him. Of course, great mathematicians cannot be linearly ordered. Nevertheless, only blind prejudice or special pleading could deny to EULER rank of solely supreme in mathematics.

Of EULER's published works only about one third concern primarily what today we should call "pure" mathematics. Roughly

[1] Cf. A. P. YOUSCHKEVITCH, "The concept of function until the middle of the nineteenth century", Archive for History of Exact Sciences 16 (1976): 37–85.

[2] Cf. H. J. M. BOS, "Differentials, higher-order differentials and the derivative in the Leibnizian calculus", Archive for History of Exact Sciences 14 (1974): 1–90.

half are devoted to various aspects of mechanics, from the foundations through celestial mechanics and down to very specific analyses of problems that today belong to mechanical engineering. If we exclude statistical mechanics, I can count ten major categories of bodies studied in mechanics as it was taught in, say, 1900. Almost single-handed, EULER discovered the basic equations for three of these: rigid bodies, ideal fluids, and flexible surfaces. To theories of these kinds of bodies and of four more he made major contributions. For example, the general equations of motion for systems of mass-points nowadays called "Newtonian" are his beyond contest; hydraulics owes to him the formulation used even today and also the analysis of rotating machines; he gave acoustics the wave equations in two and three dimensions and determined the tones of horns; the explicit solutions for the small transverse vibrations of a string arbitrarily plucked and for the propagation and reflection of pulses in tubes are his, as are those for longitudinal motion of a massless elastic cord loaded by an arbitrary number of equidistant and equal masses; he determined explicitly and classified all the forms a terminally loaded elastic bar may assume in finite deflection, and he was the first to recognize and calculate the phenomenon of buckling. [In the process, he was the first to show that the critical values of the parameters of a nonlinear theory were equal to the proper numbers of the corresponding linearized theory.] Beyond all that, he created the general principle of rotational momentum and was the first to formulate general principles of balance for classical mechanics and to obtain their explicit local forms for one-dimensional deformable continua.

Any man who attempts as much as EULER did will surely go wrong once in a while. Indeed, EULER made some errors. These, like his occasional lapses in rigor, have been magnified. A popular biography, echoing for the hundredth time a superficial obituary, asserts that EULER was justly criticized for "letting his mathematics run away with his sense of reality. The physical universe was an occasion for mathematics to EULER, scarcely a thing of much interest in itself; and if the universe failed to fit his analysis it was the universe which was in error." Anyone who has read and studied a single one of EULER's papers on physics knows that this accusation, while ever repeated with various embroidery, is a mere lie. His achievements in mechanics, added to what he did in optics and astronomy and molecular physics, make him obviously the greatest physicist of the middle half of the eighteenth century, surpassing even DANIEL BERNOULLI.

Mathematics and physics did not exhaust EULER's interest. His publications include a classic treatise on harmony and a superb paper on metaphysics as well as respectable work on statistics, navigation, and geography. His major contribution to philosophy, sneered at by the *philosophes*, is only now coming to be valued[3].

Nevertheless, it is only from his letters that we may get an idea of the range of EULER's activity. The editors of the volume under review write,

> ... the great mathematician conscientiously answered everyone who addressed him. This enormous correspondence, stretching over nearly the entire period of his scientific work—from September 1726 until September 1783—is of extraordinary interest. In it are discussed the scientific questions which concerned Euler and his correspondents: the whole range of mathematical, mechanical, astronomical, and physical problems of his time, and beyond that many questions of biology, geography, engineering, philosophy, and religion. Now and then the status of questions is merely sketched, but in many cases the letters are really small papers with the presentation of a theory, with deductions of theorems or indications of the course of proofs, with statements of new problems, with comparison of various methods and points of view *etc*. Indeed, most of the results given in Euler's letters went into his published work, but we often find in the letters the development of an idea just being born, interesting comparisons with works of other authors, and all sorts of explanations and remarks and even some results that are not to be found in his manuscripts for the press.

The editors describe also "the regular exchange of thought, which had an extraordinarily fruitful effect upon the activity of all the correspondents and is only partly reflected in their published works." They emphasize EULER's role as the great teacher of his age—in LAPLACE's words, "the master of us all".

> How many problems Euler proposes here to his correspondents, how willingly he discusses their plans of work with them, with what authority he encourages them in their undertakings, how correctly and carefully he explains their mistakes!

As the editors say, in EULER's correspondence the activities of the great academies and some of the universities of the day are reflected.

---

[3] *Cf.* A. SPEISER, *Leonhard Euler und die deutsche Philosophie*, Zürich, Orell Füssli Verlag, 1939, and R. CALINGER. "Euler's *Letters to a Princess of Germany* as an expression of his mature scientific outlook", *Archive for History of Exact Sciences* **15** (1975): 211–233.

Only from the letters may we learn what a leader EULER was in the organization of scientific work. They speak of his initiative in selecting the subjects for many an academic prize, in nominating candidates for vacant posts, in editing productions of younger men working under his personal guidance, in helping to secure books and apparatus for scientific institutions. They claim that he even engaged in diplomacy; it is one of the few stains upon his character ever alleged.

Until the correspondence of EULER is published and studied, it will remain impossible to compose a just intellectual history of Europe in the middle half of the eighteenth century. Such historical studies as we have, heavily weighted toward belles-lettres and the philosophes, concern the supporting cast, the stage hands, and the advertising men, but leave the protagonist unmentioned.

After years of hesitation, the editors of the great international edition of EULER's *Opera omnia* have decided to publish most of EULER's letters, and the volume under review, numbered Volume 1 of Series IVA, is the catalogue of the 2,948 letters to and from EULER which have come down to us through originals, copies, draughts, and extracts. We know that there must have been about 5,000 letters originally and that EULER himself set a high value upon them. Indeed, fourteen months before he finally obtained permission to leave Prussia so as to return to Petrograd he wrote to MÜLLER asking about the status of the "entire learned correspondence" he had deposited with the Academy "in a fair parcel" a quarter of a century earlier. If the Academy now found it useless, might he have his letters back? "[I]f anybody took the trouble to select these, he would find many important points in them, publication of which would be more to the taste of the public than the most profound elaborations." About one third of the letters still extant may be found in some twenty-five publications, ranging from a single extract to a complete correspondence between EULER and another person, for example BOUGUER, GOLDBACH, LAMBERT, MÜLLER. Even the most important correspondences have been published only in part: with JOHN I BERNOULLI (34 of 38), with DANIEL BERNOULLI (67 of 100), with CLAIRAUT (7 of 61), with D'ALEMBERT (8 of 39), with MAUPERTUIS (32 of 129), with LAGRANGE (33 of 37). Series IVA of the *Opera omnia*, a joint undertaking of the Swiss and the Russians, will print in full all major correspondences except the one with MÜLLER (which, having been published already in the D.D.R., would seem to involve complications) and at least EULER's side of the rest. It promises to be the major document of the intellectual history of the mid-eighteenth century to be made public in the last 100 years.

The great edition of the *Opera omnia* was planned in 1910; Series I ("Pure" mathematics) is now complete; Series II (Mechanics and Astronomy) and Series III (Physics and Miscellaneous), nearly so. These three series will make up seventy-four stately parts, carefully edited, set forth in large type on fine, tough, white paper with abundant margins. We take it for granted that anything begun today will be shabby, but the first volume of Series IVA provides a welcome exception. In format, printing, paper, and binding it differs from the old series but is not inferior. Its use of a more or less old-style type is more appropriate to a work of the eighteenth century than the cold Bodoni of Series I–III. I trust that we may presume the small font used in the catalogue is not that adopted for the words of the master himself in the succeeding six volumes planned for the series.

The reader who knows the letters already published will turn eagerly to the summaries given here for the letters not yet generally available. I am sorry to have to report that he may be disappointed. The correspondence with DANIEL BERNOULLI may serve as an example, an especially telling one because in its incomplete state as published by FUSS 135 years ago it is famous as the most interesting in the history of mechanics. Fifteen years ago I studied as much of it as I could;[4] some of the thirty-three as yet unpublished letters in this correspondence were available to me, but not all of them. I noted in particular a long gap in 1739–1740, just when the principals, because of their intense study of small transverse vibrations of an elastic bar, were having to discover how to integrate linear differential equations with constant coefficients and were discussing the criterion of minimum stored energy for the elastica—crucial topics for analysis and mechanics alike. From the volume under review I am delighted to learn that nearly all the correspondence for this period has been preserved, but the summaries of the letters that are new to me, Nos. 127–133, 135, and 137, leave me not much wiser than before about their specific content. Had EULER discovered how to handle multiple roots before 12 March 1740, when BERNOULLI sent him for publication his paper called "Excerpta ex litteris . . ."? Did BERNOULLI ever understand EULER's discovery of resonance? The summaries tell us mainly the subjects of discussion, not the theorems disclosed. The few exceptions regard studies of definite integrals and infinite series, harking quaintly back to the days of ENESTRÖM and STÄCKEL, when such things were taken as the hallmark of mathematics, but one displayed formula uses space sufficient for five lines of text, and in the case of No. 131 I think some specific statement about multiple roots

---

[4] § 24 *et seqq.* of *Opera omnia* (II) 11_2 (opus meum anglice conscriptum).

and some explanation of what is meant by "BOUGUER's theorem" would have done the user more service. Nonetheless, let us not cavil at small defects in a work so earnestly desired, so timely, and so well done as this. The catalogue will serve as the handbook for Series IVA, and let us hope that soon we may turn to the volumes of text when we wish to know the contents of the letters.

Choice of language for the descriptions must have raised a question. The two most natural solutions would have been, first, Latin, the universal language of learning (when learning was still alive) and also the language in which many of the most important letters were written; second, in each instance the language of the original. International English—stilted, circuitous, verbose, pretentious, and obscure—would have opened the work to the widest audience but would have clashed not only with EULER's own style but with the tone of the Enlightenment. The editors have selected High German. Although this choice will seal the work to most young professional historians of science, it has the appeal of being the closest to the native language of EULER and the BERNOULLIS; perhaps, also, it is the language in which EULER found himself most comfortable, for, as DANIEL BERNOULLI wrote to him in one of the letters (No. 95) that will be published for the first time in Series IVA,

> Weil ich aus dero ersterem gesehen, daß Sie sonderlich rein töutsch zu schreiben sich beflissen, als zweifle ich nicht, ich werde Dero keüsche ohren sehr mit meinem undermengten frantzösichen & lateinischen wörtern verletzt haben, weswegen sehr umb verzeihung bitte. ade noch einmahl.

Series IVB is to reprint a selection from the numerous still unpublished manuscripts of papers and books by EULER. EULER promised Count ORLOV to leave to the Academy of Petersburg enough material to fill its volumes for twenty years after his death. In fact, the regular *Acta* could not keep up with EULER's production, and so the Academy published in 1783 and 1785 the two supplements called *Opuscula analytica*, Volumes 1 and 2, containing twenty-eight of his memoirs not previously published. Nevertheless his papers made up the major part of the Academy's mathematical publication thereafter for forty-seven years regularly through 1830. By then LAPLACE, born when EULER was in his forties, had died; GAUSS had given up mathematics for astronomy and magnetism; and CAUCHY was at the height of his powers. The volume for that year contains fourteen papers by EULER, papers which he had presented to the meetings in 1780–1782, except for a lonely one of 1777, filling 137 consecutive pages. Soon thereafter, further whole volumes of previously unpublished papers or fragments appeared; the *Commentationes arithmeticae collectae*, 2 volumes, 1849, include eight new papers; the *Opera postuma*, 2 volumes, 1862, contain fifty items, none published before. All these are included in Series I–III of the *Opera omnia*. An important volume of manuscripts on mechanics, edited by G. K. MIKHAILOV, appeared in 1965; [see part b of this essay, immediately following.]

Each of these volumes provides a major source for the history of the mathematical sciences. To mention a single example, the *Opera postuma* contain a remarkable fragment on the motion of fluids in elastic tubes. Designed to analyse the flow of blood in arteries, this work stands as the earliest mathematical study in its field; in some measure it foreshadows the "new day in physiology" that DANIEL BERNOULLI expected to follow the publication of his *Hydrodynamica* in 1738, a new day that had to wait two centuries before it could dawn; but the direct contribution of this paper is to the theory of water hammer, the differential equations governing which appear in it. Delayed some eighty-seven years in publication, even so it was published a generation too soon to be appreciated, and the basic equations it obtains were rediscovered later in the nineteenth century. [Recently the fragment has been completed by the discovery of its first fourteen sections. The entire essay may be read in *Opera omnia* (II) **16**, published in 1979.]

The bulk of Series IVB will be given over to EULER's notebooks. These remarkable volumes cover most of his creative life before he became blind. Later come the volumes of *Adversaria*, which record questions he discussed with his assistants after his return to Petersburg. The two series make it possible to trace some aspects of the course of his thought over fifty years, during which, as his teacher JOHN BERNOULLI said very early, he brought the higher analysis from infancy into man's estate. The first notebook, written entirely during his student days in Basel, is the most remarkable. With some exaggeration it could be described as all of his 800 books and papers in little and in project. It can be compared with GAUSS's *Notizenjournal* and JAMES BERNOULLI's *Meditationes*; in contrast with the former, it is frank and clear, not a cryptogram; in contrast with the latter, it reveals a full-fledged, cornucopian genius at the age of nineteen, not a slow-growing, ponderous, middle-aged titan driving a divine mill. Those who have been disgusted by the decision of the BERNOULLI edition to scatter the *Meditationes* in classified snippets, destroying it as a human masterpiece, will be relieved to learn that the EULER edition will publish the notebooks as their great author wrote them.

The editors wisely do not estimate how many years it will take them to complete the edition. It was the cherished hope of ANDREAS SPEISER, the savior and refounder of the *Opera omnia*, to live to see the first three series standing upon his bookshelf, but such fulfilment was denied him. I hope that Series IV will be completed in the lifetimes of some of those who knew and loved him.

Note for the Reprinting

The volume reviewed is LEONHARDI EULERI *Commercium epistolicum. Descriptio commercii epistolici* (LEONHARDI EULERI *Opera omnia* (IVA)1), ediderunt ADOLF P. JUŠKEVIČ, VLADIMIR I. SMIRNOV, et WALTER HABICHT, impensis Societatis Scientiarum Naturalium Helveticae, venditioni exponunt Birkhäuser, Basileae, 1975.

The review is reprinted from *Archives Internationales d'Histoire des Sciences* **27** (1977): 292–296. A somewhat rougher text had appeared in *Eighteenth-Century Studies* **9** (1976): 627–634.

One more volume in Series IVA has been published: Volume **5**, containing the correspondences with CLAIRAUT (1740–1764), D'ALEMBERT (1746–1773), and LAGRANGE (1754–1775), edited by A. P. JUŠKEVIČ & R. TATON, 1980. This volume includes errata and additions for the catalogue of letters. The additions are nine newly discovered letters and descriptions in auction catalogues of nine more.

The editors of the works of JAMES BERNOULLI have wisely decided to publish as soon as possible a facsimile of the manuscript of the *Meditationes*.

b. Euler's Early Manuscripts on Mechanics (1967)

I. Catalogue

Along with his catalogue of the 873 printed works of EULER, published in 1911, ENESTRÖM issued a hastily compiled description of the unpublished manuscripts. It has long been known that this list is far from complete. The main collection of EULERian documents is now in the Archives of the U.S.S.R. Academy of Sciences. Many were loaned for over thirty years to the editors of EULER's *Opera omnia* in Basel, but little use was made of them, and in 1947–1948 they were returned to their owner. Meanwhile, in the 1930s about 1,500 more sheets had been found in the Incunabula Department of the Academy Library, disordered and packed into three bundles, in all probability after having been found in the rooms of P.-H. FUSS after his death. This material was analysed and sorted in the 1950s by G. MIKHAILOV. Photocopies were sent to Basel at that time, where they joined in the Stadtarchiv the incomplete photocopies and manuscript copies of the material that had formerly been loaned to the EULER edition.

The present volume aims at a complete description, in Russian, of the entire collection now in the Academy, although the editors allow the possibility that odd sheets may be found later. The short preface is printed also in German translation. The Archive's collection provides a nearly full record of EULER's enormous work. While his own files suffered by shipwreck and fire during his lifetime, it seems that mainly personal and family papers were affected, his scientific manuscripts and notes having been deposited with the Academy. The main permanent loss is EULER's letters, most of which seem to have been discarded by the receivers or their heirs.

First in this volume comes the list of scientific manuscripts of books and memoirs, classified according to subject and numbered from 1 to 352. Most of these have been published or are variants of published works, but many are unpublished. For example, there are two treatises on dioptrics, neither the same as the published one or the fragment published posthumously; a treatise on statics; a first draught of the *Mechanica*, broader in scope; a large treatise on elementary geometry and many fragments from an early treatise on analysis, not to mention dozens of smaller items. It turns out also that the *Opera postuma* in some cases published only portions of the manuscripts.

Nos. 353–396 on the list are evaluations of papers, projects, and machines by others, nearly all unpublished.

Nos. 397–408 are the notebooks and "Adversaria mathematica", which stretch, with gaps, from EULER's Basel years until his death. These some 1,200 manuscript pages form the bulk of the unpublished material.

The second list enumerates 260 documents concerned with EULER's relations to the Academy and some few personal matters. In the third list are references to 848 passages in the Minutes of the Academy from 1727 to 1783 which mention EULER.

The second major fund of important unpublished documents is indicated by the fourth and last list, 540 letters from EULER and 1,728 letters addressed to him, ordered alphabetically according to the name of the correspondent. The printing of selections from EULER's correspondence has been so irregular in time, nature, and place that it is hard to estimate the proportion affected. Perhaps the most important unpublished letters are the copies of twelve from EULER to DANIEL BERNOULLI, 1735–1741; publication of these is eagerly awaited, since up to now EULER's side in this famous exchange of problems, solutions, and conjectures has had to be guessed from BERNOULLI's, which was published by FUSS more than a century ago. In all cases, the index numbers in the Archive are given, so that those who wish to consult the documents at first hand or request photocopies may locate everything easily.

The volume concludes with a handy condensation of ENESTRÖM's list of published works, cross-referenced with the *Opera omnia*; some other bibliographical material; and an index of names.

An enormous labor has gone into this volume; according to the preface by SMIRNOV, the work was done by MIKHAILOV. It would have been more useful to scholars outside the U.S.S.R. if the descriptions and comments could have been translated into the language of the original or some Western language, but even without this luxury the catalogue will be of permanent value to everyone who cultivates the history of science in the eighteenth century.

Note for the Reprinting

The volume reviewed is *Manuscripta Euleriana archivi academiae scientiarum URSS*, Tomus I: *Descriptio scientifica*, ediderunt J. CH. KOPELEVIČ, M. V. KRUTIKOVA, G. K. MIKHAILOV & N. M. RASKIN (Acta archivi Academiae scientiarum URSS, Fasc. 17), Moscow & Leningrad, Izdatel'stvo Akademii Nauk SSSR, 1962.
The review is reprinted from *Isis* **58** (1967): 271–273.

II. Manuscripts on Mechanics (1967)

This volume is the first in a sequence projected to publish the main works of EULER that have remained in manuscript up to now. The two volumes of *Opera postuma*, edited by P.-H. & N. FUSS in 1862, were long thought to have completed publication of EULER's astounding output. While for some years past this misimpression has been dispelled, the mere republication of previously printed works in the seventy-three part-volumes of the *Opera omnia* has presented such a formidable task to an international committee working since 1910 that little thought has been given to the project of publishing the thousands of manuscript pages still known only to the few who have seen them in Leningrad or the imperfect copies in Basel.

The present volume owes its appearance to the double devotion of its editor, G. K. MIKHAILOV, to mechanics and to EULER. It contains twelve works, all written between 1725 and 1730, that is to say, in EULER's student days at Basel and in his first Petersburg years, before his twenty-fourth birthday. That mechanics was central in EULER's thought and a lifelong passion, is clear from his published works, but here we see evidence that it occupied most of his thoughts in his formative years.

Each paper has its own interest. No. 1, on live and dead forces, in effect calculates the integral of energy for a spring with an arbitrary law of tension. No. 2 is a textbook on statics, a subject in which EULER was a supreme master, though he published little concerning it. Nos. 3 and 4, extracted from his Basel noteboook, and No. 5, just a little later, are sketches toward the *Mechanica*, and No. 6 is the first draught of it. This last is especially valuable for its Section III, devoted to the motion of rigid bodies; first entitled "On the motion of a rigid rod", this section was developed so as to include the problem of the center of oscillation and goes as far as the theorem on moments of inertia about axes parallel to one through the center of mass. The last section shows EULER considering a body rotating simultaneously about two perpendicular axes. The difficulty of conceiving general motion of a rigid body may well have caused EULER to cancel all of Section III from the treatise as rewritten for the press.

After a note on the corpuscular nature of fluids, outlining the view EULER was to reject as "absolutely sterile" in 1752, are four articles on the motion of particles in resisting media.

The contents conclude with the fragments of a note on the motion of water in bent and inclined tubes. Here EULER uses an energy method similar to DANIEL BERNOULLI's. On 25 July 1727 EULER presented this paper to the Academy as his first communication to it. On 13 August DANIEL BERNOULLI wrote to POLENI,

> . . . at last I have happily fallen on the veritable theory of the motion of water, which is very general and can be applied to all possible cases. You know with what care it has been sought by the cleverest geometers, but in vain . . .; but what is still more remarkable is that at the same time this theory was found by a different method by Mr. Euler of Basel, student of my father, who will do him much honor.

EULER refers to BERNOULLI's results on efflux as having already been demonstrated before the Petersburg Academy. This work is incomplete from lack of any concept of fluid pressure.

In welcome departure from current practice the editor has presented his preface in Latin as well as Russian, and so those who can read the texts can also read the preface. The volume contains also a Russian translation of the texts and an index of names. The book is adequately printed, but the paper seems unlikely to last.

The enormous, hardly believable bulk of EULER's output has been an obstacle to every attempt to publish editions of any part of it. With the excuse that much of his work is "obsolete", there has always been a temptation to select only the parts of "permanent" or "immediate" interest. On the contrary, everything written by so great a man as EULER is both obsolete and of permanent value, according to taste. The greatest achievements have become part of the working tools of the common scientists of today, who will not need to consult the original presentation; to the student of the foundations of science or its history, every utterance of such a giant is of potential value, and there is no guarantee that anyone else's selection will be the right one for him. In particular, much of EULER's work that offered little interest to such men as STÄCKEL has proved to be of main importance to students two generations later and has even influenced some recent research. Although anthologies from the works of the masters have their uses for schools and for dilettantes, even to make a good anthology requires first a complete original. The title page of the volume under review suggests that it is the first in a project to publish *all* of EULER's manuscripts, arranged chronologically in series classified by general subject. Let us hope that the energy and support needed for so great an undertaking will not fail, so that men of this generation will be able to see before them all the written record of EULER's life and work.

Note for the Reprinting

The volume reviewed is *Manuscripta Euleriana archivi academiae scientiarum URSS*, Tomus II: *Opera mechanica*, Volumen I, edidit G. K. MIKHAILOV (Acta archivi academiae scientarium URSS, Fasc. 20), Moscow & Leningrad, Izdatel'stvo Akademii Nauk SSSR, 1965.

The review is reprinted from *Isis* **58** (1967): 273–274, and *Scripta Mathematica* **28** (1968): 211–212.

So far as I know, no further volumes of this collection have been published. Fortunately the task of reprinting EULER's manuscript remains has been undertaken by the editors of the *Opera omnia*. See Essay 33aV, above in this volume.

c. A Sample: Ten out of Seventy-three

Reviews of ten part-volumes of EULER's *Opera omnia*, series II, works on mechanics and astronomy

I. Principles of Mechanics (Volume 5) (1959)

This volume, despite its title, does not contain all or even the major part of EULER's work on the principles of mechanics; rather, most of its contents are related to the unfortunate quarrel between MAUPER-

TUIS and KOENIG in 1751–1752 over the principle of least action, memorable in mechanics for the researches of EULER to which it gave rise, and in literature for the satires of VOLTAIRE that it provoked. In addition to twelve papers by EULER, this volume presents the famous contested letter of LEIBNIZ, the three works of MAUPERTUIS on least action (1740–1746), and the great paper of KOENIG (1751). The brief and excellent preface of FLECKENSTEIN does not add any new sources but, on the basis of clear mastery of the voluminous published and unpublished documents presents a balanced interpretation more likely to be acceptable to a critical reader than are the hasty generalizations engraved in the ever-repeated conventional accounts. Most of this material concerns nonmathematical issues.... In any case FLECKENSTEIN's conviction, which will be shared by any reader of the unpublished correspondence between MAUPERTUIS and EULER, of the scientific sincerity of EULER's enthusiasm for the principle of least action, unlikely as it is to enlist the modern student in the cause, should induce him to perpend with diligence EULER's papers on a subject that might otherwise be dismissed as merely polemical. EULER distinguishes statics from dynamics. For the former, in E145 (1748) he shows that an arbitrary discrete system subject to arbitrary constraints obeys the MAUPERTUIS principle. In addition, he derives from it the general equation for balance of moments in a plane elastica, including as a special case the general catenary, and shows that DANIEL BERNOULLI's principle (1742) for the elastica free of distributed loads is a special case. FLECKENSTEIN remarks a fact that has escaped the notice of historians hitherto: The static principle of MAUPERTUIS is the same as the so-called DIRICHLET principle, which had been formulated and used by the elder BERNOULLIs. In E146 (1749) EULER shows that the static principle suffices to derive the conditions of equilibrium for a fluid and obtains the conditions of integrability for a "Pfaffian" form in three variables. Also, the equilibrium of a weight hung from three elastic cords is treated by a brilliant analogy to a special case of the problem for fluids. Coming to the dynamic principle (§ 30 and E197 (1752)) EULER explicitly uses, though does not explain, the fact that the varied paths obey the same integral of energy, a contribution traditionally regarded as a capital discovery of LAGRANGE (1788).

The one really important paper on the principles of mechanics, the great masterpiece E177, represents a fundamentally different approach that EULER developed in 1750 as a result of his hydraulic researches of the preceding decade; it has dominated the mechanics of extended bodies ever since. This paper contains the first proposal of the so-called NEWTON's equations, $\mathbf{f} = m\mathbf{a}$ in rectangular Cartesian co-ordinates, as a "new principle of mechanics", the common origin of all the several other principles then in use. While it is well known that these equations are not to be found in NEWTON's *Principia*, it is hard today to believe that they were not obvious from NEWTON's work and from other mechanical researches of the period, but it is an indisputable fact, borne out by many, many details, that they were not. Prior to this paper are few examples in which these equations or any fully equivalent principle are made the basis of the work. Rather, as may be seen from many entries in his unpublished notebooks, EULER gradually distilled the idea from earlier researches on special problems: a noteworthy but not quite successful attempt to treat the general compound pendulum by JOHN BERNOULLI in 1742; JOHN BERNOULLI's hydraulics (1742); EULER's own general equations for linked mass-points moving in a plane (1744); and D'ALEMBERT's partial differential equation for the vibrating string (1746). These treatments were the first to reveal the pre-eminence and sufficiency of the principle of linear momentum, resolved into perpendicular components along fixed directions, as applicable to every part of every system. . . . That the importance of this new view was seen in his own day, is shown by the nonmathematical summary published in 1754 in the *Gentleman's Magazine*, free from the sarcasms often directed by the English of that period against Continental researches on mechanics.

[EULER states in § 19,

> Commonly we find several such principles that seem worthy of being put in the rank of axioms of mechanics, since they relate to the motion of infinitely small bodies. But I remark that all these principles reduce to a single one, which can be regarded as the unique foundation of all of mechanics and the other sciences which treat of the motion of any sort of bodies. And it is on this principle alone that we should establish the other principles, not only those already received in mechanics and in hydraulics, now in use to determine the motion of solid bodies and fluids, but also those which are not yet known, and which we need in order to develop the cases noted above concerning solid bodies as well as several others occurring in fluid bodies. For in all these cases what is required is but to apply this fundamental principle adroitly . . . .

In other words, EULER regards his element of mass dM as giving rise to an integral $\int \cdots dM$ that refers to *all kinds of bodies occurring* in mechanics, not only in 1750 but *always before and forever after*.

[To understand this concept of integration, we must see how EULER interpreted $\int \cdots dM$ in the applications he himself made of his principle—in this paper, in its predecessors, and later.

1. *Rigid bodies* in E177. In earlier papers EULER had treated rigid planes; here he treats three-dimensional rigid bodies. For three-dimensional regions in general we may consult E479, "Nova methodus motum corporum rigidorum determinandi", *Novi commentarii academiae scientiarum Petropolitanae* **20** (1775): 208–238 (1776) = *Opera omnia* (II)9:99–125. In § 24 of this paper EULER uses "Lagrangean" variables X, Y, Z from his researches in hydrodynamics and interprets ∫ as integration over the "initial state", the region of space the body occupies initially. In this interpretation it is immaterial whether the body be rigid or not. This formulation is commonly used in modern continuum mechanics.

2. *Discrete systems* in E112, "Recherches sur le mouvement des corps célestes en général", *Mémoires de l'Académie des Sciences de Berlin* [**3**] (1747); 93–143 (1749) = *Opera omnia* (II)**25**, 1–44. EULER in treating finite systems interprets ∫ *fdM* for a single mass-point as *fM*.

3. *One-dimensional hydraulics* in E206, "Sur le mouvement de l'eau par des tuyaux de conduite" (1749), *Mémoires de l'Académie des Sciences de Berlin* [**8**] (1752): 111–148 (1754) = *Opera omnia* (II)**15**, 219–250. Here EULER integrates along a given curve.

4. *Two-dimensional and three-dimensional hydrodynamics* in E332, "Recherches sur le mouvement des rivières" (1751), *Mémoires de l'Académie des Sciences de Berlin* [**16**] (1760): 101–118 (1767) = *Opera omnia* (II) **12**, 272–288 and E258 "Principia motus fluidorum" (1752), *Novi commentarii academiae scientiarum Petropolitanae* **6** (1756/7): 271–311 (1761) = *Opera omnia* (II)**12**, 133–168. Here the differential elements of mass are proportional to elements of area or elements of volume.

5. *Linear continua* in general in E481, "De gemina methodo tam aequilibrium quam motum corporum flexibilium determinandi et utriusque egregio consensus," (1774) *Novi commentarii academiae scientiarum Pepropolitanae* **20** (1775): 286–303 (1776) = *Opera omnia* (II)**11**, 180–193. Here the body is an unknown curve, described with respect to a known reference curve; EULER interprets ∫ as integration along it.

[Of course a single mechanical system may consist of mass-points, wires, sheets, and space-filling bodies. This fact was recognized by LAGRANGE in his *Méchanique Analitique*, 1788. There in ¶¶ 11–12 of Première Partie, Quatrième Section, he introduces and explains his famous sign S for an integration "over all the given mass".

[LAGRANGE explicitly rejects an approach to continua through discrete systems: "...instead of considering the given mass as an assembly of contiguous points, we must follow the spirit of the infinitesimal calculus and consider it rather as composed of infinitely small elements of the same order of dimensions as the entire mass . . .." Coming back to general statements in ¶¶ 6–7 of Seconde Partie, Seconde Section, he writes the equation of dynamics in terms of "quantities . . . relative to each of the bodies of the system proposed" and denotes the sum of these "by the integral sign S, which must embrace all the bodies of the system . . . for any system of bodies regarded as points . . .". To express his meaning today we should say not "points" but "sets of points". The symbol S denotes a major instance of what is now called a three-dimensional Stieltjes integral based on a measure which may be singular at a finite number of surfaces, curves, and points.]

A different contemporary trend led to the apparently invariant formalism of LAGRANGE (1788), which is more appreciated by historians and physicists. It almost but not quite succeeds in burying the necessarily Euclidean character of the space of classical mechanics. It has never led to anything useful in the mechanics of space-filling bodies; rather, by obscuring the role of finite transport and rigidity in classical mechanics it has allowed confusion between co-ordinate invariance and dynamical invariance that is only being resolved, at last, by the most recent attempts toward an axiomatic foundation of mechanics. These attempts start from the viewpoint of EULER and CAUCHY. [See Essay 39, below.]

EULER's reformulation of the foundations of mechanics bore immediate fruit—indeed, it was designed specifically to yield general equations for fluids and rigid solids. The same paper contains both the axiom and its use [to solve the problem that, more than any other, had driven EULER to seek the principle itself, namely to discover differential equations for the general motion of a rigid body. See Parts bII and cV of this essay, above and below]. In this application EULER simply assumes, because whatever internal forces there may be within a rigid body are powerless to change its shape, that in calculation of the resultant torque on the body those forces may be left out. By appeal to this additional assumption he arrives at what have ever since been called "the EULER equations" of rigid dynamics (§ 49), with the angular velocity vector (§ 40) and the tensor of inertia (§ 48) appearing as necessary incidentals.

This paper, obscured by partially unsymmetric notations and exploration of special, preliminary cases, is the beginning, not the end, of EULER's work on the general motion of a general rigid body. Here (§ 15) he finds necessary and sufficient conditions for permanent rotation but does not investigate their solution; on the basis of this paper, SEGNER (1755) and EULER in E292 (the correct date of which is 9 November 1758, not the ENESTRÖM-JACOBI date of 7 October 1751), were soon to show that every body has at least one set of three orthogonal principal axes, and from this work a branch of algebra begins.

In respect to the principles of mechanics, this paper is as important a turning point as NEWTON's *Principia*. Nearly every prior treatment of a mechanical problem is rendered obsolete. Problems that before seemed intricate become at once easy to formulate in general equations of motion. In his remaining thirty-three years of life EULER found time to do over again on the basis of his "New Principle", or, as he later called the linear and angular momentum equations together, "the first principles of mechanics", nearly every one of his earlier mechanical researches, each time plucking new fruit from a seemingly endless supply. This cornucopia poured out the wave equation, the equations of hydrodynamics, the hydraulic theory of motions of finite amplitude, the partial-differential equations for vibrating rods and membranes. Few indeed are works contributing so much to mechanics as this one paper.

Note for the Reprinting

The volume reviewed is *LEONHARDI EULERI Commentationes mechanicae [ad] principia mechanica [pertinentes]* (LEONHARDI EULERI *Opera omnia* (II)**5**), edited by JOACHIM OTTO FLECKENSTEIN, Zürich, Orell Füssli Verlag, 1957.

The review was first published in *Mathematical Reviews* **20** (1959): 620–621.

The passages within square brackets replace brief, oversimplified, and inaccurate phrases in the original.

II. Mechanics of Mass-Points (Volume 6) (1959)

The volume under review is the first of two collecting EULER's papers on the mechanics of mass-points. The preface by BLANC consists of short, clear summaries of the contents of the papers and enables the modern reader to see at once the domain to which the work belongs and to estimate the results obtained.

While these volumes contain the least interesting of EULER's researches on mechanics, being concerned with the kind of "Analytical Mechanics" nowadays associated with examination problems of the last century and reducing, at bottom, to little more than investigation of explicitly integrable cases of certain ordinary differential equations, nevertheless a major paper on the principles of mechanics is included.

That is E86, published in 1746; it concerns the motion of bodies constrained to move on a rigid curve which itself may be in free or constrained motion. The posthumously published pieces E826, E827, E828, and E829 contain earlier treatments of special cases. The problem was raised by JOHN BERNOULLI about 1730; a decade later, it attracted the attention of DANIEL BERNOULLI and EULER, and the latter communicated it to CLAIRAUT. An early fruit, as BLANC remarks, was EULER's and DANIEL BERNOULLI's recognition of the angular momentum and proof that it is conserved in certain cases. More important is the realization by EULER and CLAIRAUT of the nature and role of the principle of relative motion, which had major repercussions in analyses of invariance a century later.

So far as I can learn, no adequate history of the problem of relative motion has appeared. CLAIRAUT[1] first achieved a correct general statement, in words, of the laws of mechanics in noninertial frames, but his calculation of the relative acceleration in special cases is faulty. DANIEL BERNOULLI, characteristically, was able to solve very special problems correctly by special devices but made no attempt to face the general situation. EULER stood between. In the paper E86 he succeeded in solving, by a method which is general in principle but mathematically complex, some extremely difficult cases; for example, that of a particle mobile within a curved tube free in space. Application to the case when the tube is given an assigned motion foreshadows EULER's theory of rigid bodies.

For EULER's further development of the principle of relative motion, we must turn to his [basic paper E177, discussed in the preceding part of this section, and to his] papers of 1751–1753 on hydraulics, republished in *Opera omnia* (II) **15**. There he gives a complete and correct verbal statement, followed by a complete and correct mathematical theory of one-dimensional motion expressed in the angular variables appropriate to rotating hydraulic machines. This brilliant analysis contains the first explicit appearance of the "Coriolis acceleration"....

Note for the Reprinting

The volume reviewed is *LEONHARDI EULERI Commentationes mechanicae ad theoriam motus punctorum pertinentes*, Volumen Prius (LEONHARDI EULERI *Opera omnia* (II) **6**), edited by CHARLES BLANC, Zürich, Orell Füssli Verlag, 1957.

The review was first published in *Mathematical Reviews* **20** (1959): 1138.

---

[1] A.-C. CLAIRAUT, "Sur quelques principes qui donnent la solution d'un grand nombre de problèmes de dynamique", *Histoire de l'Académie des Sciences avec les Mémoires pour 1742* (Paris): 1–42 (1745).

III. Elastic and Flexible Bodies (Volumes 10 and 11) (1959)

The universal ignorance of the history of mechanics in the eight-eenth century is reflected in the contents of this volume, which is substantially that planned for it when the EULER edition was initiated in 1910. It seems unlikely that G. H. ENESTRÖM, from whose cata-logue[1] STÄCKEL adopted almost unchanged the classification used in distributing the material on mechanics and physics, can have penetrated much further than the titles of the works he regarded as "applied". Volumes 10 and 11 in Series II, "Mechanics and Astronomy", contain forty-one papers "pertaining to the theory of flexible and elastic bodies". In Volume 10 appears E268 which is an abstract of E306, published as "physics" in Volume 1 of Series III. In fact, both papers are pure hydrodynamics, being the sources of the so-called Lagrangean equations of motion and of the wave equation in two and three dimensions. Every paper whose title contains the world "oscillation" seems to have been put into the volumes on elas-ticity. E.g., E455 concerns the compound pendulum, and the remark-able paper E126, which seems to have escaped the notice of his-torians, contains the first analysis of a single harmonically driven oscillator. This latter paper, written in 1739, proceeds by slow trials and transformations of cases to discover mathematically the phenomenon of resonance. EULER does not recognize the result as being a precise mathematical counterpart to the qualitative explana-tion published by FRACASTORO in 1546 and, in brilliant style, by GALILEO in 1638, besides having been found independently by BEECKMAN (1616–1618) and published by MERSENNE in 1635. Rather, EULER wrote on 5 May 1739 to JOHN BERNOULLI that he had found "such various and wonderful motions as would surely fail to be suspected until the calculation was completed." He recognized the case where the driving frequency equals the natural frequency as "the one . . . which deserves the greatest notice", since the amplitude "increases continually and finally grows out to infinity", even though the effect "arises from finite forces". Thus, to produce a perpetual motion, one has only to drive a cycloidal pendulum by an "auto-maton" having the same period and to overcome resistance and friction sufficiently that the oscillations, though not increasing, per-petually conserve the same amplitude. This paper furnishes a brilliant example of purely mathematical discovery (in this case, rediscovery) of a major physical phenomenon. Also, it reminds us that the EULER

---

[1] G. H. ENESTRÖM, *Verzeichnis der Schriften Leonhard Eulers, Jahresbericht der Deut-schen Mathematiker-Vereinigung*, der Ergänzungsbände **IV**. Band, 1. Lieferung, Leipzig, Teubner, 1910.

who could effect dazzling calculations was also, in his early years, the discoverer of some of the simplest and most useful elementary analysis, for it was only four months later that he obtained the general solution for linear differential equations with constant coefficients, to which he had been led by encountering two major instances in the theory of vibration: the forced harmonic oscillator and the rod in infinitesimal transverse motion. [Using the incomplete mechanical principles then accepted as providing the only handle upon the motion of systems with many degrees of freedom,] EULER and DANIEL BERNOULLI independently in 1734–1735 had approached the transverse vibrations of rods through these simple modes and proper frequencies, which are determined by the proper numbers and proper functions of the ordinary differential equation $k^4 y'''' = y$. In EULER's first paper on vibrating bars (E40, written in 1735) he can integrate this special equation only in power series. It is a thoughtworthy example that the greatest of all manipulators pondered over these series for four years before recognizing their finite expression in circular and exponential functions. The papers E443 and E526, of 1772 and 1774, present the fully explicit theory, including accurate calculations of frequencies and nodal ratios. Neither here nor in any other paper of the eighteenth century is there any hint of the orthogonality of proper functions.

The modern concept of function as a single-valued correspondence arose in connection with EULER's researches on the vibrating string, reported in E213, E317, E318, and many other memoirs, which make for confusing reading because of polemics over trigonometric series and personalities. While the popular histories of mathematics still parrot the easy generalizations of Victorian dilettantes who singled out EULER as the scapegoat for all the sins of formalism in his century, in fact he was not only the first but also the only mathematician of that time to see the inadequacy, though apparently not the vagueness, of the definition of function as an analytic expression, a definition he had transcribed from common practice in 1745 and to which the formalists D'ALEMBERT and LAGRANGE adhered, with their respective pugnacity and dryness, to the ends of their lives. These same works of EULER contain the discovery of wave propagation and wave reflection as consequences of the partial-differential equation. EULER, having taken a dislike to trigonometric series, developed by use of the functional equation alone all of the properties of the uniform string that are now usually demonstrated by application of trigonometric solutions obtained a half century later by the French school. EULER was alone in his century in seeing that physical problems, especially those governed by what we now call hyperbolic equations, require a concept of "solution" general enough to include certain kinds of propagating discontinuities. The remarkable paper E287 gives EULER's explicit solution of the initial-value problem for a string composed of two uniform parts of different densities. Here the differential condition of continuous slope at the junction cannot always be satisfied; in replacing it by an integral condition that reduces to it for differentiable functions, EULER shows himself the father of "generalized" solutions.

E410, written in 1770, introduces the concept of resultant shear and obtains the general differential equations for a deformable line bent in its own plane, no constitutive equation being assumed. This paper and its sequel E481 are the culmination of EULER's struggle for fifty years to embrace the elastica and the catenary within a single conceptual framework. Apart from mathematically rudimentary work by PARENT (1713) and COULOMB (1773), shear stress is not to appear elsewhere until the great researches of CAUCHY in 1821–1823.

The last paper printed, E831, was the first to be written. Dating from 1727, EULER's time as a student of JOHN BERNOULLI in Basel, it contains the faulty theory of elastic rings that EULER published in developed form in E303 (1760). The flaw in this theory and in later attempts by EULER and other writers of the eighteenth century lies, not in incorrect or inadequate physical hypotheses or information, but in the lack of sufficient differential geometry to describe the deformation of a curved line or surface. While EULER's theory of bending of the skew elastica, obtained in 1774–1775 and published in E471 and E608, led him to much of the vectorial theory of a single curve, including the concept of the second-order magnitudes, the binormal and one of the "SERRET-FRENET" formulæ, the reader is struck by the generally primitive state of geometry reflected.

But a more important contribution lies in E831: the so-called YOUNG's modulus of linear elasticity. While the use of such a modulus to characterize a material rather than a particular body of material was foreshadowed in a more general investigation of JAMES BERNOULLI (1704), EULER was the first theorist to put enough faith in HOOKE's law to develop its consequences with any attention. His first achievement was to derive by its aid the celebrated formula $M = EI/r$ for the bending moment acting upon a beam, his method being that now found in any engineering text. EULER's definitive treatment of this idea and of "YOUNG's modulus" is given in E508, but for this paper we have to wait for the projected Volume 17 ("Theory of machines") of Series II, and to understand that paper we must turn back to Volume 24 ("Calculus of variations") of Series I.

The unfortunate classification should not be judged too harshly. While only thirty-one volumes of EULER's works have been labeled as "mechanics", examination shows that mechanics was the dominating interest of his life and gave rise to most of his researches on "pure" mathematics. His accidental discovery of a property of the rectangular elastica helped lead him finally to the addition theorem for elliptic integrals. At least two of the twelve volumes classified as "physics" concern mechanics primarily, and of the researches in the twenty-nine volumes on pure mathematics, now completely in print, at least one half concern problems growing from his researches on mechanics. The interconnections of all these works are intricate and strong.

This review mentions only a few of the important results contained in the volume.

Note for the Reprinting

The volume reviewed is *LEONHARDI EULERI Commentationes mechanicae ad theoriam corporum flexibilium et elasticorum pertinentes, Volumen Posterius, Sectio Prima* (LEONHARDI EULERI *Opera omnia* (II)11_1, edited by FRITZ STÜSSI & ERNST TROST, Zürich, Orell Füssli Verlag, 1957.

The review was first publshed in *Mathematical Reviews* **20** (1959): 622–623.

The "projected Volume 17" appeared in 1982, edited by C. BLANC & P. DE HALLER.

IV. Naval Science (Volumes 18–21) (1977, 1981)

WALTER HABICHT has completed the edition of EULER's works on naval science in the *Opera omnia*. He has edited Volumes 20 and 21, which contain memoirs and a short book, and has provided introductions in German for them; his introduction to Volume 21 contains also an analysis of EULER's great treatise, *Scientia navalis*, first published in 1749. These introductions maintain the standard and custom prevailing in the more recent volumes of the edition: The works are compactly described in detail sufficient that any reader with reasonable knowledge of mechanics and calculus can follow everything and easily locate such passages as he wishes to read in the original. Such analyses, untainted by the jejune isms and slants of modern historiography of science, are solid food for the student who loves science as science.

Mr. HABICHT's excellent summaries make anything further of that kind superfluous in a review. I will merely comment on some points of particular interest.

The first essay in Volume 20 is also EULER's first attempt to pluck a solid gold apple from the orchard of the sciences; it is his entry for the Paris prize of 1727. The topic set was the masting of ships. EULER was twenty years old and had never left Basel; thus he could not yet have seen a seagoing ship. Mechanics applies just as well to things far off and unseen as to things nearby and familiar. It was mechanics, expressed in mathematical form, on which EULER chose to base the design of ships. The prize was awarded, and justly, to BOUGUER, but the boy EULER received an "accessit", and his essay was printed immediately. It is No. 4 in ENESTRÖM's catalogue of EULER's works.

The reader already familiar with any of EULER's calculations of the positive forces exerted on obstacles by fluids will know that he adopted "NEWTON's Law of Resistance": The pressure exerted by a fluid impinging upon a plane surface is proportional to the square of the speed, and back pressure is neglected. To calculate the forces exerted by wind and wave on sail and hull, EULER interprets "NEWTON's Law" as a statement regarding differential elements of surface. This assumption, which he calls "the common hypothesis", enables him to calculate the total resistance by integration. Thus he can and does obtain, time after time, a definite answer, often in an elegant, explicit form that allows quantitative as well as qualitative conclusions, which he takes great pains to develop and interpret clearly. Today the resistance experienced by a ship and the reaction of a sailing ship to its rudder and the winds are regarded as intractable except through basic partial-differential equations and methods of solving them that lay many years ahead of EULER's time, equations some of which were to be discovered by EULER himself in his later researches on hydrodynamics. These equations presume a more detailed specification of the ship and its circumstances than anyone could have had when EULER was writing his early papers. EULER himself[1] was to explain as follows his position regarding "the common hypothesis":

> When I treated this subject some years ago, I founded my calculations on the common hypothesis..., not that I believed that hypothesis to be entirely conformable with the truth, but rather because the true law of these forces was still unknown. I even agree that in determination of the force of the wind this hypothesis can diverge considerably from the truth..., while the same hypothesis in regard to the impulse of water agrees better with experiment, although there often enough the disagreement is rather noticeable. Thus if I have employed this defective hypothesis in my researches on the effects of windmills, to it only should be attributed such errors as comparison of the calculation with experiments shall reveal.

---

[1] § 1 of E233, "Recherches plus exactes sur l'effet des moulins à vent", *Mémoires de l'académie des sciences de Berlin* [12] (1756): 165–234 (1758) = pages 65–125 of LEON-HARDI EULERI *Opera omnia* (II)16.

... [U]ntil now the true theory has not been discovered. It should come as no surprise that in calculation we stay with this hypothesis, which we do not fail to recognize as being unsatisfactory.

Here we see EULER taking an attitude common and reasonable in engineering today: Because some answer may be better than no answer, make plausible if crude guesses to exploit what little knowledge you have. We know now that "the common hypothesis" should not be interpreted locally; if it is applied to the total surface exposed to flow of a highly rarefied body of gas, it can square fairly well with the facts.

To determine the propelling force of the wind upon a ship with several sails, EULER simply assumes that each sail stops all the wind upon it and so produces a wake of dead air behind it; any portion of a sail that is downwind in this wake is useless. Unfortunately nature is not always so simple as the great explorers of science in the eighteenth century hoped. Their dauntless courage led to victories upon which all later progress in mathematical physics rests; we must not find cause for astonishment in their failure to achieve the impossible.

In the remainder of this review I will emphasize portions of EULER's work on naval science that are largely or entirely free of appeal to "NEWTON's Law of Resistance".

Mr. HABICHT remarks upon one paper that gives a pretty solution of a purely kinematical problem. That is E94: A boat with a perfect rudder moves at constant speed across a stream flowing in straight fillets at assigned speeds. This problem may well have been suggested to EULER by the Basel ferries, which still today cross a broad river by exploiting the power of its swift current. On the assumption that the downstream velocity of the boat equals that of the river, how should the rudder be set so as to make the boat traverse a given path? Various instances, one of which gives rise to a variational problem, are worked out in detail.

Mr. HABICHT calls attention also to E413, which won the Paris prize of 1753 but was not published until 1771. Here EULER analyses various means of propelling ships: paddle wheel, screw propeller, and water jet. He composed this paper just while he was developing his field theory of hydrodynamics, and toward the end he brings that theory to bear upon the practical problems here considered.

Volume 21 opens with the essays which were awarded the Paris prizes of 1759 and 1761, the former on pitch and roll and the latter on lading. The bulk of the text is used by the book called *Complete Theory of the Construction and Manœuver of Ships,* published in 1773 in

Russia; a second edition in corrected French was prepared at TUR-
GOT's request in 1776 for use in teaching at all naval academies in
France. Of course it is the first edition that is reprinted here, without
LEXELL's supplements subjoined to the second edition. The volume
concludes with two short notes, one of which calculates the resistance
that the prow of a ship encounters, and the other proposes and analyses a device for pulling a ship upstream by means of a sail attached to
a roller.

At the time I edited Volumes 18 and 19, which contain the *Scientia
navalis,* I expected to be able to write a preface for it, but later circumstances prevented me. Although H. E. TIMERDING in a brief article
published in 1908 distinguished its central importance in the history
of mechanics, the book seems to have remained closed to historians.
Mr. HABICHT's excellent survey, followed by his clear and detailed
analysis, section by section and with the main equations written in
modern notation, renders the contents accessible to mathematicians
and engineers today.

Any reader who can put himself in the position of the times will be
astounded at the breadth, depth, and originality of EULER's great
book. In Mr. HABICHT's words, "here for the first time the principles
of hydrostatics are established in full clarity, on the basis of which a
scientific foundation for the theory of naval architecture is provided . . . ." Here we find the concepts of centroid and metacenter as
distinct from center of gravity; a theory of stability based upon the
direction of the restoring torque in a small displacement; the earliest
treatment of three-dimensional motion of a rigid body of general
form in response to applied torque; the theory of small oscillations of
floating bodies; and a cornucopia of solutions of specific problems
based upon local use of "NEWTON's Law of Resistance". The astonishing courage of the work may be guessed from the titles of the
chapters.

Volume I: General theory of the location and motion of bodies
floating on water

(1) Equilibrium of floating bodies

(2) Restitution of floating bodies to equilibrium

(3) Stability of floating bodies in equilibrium

(4) Effects of external forces upon floating bodies

(5) The resistance of water to moved plane figures

(6) The resistance of water to moved bodies

(7) The progressive motion of floating bodies

Volume II: Rules and precepts for constructing and steering ships

 (1) Ships in general

 (2) Equilibrium of ships

 (3) Stability of equilibrium

 (4) Oscillations of ships

 (5) Inclination under the influence of arbitrary forces

 (6) Effect of rudders

 (7) Effect of oars

 (8) Construction of rowed ships

 (9) Force exerted by the wind on a sail

(10) Masting of sailing ships

(11) Ship on a skew course

Publication of this masterpiece of fundamental and applied mechanics was delayed for nine years by the instability of Russian circumstances. The French geometers were particularly active in naval science then, and for several aspects of the subject priority in publication belongs to BOUGUER's great work, *Traité du navire, de sa construction et de ses mouvemens*, which appeared in 1746, five years after EULER had delivered his complete manuscript to the Academy at St. Petersburg but three years before it was published. An example of reversed priority is provided by the centrally important concept of metacenter, first discovered (as far as we can now tell) by EULER, rediscovered and first published by BOUGUER.

While EULER's table of contents reflects his strictly projected organization, he wrote the second volume after he had completed the first, and so in developing applications he could avail himself of the opportunity to improve the basic theory. Taking account of this duplication, Mr. HABICHT divides the whole work into subjects as follows:

A. Hydrostatics

   a) Equilibrium (I1, II1, and II2)

   b) Stability (I2, I3, and II3)

   c) Change of situation and stability due to lading and to the action of external forces (I4 and II5)

   d) Considerations of strength and scaling (II6 and II8)

B. Dynamics and kinetics

   a) Vibrations (II4)

b) Progressive motion and rotation about a vertical axis (I5–I7, II6–II11)

Although EULER solves all the problems he sets himself, the portion of his results that rests on "NEWTON's Law of Resistance" can have served little in the design and handling of ships.

All of the book and each of the memoirs demonstrate EULER's intense interest in practical problems. As Mr. HABICHT remarks on page L of his preface to Volume 20, the character of EULER's work on naval science belies the often repeated claim of secondary writers who have reproached him for putting all his trust in mathematics, neglecting facts of experience:

In truth Euler was an astonishing discoverer and as such had insights that often were not rediscovered for a century. That he was not understood, is altogether the fault of the "practical men" of the time, to whom his writings were addressed. Far from having any idea of analysis, they were not even accustomed to calculate anything.

Note for the Reprinting

The volumes here reviewed are *LEONHARDI EULERI Scientia navalis* (LEONHARDI EULERI *Opera omnia* (II)**18–19**), edited by C. TRUESDELL, and *Leonhardi Euleri Commentationes mechanicae et astronomicae ad scientiam navalem pertinentes* (LEONHARDI EULERI *Opera omnia* (II)**20–21**), edited by W. HABICHT, Basel, Orell Füssli Verlag, 1967, 1972, 1974, 1978. For Volumes 18 and 19 Mr. HABICHT has provided an excellent analysis in his preface to Volume 21.

The text is condensed and recast from reviews published in *Centaurus* **21** (1977): 76–77, and **26** (1983): 323–335.

V. The *Mechanica* and Euler's Program
in Mechanics (1981)

PAUL STÄCKEL wrote as follows of EULER's numerous works on mechanics: "Collected and ordered in the *Opera omnia* LEONHARDI EULERI, they will awaken to new life; this new edition will make it for the first time possible to gain a clear picture of the enormous activity of EULER in the entire field of mechanics." Those words were written in 1912; they were printed in the preface to the first volume of the *Mechanica*, which appeared as Volume 1 of Series II (Works on

Mechanics and Astronomy). STÄCKEL was right. Even though he was an expert on "classical mechanics" as it was then conceived, the bulk of EULER's publications was more than he could face effectively: His disposition into volumes concerning particular topics betrays his ignorance of the contents of the works he was classifying. Some specific examples are noted above in Part bIII of this essay. Now that the volumes on naval science have appeared, we can indeed, as STÄCKEL projected, form a picture of EULER's whole activity in mechanics, even though five of the thirty-two part-volumes of Series II and one of the twelve volumes of Series III are still to come. While it is risky to dismiss any of EULER's papers as being irrelevant to some particular subject, I hazard the guess that only one of the still wanting volumes, namely the third "on the theory of machines", will bear on a general estimate of his contribution to basic mechanics. Also I say nothing of the ten volumes on celestial mechanics and astronomy (Volumes 22–31), four of which are still outstanding.

While EULER was a student of JOHN BERNOULLI in Basel, he projected a great treatise on mechanics as a whole. Mr. MIKHAILOV[1] has published two early draughts written then by EULER concerning motion subject to central forces, followed by a more mature draught *Mechanica seu scientia motus* written soon after he had arrived in Petersburg at the age of twenty. The 135 pages of this manuscript are divided into three sections:

I. On the motion produced by forces acting on a free point

II. On the motion produced by forces acting on a constrained point

III. On the motion of rigid bodies loaded by arbitrary forces

EULER was stopped by the problem he had set himself in Section III. He was not to master the theory of rigid motions for nearly a quarter of a century. He decided instead to revise, clarify, and extend the first two sections, which concern (§ 9) the motion of "infinitely small bodies or points, for it cannot so easily be said what sort of motion there be in bodies having a size, because the various parts can have various velocities." The first two sections grew into the two volumes of his *Mechanica*, the full title of which is *Mechanics, or the Science of Motion set forth Analytically*. EULER had finished the manuscript of the first volume by the end of 1734, when he was twenty-seven years old. Both

---

[1] G. K. MIKHAILOV, *Manuscripta Euleriana Archivi Academiae scientiarum URSS.* Tomus II, *Opera mechanica* Volume 1, Moscow & Leningrad, Nauka, 1965. See part bII, above, of this essay.

parts were published in 1736 "by way of a supplement" to the *Commentarii* of the Petersburg Academy of Sciences. In the preface EULER tells us that the *Mechanica* is only the beginning, and he speaks of "the following books, in which the motion of finite bodies will be determined." In § 98 of the text, the title of which is "general scholium", he sets out his program in detail. The motion of a finite body "is compounded of the efforts of its several tiny parts"; from "lack of sufficient principles" such motions "cannot yet be determined" and so must be deferred to succeeding volumes. "The diversity of bodies . . . will provide us the primary division of the work." EULER will treat in turn motions of bodies of finite size that are

(1) Rigid

(2) Flexible

(3) Elastic

(4) Subject to impacts with each other

(5) Fluid

EULER was to devote much of his life to this program. His treatises and memoirs on naval science and his book on ballistics fall only partly within it; much of their contents could be classified in modern terms as applied mechanics or mechanical engineering. While his ambitions in mechanics seem grand enough to consume the whole life's work of anyone, no matter how potent his genius, EULER found time to cultivate also every other part of mathematics.

Although it was the original policy of the *Opera omnia*, which were organized in 1910, to let the texts speak for themselves, STÄCKEL saw even then, when nearly every trained scientist could read Latin, French, and German easily, that the styles and criteria of science had changed so much since EULER's day as to render comprehension word-by-word insufficient to ensure understanding of contents. He provided an excellent though short preface to explain the circumstances giving rise to the *Mechanica* and to describe its reception by the community of learning. He also pointed out some of its limitations and virtues.

While historical works continue even today to write of the *Mechanica* as if it were neither more nor less than a translation of NEWTON's *Principia* into the language of infinitesimal calculus, it is nothing of the kind. Much narrower in scope, it concerns only what are now called "mass-points". EULER here (Volume 1, § 98 and Preface) introduces these "infinitely small bodies, which can be considered as points" because the assumptions and conclusions he develops in Chapter I regarding free bodies "belong properly" to punctiform bodies and cannot be extended in all aspects to "bodies of finite size". More than that, the only system it treats is *a single mass-point*, except for a few pages at the end of Chapter I on the motion of one point relative to another moving point. The remainder of Chapter I concerns the nature of rest and uniform motion; at Proposition 20 in Chapter II EULER at last reaches NEWTON's Second Law; and because his scope is limited to motions of a single point, he does not need the Third Law.

Volume 1 concerns motion "free" in the sense that the mass-point is subject to an entirely known force. Mathematically speaking, the acceleration is prescribed to within an arbitrary multiplicand; in all the numerous instances EULER considers, the arguments of the force function are place and speed only. Thus the whole volume is devoted to integration of particular differential equations of second order and to interpretation of the results.

Motion along a straight line occupies the reader for nearly half of the volume. Most of the rest concerns motion in a plane, and at the end are a few pages on motion along a skew curve. EULER introduces fixed rectangular Cartesian co-ordinates for the position of the mass-point, but to set up the differential equations of motion he takes arc length as the independent variable and resolves the enforced acceleration into components along the tangent and the normal to the path. In three-dimensions he uses two orthogonal normals, one of which he requires to be parallel to some fixed plane. Thus he does not here obtain either the "NEWTONian" differential equations of dynamics or the SERRET–FRENET formulæ of differential geometry. STÄCKEL in his preface notes the former fact and then repeats from LAGRANGE's *Méchanique Analitique* the false claim, still frequently repeated today, that MACLAURIN in 1742 was the first to take the final step, while in truth the "NEWTONian" equations are not to be found in MACLAURIN's book but were first given by EULER himself in a paper presented in 1747 and published in 1749: "Recherches sur le mouvement des corps célestes en général", republished in *Opera omnia* (II) **25**.

Volume 2 of the *Mechanica* concerns the motion of a single point constrained to lie upon a given curve or a given surface. In one of his earliest papers EULER had found the differential equations of the geodesics on an arbitrary surface. He obtains the same differential equations again as governing the problem of free motion on a surface (§§ 58–63). Thus he shows that the path of a mass-point free to move upon a fixed surface is the shortest possible (locally) between its initial and final points.

In both volumes there are chapters on motion in resisting media of various kinds, a topic to which NEWTON had devoted much of his

Principia. In this regard, but in no other, EULER's scope here is broader than LAGRANGE's in the *Méchanique Analitique*, 1788.

In neither volume does EULER treat as such the problem of two bodies subject to their own mutual forces. Of course, since it can be reduced to determining the motion of a single body, we may regard that problem as being included by implication, and in his extensive and thorough treatment of KEPLER's Laws and related problems (Chapter V of Volume 1) EULER seems to presume that his readers already know such a reduction to be possible.

EULER wrote in the preface to the *Mechanica*, "I have explained by the analytic method and in a convenient order both what I have found in the writings of others and what I myself have thought out . . . .." As DUGAS states in Chapter 3 of Book III of his *Histoire de la Mécanique* (see Essay 17, above)

> Euler seeks to develop dynamics as a rational science, starting from definitions and with propositions logically ordered. He intends to prove the laws of mechanics in such a way as to make us understand that they are not only certain but even of *necessary truth.*

It was these aspects that brought the book and its author immediate fame and blame. EULER's teacher, the old JOHN BERNOULLI, was as delighted with it as LEIBNIZ had been with HERMANN's *Phoronomia*. Mathematicians for the most part welcomed the systematic development using differential equations. The modern student, unlikely to be familiar with any other way of looking at the mechanics of mass-points, may fail to see that here lay one of EULER's great innovations. In his preface EULER writes of the two preceding treatises in which dynamics was developed by infinitesimal methods, namely the books of NEWTON and HERMANN, that although in reading them he found numerous problems well enough solved, the method presented did not enable him to solve other problems that were just a little different. The earlier works were neither systematic nor general.

The CARTESian physicists disliked EULER's treatment through differential equations for the reason they disliked NEWTON's geometrical arguments in his *Principia*: too mathematical. The Englishman ROBINS, expressing himself with what STÄCKEL described as "impudent arrogance", rejected the philosophical groundwork because it made no appeal to experiment; he disliked the systematic use of differential equations because simpler and shorter arguments would have sufficed to get some of the special instances. LAGRANGE, writing 75 years later, bestowed upon the *Mechanica* a few of his spare words of faint praise for EULER: "the first large work in which analysis is applied to the science of motion". This statement is unjust to HERMANN, whose book is both analytic in style and large.

In the first half of the eighteenth century mechanics was the center of scientific interest. Every great geometer had to prove his rank by what he could do in research on mechanics. With the *Mechanica* EULER gained entry into the circle of grand masters of his day. The two volumes are the fifteenth and sixteenth of EULER's more than 800 publications. He was twenty-nine when they appeared.

The *Mechanica* itself has joined EUCLID's *Elements* and COPERNICUS's *De revolutionibus* as one of those works that must be cited and are easy to praise for their first few pages but otherwise are revered from the outside, covers shut. I doubt there be a man alive who has studied the *Mechanica* carefully, straight through. I hope that someone expert in analytical dynamics will soon do so and provide reasonably compact summaries and analyses of the contents, thus closing the greatest remaining gap[2] in our picture of the development of basic mechanics from NEWTON's time until LAGRANGE's.

With EULER's other work on the motions of mass-points and with his researches on other kinds of bodies we are much better off because of the prefaces to several volumes in the *Opera omnia*, a list of which is given below, at the end of Part d, the final portion of this essay.

While the *Mechanica* brought EULER his most immediate and greatest fame in his own time, today what it contains seems the least interesting of his research on mechanics. It is dated. As we easily learn from WHITTAKER's *Analytical Dynamics*, the Cambridge tripos examinations taught generations of British mathematicians, among

---

[2] The next-greatest gap concerns the contents of HERMANN's *Phoronomia*. Much of what is said of it today derives from LAGRANGE's few and vague remarks about it in his *Méchanique Analitique*. On pages 12–13 of the preface to the work cited above in Footnote 1 Mr. MIKHAILOV has lamented the "undeserved oblivion" bestowed upon the *Phoronomia*. Mr. FELLMANN in the course of his sympathetic article on HERMANN in the *Dictionary of Scientific Biography* mentions that the book still awaits analysis. So far as I know, the only published studies of it that enter into details are as follows:

1. Kinetic theory of gases, pages 272–273 of my *Essays on the History of Mechanics*, New York, Springer-Verlag, 1968, and § 1 of "Early kinetic theories of gases", *Archive for History of Exact Sciences* 15 (1975/1976): 1–66 (1975).

2. Flexible lines, pages 80, 82, and 86–87 of my *The Rational Mechanics of Flexible or Elastic Bodies, 1632–1788*, LEONHARDI EULERI *Opera omnia* (II)11$_2$, Zürich, Orell Füssli Verlag, 1960.

3. Pressure waves and the vibrating string, pages 28–32 of J. T. CANNON & S. DOSTROVSKY, *The Evolution of Dynamics, Vibration Theory from 1687 to 1742*, New York *etc.*, Springer-Verlag, 1981. The reader of this work must take care not to rely on the authors' translations from Latin.

HERMANN's work on geometry is described sympathetically by C. B. BOYER, pages 170–174 of his *History of Analytic Geometry*, New York, Scripta Mathematica, 1956.

them STOKES and KELVIN and MAXWELL, how to solve explicitly special problems more difficult and tricky than any in the *Mechanica*. I do not know whether any of the methods and devices of integration that EULER presented there have proved valuable in later research on differential equations as such. The *Mechanica* furnishes an example to show that a book upon which its own time justly sets supreme value may lose that rank in later ages. It exemplifies also the inertia of fame. Lists of "great books" include it; the tributes to EULER that senior scientists deliver at official celebrations usually praise it; and merchants of books for collectors dilate its fame by making it one of the two or three most costly of EULER's some twenty treatises. These sources of acclaim spring only from earlier ones of the same fluffy kind and can be traced back to the two obituaries of EULER.

Note for the Reprinting

EULER's *Opera omnia* reprint the *Mechanica* in Volumes (II) **1–2**, edited by P. STÄCKEL. The text above derives in large part from a review published in *Centaurus* **26** (1983) 323–335.

d. Leonard Euler, Supreme Geometer (1972, 1982)

On 23 August 1774, within a month of his appointment as Ministre de la Marine and the day before he was made Comptrolleur Général of France, TURGOT wrote as follows to LOUIS XVI:

> The famous Leonard Euler, one of the greatest mathematicians of Europe, has written two works which could be very useful to the schools of the Navy and the Artillery. One is a *Treatise on the Construction and Manœuver of Vessels*; the other is a commentary on the principles of artillery of Robins . . . I propose that Your Majesty order these to be printed; . . . .
>
> It is to be noted that an edition made thus without the consent of the author injures somewhat the kind of ownership he has of his work. But it is easy to recompense him in a manner very flattering for him and glorious to Your Majesty. The means would be that Your Majesty would vouchsafe to authorize me to write on Your Majesty's part to the lord Euler and to cause him to receive a gratification equivalent to what he could gain from the edition of his book, which would be about 5,000 francs. This sum will be paid from the secret accounts of the Navy.

"The famous Leonard Euler", then sixty-nine years old and blind, was the principal light of CATHERINE II's Academy of Sciences in Petersburg. His name had figured before in the correspondence between TURGOT, the economist and politician, and CONDORCET, the prolific if rather superficial mathematician and littérateur soon to become Perpetual Secretary of the Paris Academy of Sciences, and later first an architect and then a victim of the Revolution. Just twenty years afterward CONDORCET was to die because his hands had been found to be uncalloused and his pocket to contain a volume of HORACE, but in 1774 equality, while already advocated and projected by TURGOT, had not progressed so far. In a France threatened by bankruptcy a minister of state could still find time to write in letters to a friend his opinions and doubts and conjectures about everything from literature to manufacture, and by the way the solution of algebraic equations. It was such a minister who asked whether "this EULER, who lets nothing slip by unnoticed, might have treated in his mechanics or elsewhere" the most advantageous height for wagon wheels[1].

In a time when intelligence was the highest virtue, when even men and women then thought to be lazy and stupid (and today proved by their words and deeds to have been lazy and stupid) were portrayed with little wrinkles of alertness around their sparkling, comprehending eyes, the name of LEONARD EULER, the greatest mathematician of the century in which mathematics was almost unexceptionally regarded as the summit of knowledge, was better known than those of the literary and musical geniuses, for example SWIFT and BACH. In the firmament of letters only VOLTAIRE outshone EULER. True, in all the world there were but seven or eight men who could enter into discourse with him, VOLTAIRE certainly not being one of them, and most of what he wrote could be understood in detail by only two or three hundred, VOLTAIRE not being one of these either, but pinnacles could then still be admired from below. In the volume for 1754 of *The Gentleman's Magazine*, a British periodical of general interest the contents of which ranged from heraldry to midwifery, we find an article

---

[1] This remark is enlightening. The book to which TURGOT refers is EULER's famous *Mechanica*, published in 1738. One of the most abstract works of the century, it never comes near anything concerning a wheel, let alone a wagon. Respect unsupported by even vague familiarity with the contents of this book is not limited to statesmen but is shown even by modern general histories of science or mathematics, which regularly and in positive terms provide it with a purely imaginary description as the "analytical translation" of NEWTON's *Principia*. In fact, as I have made clear above in Part cV of this essay, it is a treatise on the motion of a single point whose acceleration is induced by a rule of one of several simple kinds. Were it not for the headings, only an initiate would be able to recognize the contents as being mechanics.

entitled "Of the general and fundamental principles of all mechanics, wherein all other principles relative to the motion of solids or fluids should be established, by M. Euler, extracted from the last Berlin Memoirs." The anonymous extractor concludes that EULER's principle "comprises in itself all the principles which can contribute to the knowledge of the motion of all bodies, of what nature soever they be." This principle we call today the *principle of linear momentum*. There are in fact two further general principles of motion, the *principle of rotational momentum* and the *principle of energy*. The former of these EULER himself evolved and enounced twenty-five years later; it was the culmination of his researches on special cases of rotation that had extended over half of the eighteenth century. The latter principle was left for physicists of the next century to discover.

An entire volume is required to contain the list of EULER's publications. Approximately one third of the entire corpus of research on mathematics and mathematical physics and engineering mechanics published in the last three quarters of the eighteenth century is by him. From 1729 onward he filled about half of the pages of the publications of the Petersburg Academy, not only until his death in 1783 but on and on over fifty years afterward. (Surely a record for slow publication was won by the memoir presented by him to that academy in 1777 and published by it in 1830.) From 1746 to 1771 EULER filled approximately half of the scientific pages of the proceedings of the Berlin Academy also. He wrote for other periodicals as well, but in addition he gave some of his papers to booksellers for issue in volumes consisting wholly of his work. By 1910 the number of his publications had reached 866, and five volumes of his manuscript remains, a mere beginning, have been printed in the last ten years. There is almost no duplication of material from one paper to another in any one decade, and even most of his expository books, some twenty-five volumes ranging from algebra and analysis and geometry through mechanics and optics to philosophy and music, include matter he had not published elsewhere. The modern edition of EULER's collected works was begun in 1911 and is not yet quite complete; although mainly limited to republication of works which were published at least once before 1910, it will require seventy-four large quarto parts, each containing 300 to 600 pages. EULER left behind him also 3000 pages of clearly and consecutively written mathematical notebooks and early draughts of several books[2]. A whole

---

[2] There are also four classes of manuscripts of memoirs and books:

1. Manuscripts from which, perhaps with some correction, the works were set in type in EULER's lifetime.

volume is filled by the catalogue of the manuscripts preserved in Russia. EULER corresponded with savants and administrators all over Europe; the topics of his letters range more widely than his papers, going into geography, chemistry, machines and processes, exploration, physiology, and economics. About 3000 letters from or to EULER are presently known; the catalogue of these, too, occupies a large volume; nearly one-third of them have been printed, usually in volumes consisting of particular correspondences. The first such volume, published in 1843, was of great importance for its impetus to developments in the theory of numbers in the nineteenth century, more than fifty years after all the principals in the correspondence had died. This kind of permanence, difficult for literary men and historians and physicists to comprehend, is typical of sound mathematics.

In modern usage EULER's name is attached as a designation to dozens of theorems scattered over every part of mathematical science cultivated in his time. Even more astonishing than this broad though vague and incomplete tradition is the influence EULER's own writings continue to exert upon current research. The *Science Citation Index* for 1975 through 1979 lists roughly 200 citations of some 100 of EULER's publications; most of the works in which these citations occur are contributions to modern science, not historical studies.

It was EULER who first in the western world wrote mathematics openly, so as to make it easy to read. He taught his era that the infinitesimal calculus was something any intelligent person could learn, with application, and use. He was justly famous for his clear style and for his honesty to the reader about such difficulties as there were. While most of his writings are dense with calculations, four of his books are elementary. One of these is a textbook for the Russian schools; one is the naval manual which TURGOT caused to be reprinted in France; one is a treatise on algebra which begins with counting and ends with subtle problems in the theory of numbers; and the fourth, called *Letters to a Princess of Germany on Different Subjects in Natural Philosophy*, is a survey of general physics and metaphysics. This last is the most widely circulated book on physics written before the recent explosion of science and schooling. It was translated into eight languages; the English text was published ten times, each time

2. Manuscripts intended for publication and published in the regular volumes of the Petersburg Academy after EULER's death.

3. Manuscripts which EULER withheld from publication but which were published in the collections entitled *Commentationes arithmeticae collectae* (St. Petersburg, 1849) and *Opera postuma*, 2 vols. (St. Petersburg, 1862).

4. Manuscripts of works not published before 1966. Many of these remain unpublished.

revised so as to bring the contents somewhat up to date; six of the editions were American, the last one in 1872, a date only a little further from the present day than from 1768, when the original first appeared.

While EULER is known today primarily as a mathematician, he was also the greatest physicist of his era, a rank which was obscured for 200 years but has been re-established by the recent studies of Mr. DAVID SPEISER. EULER was the first person to derive an equation of state for a gas from a kinetic-molecular theory. In geometrical optics he invented the achromatic lens. His design for it required glasses of high, distinct, and reproducible quality; attempts to construct lenses according to his prescriptions have been adduced as impulses to the rise of the optical industry in Germany, which was supreme in precision for at least a century. He designed and caused to be built and tried an apparatus for measuring the refractive index of a liquid; it worked, and it remained in use for a century and a half. EULER's hydrodynamics was the first field theory. Perhaps his most important progress in physics other than mechanics is his having taken the observed fact that beams of light pass through each other without interference as justifying use of his linear field theory of acoustic waves to describe waves of light in a luminiferous aether, which he visualized as a subtle fluid.

To study the work of EULER is to survey all the scientific life, and much of the intellectual life generally, of the central half of the eighteenth century. Here I will not even list all the fields of science to which EULER made major additions. The most I attempt is to give some idea what kind of man he was.

LEONARD EULER was born in Basel in 1707, the eldest son of a poor pastor who soon moved to a nearby village. The parsonage there had two rooms: the pastor's study and another room, in which the parents and their six children lived. EULER in the brief autobiography he dictated to his eldest son when he was sixty wrote that in his tender age he had been instructed by his father;

> as he had been one of the disciples of the world-famous James Bernoulli, he strove at once to put me in possession of the first principles of mathematics, and to this end he made use of Christopher Rudolf's *Algebra* with the notes of Michael Stiefel, which I studied and worked over with all diligence for several years.

This book, then some 160 years old, only a gifted boy could have used. Soon EULER was turned over to his grandmother in Basel,

> so as partly by attendance at the gymnasium and partly by private lessons to get a foundation in the humanities [*i.e.* Greek and Latin

languages and literatures] and at the same time to advance in mathematics.

Documents of the day picture the gymnasium in a lamentable state, with fist-fights in the classroom and occasional attacks of parents upon teachers. Mathematics was not taught; EULER was given private lessons by a young university student of theology who was also a tolerable candidate in mathematics.

At the age of thirteen EULER registered in the faculty of arts of the University of Basel. There were approximately 100 students and nineteen professors. Instruction was miserable, and the faculty, underpaid, was mediocre with one exception. The Professor of Mathematics was JOHN BERNOULLI, the younger brother of the great JAMES, by that time deceased. JOHN BERNOULLI, a mighty mathematician and ferocious warrior of the pen, was universally feared and admired as a geometer second only to the aged and long silent NEWTON. BERNOULLI had returned, reluctantly, to the backwater of Basel despite brilliant offers of chairs in the great universities of Holland; he had had to return because of pressure from his patrician father-in-law. Single-handed, he had made Basel the mathematical center of Europe. Three of the four principal French mathematicians of the first half of the century had sought and received instruction from him; his sons and nephews became mathematicians, some of them outstanding ones. He hated the "English buffoons", as he called them, and like Horatius at the bridge he had defeated every British champion who dared challenge him.

BERNOULLI discharged his routine lecturing on elementary mathematics at the University with increasing distaste and decreasing attention. Those few, very few, students whom he regarded as promising he instructed privately and sometimes gratis. EULER recalled,

> I soon found an opportunity to gain introduction to the famous professor John Bernoulli, whose good pleasure it was to advance me further in the mathematical sciences. True, because of his business he flatly refused me private lessons, but he gave me much wiser advice, namely to get some more difficult mathematical books and work through them with all industry, and wherever I should find some check or difficulties, he gave me free access to him every Saturday afternoon and was so kind as to elucidate all difficulties, which happened with such greatly desired advantage that whenever he had obviated one check for me, because of that ten others disappeared right away, which is certainly the way to make a happy advance in the mathematical sciences.

When he was fifteen, EULER delivered a Latin speech on temperance and received his *prima laurea*, first university degree. In the same year he was appointed public opponent of claimants for chairs of logic and of the history of law. In the following year he received his master's degree in philosophy, and to the session of 8 June 1724, at which the announcement was made, he gave a public lecture on the philosophies of DESCARTES and NEWTON. Meanwhile, he remembered, for the sake of his family

> I had to register in the faculty of theology, and I was to apply myself besides and especially to the Greek and Hebrew languages, but not much progress was made, for I turned most of my time to mathematical studies, and by my happy fortune the Saturday visits to Mr. John Bernoulli continued.

At nineteen EULER published his first mathematical paper, an outgrowth of one of BERNOULLI's contests with the English; EULER had found that his teacher's solution of a certain geometrical problem, while indeed better than the English one, could itself be greatly improved, generalized, and shortened. In the case of his own sons, such turns aroused BERNOULLI's jealousy and competition, but EULER at once became and remained his favorite disciple.

The next year, at the age of twenty, EULER competed for the Paris prize. These prizes were the principal scientific honors of the century; golden honors they were, too, 2500 livres or even twice or thrice that much, not the empty titles of our time. JOHN BERNOULLI himself won the prize twice; his son DANIEL, ten times; EULER was to win it twelve times, or about every fourth year of his working life. The assigned topics were usually dull or vague or intricate matters of celestial mechanics, nautics, or physics, never mathematics as such. Often they were directed toward the interests of a specific Frenchman who had something ready and was expected therefore to win, but the competitions were administered fairly, and when an outsider sent in a fine essay, as a rule he was given the prize. The Basler mathematicians had a knack of twisting a promiseless subject into something more fundamental, upon which mathematics could be brought to bear. The prize essays themselves rarely solved the problem announced and usually were works of second class in their authors' total outputs, but the competitions caused the great savants to take up and deepen inquiries they might otherwise never have begun, and so the competitions tended indirectly to broaden the range of mathematical theories of physics. Thus they played, though at a more individual and aristocratic height, a role like that of military support for science in our time. The subject of 1727 was the masting of ships. EULER had never seen a seagoing ship, but his entry received honorable mention and was published forthwith. The winner was BOUGUER, for whom the prize had been designed, and who had submitted an entire treatise he had been writing for some years; this treatise immediately became the standard work on the subject. The other two classics of the eighteenth century on naval science, one being much more general and mathematical and profound, and the other being the little handbook to which TURGOT referred, were both to be written later by EULER.

In the same year, his twenty-first, EULER on BERNOULLI's advice competed for the chair of physics. While he was quickly eliminated as a candidate, he published his specimen essay, *A Physical Dissertation on Sound*. With the clarity and directness that were to become his instantly recognizable signature, in sixteen pages he laid out in order and in simple words, without calculations, all that was then known about the production and propagation of sound, added some details of his own, and listed a number of open problems. This work became a classic at once; it was read and cited for over a hundred years, during which it served as the program for research on acoustics. EULER himself later wrote at least 100 papers directly or indirectly related to the problems set here, and many of these he solved once and for all. The last page lists six annexes. The first denies the principle of pre-established harmony; the second asserts that NEWTON's Law of gravitation is indeed universal; the fourth affirms that kinetic energy is the true measure of the force of bodies; while the remaining three announce solutions of problems concerning oscillation through a hole in the earth, the rolling of a sphere, and the masting of ships. The professorship was given to a man never heard of again, who in fact was interested primarily in anatomy and botany. EULER at twenty had entered the field of mechanical physics and philosophy as a challenger with firm positions, openly avowed, on every main question then under debate. At the same time, and in equal measure, he was able to announce definite and final solutions to several specific problems. When he died, fifty-five years later, his mastery of all physics as it was then understood, and his ability to solve special problems, were just the same. Indeed, most of the main general advances of the entire century had been made by him, and in addition he had solved many key-problems and hundreds of examples. On the day of his death he had discussed with his disciples the orbit of the planet Uranus, which HERSCHEL had discovered two years before. On his slate was a calculation of the height to which a hot-air balloon could rise. The news of the MONTGOLFIERS' first ascent had just reached St. Petersburg, where EULER had been residing for most of his life.

Having had the good luck not to win the chair of physics at Basel, EULER went to Petersburg in 1727. JOHN BERNOULLI had been invited but felt himself too old; instead he offered one of his two sons, DANIEL and NICHOLAS, and then adroitly required that neither should go unless the other went too for company and comfort. One was a professor of law and the other was studying medicine in Italy; both were pleased to accept chairs of mathematics or physics. They promised the young EULER the first vacant place, but Russia's thirst for the mathematical sciences was slaked at the moment, and so they suggested he take a position as "Adjunct in Physiology". To this end they advised him to read certain books and learn anatomy; accordingly

> I matriculated in the medical faculty of Basel and began to apply myself with all industry to the medical course of study . . ..

EULER arrived in Petersburg on the day the empress died and the Academy fell into

> the greatest consternation, yet I had the pleasure of meeting not only Mr. Daniel Bernoulli, whose elder brother Mr. Nicholas had meanwhile died, but also the late Professor Hermann, a countryman and also a distant relative of mine, who gave me every imaginable assistance. My pay was 300 rubles along with free lodging, heat, and light, and since my inclination lay altogether and only toward mathematical studies, I was made Adjunct in Higher Mathematics, and the proposal to busy me with medicine was dropped. I was given liberty to take part in the meetings of the Academy and to present my developments there, which even then were put into the *Commentarii* of the Academy.

The Academicians were all foreigners—Germans, Swiss, and a Frenchman, not only the professors but also the students. Thus language was not a problem, but the senior colleagues were. To a man the chiefs, like university officials today, were tumors, the only question being whether benign or malignant. The most promising mathematician, NICHOLAS II BERNOULLI, had died of a fever before EULER arrived. EULER's friends were DANIEL BERNOULLI, seven years older and already a famous mathematician and physicist, and GOLD-BACH, an energetic and intelligent Prussian for whom mathematics was a hobby, the entire realm of letters an occupation, and espionage a livelihood. The Academy fell on evil days; its effective director was an Alsatian named SCHUMACHER, whose main interest lay in the suppression of talent wherever it might rear its inconvenient head. SCHUMACHER was to play a part in EULER's life for more than a quarter century.

Soon most of the old tumors had been excised by departure or death. So had most of the capable men. DANIEL BERNOULLI, after having competed for every vacancy in Basel, in 1733 finally obtained the chair of anatomy. Once back, he felt himself a new man in the good Swiss air, but in the rest of his long life he never again reached the level and the fruitfulness of his eight years in Petersburg, six of which were enlivened by friendly competition with EULER.

EULER stayed on. For him, these were years of growth as well as production. While he never lost his love for mechanics and the "higher analysis", he steadily enlarged his knowledge and power of thought to include all parts of mathematics ever before cultivated by anyone. He was able to create new synthetic theorems in the Greek style, such as his magnificent discovery and proof that every rotation has an axis. He sought and read old books such as FERMAT's commentary on DIOPHANTOS. On the basis of such antiquarian studies he recreated the arithmetic theory of numbers, which had been scarcely noticed by the BERNOULLIs and LEIBNIZ, in whose school of thought he had been trained. He gave this subject new life and discovered more major theorems in it than had all mathematicians before him put together. He was equally at home in the algebra of the seventeenth century, a field neither easy nor elementary, tightly wed to the theory of numbers. He also probed new subjects which were to flower only much later. One of these is combinatorial topology, in which he conjectured but was not able to prove what later became a key-theorem, now called the EULER polyhedron formula[3]. Unifying and subjecting to system the work of many predecessors, he created analytic geometry[4] as we know that discipline today; from his textbook,

---

[3] Namely, in any simple polyhedron the number of vertices plus the number of faces is greater by two than the number of edges. EULER could not have known that the same assertion lay in an unpublished manuscript of DESCARTES. EULER did publish a proof, but it is false as it stands; the basic idea of it, nevertheless, is sound and has been applied in countless later researches.

[4] Analytic geometry is ordinarily attributed to DESCARTES and FERMAT. Of course, like any other mathematical innovation, it was neither without antecedents nor beyond improvement. The reader who doubts my statement should draw his own conclusion by comparing DESCARTES' *La Géométrie*, Volume 2 of EULER's *Introductio in analysin infinitorum*, and a textbook of the 1930s.

EULER's development of analytic geometry is described by C. B. BOYER on pages 180–181 of his *History of Analytic Geometry*, New York, Scripta Mathematica, 1956. Of EULER's *Introductio in analysin infinitorum* BOYER writes

The *Introductio* of Euler is referred to frequently by historians, but its significance generally is underestimated. This book is probably the most influential textbook of modern times. It is the work which made the function concept basic in mathematics. It popularized the definition of logarithms as exponents and

and from others based upon it, and still others based on them, and so on, students of mathematics learned the subject from 1748 until the 1930s, when it was largely superseded by the rise of modern linear algebra. Students of natural science even today learn it in essentially EULER's way. EULER was the first man to publish a paper on partial-differential equations, and the world has learnt most of the elementary calculus of partial derivatives from his books, although some of the rules had been known to NEWTON and LEIBNIZ but not published by them. It was mainly in his first Petersburg years that EULER developed his taste for pure mathematics, which has remained forever after, in a tradition deriving from him and unbroken by the most violent political changes, a Russian specialty. About one-third of his total product was regarded as "pure" mathematics in his own day; in the classification of our time, this term would apply to only about one-fifth of it; but that small fraction includes many of his deepest and most permanent contributions. One of these is the concept of real function: namely, a rule assigning to each real number in some interval another real number. In his earlier years EULER, like his predecessors, had used a concept of function both narrow and vague, but his own discoveries in the theory of partial-differential equations and wave propagation had shown him the clear way[5], which every mathematician since 1850 has

---

the definitions of the trigonometric functions as ratios. It crystallized the distinction between algebraic and transcendental functions and between elementary and higher functions. It developed the use of polar coordinates and of the parametric representation of curves. Many of our commonplace notations are derived from it. In a word, the *Introductio* did for elementary analysis what the *Elements* of Euclid did for geometry. It is, moreover, one of the earliest textbooks on college level mathematics which a modern student can study with ease and enjoyment, with few of the anachronisms which perplex and annoy the reader of many a classical treatise.

BOYER states that EULER's "treatment of the linear equation is characteristic for its generality, but it is startlingly abbreviated." By the standards of modern textbooks for freshmen EULER's book is rather advanced. For example, he stated "the geometry of the straight line is well known."

Finally, writes BOYER,

The *Introductio* closes with a long and systematic appendix on solid analytic geometry. This is perhaps the most original contribution of Euler to Cartesian geometry, for it represents in a sense the first textbook of algebraic geometry in three dimensions.

By "Cartesian geometry" BOYER refers more or less to what is usually called "analytic geometry"; by "algebraic geometry", to what is usually called "co-ordinate geometry".

[5] The "clear way" is commonly attributed to DIRICHLET or other mathematicians of the nineteenth century.

followed. Other great discoveries were the law of quadratic reciprocity[6] in number theory and the addition theorem for elliptic functions[7], but these came later than the time of which I am now speaking.

What EULER did for mechanics blanks superlatives. The contents of any one of the two dozen volumes of his *Opera* that concern mechanics primarily would have sufficed to earn its author a place at or near the summit of the field. There is no aspect of it as it stood before his day that he did not change essentially; he solved problems set by his predecessors, applied existing theories to important new instances, simplified ideas while making them more general, unified domains that before him had seemed separate. He created new concepts and new disciplines to embrace phenomena of nature that previously were not understood. Sometimes he worked with the most abstruse mathematics known in his day; he was equally ready to explain his results and their applications by simple rules of practice; he regularly furnished numerical methods and worked-out instances. Above all, he sought and achieved clarity.

Analysis was the key to mechanics, and in turn mechanics suggested most of the problems of analysis that mathematicians of the eighteenth century attacked. Astronomy and physics were mainly applications of mechanics. Over half of the pages EULER published were expressly devoted to mechanics or closely connected with it.

Nonetheless, there is no evidence that EULER preferred any one part of mathematics to the rest[8]. The only sure conclusion we can draw from his prodigious output is that he sought to enlarge the

---

[6] That is, in the notation of GAUSS, of the two congruences $x^2 \equiv q \pmod{p}$ and $x^2 \equiv p \pmod{q}$, p and q being prime numbers, either both are soluble or neither is except if $p \equiv q \equiv 3 \pmod 4$, in which case one is soluble and the other is not.

[7] That is, in the notation of JACOBI,

$$\mathrm{sn}(u+v) = \frac{(\mathrm{sn}u)(\mathrm{cn}v)(\mathrm{dn}v) + (\mathrm{cn}u)(\mathrm{sn}v)(\mathrm{dn}u)}{1 - k^2(\mathrm{sn}^2u)(\mathrm{sn}^2v)}$$

and related formulæ.

[8] In his beautiful book *Fermat's Last Theorem*, New York *etc.*, Springer-Verlag, H. M. EDWARDS writes as follows:

It is a measure of Euler's greatness that when one is studying number theory one has the impression that Euler was primarily interested in number theory, but when one studies divergent series one feels that divergent series were his main interest, when one studies differential equations one imagines that actually differential equations was his favorite subject, and so forth. ... Whether or not number theory was a favorite subject of Euler's, it is one in which he showed a lifelong interest and his contributions to number theory alone would suffice to establish a lasting reputation in the annals of mathematics.

domain of mathematics and its applications with a dediction as eager as that which led Don GIOVANNI to seduce even ugly girls *pel piacer di porle in lista*, but EULER's outposts, even those ridiculed by some of his contemporaries, have been bridgheads to future and permanent, total conquests.

The first Petersburg years brought EULER success, instruction in the facts of life, and misfortune.

> In 1730, when Professors Hermann and Bülfinger returned to their native land, I was named to replace the latter as Professor of Physics, and I made a new contract for four years, granting me 400 rubles for each of the first two and 600 for the next two, along with 60 rubles for lodging, wood, and light.

Then EULER had the experience, not uncommon in the Enlightenment, of being unable to collect all of his contracted salary. In 1731 there was a matter of promotion: Four little men, who up to that time had been receiving less than he, were set equal to him. In a formal protest EULER wrote [*cf.* above, Parts aIII and aIV of this essay],

> That we shall each be treated on the same footing is something I can't get through my head at all. .... It is true that I have never applied myself so much to physics as to mathematics, but nevertheless I doubt much that you can get from the outside such a person as I for any 400 rubles. In the matter of mathematics, I think the number of those who have carried it as far as I is pretty small in the whole of Europe, and none of those will come for 1000 rubles.

(We should take note of EULER's estimated difference of salaries: 400 for a physicist, 1000 for a mathematician. In those days physics was a speculative or experimental science, not a mathematical one.)[9] BÜLFINGER, whose talent was modest at best and for mathematics naught, had been Professor of Physics; DANIEL BERNOULLI, whose lifelong passion was what he himself called physics, was Professor of Higher Mathematics. SCHUMACHER advised the President of the Academy not to grant EULER the least concession, since otherwise he would straightway grow impudent. EULER learned a lifelong lesson

---

[9] This difference in their predecessors is recognized by both mathematicians and physicists today, since the latter are wont to say that the greatest discoveries in mathematics were made by (theoretical) physicists, while the former often remark that most of the major discoveries in theoretical physics were made by mathematicians (until very recently). Usually they are speaking of the same persons, *e.g.*, HUYGENS and NEWTON and EULER and LAGRANGE and CAUCHY and FOURIER.

from this experience: It is futile to argue with administrators but easy to outwork and forget them.

In 1733, EULER states,

> when Professor Daniel Bernoulli, too, went back to his native land, I was given the professorship of Higher Mathematics, and soon thereafter the directing senate ordered me to take over the Department of Geography, on which occasion my salary was increased to 1200 rubles.

Earlier in the same year, even before this splendid increase in his salary, EULER had married, of course choosing a Swiss wife, the daughter of a court artist; in this way he continued the tradition of the BERNOULLIS, all of whom were either professors or painters, and his younger brother also became a painter. The first of EULER's many children was born the next year. In 1738 a violent fever destroyed the sight of one of EULER's eyes. The work in the geographical department strained his eyesight severely, but he was really interested in constructing a good general map of Russia, and he succeeded in doing so. He wrote to order a school arithmetic text and a great treatise on naval science, receiving for this latter 1200 rubles, in this way doubling his salary one year. EULER's precise recollection of the dates and salaries of his early appointments reflects his Swiss talent for making and saving money. On at least one occasion even Tyche smiled upon him: In the spring of 1749 he wrote to GOLDBACH that he had received 600 Reichsthaler from a lucky ticket in a lottery, "which was just as good as if I had won a Paris prize this year."

In 1740 EULER was requested to cast the horoscope of the new Czar, who was only a few weeks old. While such a task would have been normal a century earlier, for the Enlightenment it was *retardataire*. EULER smoothly passed the honor on to the Professor of Astronomy. The contents of the horoscope is not known, but in less than a year the child Czar was deposed and hidden; twenty-four years later, still in prison, he died.

In 1740 FREDERICK II ascended the throne of Prussia. This eccentric and semi-educated general, flute player, and homosexual lay under the spell of France and French men. He wished to create in Berlin a mingled French Académie des Sciences and Académie Française. VOLTAIRE was his Apollo, and VOLTAIRE recommended as director a trifling but extremely eminent French scientist named MAUPERTUIS, whom he dubbed "Le Grand Aplatisseur" for his having led an expedition to Lapland to measure the length of one degree of a meridian, whence he had concluded that the earth was flatter at the poles than at the equator. For VOLTAIRE, who endorsed mathematical philosophy but did not understand it, this proved DESCARTES wrong and NEWTON right about everything. The later *philosophes* followed his judgment; the British gleefully followed them; and somehow this minor and precarious if not puerile side issue has assumed in the folklore of science an importance it never for a moment deserved or enjoyed among those who knew what was what in rational mechanics. In addition to being an argonaut, MAUPERTUIS was an *héros de salon* and a *causeur*, a fit table companion for the king; notwithstanding that, he had been a disciple of JOHN BERNOULLI, and though no geometer himself, he knew mathematics when he saw it. He proposed to bring all the BERNOULLIs and EULER to Berlin.

Only EULER was seduced, and at that only because, as he put it, in the regency following the death of Empress ANNA "things began to look rather awkward." That the prospect in Russia was bad indeed, is proved by EULER's consenting to move at no increase in pay. Even so, the Prussian king did not feel himself compelled to discharge his promise in full. After his return to Petersburg, EULER's dictated summary of his twenty-five years in Berlin was "What I encountered there, is well known."

No sooner did EULER arrive in Berlin but the king's wars overturned everything and endangered MAUPERTUIS, who withdrew from Prussia until he was sure FREDERICK's seat was firm. EULER, meanwhile, was writing mathematical papers. Every associate member of the Academy was required to compose for publication at least one memoir per year; every pensioner, at least two; EULER never presented fewer than ten.

The keys to the treasurehouse of learning in the eighteenth century—I should be tempted to say also today, were it not that any such statement would be empty because "learning" has been taken off the gold standard—were the Latin language and the infinitesimal calculus. FREDERICK II understood neither; he detested both. He ordered his Academy to speak and publish only in French, and he encouraged it to cultivate the sciences useful in promotion of trades and manufactures, in the restraint of savage passions, and in the development of a subject's duties. EULER, despite his thoroughly Classical training and his consummate mastery of the new "analysis of curves", easily accepted these conditions. He continued his connection with the Academy of Petersburg, not only sending it a stream of papers, mainly on pure mathematics, but also serving as editor of its publications; in addition, he conveyed to SCHUMACHER information of all sorts regarding the scientific life of the West. In return, of course, he received a salary. These relations continued even through the Seven Years' War, during which Russia joined the alliance against Prussia and at one time overran Berlin. When a farm belonging to

EULER[10] was pillaged by the Russians, their commander, General TOTLEBEN, saying he did not make war upon the sciences, indemnified EULER for more than the damage sustained, and the Empress ELIZABETH added a further gift, finally turning the loss into a handsome profit. EULER also lodged and boarded in his house Russian students sent by the Petersburg Academy, one of these being RASUMOVSKI, hetman of the Cossacks, who later became president of the Academy. EULER gave these students instruction in mathematics, this being as close as he ever came to what is called "teaching" in American universities. EULER taught mathematics and physics to the whole world, and down to the present time his influence on instruction in the exact sciences has been second only to EUCLID's. In person, had he held a chair in a university, he might have reached a few hundred students at most; like EUCLID, by writing EULER has taught mathematics to millions.

By no means all of EULER's books were popular ones. Until about fifteen years ago unopened copies of his more advanced works turned up at low prices on the book market. At least five of these were the first treatises ever published on their subjects, and while easy for a dedicated reader to study, they seemed abstruse to the laity. Few as were the copies sold in EULER's own day[11], they fell into the right hands. His treatises on rigid-body dynamics, infinite series, differential and integral calculus, and the calculus of variations were mother's milk to three or four generations of mathematicians and theoretical physicists, including the great Frenchmen of the NAPOLEONic revival, as well as the less eminent but equally influential German and Italian professors of the same period; from the teaching of these three schools the basic core of EULER's work has passed into the common tradition of the mathematical sciences[12]. While it is a rare young Doctor of Philosophy in America today who can decipher a page of JOHNSON's *London* without a dictionary if not a crib or coach, and

---

[10] The episode has come down to us only through CONDORCET's *Eloge*; we do not know whether EULER had more than one farm.

[11] EULER's correspondence with KARSTEN shows that the printing of his book on the motion of rigid bodies, an acknowledged masterpiece of mechanics, was delayed four years for lack of interest. The publisher demanded subscriptions for 100 copies, but after waiting eighteen months he had received only thirty. EULER finally waived royalties; instead, he requested twenty free copies but said he would be satisfied with twelve. It seems this latter number was what he did in the end receive. Twenty-five years later, and after EULER's death, the same publisher found it worthwhile to issue the work in a second edition, adding some of EULER's major papers on the subject as an appendix.

[12] It is well known that the British school of the mid-nineteenth century, the greatest representatives of which were GREEN, STOKES, KELVIN, and MAXWELL, learnt mathematics and mathematical physics primarily from French books.

while in another academic generation we can confidently expect that *Robinson Crusoe* will have to be translated into "modern English", even the mediocre juniors in engineering the world over have learnt and are able to use a dozen of EULER's discoveries. With the music of the same period, the contrast is more striking. For example, in the eighteenth century no-one outside Hamburg can have heard TELEMANN's *Der Tag des Gerichtes*; few can have been those who heard even some part of BACH's *Messe in H-moll,* and no-one, certainly, had heard the whole of it or any part at all of *Die Kunst der Fuge.* While these works seem to us now to stand at the summit of the Enlightenment, even their authors had in their own day merely national or local reputations. Not so with EULER, who was famous far, far beyond the tiny though international circle of those who could understand what he wrote. He was one of those favored few who achieved even from their own contemporaries the respect of which posterity has judged them worthy. EULER won his later fame by the usual method: merciless trials by the fire and water of time. In his own day, from his twenty-fifth year onward, he was a senior academician, and he used well the advantages his position gave him.

The academies of the eighteenth century, although few in number, dominated its science, which had become professional. While in the earlier Baroque period there had been many savants, mainly private, who had contributed in some degree to the spring tide of the new natural philosophy, by the time of the Enlightenment science had become a serious business, valued and rewarded though little understood. The high positions were paid well. EULER's initial salary in Berlin was 1600 talers; MAUPERTUIS received nearly twice as much; the junior members, about 300. Paid positions were few, and they were hotly sought. A senior professor in Basel and "the ARCHIMEDES of his age", as he justly regarded himself, old JOHN BERNOULLI at the end of his life received only about as much as a "student" or "adjunct" in one of the great academies. It is difficult to estimate equivalents in modern currency, but in terms of goods and services in 1982 I think the value of EULER's 1600 talers was around $80,000, tax-free. For example, in 1742 he bought a fine house with a large garden for 2000 taler, one and one quarter year's salary, while the wages of a professor today for the same length of time, after income taxes, would fall well below the price of a run-down row house. Nonetheless we must not be misled by today's social-democratic guilt syndrome, which dictates that the greatest genius of the age must not be paid more than twice the wages of idleness for a congenital fool. The Enlightenment, as its name might suggest, was a period of economic variety, in which EULER found himself further from the top than from the bottom. It would have cost a whole Paris prize to buy a

Savonnerie carpet fourteen feet square, had the royal monopoly let any be sold. This is one point where comparisons might be thought simple, since the factory still exists. A carpet of the same size, presumably one of the garish sprays of splotches now regarded as art, cost $36,000 in 1972; as in the Enlightenment, today the total product of the factory is reserved, though no longer for the splendid galleries of kings and their pretty mistresses' bedrooms but rather for the upper beaches washed by the flux and reflux of interchangeable functionaries of the Nth Republic.

EULER practised the thrift for which the Swiss are justly famous. In 1753 he bought a farm for 6000 talers, and with its produce of hay, grain, vegetables, and fruit he cut his household expenses in half. He lodged there his widowed mother, his younger children, and their private tutors. A portrait of the time, a fine pastel by HANDMANN, shows him in an elegant nightcap and a dressing gown of light and dark blue strips of satin, presumably his working clothes. In this portrait, his blind right eye is turned aside from the beholder. The somewhat confused expression of the mouth is due only to damage to the pastel and does not reflect the ease and decision visible in other portraits of EULER. One of these, unpublished, shows him in a scarlet velvet morning coat. In another, reproduced as frontispiece to this volume, he sits on a curvilinear chair with vasiform splat, writing at a carved and gilt table in rococo style.

To learn what an Academy of the eighteenth century was, we may begin with Gulliver's third voyage, published the year before EULER first went to Petersburg. SWIFT had the Royal Society of London in mind, but the glove fits the more formal academies of the Continent almost as well. First there was the mathematical class:

> ... a race of mortals ... singular in their shapes, habits, and countenances. Their heads were all reclined either to the right or the left; one of their eyes turned inward, and the other directly up to the zenith. Their outward garments were adorned with the figures of suns, moons, and stars, interwoven with those of fiddles, flutes, harps, trumpets, guitars, harpsichords, and many other instruments of music . . . . I observed here and there many in the habit of servants, with a blown bladder fastened like a flail to the end of a short stick, which they carried in their hands. In each bladder was a small quantity of dried pease, or little pebbles . . .. With these bladders they now and then flapped the mouths and ears of those who stood near them . . .; it seems the minds of these people are so taken up with intense speculations, that they neither can speak, nor attend to the discourses of others, without being roused by some external taction upon the organs of speech and hearing; for which reason those persons who are able

to afford it always keep a flapper . . . . And the business of this officer is, when two or more persons are in company, gently to strike with his bladder the mouth of him who is to speak, and the right ear of him or them to whom the speaker addresseth himself. This flapper is likewise employed diligently to attend his master in his walks, and upon occasion to give him a soft flap on his eyes, because he is always so wrapped up in cogitation, that he is in manifest danger of falling down every precipice, and bouncing his head against every post, and in the streets, of justling others, or being justled himself into the kennel. . . .

At last we entered the palace, and proceeded into the chamber of presence, where I saw the King seated on his throne, attended on each side by persons of prime quality. Before the throne was a large table filled with globes and spheres, and mathematical instruments of all kinds. His Majesty took not the least notice of us, although our entrance was not without sufficient noise, by the concourse of all persons belonging to the court. But he was then deep in a problem, and we attended at least an hour, before he could solve it. . . . My dinner was brought . . . . In the first course there was a shoulder of mutton, cut into an equilateral triangle, a piece of beef into a rhomboides, and a pudding into a cycloid. The second course was two ducks, trussed up into the form of fiddles; sausages and puddings resembling flutes and hautboys, and a breast of veal in the shape of a harp. The servants cut our bread into cones, cylinders, parallelograms, and several other mathematical figures. . . .

The knowledge I had in mathematics gave me great assistance in acquiring their phraseology, which depended much upon that science and music; and in the latter I was not unskilled. Their ideas are perpetually conversant in lines and figures. If they would, for example, praise the beauty of a woman, or any other animal, they describe it by rhombs, circles, parallelograms, ellipses, and other geometrical terms, or by words of art drawn from music, needless here to repeat.

(We remark that the mathematicians of the Enlightenment shared the common passion for music. EULER himself wrote a major treatise on harmony, which as far as it goes has never been superseded; he projected a treatise on composition; and he published some short papers concerning the function of dissonances. D'ALEMBERT likewise wrote a treatise on music. Some musicians returned the compliment: RAMEAU wrote,

Music is a science which should have secure rules; these rules should be drawn from an evident principle, and this principle can scarcely be known to us without the aid of mathematics. Thus I must admit that despite all the experience I could get from music in practising it for so long a time, nevertheless it is only by the help of mathematics that my ideas have grown clear.

Whatever RAMEAU's study of mathematics may have been, no sign of it may be detected in his book, in which even the experimental facts of acoustics as they were then known are partly misrepresented.)

In Laputa

[t]heir houses are very ill built, the walls bevil, without one right angle in any apartment, and this defect ariseth from the contempt they bear to practical geometry, which they despise as vulgar and mechanic, those instructions they give being too refined for the intellectuals of their workmen, which occasions perpetual mistakes. And although they are dexterous enough upon a piece of paper in the management of the rule, the pencil, and the divider, yet in the common actions and behaviour of life, I have not seen a more clumsy, awkward, and unhandy people, nor so slow and perplexed in their conceptions upon all other subjects, except those of mathematics and music. They are very bad reasoners, and vehemently given to opposition, unless when they happen to be of the right opinion, which is seldom their case. Imagination, fancy, and invention, they are wholly strangers to, nor have any words in their language by which those ideas can be expressed; the whole compass of their thoughts and mind being shut up within the two forementioned sciences.

Most of them, and especially those who deal in the astronomical part, have great faith in judicial astrology, although they are ashamed to own it publicly. But what I chiefly admired, and thought altogether unaccountable, was the strong disposition I observed in them towards news and politics, perpetually enquiring into public affairs, giving their judgments in matters of state, and passionately disputing every inch of a party opinion. I have indeed observed the same disposition among most of the mathematicians I have known in Europe, although I could never discover the least analogy between the two sciences; unless those people suppose, that because the smallest circle hath as many degrees as the largest, therefore the regulation and management of the world require no more abilities than the handling and turning of a globe. But I rather take this quality to spring from a very common infirmity of human nature, inclining us to be more curious and conceited in matters where we have least concern, and for which we are least adapted either by study or nature.

These people are under continual disquietudes, never enjoying a minute's peace of mind; and their disturbances proceed
from causes which very little affect the rest of mortals. Their
apprehensions arise from several changes they dread in the celestial bodies. For instance, that the earth, by the continual
approaches of the sun towards it, must in course of time be absorbed or swallowed up. That the face of the sun will by degrees
be encrusted with its own effluvia, and give no more light to the
world. That the earth very narrowly escaped a brush from the tail
of the last comet, which would have infallibly reduced it to ashes;
and that the next, which they have calculated for one and thirty
years hence, will probably destroy us. For if in its perihelion it
should approach within a certain degree of the sun (as by their
calculations they have reason to dread) it will conceive a degree of
heat ten thousand times more intense than that of red-hot
glowing iron; and in its absence from the sun, carry a blazing tail
ten hundred thousand and fourteen miles long; through which if
the earth should pass at the distance of one hundred thousand
miles from the nucleus or main body of the comet, it must in its
passage be set on fire, and reduced to ashes. That the sun daily
spending its rays without any nutriment to supply them, will at last
be wholly consumed and annihilated; which must be attended with
the destruction of this earth, and of all the planets that receive their
light from it.

They are so perpetually alarmed with the apprehensions of
these and the like impending dangers, that they can neither sleep
quietly in their beds, nor have any relish for the common pleas-
ures or amusements of life.

The mathematicians Swift described lived upon an island mag-
netically suspended in the air. They were able to control its motions
perfectly and so dominate the low earth beneath them. At this baser
level lay the practitioners of applied and natural science, who
inhabited the Grand Academy of Lagado:

This Academy is not an entire single building, but a continuation of several houses on both sides of a street, which growing
waste was purchased and applied to that use.

I was received very kindly by the Warden, and went for many
days to the Academy. Every room hath in it one or more projectors, and I believe I could not be in fewer than five hundred
rooms.

The first man I saw was of a meagre aspect, with sooty hands
and face, his hair and beard long, ragged and singed in several

places. His clothes, shirt, and skin were all of the same colour. He had been eight years upon a project for extracting sun-beams out of cucumbers, which were to be put into vials hermetically sealed, and let out to warm the air in raw inclement summers. He told me he did not doubt in eight years more he should be able to supply the Governor's gardens with sunshine at a reasonable rate; but he complained that his stock was low, and entreated me to give him something as an encouragement to ingenuity, especially since this had been a very dear season for cucumbers. I made him a small present, for my lord had furnished me with money on purpose, because he knew their practice of begging from all who go to see them.

I went into another chamber, but was ready to hasten back, being almost overcome with a horrible stink. My conductor pressed me forward, conjuring me in a whisper to give no offence, which would be highly resented, and therefore I durst not so much as stop my nose. The projector of this cell was the most ancient student of the Academy; his face and beard were of a pale yellow; his hands and clothes daubed over with filth. When I was presented to him, he gave me a close embrace (a compliment I could well have excused). His employment from his first coming into the Academy, was an operation to reduce human excrement to its original food, by separating the several parts, removing the tincture which it receives from the gall, making the odour exhale, and scumming off the saliva. He had a weekly allowance from the society, of a vessel filled with human ordure, about the bigness of a Bristol barrel.

I saw another at work to calcine ice into gunpowder, who likewise showed me a treatise he had written concerning the malleability of fire, which he intended to publish.

There was a most ingenious architect who had contrived a new method for building houses, by beginning at the roof, and working downwards to the foundation, which he justified to me by the like practice of those two prudent insects, the bee and the spider.

There was a man born blind, who had several apprentices in his own condition: their employment was to mix colours for painters, which their master taught them to distinguish by feeling and smelling. It was indeed my misfortune to find them at that time not very perfect in their lessons, and the professor himself happened to be generally mistaken: this artist is much encouraged and esteemed by the whole fraternity.

In another apartment I was highly pleased with a projector, who had found a device of ploughing the ground with hogs, to save the charges of ploughs, cattle, and labour. The method is this: in an acre of ground you bury, at six inches distance and eight deep, a quantity of acorns, dates, chestnuts, and other mast or vegetables whereof these animals are fondest; then you drive six hundred or more of them into the field, where in a few days they will root up the whole ground in search of their food, and make it fit for sowing, at the same time manuring it with their dung. It is true, upon experiment they found the charge and trouble very great, and they had little or no crop. However, it is not doubted that this invention may be capable of great improvement.

I went into another room, where the walls and ceiling were all hung round with cobwebs, except a narrow passage for the artist to go in and out. At my entrance he called aloud to me not to disturb his webs. He lamented the fatal mistake the world had been so long in of using silk-worms, while we had such plenty of domestic insects, who infinitely excelled the former, because they understood how to weave as well as spin. And he proposed farther that by employing spiders the charge of dyeing silks should be wholly saved, whereof I was fully convinced when he showed me a vast number of flies most beautifully coloured, wherewith he fed his spiders, assuring us that the webs would take a tincture from them; and as he had them of all hues, he hoped to fit everybody's fancy, as soon as he could find proper food for the flies, of certain gums, oils, and other glutinous matter to give a strength and consistence to the threads. . . .

I was complaining of a small fit of the colic, upon which my conductor led me into a room, where a great physician resided, who was famous for curing that disease by contrary operations from the same instrument. He had a large pair of bellows with a long slender muzzle of ivory. This he conveyed eight inches up the anus, and drawing in the wind, he affirmed he could make the guts as lank as a dried bladder. But when the disease was more stubborn and violent, he let in the muzzle while the bellows were full of wind, which he discharged into the body of the patient, then withdrew the instrument to replenish it, clapping his thumb strongly against the orifice of the fundament; and this being repeated three or four times, the adventitious wind would rush out, bringing the noxious along with it (like water put into a pump), and the patient recover. I saw him try both experiments upon a dog, but could not discern any effect from the former. After the latter, the animal was ready to burst, and made so violent a discharge, as was very offensive to me and my companions. The dog died on the spot, and we left the doctor endeavouring to recover him by the same operation. . . .

I had hitherto seen only one side of the Academy, the other being appropriated to the advancers of speculative learning, of whom I shall say something when I have mentioned one illustrious person more, who is called among them *the universal artist*. He told us he had been thirty years employing his thoughts for the improvement of human life. He had two large rooms full of wonderful curiosities, and fifty men at work. Some were condensing air into a dry tangible substance, by extracting the nitre, and letting the aqueous or fluid particles percolate; others softening marble for pillows and pin-cushions; others petrifying the hoofs of a living horse to preserve them from foundering. The artist himself was at that time busy upon two great designs; the first, to sow land with chaff, wherein he affirmed the true seminal virtue to be contained, as he demonstrated by several experiments which I was not skillful enough to comprehend. The other was, by a certain composition of gums, minerals, and vegetables outwardly applied, to prevent the growth of wool upon two young lambs; and he hoped in a reasonable time to propagate the breed of naked sheep all over the kingdom.

So much for the Department of Doing Material Good to Humanity. You may think that these long quotations from *Gulliver's Travels* are no more than a fantastic parody and so digress from my subject; on the contrary, for each episode in the Third Voyage a specific source, either in the *Philosophical Transactions* or in other scientific literature available to SWIFT, has been traced[13]. The truth was so

---

[13] A colleague in literature has kindly brought to my attention the fascinating, learned article in which these sources are discovered and quoted: MARJORIE NICOLSON & NORA M. MOHLER, "The scientific background of Swift's *Voyage to Laputa*", *Annals of Science* **2** (1937); 299–334. [*cf.* also their "Swift's 'Flying Island' in the *Voyage to Laputa*", *ibid.* 405–430.] The reader of the article of NICOLSON & MOHLER cannot fail to notice also their opening comments on the third voyage: "There is general agreement that in interest and literary merit it falls short of the first two voyages. It is marked by multiplicity of themes; it is episodic in character. In its reflections upon life and humanity, it lacks ... philosophic intuition. ... Any reader sensitive to literary values must so far agree with the critics who disparage the tale." On the contrary, I think that such critics have approached it without first learning the language in which it is written. They have forgotten that in the eighteenth century the cultivators of literature, unlike most of their successors today, did not despise science and vaunt their ignorance of it, but rather did their best to understand it, or at least pretended they did, as may be noted for instance in Uncle Toby's reference to the *Acta eruditorum* and in SAMUEL JOHNSON's disturbingly precise recollection of the contents of volume upon volume of popularized science (*cf.* also R. B. SCHWARTZ, *Samuel Johnson and the New Science*, Madison: University of Wisconsin Press, Wisconsin, 1971). Even pragmatically, how can the literary critics have for a moment fancied that so accomplished, artful a writer

bizarre as to need only recounting to serve as satire of itself. Today we sometimes forget that the abuse which accompanied the rise of experiment amounted to a second childhood of the human mind.

Crossing the walk, Gulliver arrived at the part where resided "the projectors in speculative learning", that is, to the Department of Moral and Humanitarian Studies.

> The first professor I saw was in a very large room, with forty pupils about him. After salutation, observing me to look earnestly upon a frame, which took up the greatest part of both the length and breadth of the room, he said perhaps I might wonder to see him employed in a project for improving speculative knowledge by practical and mechanical operations. But the world would soon be sensible of its usefulness, and he flattered himself that a more noble exalted thought never sprang in any other man's head. Every one knew how laborious the usual method is of attaining to arts and sciences; whereas by his contrivance the most ignorant person at a reasonable charge, and with a little bodily labour, may write books in philosophy, poetry, politics, law, mathematics, and theology, without the least assistance from genius or study. He then led me to the frame, about the sides whereof all his pupils stood in ranks. It was twenty foot square, placed in the middle of the room. The superficies was composed of several bits of wood, about the bigness of a die, but some larger than others. They were all linked together by slender wires. These bits of wood were covered on every square with paper pasted on them, and on these

---

as Swift would have published a "pointless" book, a satire of "slight importance", which would not strike home to the general reader of his own day?

The critics, few of whom have written great satires themselves, here remind me of the man born blind who engaged himself to mix colors for painters. Him who can read all four books, the satire bites deeper from each voyage to the next. The practitioners of music, mathematics, applied natural philosophy, and projective humanitarianism were the intellectual elite of the Enlightenment. To smite such men, who would have felt themselves little touched by the pettiness and brutality encountered in the first two voyages, Swift rose in the third. In the fourth the moral philosophers, superior both to the arrogant desiccation of the Laputan judicial astrologers and to the sordid scheming of the Lagadian quacks, found themselves revealed as being no more than Yahoos.

To strike the mathematicians and projectors where they were weakest, Swift paraphrased their own writings. The truth was more bizarre than any imagination. Nevertheless, Swift's deadly penetration selected examples so representative as to picture not only the Royal Society but also any other academy of the day and, with weird vision, even a great range of professional science, pseudoscience, and progressive learning 250 years later. When the third voyage is read out loud to an audience of scientists today, though their professionally glazed eyes have never seen nor ever will see a Lilliputian, Brobdingnagian, or Houyhnhm, they instantly recognize in it a harsh picture of their enemies, their friends, and themselves.

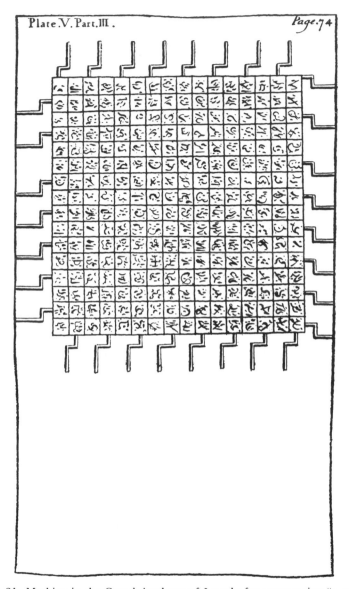

Figure 21. Machine in the Grand Academy of Lagado for constructing "a complete body of all arts and sciences".

papers were written all the words of their language, in their several moods, tenses, and declensions, but without any order. The professor then desired me to observe, for he was going to set his engine at work. The pupils at his command took each of them hold of an iron handle, whereof there were forty fixed round the edges of the frame, and giving them a sudden turn, the whole disposition of the words was entirely changed. He then commanded six and thirty of the lads to read the several lines softly as they appeared upon the frame; and where they found three or four words together that might make part of a sentence, they dictated to the four remaining boys who were scribes. This work was repeated three or four times, and at every turn the engine was so contrived that the words shifted into new places, as the square bits of wood moved upside down.

Six hours a day the young students were employed in this labour, and the professor showed me several volumes in large folio already collected, of broken sentences, which he intended to piece together, and out of those rich materials to give the world a complete body of all arts and sciences; which however might be still improved, and much expedited, if the public would raise a fund for making and employing five hundred such frames in Lagado, and oblige the managers to contribute in common their several collections.

He assured me, that this invention had employed all his thoughts from his youth, that he had emptied the whole vocabulary into his frame, and made the strictest computation of the general proportion there is in books between the numbers of particles, nouns, and verbs, and other parts of speech. . . .

We next went to the school of languages, where three professors sat in consultation upon improving that of their own country.

The first project was to shorten discourse by cutting polysyllables into one, and leaving out verbs and participles, because in reality all things imaginable are but nouns.

The other project was a scheme for entirely abolishing all words whatsoever; and this was urged as a great advantage in point of health as well as brevity. For it is plain that every word we speak is in some degree a diminution of our lungs by corrosion, and consequently contributes to the shortening of our lives. An expedient was therefore offered, that since words are only names for *things*, it would be more convenient for all men to carry about them such things as were necessary to express the particular business they are to discourse on. And this invention would certainly have taken place, to the great ease as well as health of the subject, if the women, in conjunction with the vulgar and illiterate, had not threatened to raise a rebellion, unless they might be allowed the liberty to speak with their tongues, after the manner of their ancestors; such constant irreconcilable enemies to science are the common people. However, many of the most learned and wise adhere to the new scheme of expressing themselves by things, which hath only this inconvenience attending it, that if a man's business be very great, and of various kinds, he must be obliged in proportion to carry a greater bundle of things upon his back, unless he can afford one or two strong servants to attend him. . . .

Another great advantage proposed by this invention was that it would serve as an universal language to be understood in all civilised nations, whose goods and utensils are generally of the same kind, or nearly resembling, so that their uses might easily be comprehended. And thus ambassadors would be qualified to treat with foreign princes or ministers of state, to whose tongues they were utter strangers.

I was at the mathematical school, where the master taught his pupils after a method scarce imaginable to us in Europe. The proposition and demonstration were fairly written on a thin wafer, with ink composed of a cephalic tincture. This the student was to swallow upon a fasting stomach, and for three days following eat nothing but bread and water. As the wafer digested, the tincture mounted to his brain, bearing the proposition along with it. But the success hath not hitherto been answerable, partly by some error in the *quantum* or composition, and partly by the perverseness of lads, to whom this bolus is so nauseous, that they generally steal aside, and discharge it upwards before it can operate; neither have they been yet persuaded to use so long an abstinence as the prescription requires.

Thus we see that "relevant" studies were subsidized by the governments of the Enlightenment, that they employed large staffs and needed costly apparatus, and that of modern educational tools only television and computerized dating were yet to be discovered. None of the products of these gossamer schemes for human betterment led to anything we now value. On the other hand, the military projects rarely if ever brought any improvement in the arts of warfare, but they did yield as by-products much basic science which every man curious to understand the world around him must learn today, science upon which rests much of our ordinary technology, that ubiquitous and supremely ugly technology whose products the most humanitarian of humanists insist upon having, and at low cost, however much they may despise the kind of learning that has produced them. For example, EULER's treatise on naval science was based largely on assumptions about the inertial and frictional resistances of water and air which were later shown to be false, and so his tediously scrupulous calculations of the efficiency of sails, oars, and paddle wheels, the design of hulls, and the courses of sailing ships, while correct as calculations, can have been little but useless to the Russian navy, yet his book contains also the first analysis of the stability of floating bodies in general and of the motion of rigid bodies about a variable axis. One device based upon EULER's basic theory but not invented until over 150 years after his death is the gyrocompass, which has saved a thousand times the number of lives it has helped to destroy.

SWIFT did not mention the disputes of the academicians and the precarious finances of the academies. Although by disposition somewhat irascible, EULER was not quarrelsome; he was exceptionally generous, never once making a claim of priority and in some cases actually giving away discoveries that were his own. He was the first to cite the works of others in what is now regarded as the just way, that is, so as to acknowledge their worth. Up to his time citation had been little more than a weapon of attack, to show where predecessors went wrong. EULER's intellectual generosity can hardly be set as an example, any more than a rich man's scale of giving can be imitated by a poor one: EULER was so wealthy in theorems that loss of a dozen more or less would not be noticed.

It was a different matter with religious issues. EULER maintained throughout his life the simple Protestant faith his father had preached. It had no pretensions in science, and science for EULER had no just pretensions in morality and religion. Thus for EULER the atheism or deism or agnosticism of the French *philosophes* was devilish. King FREDERICK, on the other hand, while regarding organized religion as desirable for the ignorant, upheld the supremacy of the human intellect so long as it impinged only upon GOD's rights, not those of earthly kings. A Swiss Protestant was ready to bow to his king, but not to the DEVIL. EULER published anonymously a booklet called *The Rescue of Divine Revelation from the Objections of the Freethinkers.*

In addition, EULER was a philosopher in his own right. Whereas the *philosophes* ridiculed him as naive, KANT later was to derive his own metaphysics from his study of EULER's writings, but he was not able enough in mathematics to understand EULER's major metaphysical paper, *Reflections on Space and Time.* The ridiculously narrow doctrine of the physical universe we are accustomed to associate with KANT and his successors in German philosophy was evolved after EULER's death, and EULER's point of view did not come into its own until the rise of non-Euclidean geometries and relativity, one and two centuries later[14].

---

[14] EULER did not anticipate these much later specific theories, but they are in no way contradictory or repugnant to the general conceptions of space and time he formulated.

MAUPERTUIS, President of the Berlin Academy, was not precisely a *philosophe*. EULER was loyal to him, and he stood between EULER and the dislike, even contempt, of the king. MAUPERTUIS had sputtered an overriding law of nature, the Law of Least Action, according to which all natural operations rendered something the smallest it could possibly be. MAUPERTUIS' attempt to phrase this law in its application to mechanics was wrong, and ridiculously so. A year earlier EULER had found a correct statement for the motion of a single particle, greatly more special than MAUPERTUIS' pronouncement, but, as far as it went, right. When he heard of MAUPERTUIS' principle, far from claiming any credit, EULER published his own result as being a confirmation of MAUPERTUIS's grand idea, which he praised beyond measure.

Not so the rest of the world. A distinguished nonresident member of the Academy named KOENIG, a good mathematician and a friend and former protégé of MAUPERTUIS, had some objections, which he confided to MAUPERTUIS in a private conversation. A break followed, for MAUPERTUIS tolerated no criticism. The next year KOENIG published his objections, along with counterexamples, and he mentioned that in any case the idea had been sketched in a letter of LEIBNIZ, long dead, an extract from which he included. A dreadful rumpus ensued in Berlin. KOENIG could not produce the letter, which he said he had seen in the possession of his unfortunate friend HENZI, whom the fathers of the Canton of Bern had beheaded because he had accepted their invitation to make some suggestions regarding the government. EULER came to the defense of Least Action and MAUPERTUIS. Having handed over to MAUPERTUIS as a gift his own discovery of the one case in which the principle could then be proved right, he was sure MAUPERTUIS could not have stolen it from LEIBNIZ, and he had shown that something could be done with the principle if properly corrected. Unfortunately he chose to launch a counterattack against KOENIG, claiming that the letter was forged[15].

Meanwhile VOLTAIRE, who after the death of his mistress the Marquise DU CHÂTELET had no agreeable lodging, came to visit King FREDERICK at Potsdam. Formerly VOLTAIRE had been a great admirer of MAUPERTUIS and had written:

---

[15] In EULER's entire life this episode is the only one that has given rise to any *suspicion* of wrongdoing. With the gleeful desire now in fashion to show that *everyone* is as evil as everyone else—or conversely, that nobody is better than anybody—so that no moral or intellectual values can have any but transitory and subjective, and hence meaningless, meaning, every biographical notice on EULER, no matter how meagre or slipshod, manages to mention his unfairness to KOENIG.

Héros de la physique, Argonautes nouveaux
Qui franchissez les monts, qui traversez les eaux,
Dont le travail immense et l'exact mesure
De la terre étonnée ont fixé la figure.

Heroes of physics, new Argonauts,
Who cross the mountains and the seas,
Whose immense labor and exact measurement
Have fixed the figure of the astonished earth.

After having sat for a while as the rival of MAUPERTUIS at the king's table, VOLTAIRE changed his mind and republished the quatrain with "hero" replaced by "courier" and with the couplet about immense work and exact measurement replaced by:

Ramenez des climats, soumis aux trois couronnes
Vos perches, vos secteurs, et surtout deux Lapones!

You bring back from climes subject to the three crowns
Your poles, your sectors, and above all two Lapp girls.

Indeed MAUPERTUIS had a strange household, which his Lapp mistress had to share with tropical birds, exotic dogs, and a black man, but this was only the beginning. Just at that time MAUPERTUIS published a medley called *Letter on the Progress of the Sciences*, in which he proposed numerous things worthy of the Academy of Lagado: investigations of the Patagonian giants, methods of prolonging life, a college composed of perfectly educated representatives of all nations, vivisection of criminals, a town where only Latin would be spoken, boring a study hole into the earth, use of drugs to allow experiments on the brain, and other metaphysical matters. VOLTAIRE was thus well prepared to regard the treatment of KOENIG by MAUPERTUIS as unjust, and MAUPERTUIS' eccentricities and pretensions furnished an immediate subject for a satire: *Dr. Akakia, Physician of the Pope*. The doctor's mission was to cure MAUPERTUIS of his dreadful case of insufferable arrogance.

The king, while presumably amused by the wit displayed, was insulted by the attack on his own President. It must be remembered that the king himself regularly participated in the doings of his Academy by composing essays on moral philosophy for its memoirs. He forbade VOLTAIRE's satire to be printed. VOLTAIRE printed it anyway, using a permit issued for another work. The king, doubly insulted, had the edition burnt by the hangman. The satire was reprinted in Holland, and Berlin was flooded with copies. VOLTAIRE, in increasing disgrace, left town as quickly as he could gain permission to do so. On his slow progress to Switzerland he was in fact arrested and detained for a while by the king's officers. MAUPERTUIS, already sick to death with tuberculosis, also left Berlin to take refuge in the home of one of the BERNOULLIS in Basel, where in a few years he died. VOLTAIRE published a sequel, in which Akakia induced MAUPERTUIS and KOENIG to sign a treaty of peace. Article 19 concerns EULER:

> ... our lieutenant general L. Euler hereby through us openly declares
>
> I. That he has never learnt philosophy and honestly repents that by us he has been misled into the opinion that one could understand it without learning it, and that in future he will rest content with the fame of being the mathematician who in a given time has filled more sheets of paper with calculations than any other ...."

Unfortunately the further sections of this article of the treaty, while equally witty, repeat some of the specific objections of the Englishman ROBINS about mathematical points, objections which reflect only the inability of ROBINS to understand the advanced mathematics of his day. In a typical effusion of literary philosophy, VOLTAIRE did no more than blindly copy passages of bad science.

After MAUPERTUIS' departure all the duties of the presidency fell on EULER, but the king would not have a German (for as such he regarded EULER) assume the title, be given the powers, or receive the pay of the office. The Academy had to finance itself from the sale of almanacs, and EULER had to direct their production and marketing. The depression caused by the Seven Years' War was severe. Serious disputes with the king ensued. Meanwhile, the Academy grew smaller from attrition, until besides EULER there was only one other man of any capacity, namely, the lately arrived, self-taught Genevan genius LAMBERT, whom FREDERICK regarded as a bear and could only with great difficulty and after long delay be persuaded to accept.

Almost as soon as he had arrived in Berlin, EULER came to realize that in leaving Russia he had made a grave mistake (see Part aIV of this essay, above). He found neither the leisure to work, for he was immediately engulfed in the administration of the academy, nor the stimulation from gifted friends and acquantances he had enjoyed in Petersburg. After having been in Berlin for eight years he wrote

> I and all those who have had the good fortune to spend some time in the Imperial Russian Academy must admit that we owe all we

are to the advantageous circumstances in which we found ourselves there. For my part, had I not had that splendid opportunity, I should have had to devote myself primarily to some other
field of study, in which by all appearances I should have become
only a bungler.

Such vehemence of expression may be due to its having been directed
to SCHUMACHER, on whose good will EULER's pension depended, yet
because all evidence confirms his truthfulness at other times and in
other matters, it is unlikely that what he wrote here differed much
from what he felt.

While throughout his long life FREDERICK again and again
expressed his contempt for the infinitesimal calculus, the elements of
which, it seems, he had tried to learn several times but in vain, he
insisted upon having a mathematician as President of his Academy.
At the same time this mathematician had to be French, a man of the
world, a lion of society. Few indeed have been the mathematicians of
this kind, but FREDERICK found one.

In 1759, when MAUPERTUIS died, there were besides EULER
and LAMBERT only two other major mathematicians in Europe:
DANIEL BERNOULLI and D'ALEMBERT. The former did not fit
any of FREDERICK's qualifications. The latter, a Frenchman ten
years younger than EULER, was at the height of his fame; he
was FREDERICK's ideal, being a man of wit, a *philosophe*, a major collaborator on DIDEROT's *Encyclopédie*, and a light of literature. Even
seven years earlier the king had offered him a salary of 12,000 francs,
which was seven times what he was receiving in Paris, and also free
lodging in the royal château and meals at the royal table, but D'ALEM-
BERT had preferred freedom in poverty to the dangerous vicinity of a
king. Moreover, D'ALEMBERT had quarreled with the Berlin Academy
over one of its prizes, and for a time he seemed to be a rival of EULER
in mechanics and in some parts of analysis. The major scientific dispute of the mid-century, which concerned the tones and motions of
the monochord, was at its hottest; the disputants were D'ALEMBERT,
EULER, and DANIEL BERNOULLI, three powerful parties each consisting in just one man, since there was no-one else who could understand the mathematics enough to form a founded opinion, let alone
take part. Here[16], as in several other circumstances of science, the

---

[16] While it had antecedents going back for over a century, the dispute began with a
paper by D'ALEMBERT published in 1749 and continued through D'ALEMBERT's
remaining life. HANKINS on page 48 of his biography of D'ALEMBERT, reviewed
above in Essay 27, states that D'ALEMBERT conceded defeat in a final volume of his
Opuscules, which exists in manuscript but was never published. On the whole, the
controversy was not resolved during the lifetimes of any of the main disputants but

eighteenth century is unique: never before had mathematics been so highly regarded by the community of learning, but never before or after were there so few persons able to enter the arena of mathematical research.

D'ALEMBERT came to visit FREDERICK at Potsdam in 1763. The Academicians, most of whom were Swiss, feared the worst. D'ALEMBERT spoke graciously to them and recommended them to the king. In particular, he declined the presidency and recommended EULER for it; the king positively refused, and indeed all along he had spoken contemptuously of EULER, written to him with harsh disrespect, and declined to grant him the least of the requests he had submitted from time to time on behalf of his family and friends. After D'ALEMBERT had returned to Paris, FREDERICK wrote for his advice on all matters concerning the Academy of Berlin, to the extent that when the Academicians wished to suggest something to the king, they found it best to convey the message first to D'ALEMBERT in Paris, who thereupon, if he agreed, offered it to the king as his own idea.

EULER then found the position intolerable. For a long time he had been negotiating intermittently regarding return to the Petersburg Academy. With the accession of a German princess as CATHERINE II of Russia in 1762, the auspices for the arts and sciences there improved greatly, and EULER succeeded in obtaining an excellent appointment. He tendered his resignation to King FREDERICK, who brusquely told him to stop petitioning. EULER desisted from taking part in any activity of the Academy. D'ALEMBERT, meanwhile, had found a replacement for him, the young LAGRANGE, a Piedmontese who had begun in 1760, at the age of twenty-four, to pour forth brilliant research on analysis and mechanics at EULER's own level and

---

rather just died out. EULER solved all the central problems concerning a homogenous string correctly and in generality. DANIEL BERNOULLI's point of view has been used more often subsequently and is susceptible of greater generalization, but he himself was unable to do much on the basis of it, since the mathematical theory essential for exploiting it was not developed until the middle of the next century. LAGRANGE also took part from 1760 onward, but his work is largely incomplete or incorrect. While it made a great stir in its day and drew high praise from both EULER and D'ALEMBERT, it stands up but ill under critical scrutiny. For a review of the whole matter, see pages 237–300 of my *Rational Mechanics of Flexible or Elastic Bodies, 1638–1788*, LEONHARDI EULERI *Opera omnia* (II) 11_2, 1960. Although various historians of science have protested that my estimates of LAGRANGE's work in mechanics and analysis (for I have never formed any judgment whatever concerning his work in algebra and number theory) are too harsh, those estimates are induced from detailed examination of the sources, page by page and line by line, and so I will not revise them until such time as I be shown specific errors in my evaluations of specific passages. Anyone who has read older essays on the history of mathematics will be accustomed to sweeping generalities based on a glancing acquaintance with a few of the more elementary parts of works cited, but I see no reason to respect utterances of this kind today.

speed. EULER had tried to induce him to come to Berlin, but LAGRANGE, seeing that he had to choose between EULER and D'ALEMBERT, took D'ALEMBERT as his foster father in the politics of science, though in research he always followed tacitly in EULER's footsteps. The choice reflected LAGRANGE's sagacity. D'ALEMBERT, though not old, had ceased to produce anything worthwhile and had become merely a conniver; he had quarreled with all mathematicians of his own age or older, and he was detested by his fellow academicians in Paris; vain, he badly needed an admirer at the highest echelon of mathematics. EULER was at the summit and plateau of his creative powers, was on excellent terms with everyone except D'ALEMBERT, KOENIG, and King FREDERICK, and needed nothing but money and rank. D'ALEMBERT arranged that LAGRANGE go to Berlin as EULER's successor[17]. In order to do so, D'ALEMBERT had to tell FREDERICK a white lie, namely, that LAGRANGE was a *philosophe* and man of the world. In fact he was neither; he had no interests outside mathematics and a narrow outlook within it, but in society he knew how to keep his mouth shut when not expressing deference to the views of his seniors. In addition, he could pass more or less for a Frenchman, and he later became one[18], but he never lost his heavy Piedmontese accent.

In all of EULER's vast correspondence there is no mention of politics and little reference to social conditions. Evidently one country, government, or party was the same as another for him, provided it allowed free worship in the Protestant faith his father had taught him and the chance to do a mountain of mathematics for a good salary. Like many other men of the Enlightenment, EULER expressed a general interest in human wellbeing and in good works such as widows' pensions, charity for orphans and cripples, and common measures for prevention of disease and promotion of trades and manufactures, but his own contribution to these estimable objectives seems to have been confined, beyond a few special mathematical

---

[17] The relations between EULER and D'ALEMBERT in 1763–1766 are too complicated to trace here. Like most other savants of the period, EULER despised D'ALEMBERT's character, and he did not wish to remain in the Academy if D'ALEMBERT were to become its president. By the time D'ALEMBERT came to decline the presidency, EULER wished only to leave Berlin and feared that D'ALEMBERT's recommendation of him might result in his being retained against his will; and by the time it came to persuading FREDERICK to accept LAGRANGE as EULER's successor, D'ALEMBERT's actions were in EULER's best interest, because without a replacement EULER would not have succeeded in getting permission to go. *Cf.* Parts aII and aIII of this essay, above in this volume.

[18] LAGRANGE's mother tongue was the Piedmontese dialect; his first publication was in Latin. The errors of language in his earliest papers in French have been silently corrected in the reprints in his *Œuvres Complètes*, the editors of which, unfortunately, for the most part have not taken similar pains with the numerous errors in mathematics.

studies, to an exemplary personal life and a miraculously creative and ageless exercise in mathematical science. Again and again he stated that truth of all kinds, knowledge in general, and mathematics in particular led to the betterment of man's condition, and he never showed evidence of seeing any conflict between service to his prince and service to humanity. While obviously neither a Prussian nationalist nor a Russian one, EULER served both countries with the total loyalty which in those days was regarded as the ordinary, moral duty of a servant to his master. The personal failings of FREDERICK II as a candidate for GOD's lieutenant on earth must have been more than obvious to EULER, but it was not those that drove him from Berlin. Rather, he sought a social and financial position worthy of himself and, above all, advancement for his children.

Finally FREDERICK granted EULER leave to depart with most of his family and some of his servants, eighteen persons all told. EULER, then in his sixtieth year, was entertained en route by the King of Poland and the eminent nobility, and upon arrival in Russia was received by the empress. In addition to his salary of 3000 rubles he was given 8800 rubles to buy a good house and 2000 rubles for furniture. He was not burdened with duties; his counsels were requested regularly and often followed. His greatest reward was that good places in the Academy or the imperial service were found for his sons, and marriages into the nobility were arranged for his daughters.

In his last years in Petersburg EULER had more time free for mathematics than ever before. He soon lost the sight of his one remaining eye. Like BACH, he underwent the torment of an operation for cataract, which was unsuccessful and rendered him almost totally blind. If anything, this enforced end to most of the ordinary duties of life left him still freer to work. About half of his 800 publications were written in these, the last seventeen years of his life. In 1766, the year he moved, EULER composed the first general treatise on hydrodynamics; it was to be about 100 years before anyone wrote another. The next year EULER wrote his famous *Complete Introduction to Algebra*. After EUCLID's *Elements*, this is the most widely read of all books on mathematics, having been printed at least thirty times in three editions and in six languages; selections were being used as textbooks in the Boston schools in the 1830s. The next year, 1768, EULER wrote his treatise on geometrical optics in three volumes and his tract on the motion of the moon; both of these are filled with colossal calculations, and the latter contains a single table 144 pages long, calculated under EULER's direction "by the tireless labor" of his son, KRAFFT, and LEXELL, all of them academicians. In 1770 he wrote a monograph on the difficult orbit of a comet which had appeared the year before.

EULER's total blindness put an end to composition of such long treatises, and the great increase in the annual number of his publications reflects the change in his method of work. In the middle of his study he had a large table with a slate top. Being barely able to distinguish white from black, he could write a few large equations. Every morning a young Swiss assistant read him the post, the newspaper, and some mathematical literature. EULER then explained some problem he had been sleeping on and proposed a method of attacking it. The assistant was usually able to produce the outline for a draught of a short memoir, or part of one, by the next morning. In 1775, for example, EULER composed more than one complete paper per week; these run from ten to fifty pages in length and concern widely different special problems.

Two years before his death EULER presented to the Petersburg academy a pair of papers suggested by VERGIL's line

anchora de prora jacitur, stant littore puppes.

The problem is to find the motion of a ship whose prow is anchored. The title of the first paper tells us that the problem is "commonplace enough, but very difficult to solve"; EULER derives the differential equation of motion for a much simplified model and obtains some integrals of the motion but despairs of proceeding further; in the second paper he presents and analyses the general solution. The *Acta* for that year include five further papers by EULER, but his output was become too great for the ordinary channels, and in the year of his death the Academy issued in addition to nineteen memoirs in the *Acta* an extra volume called *Opuscula analytica*, which consists in thirteen of his papers composed and presented to the Academy nine to twelve years earlier.

EULER's memory, always extraordinary, had by then become prodigious. He could still recite the *Æneid* in Latin from beginning to end, remembering also which lines were first and last on each page of the edition from which he had learnt it some sixty or seventy years earlier. Enormous equations and vast tables of numbers were ready on demand for the eye of his mind. He became one of the sights of the town for distinguished visitors, with whom he usually spoke on nonmathematical topics. Amazed by the breadth and immediacy of his knowledge concerning every subject of discourse, they spread fairy tales about what he could do in his last years.

Only recently have we been able, by study of the manuscripts he left behind, to determine the course of EULER's thought. We now know, for example, that many of the manuscript memoirs published in the two volumes of posthumous works in 1862 he wrote while still a student in Basel and himself withheld from publication for a reason—

which usually was some hidden error or an unacceptable or unconvincing result. The first page of one of these memoirs is reproduced here as Figure 22. The memoir it opens is the one that served to introduce EULER to DANIEL BERNOULLI and was important in securing him his first post in Petersburg. There can be only one reason EULER did not publish it: DANIEL BERNOULLI had obtained the same result at about the same time by somewhat different means, and EULER did not wish to detract from his friend's glory. The result itself, the solution of the problem of efflux of water from a vessel, became known through DANIEL BERNOULLI's book, published twelve years later.

The manuscript is a typical one. The spots are ink from the other side showing through. There are few corrections in the smooth, easy writing. The manuscripts of the books EULER wrote in later life are much the same, but for some remain one or even two complete earlier manuscripts of the whole, showing many differences from the final one. When EULER wished to revise a work, he wrote it all out afresh, neat and clean. Like MOZART, he revised in his head and did not begin to use paper until the revision was complete.

The most interesting of all EULER's remains is his first notebook, written when he was eighteen or nineteen and still a student of JOHN BERNOULLI. It could nearly be described as being all his 800 books and papers in little. Much of what he did in his long life is an outgrowth of the projects he outlined in these years of adolescence. Later, he customarily worked in some four domains of mathematics and physics at once, but he kept changing these from year to year. Typically he would develop something as far as he could, write eight or ten memoirs on various aspects of it, publish most of them, and drop the subject. Coming back to it ten or fifteen years later, he would repeat the pattern but from a deeper point of view, incorporating everything he had done before but presenting it more simply and in a broader conceptual framework. Another ten or fifteen years would see the pattern repeated again. To learn the subject, we need consult only his last works upon it, but to learn his course of thought, we must study the earliest ones, especially those he did not himself publish.

In an age when genius, intellectual ambition, and drive were common, no man surpassed EULER in any one, and none came near him in combination of all three. Nevertheless, histories of the eighteenth century and social or intellectual histories in general rarely mention him. The explanation was written by FONTENELLE, before EULER was born:

> We like to regard as useless what we do not know; it is a kind of revenge; and since mathematics and physics are rather generally

De Effluxu Aquae ex Tubis Cylindricis utcunque inclinatis ex inflexis

§ I.

Quae hucusque contemplati sumus vasa, e quibus aqua effluat, ejusmodi omnia fuere, ut circa axem verticalem quaquaversus aequaliter fuerit diffusa. Jam quâ lege aqua erumpat ex tubis vel inclinatis ad horizontem vel inflexis aut incurvatis. Quod quidem attinet ad inclinatos tubos, facile posse colligi videtur ex motu de descensu super plano inclinato principiis, aquam inde effluxuram, eâ prorsus velocitate quâ ex tubo verticali ejusdem altitudinis, cujusque foramen ad basin eandem habet. Ne autem temere aliquid admittere videar, tubos inclinatos et ea Theoriae subjiciam.

§ II.

Sit itaque tubus inclinatus cylindricus $ABCD$ sit ejus sectio recta AB. Sectio autem horizontalis fundum DC, quod diametro EF quoque elliptico. Dum est pertusa. Sit vena effluens $Eefe$, cujus recta recta, eg, sit verticalis BG, et horizontalis CG, ut habeatur angulus inclinationis BCG, et ratio sinus totius BCG sinum anguli inclinationis BG, quem dic α ad β. Sit ratio sectionis cylindri rectae AB ad sectionem venae rectam, eg seu quod eodem

unknown, they rather generally pass for useless. The source of their misfortune is plain; they are prickly, wild, and hard to reach . . . .

Such is the destiny of sciences handled by few. The usefulness of their progress is imperceptible to most people, especially if they are practised by professions not particularly illustrious.

ACKNOWLEDGMENT

I am grateful to Dr. MARTA REZLER for correction of some details regarding VOLTAIRE.

ANNOTATED BIBLIOGRAPHY

Biography:

Article X, pages 32–60 of *Adumbratio eruditorum Basiliensium meritis apud exteros olim hodieque celebrium*, published as "Adpendix" to *Athenae Rauricae, sive catalogus professorum academiae Basiliensis ab anno 1770 ad annum 1778, cum brevi singulorum biographia*, Basileae, 1778. I know this work only through the article by F. MÜLLER, "Über eine Biographie L. Eulers vom Jahre 1780 und Zusätze zur Euler-Literatur", *Bericht der Deutschen Mathematiker-Vereinigung* **17** (1908): 36–39.

NICOLAUS FUSS, *Lobrede auf Herrn Leonhard Euler . . . 23 Octob. 1783 vorgelesen . . .*, Basel, 1786 = pages XLIII–XCV of LEONHARDI EULERI *Opera omnia* (I)**1**, Leipzig & Berlin, Teubner, 1911.

M.-J.-A.-N. CARITAT, Marquis de CONDORCET, "Eloge de M. Euler", *Histoire de l'Académie Royale des Sciences* (Paris) 1783: 37–68 (1786) = pages 287–310 of LEONHARDI EULERI *Opera omnia* (III)**12**, Zürich, Orell Füssli, 1960.

O. SPIESS, *Leonhard Euler, Ein Beitrag zur Geistesgeschichte des XVIII. Jahrhunderts*, Frauenfeld/Leipzig, 1929.

Note: FUSS did not meet EULER until 1773, EULER's sixty-seventh year; CONDORCET never met him at all. Neither was competent in more than a small part of the range of science enriched by EULER; both were younger than he by more than thirty years, and neither showed evidence of having studied EULER's early work in detail. Their necrologies of EULER are heavily weighted by hearsay and treat his youth as already legendary. The accounts of EULER's life and work in the general histories of mathematics or collected biographies of mathematicians are mainly if not entirely their authors' personal embroideries upon odds and ends pecked out of the two necrologies. The biography by SPIESS, in welcome contrast, is based upon extensive study of unpublished letters and documents as well as all published sources concerning EULER's life. Nevertheless, it is a biography in the literary sense; while SPIESS made some attempts to write what is now called intellectual history, his understanding of the contents of EULER's researches was limited not only to what in SPIESS's day was called pure mathematics but even to elementary matters such as quadratures, properties of particular curves, explicit sums of series, *etc.* Thus, inevitably, EULER appears in SPIESS's pages as the most dazzling of mathematical jugglers but not as the great creator of concepts and organizer of doctrines he really was. In general, the critical reader who would understand EULER's conceptual frame and intellectual achievement can find today no intermediary between himself and EULER's own writings except the prefaces to some volumes of the *Opera omnia*, for which see below, "EULER's place in the history of science".

A. P. YOUSCHKEVITCH, article "Euler", *Dictionary of Scientific Biography,* Volume **4**, 1971.

Portraits:

H. THIERSCH, "Zur Ikonographie Leonhard und Johann Albrecht Euler's", *Gesellschaft der Wissenschaften zu Göttingen, Nachrichten der Philosophisch-Historischen Classe* 1928: 264–289 + 4 plates.

H. THIERSCH, "Leonhard Euler's 'verschollenes' Bildnis und sein Maler", *ibid.* 1930: 193–217 + Nachtrag + 2 plates.

H. THIERSCH, "Weitere Beiträge zur Ikonographie Leonhard und Johann Albrecht Euler's", *ibid.* 219–249 + 3 plates.

Lists of publications, of manuscripts, and of letters:

G. ENESTRÖM, "Verzeichnis der Schriften Leonhard Eulers", *Jahresbericht der Deutschen Mathematiker-Vereinigung* **4**. *Ergänzungsband* (2 Lieferungen), 388 pages (1910) and **22**: 191–205 (1910).

Manuscripta Euleriana Archivi Academiae Scientarum URSS, **1**, Moscow & Leningrad, 1962. (This volume describes the scientific manuscripts preserved in Russia. It is reviewed above in Part bI of this essay. According to ENESTRÖM, the manuscripts left in the Archives of the Academy in Berlin were once described by JACOBI. I have not seen his description and do not know if it was ever published or if the manuscripts still exist.)

LEONHARDI EULERI *commercium epistolicum. Descriptio commercii epistolici.* LEONHARDI EULERI *Opera omnia* (IVA)**1**, ediderunt A. P. JUŠKEVIČ, V. I. SMIRNOV, & W. HABICHT, Basel, Birkhäuser, 1975.

Works:

Memoirs, books, and manuscripts, mainly those published at least once before 1911:

LEONHARDI EULERI *Opera omnia*, at first Leipzig, then Zürich or other cities of Switzerland, 1911–:
Series I. *Opera mathematica* (complete, 29 volumes issued in 30 parts).
Series II. *Opera mechanica et astronomica* (27 of 32 part-volumes published by the end of 1982).
Series III. *Opera physica et miscellanea* (11 of 12 volumes published by the end of 1982).

Manuscripts not published before 1911:

Manuscripta Euleriana Archivi Academiae Scientiarum URSS, Volume **2**, Moscow & Leningrad, 1965. This volume is reviewed above in Part bII of this essay.

Letters:

LEONHARDI EULERI *Opera omnia* (IVA)**1**, the catalogue of the letters, gives references to the some thirty publications in which one or more letters appear. Other volumes in this series are to publish the letters in full. Volume **5** was published in 1980. It includes errata and addenda for Volume **1**.

Euler's place in the history of science:

Although it would be hard to find any history of mathematics or physics that does not say something about one or more aspects of EULER's work, and although his name is used as a label for a dozen or more of the commonest and most useful theorems in the mathematical sciences, the bulk and level of his works seem to have discouraged critical study of them. Even volumes of essays devoted to celebrations of EULERian anniversaries often contain no more than musings by senior scientists who have glanced at a few pages before composing variants of the generalities imparted to them by their teachers in elementary courses half a century earlier. In regard to eighteenth-century mathematics and physics the general histories of science or mathematics or physics are grossly unreliable because they are based largely on tale-bearing or caprice or both. Some of the prefaces to individual volumes of LEONHARDI EULERI *Opera omnia* explain succinctly some part of EULER's work, especially those in Volumes (I)**4** and **5** (by FUETER), (I)**9** (by A. SPEISER), (I)**24** (by CARATHÉODORY), (I)**25** through **29** (by A. SPEISER), (II)**3** (by BLANC), (II)**5** (by FLECKENSTEIN), (II)**6**, **7**, and **9** (by BLANC), (II)**11**$_2$ through **13** (by TRUESDELL), (II)**14** (by SCHERRER), (II)**15** (by ACKERET), (II)**16** and **17** (by BLANC & DE HALLER), (II)**20** and **21** (by HABICHT), (II)**22** (by COURVOISIER), (II)**23** (by FLECKENSTEIN), (II)**25** (by SCHÜRER), (II)**28** (by A. SPEISER), (II)**29** and **30** (by COURVOISIER), (III)**5** (by D. SPEISER), (III)**6** (by A. SPEISER), (III)**7** (by HABICHT), (III)**8** (by HERZBERGER), (III)**9** (by HABICHT), (III)**10** (by D. SPEISER), (III)**11** and **12** (by A. SPEISER). A few of these also place EULER's work in the setting of its antecedents and its time. For mechanics there is also my book, *Essays in the History of Mechanics*, New York, Springer-Verlag, 1968, and SZABÒ's, described above in Essay 31; both treat EULER merely incidentally.

The only other occasional yet solid analyses of EULER's work I have found in languages other than Russian are included in Chapter VII of C. R. BOYER's *History of Analytic Geometry*, New York, Scripta Mathematica, 1956, and in six articles in the *Archive for History of Exact Sciences:*

J. E. HOFMANN, "Über zahlentheoretische Methoden Fermats und Eulers, ihre Zusammenhänge und ihre Bedeutung", **1**(1960/1962): 122–159 (1961).

O. B. SHEYNIN, "On the mathematical treatment of observations by L. Euler", **9** (1972): 45–56.

H. J. M. BOS, "Differentials, higher-order differentials and the derivative in the Leibnizian calculus", **14** (1974/1975): 1–90 (1974).

R. CALINGER, "Euler's 'Letters to a Princess of Germany' as an expression of his mature scientific outlook", **15** (1975/1976): 211–233 (1976).

A. P. YOUSCHKEVITSCH, "The concept of function up to the middle of the 19th century", **16** (1976/1977): 37–85 (1976).

C. A. WILSON, "Perturbations and solar tables from Lacaille to Delambre: the rapprochement of observation and theory", **22** (1980); 53–304.

Note also the chapter in EDWARDS' book cited above in Footnote 8.

A distinguished mathematician of our day, GEORG PÓLYA, has composed a treatise on methods of discovery in mathematics which refers to EULER so often, even including analyses and schemas of some of his papers, that EULER might be said to be the hero of the work. This treatise is *Mathematics and Plausible Reasoning*, 2 volumes, Princeton, Princeton University Press, 1954. PÓLYA's estimate of EULER, on page 90 of Volume **1**, is as follows:

> Yet Euler seems to me almost unique in one respect: he takes pains to present the relevant inductive evidence carefully, in detail, in good order. He presents it convincingly but honestly, as a genuine scientist should do. His presentation is "the candid exposition of the ideas that led him to those discoveries" and has a distinctive charm. Naturally enough, as any other author, he tries to impress his readers, but, as a really good author, he tries to impress his readers only by such things as have genuinely impressed himself.

We await with great eagerness the first volume of ANDRÉ WEIL's history of number theory, which will concern EULER's work primarily.

Note for the Reprinting

This essay is reprinted, after restoration of a few passages emended by the editor, from pages 51–95 of *Studies in Eighteenth Century Culture*, Volume **2**: *Irrationalism in the Eighteenth Century*, Cleveland and London, Press of the Case Western Reserve University, 1972. After Dr. FELLMANN had kindly provided me with a copy of the manuscript of EULER's brief autobiography, I conformed my account of several details of his early life with what he himself had recounted. Also I corrected or made more precise some statements here and there, and I brought the bibliography up to date.

34. THE ESTABLISHMENT STIFLES GENIUS: HERAPATH AND WATERSTON (1968, 1982)

1. ANTECEDENTS

That matter is composed of molecules in violent agitation, and that heat is a manifestation of that motion, are ideas of great age. The first definite measure of the heat of intestine motion was proposed by HERMANN in 1716. He identified heat with the kinetic energy of the molecules, supposed to be in translatory motion. EULER in 1729 proposed a molecular model in which variously composed globular molecules, closely pressed together, whirled at a uniform linear speed upon their surfaces. From this assumption, which he revived and simplified fifty years later, he derived a simple "equation of state" that seemed to fit fairly well the facts then known. DANIEL BERNOULLI in his *Hydrodynamica* (1738) proposed a model of particles of random size undergoing purely translational motion. He, too, derived an equation of state that agreed with the TOWNELEY-POWER-[BOYLE] Law for gases that are not too dense, and he remarked that for a dense gas deviations would be expected.

For analysis of the mathematics and physics of this and other early work, the reader may turn to my "Early kinetic theories of gases", *Archive for History of Exact Sciences* **15** (1975): 1–66. Here we are concerned with the men of the early nineteenth century who tried to revive the kinetic model of the gaseous state.

2. HERAPATH (1816–1821)

The kinetic theory of gases remained dormant, indeed forgotten, until 1816, when HERAPATH[1] proposed a theory which is essentially BERNOULLI's. Unfortunately he chose to define temperature as being proportional to the momentum rather than the kinetic energy of the molecules. HERAPATH was the first to show, more or less, that the kinetic theory can provide crude explanations for changes of state, diffusion, and the propagation of sound. While his ideas were not sufficiently developed or happy, the hostility they aroused seems to have arisen not from any defect in them but rather from the difference between the scene in which he found himself and that of a century earlier. By the 1800s, in England at least, original speculation of any kind and mathematical theory in particular were distrusted. Mechanics had been established and exhausted by NEWTON. Heat was explained in gross by the caloric theory, and on the molecular scale by the vibratory motion of atoms within an ætherial substance, not a fit subject for mathematics.

HERAPATH was self-educated, having learnt mathematics and natural philosophy from the writings of NEWTON and the great French mathematicians of his own day. He submitted his main paper to the Royal Society in 1820 after having made the necessary tactful inquiries of influential members. DAVIES GILBERT, the Fellow to whom he submitted it, replied within a fortnight: "[while] I was much pleased with the great ingenuity displayed throughout the whole, . . . I entertained strong doubts on the propriety of laying before the Royal Society, anything so abstruse and metaphysical. I, therefore, desired two of the best mathematicians in London to look at the premises; and their opinion confirmed my doubts. They say, that such a work should be laid before the public in a separate form."

HERAPATH offered to explain the mathematics (in fact merely elementary, even primitive) and so as to confirm his predictions went on to perform some experiments on mixing liquids at various temperatures. GILBERT and his august colleagues stood fast; "some members of the Council, who are usually looked up to on such occasions, . . . considered the investigations too theoretical for the

---

[1] J. HERAPATH, "On the physical properties of gases", *Annals of Philosophy* (1) **8** (1816): 56–60; "A mathematical investigation into the causes, laws, and principal phænomena of heat, gases, gravitation, etc.", *Annals of Philosophy* (2) **1** (1821): 273–293, 340–351, 401–416; "On Mr. Tredgold's 'refutation' of Mr. Herapath's theory", *ibid.* (2) **1** (1821): 303–307; *Mathematical Physics . . .*, 2 volumes, London, Whittaker, 1847. The second and fourth of these works are reprinted in HERAPATH's *Mathematical Physics . . . and Selected Papers*, edited by S. G. BRUSH, New York & London, Johnson Reprint Co., 1972. The preface by BRUSH contains an account of HERAPATH's life and works.

Figure 23. JOHN HERAPATH (1790–1869), by courtesy of SPENCER D. HERAPATH, Esq., after a portrait of about 1815, destroyed in 1951.

Transactions, without taking on themselves to judge of the mathematics." GILBERT proposed that the revised paper, including the experiments, be put in the hands of the new president and council, to be elected a month later.

HERAPATH expected to find a backer in the new president, Sir HUMPHRY DAVY, who was an avowed atomist. To support himself, HERAPATH was just then setting up "a school for mathematics, and for preparing young men for the navy." He wished "to strain every nerve to establish my scientific reputation, as well for the sake of maintaining myself in the good opinion of my friends, as for the benefit which it may be of to my family, in my business." He was neither the first nor the last to learn how fragile is the hope of help from nabobs of science.

DAVY wrote to HERAPATH that he had

> read those parts of it which are intelligible, without profound mathematical study, with attention; and highly ingenious as I find your views, I must say I am not impressed with a conviction of their truth . . . . Before I return the paper, I shall take the liberty of submitting it to the most acute philosopher (I know) in this country. . . .
>
> There is so much ingenuity and so minute an acquaintance with the progress of discovery displayed in your paper, that I cannot help wishing that its views and objects had been limited to matters of pure experimental inquiry. For instance, the doctrine of heat, and the investigation of its laws, supposing it to be motion. Such a preliminary paper, if satisfactory, would prepare the philosophical world for greater and more abstruse researches. You refer, in your last letter, to some experiments which you have concluded, on heat, and which confirm your general views. Any investigation of this kind, I shall have great pleasure in communicating to the Royal Society. Should anything call you to London, I hope you will favour me with a call; I shall be glad to see you at my house any Wednesday evening, after nine o'clock. You will always find some of our most distinguished men of science with me on that day. I am, Sir, with much consideration,
>
> Your obedient humble servant,
> H. Davy

HERAPATH paid the visit at the very next opportunity, four days later. He was introduced to THOMAS YOUNG, "who, on my asking him, said, he had not seen or heard of the paper, yet, in a few minutes after, discussing the mode of making some experiments, he observed, 'but you, in your paper,' (referring to this very one) 'say . . . .' "

DAVY wished to reject the theory and publish the experiments. HERAPATH refused but got into an argument about the experiments, resting on his unfortunate definition of temperature. Here HERAPATH was in the wrong, but DAVY, with a talent common among great and important officers, chose to attack him instead on a point where he was right and invincible. DAVY wrote, "... having considered a good deal the subject of the supposed real zero, I have never been satisfied with any conclusions respecting it. I cannot see any necessary connexion between the capacity of bodies for heat, and the absolute quantity they contain; and temperature does not measure a quantity, but merely a property of heat."

HERAPATH's paper, withdrawn from the Royal Society, was published in a minor journal in 1821, but since it had been damned already by the great of British science, it attracted little notice in Britain and none whatever abroad. The little British notice was hostile, and HERAPATH in the course of answering some objections published an extract from the letter in which DAVY had referred to comprehending the paper "without profound mathematical study". DAVY, self-condemned but self-righteous, wrote to HERAPATH not to expect anything kind of him in the future. The Royal Society refused to witness HERAPATH's experiments, and nonofficial journals began to reject his papers about them. HERAPATH considered himself persecuted by "the insidious propagation of reports". Five years later, in 1826, he laid out the whole matter in letters to *The Times* (London), publishing most of the correspondence, connected by his account of the whole affair. No answer. On the 10th of January, 1826, HERAPATH challenged DAVY:

> Problem of Defiance to Sir H. Davy and the Royal Society.
> From two postulates only, of less than fifty words, the one on the nature of heat, the other on that of a gas, to determine from theory alone in a function of the altitude, the law of the diminution of temperature and elasticity of any air surrounding a globe of uniform temperature, whose action varies according to any law whatever.

While this is but the beginning, I will not detail the rest. If you are expecting a glorious Hollywood finale in which HERAPATH is given a golden Base Prize and then elected Fellow of the Royal Society, installed at a special ceremony in which his chastened and converted old detractors rise to applaud him, while he is toasted from the galleries packed with beautiful women in décolletage designed to reveal enough of their best upper parts as to suggest that the lower would abundantly repay exploration, that finale never occurred. Official silence greeted HERAPATH's challenge problem. A second and a third met the same response. DAVY resigned his command, but the war went on. HERAPATH complained that the Fellows of the Royal Society whispered he could not solve his own problems, and he ought to enter the field in pure mathematics as being fairer to all. (Since he had done original work in the theory of gases, in it he plainly had a head start on the Fellows, most of whom had done no original work of any kind, and so it was only decent, in a sporting trial of skill, to choose from the common school background, in which HERAPATH had not shown any special distinction.) HERAPATH was drawn in. He proposed a mathematical problem and backed his stand with a substantial bet. No reply. In 1828 HERAPATH wrote a final letter to *The Times*, mentioning that the time for solving all his problems had expired without a single written reply or comment. His letter concludes, "I hope never to appear before the public again on such business."

Perhaps DAVY's attitude is reflected in his brother's biography of him, published not long afterward. In discussing the onerous burdens of a President of the Royal Society, JOHN DAVY wrote in regard to disappointed authors, "The man of real ability or of true dignity would be above the Royal Society, and would not condescend to resent any act of injustice toward him. . . .. He has the world for his tribunal; and it is only necessary for him to publish the results of his inquiries, and he is sure to have justice done to him . . .." This cozy attitude seems not to have been confirmed in HERAPATH's case. Today, DAVY remains a general hero of science, while few specialists in statistical mechanics even recognize HERAPATH's name. Time has proved that the opinion of a President of the Royal Society, in a case when he was not only wrong but also foolish, outweighs pages of sound, logical calculation by an original but unknown outsider. Neither DAVY nor anyone else then in the Royal Society, it seems, knew enough mathematical physics to appreciate the earlier work of HERMANN, BERNOULLI, and EULER, which was never mentioned in the course of the controversy. The British establishment ignored altogether the marvellous discoveries the French were just then making in mechanics and mathematical physics; since they did not use the notation of fluxions, their work was *a fortiori* contemptible.

3. JOULE (1851), THOMSON (1852/1853), AND KRÖNIG (1856)

HERAPATH's theory joined its predecessors in oblivion (Figure 24). If we are to follow the record in the order of printing the next step

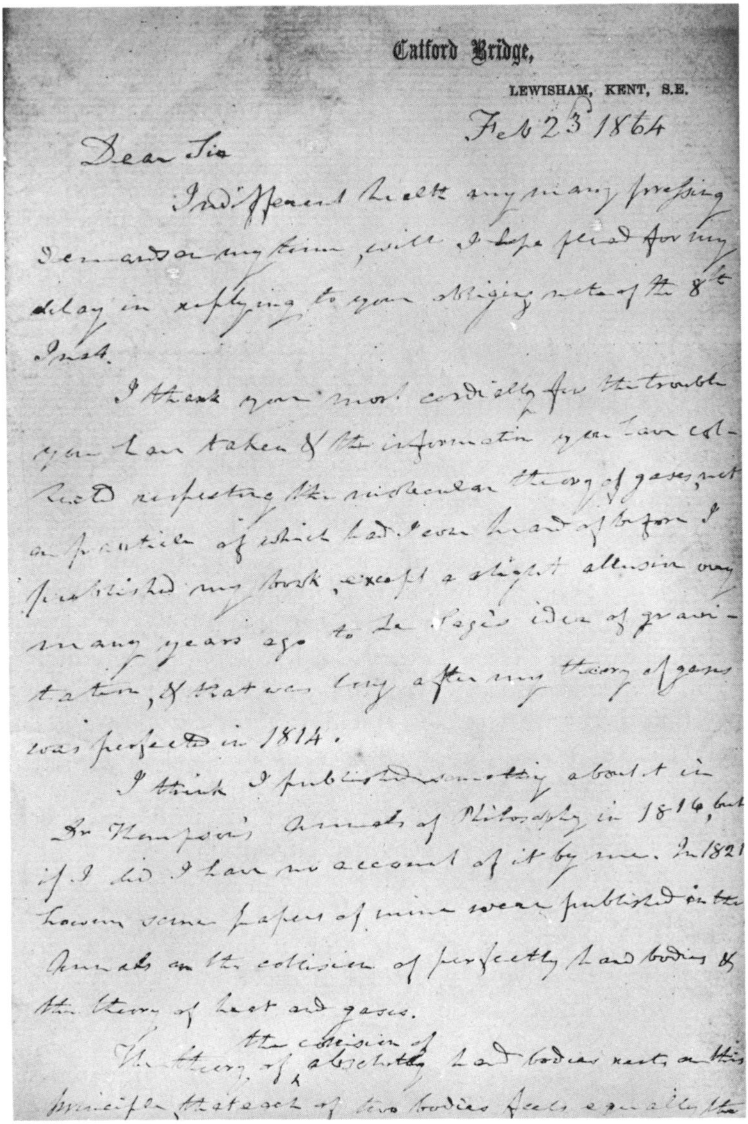

Catford Bridge,
LEWISHAM, KENT, S.E.

Feb 23 1864

Dear Sir

Indifferent health my many pressing demands on my time, will I hope plead for my delay in replying to your obliging note of the 8th Inst.

I thank you most cordially for the trouble you have taken & the information you have collected respecting the molecular theory of gases, not a particle of which had I ever heard of before I published my book, except a slight allusion many years ago to Le Sage's idea of gravitation, & that was long after my theory of gases was perfected in 1814.

I think I published something about it in Dr Thomson's Annals of Philosophy in 1816, but if I did I have no account of it by me. In 1821 however some papers of mine were published in the Annals on the collision of perfectly hard bodies & the theory of heat and gases.

The theory of the collision of absolutely hard bodies rests on this principle that each of two bodies feels equally the

Figure 24. First and last pages of a letter from HERAPATH to an unknown correspondent and dated February 23, 1864.

HERAPATH states that his theory was "perfected in 1814", before he had any idea that others had touched the subject. He concludes, "there are other parts of my labours in relation to gases and their chemical combinations, &c which I think, if I can ever find leisure to publish what I have had by me for many years, will change the ideas entertained respecting elastic particles in gases and some other points of Physics. . . . [M]ine is strictly a mechanical system of Physics, based on the property of absolute hardness in the component atoms or particles." The two pages not shown explain

HERAPATH's views on the impact of bodies. The letter belongs to C. TRUESDELL. The concluding remarks about electricity suggest that it may be addressed to MAXWELL, who took up the kinetic theory between the appearance of the February, 1859, issue of the *Philosophical Magazine*, and 30 May, 1859, on which date he mentioned in a letter to STOKES "Clausius' (or rather Herapath's) theory". In his paper of 1860 MAXWELL had mentioned DANIEL BERNOULLI and HERAPATH as having derived equations of state from the hypothesis of uniform rectilinear motions, and then CLAUSIUS in 1862 had given a list of early authors on the kinetic theory, such as that which HERAPATH in this letter acknowledges having received in a letter of 8 February 1864, to which he is replying.

was made in 1851, when JOULE[2] published a crude calculation of molecular speeds based on HERAPATH's ideas. Like HERAPATH, JOULE was self-educated and unspoilt by the rotten Cambridge tradition of his day. Unlike HERAPATH, JOULE found within the stream of new men who were enthusiastically transplanting to Britain the French way of looking at mathematical physics a strong and unswerving defender, adherent, and collaborator: WILLIAM THOMSON, later Lord KELVIN.

THOMSON set as an examination question in 1852 an improved calculation of the essential step in JOULE's result, and in the next year he published his solution.

There are persons who claim that work which is without influence does not belong in a history of science. Perhaps they transfer the operationalist philosophy, which until recently was popular enough among physicists to annihilate any attempt at theory that did not conform with the prejudices of the crowned heads at the moment, to the historical domain, regarding papers that do not happen to be read by physicists, like events that do not happen to be observed by physicists, as not existing at all. For such persons the history of the kinetic theory begins in 1856 with a paper by KRÖNIG. KRÖNIG's ideas, though much the same as BERNOULLI's and JOULE's, are a little easier to follow and a little less correct.

While the work of HERMANN, EULER, and BERNOULLI had been forgotten, and HERAPATH was ignored as a crank, KRÖNIG's rudimentary paper, not even correct in detail, was published, read, and respected, and to this day it is commonly regarded as the starting point of the modern kinetic theory. Indeed, KRÖNIG's paper made no advance. To account for the favor bestowed upon it, we can remark that in the years just before it was printed the mechanical theory of heat had been established experimentally, and that KRÖNIG, however small his name may sound today, in his own time was a person important enough for other scientists to pay attention to him.

We are left with the question, what prompted KRÖNIG, an otherwise undistinguished man, to take up gas theory just at this time, never before and never again? A plausible answer has been suggested by DAUB in his article on KRÖNIG in the *Dictionary of Scientific Biography*:

> In the absence of any other significant theoretical work, the paper on the kinetic theory seems strangely anomalous. It is likely that Krönig's theory was not altogether original with him. During

---

[2] Details and full references may be found in my history of the early kinetic theories, cited above in Footnote 1.

his years as editor of *Die Fortschritte der Physik*, an abstract of Waterston's ideas came under review in which every one of these conclusions was succinctly stated, and Krönig may well have inadvertently received general guidelines of his theory from reading that brief résumé.

4. WATERSTON (1843–1846)

If the work of HERMANN, EULER, BERNOULLI, HERAPATH, and JOULE was unnoticed, what shall we say of WATERSTON's? While his name is well known today to historians of nineteenth-century science, most specialists in statistical mechanics have never heard it even now.

On 11 December, 1845, six years before JOULE's publication and eleven years before KRÖNIG's, the Royal Society received from Bombay a memoir by JOHN JAMES WATERSTON (1811–1883) entitled "On the physics of media that are composed of free and perfectly elastic molecules in a state of motion". The next year, the Royal Society published a half-page abstract[3] of the paper, with the author's name misspelt; in it we read only that WATERSTON used a molecular model like BERNOULLI's and that he discussed the gas laws, the effect of different molecular masses, condensation and dilatation and the value of the *vis viva*, the resistance of media to a moving surface, the equilibrium of the atmosphere, and the speed of sound. In 1851 in connection with a meeting of the British Association WATERSTON published a more descriptive one-page abstract[4]. Among other things it states that "the *vis viva* of those [random] motions in a given portion of gas constitutes the quantity of heat contained in it", and "equilibrium of pressure and heat between two gases takes place when the number of atoms in unity of volume is equal, and the *vis viva* of each atom equal. Temperature, therefore, in all gases, is proportional to the mass of one atom multiplied by the mean square of the velocity of the molecular motions, being measured from an absolute zero[5] 491° below the zero of Fahrenheit's thermometer." Thus WATERSTON was the first to publish the modern kinetic-theory definition of temperature and the simplest case of the law of equipartition.

---

[3] J. J. WATERS[T]ON, "On the physics of media that are composed of free and perfectly elastic molecules in a state of motion" (abstract), *Proceedings of the Royal Society* (London) **5** (1846): 604 = *Philosophical Transactions of the Royal Society* (London) **A183** (1893): 78 = pages 317–318 of WATERSTON's *Papers*. This abstract can scarcely have been composed by WATERSTON himself.

[4] J. J. WATERSTON, "On a general theory of gases", *Report of the Association for the Advancement of Science 1851*: 6 = *Philosophical Transactions of the Royal Society* (London) **A183** (1893): 79 = pages 318–319 of WATERSTON's *Papers*.

[5] Other work of WATERSTON suggests that 491° is a misprint of 461°.

Figure 25. JOHN JAMES WATERSTON (1811–1883), after a photograph at the age of 46, taken by his nephew, GEORGE WATERSTON, reproduced by courtesy of the late ROBERT WATERSTON, Esq.

solids; seeing that any amount of rigidity, however small, will account for the phænomena, if we adopt certain suppositions as to molecular forces.

V. Our knowledge of molecular forces is not as yet sufficiently advanced to enable us to use experiments on the velocity of sound as a means of determining accurately the coefficients of elasticity of solids.

If we adopt for them the hypothesis already stated with respect to liquids, a theoretical investigation given in an Appendix shows that the velocity of sound in a cylindrical rod of an uncrystallized substance, whose surface is absolutely free, will be less than that in an unlimited expanse in a ratio which is sensibly $1 : \sqrt{2}$ for a very slender filament, and approaches $\sqrt{2} : \sqrt{3}$ as the diameter of the rod increases; but the absolute freedom of the surface cannot be realized in practice; the means used in fixing the rod tend to restrain the lateral oscillations and accelerate the velocity of sound. The ratios ascertained by experiment range from $\sqrt{2} : \sqrt{3}$ to near equality; but they are not sufficiently numerous to form data for any definite conclusions.

The oscillations treated of in the special problems of the body of the paper being of a kind called *nearly longitudinal*, a second Appendix is added, containing the general equations of another kind, called *nearly transverse*, in uncrystallized bodies.

On a General Theory of Gases. By J. J. WATERSTON, Bombay.

The author deduces the properties of gases, with respect to heat and elasticity, from a peculiar form of the theory which regards heat as consisting in small but rapid motions of the particles of matter. He conceives that the atoms of a gas, being perfectly elastic, are in continual motion in all directions, being restrained within a limited space by their collisions with each other, and with the particles of surrounding bodies. The *vis viva* of those motions in a given portion of gas constitutes the quantity of heat contained in it.

He shows that the result of this state of motion must be to give the gas an elasticity proportional to the mean square of the velocity of the molecular motions, and to the total mass of the atoms contained in unity of bulk; that is to say, to the density of the medium. This elasticity, in a given gas, is the measure of temperature. Equilibrium of pressure and heat between two gases takes place when the number of atoms in unity of volume is equal, and the *vis viva* of each atom equal. Temperature, therefore, in all gases, is proportional to the mass of one atom multiplied by the mean square of the velocity of the molecular motions, being measured from an *absolute zero* 491° below the zero of Fahrenheit's thermometer.

If a gas be compressed, the mechanical power expended in the compression is transferred to the molecules of the gas increasing their *vis viva*; and conversely, when the gas expands, the mechanical power given out during the expansion is obtained at the expense of the *vis-viva* of the atoms. This principle explains the variations of temperature produced by the expansion and condensation of gases—the laws of their specific heat under different circumstances, and of the velocity of sound in them. The fall of temperature found on ascending in the atmosphere, if not disturbed by radiation and other causes, would correspond with the *vis viva* necessary to raise the atoms through the given height.

The author shows that the velocity with which gases diffuse themselves is proportional to that possessed by their atoms according to his hypothesis.

LIGHT, HEAT, ELECTRICITY, MAGNETISM.

On the Conduction of Electricity through Water. By F. C. BAKEWELL.

Mr. Bakewell stated the results of some experiments on the conduction of electricity by water, made with a view to prove that an electric current may be transmitted for a considerable distance through unprotected wires immersed in water. The experiments were conducted on Saturday last in one of the Hampstead ponds. A thin copper wire (No. 20), three hundred and twenty feet long, was stretched across the pond, and two copper plates ten inches square, to which wires were soldered,

Figure 26. WATERSTON's abstract of 1851, giving his statement of equipartition.

The original memoir of WATERSTON is a masterly exposition of all the kinetic-theory notions we have explained thus far and of related ones. It is easy and persuasive reading. The results are summarized in sweeping italicized statements at the end of each argument. By a slip in calculation WATERSTON obtained $\gamma = \frac{4}{3}$ rather than $\frac{5}{3}$, rendering invalid certain other numbers, but the reasoning is right. WATERSTON was familiar with all the experimental data on gases known at the time, and his paper contains many numerical results and comparisons with experiment. Included is a faulty[6] calculation of the mechanical equivalent of heat. The treatment of the resistance encountered by a moving surface is defective for the same reason as was NEWTON's, *viz* neglect of the action of reflected molecules on those about to impinge, and naturally the solution to the problem of equilibrium of the atmosphere, an elaborate matter still imperfectly understood, is not satisfactory. Finally in an appended note WATERSTON constructed a theory of chemical combination of gases differing from that now accepted only in his not recognizing molecules as composed of indivisible atoms. At the end of the note is a great chart of the molecular properties of specific substances. To say that WATERSTON in 1845 had worked out all of the kinetic ideas and results that were to be published by others in the next eleven years, thus falls short of justice to his great paper.

In submitting it to the Royal Society, he wrote:

> Whether gases do consist of such minute elastic projectiles or not, it seems worth while to enquire into the physical attributes of media so constituted, and to see what analogy they bear to the elegant and symmetrical laws of aeriform bodies. Some years ago I made an attempt to do so, proceeding synthetically from this fundamental hypothesis.... The results have appeared so encouraging, although derived from very humble applications of mathematics, that I have been led to hope a popular account of the train of reasoning may not prove unacceptable to the Royal Society.

In 1845 British physicists believed heat to be either a subtle fluid which can pass through continuous matter or a vibratory motion of molecules. They did not wish to think about other possibilities. Moreover, as there developed, first in Britain and then on the Continent, a straitening of the channel of physics until its walls nearly met

---

[6] According to HOYER, the nearness of WATERSTON's result to that now accepted is due to compensating errors. See his "Über Waterstons mechanisches Wärmeäquivalent", *Archive for History of Exact Sciences* **19** (1978): 371–384.

at the center-line of tables of data from experiments, anything specu-lative, unless written by a recognized authority, had already become distasteful, being relegated to mere philosophy. Then as now, docile and laborious calculation of details within the accepted views was preferred above search of principle. The first referee objected to the underlying assumption that the gas pressure results from molecular impacts, which he asserted was "by no means a satisfactory basis for a mathematical theory." While he conceded that the paper "admits much skill and many remarkable accordances with the general facts as well as numerical values furnished by observation," he did not think it worthy to be published in the *Philosophical Transactions*. The second referee wrote that "the paper is nothing but nonsense, unfit even for reading before the Society." The paper was rejected. The system of secret refereeing, still in almost exclusive use today, has protected the anonymity of the eminent referees, while burying the genius of WATERSTON[7] in permanent oblivion.

According to the Society's rules, manuscripts sent to it became its property forever. WATERSTON submitted a written request that his manuscript be returned, but it was not. In 1891 Lord RAYLEIGH found it in the archives; he caused it to be published in 1893 in the journal to which it had been submitted forty-eight years earlier[8]. He furnished it with an introduction, in which he defended the conscien-tiousness of the second referee, "one of the best qualified authorities of the day", adding the counsel that "highly speculative investigations, especially by an unknown author, are best brought before the world through some other channel than a scientific society...," and "a young author who believes himself capable of great things would usually do well to secure the favourable recognition of the scientific

---

[7] The names of the referees were revealed in 1965: The Reverend Mr. BADEN POWELL, Savilian Professor of Geometry at Oxford, and Sir J. LUBBOCK, who was considered to be an expert in the field of gas theory. POWELL had done experimental work on heat and light in the 1830s and 1840s. LUBBOCK was a banker, entomologist, anthropologist, botanist, politician, and public educator. Both referees were of out-standing incompetence in mathematical theory of physics.

Priority fifty years after the fact is legal priority honored only in the breach. Though the work of WATERSTON—including his *unquestionable* priority not only in discovery but also in publication of the principle of equipartition—has been discussed again and again by writers on the history of physics, his name remains unknown even among specialists on the kinetic theory and statistical mechanics. Physicists, apparently, are as little likely to take note of what historians write as they are to study the sources of their science.

[8] J. J. WATERSTON, "The physics of media that are composed of free and perfectly elastic molecules in a state of motion" (with an introduction and notes by Lord RAY-LEIGH), *Philosophical Transactions of the Royal Society* (London) **A183** (1893): 1–79 = pages 207–331 of WATERSTON's *Papers*.

world by work whose scope is limited, and whose value is easily judged, before embarking upon higher flights." We are to presume that these words, while if published today they would be read as sarcasm, were written in Victorian earnest. RAYLEIGH praised the "marvellous courage" of the work and the great advance it attained, according WATERSTON priority in the law of equipartition on the basis of the abstract from 1851. By this time WATERSTON had been dead for ten years. That he made no claim of priority later when the kinetic theory became a recognized field, is not hard to explain. First, he learned that a part of his views had been put forth earlier by BERNOULLI and by HERAPATH. Second, he felt himself bound by the rule of the Royal Society in impounding as its own his manuscript, since he had by implication acknowledged this rule in sending anything to the Society. Third, his Sandemanian faith forbade contention for personal eminence, and his sole object was the truth. His critical attitude toward the acknowledged leaders of science, such as Lord KELVIN (whose age and station must have kept him free of taint by the affair of WATERSTON's great paper), caused him in later life to be regarded as a crank. One of his relatives reported that

> any mention of the Royal Society led to very strong and contemptuous language. . . . [H]e showed a restlessness and dislike at the mention of scientific men, except Faraday, and he used very strong language in respect to some who bulk largely in public estimation.

RAYLEIGH's suggestion of "some other channel" calls for a further remark. Some thirty years before WATERSTON, HERAPATH, after his similar experience with the Royal Society, had gained oblivion through publishing in the channel of a minor journal. In 1843 WATERSTON had employed the channel of a minor press for a book on the physics of the mind; at the end is a note which is virtually an abstract of the long paper afterward rejected by the Royal Society. We are not surprised that publication in this channel attracted no notice whatever; if the leaders of science are hostile to a new idea when it is called to their official attention, there is little likelihood they will choose to go out of their ways to search for it in obscure places. In this note of 1843 WATERSTON introduced the concept of *mean free path*[9] (Figure 27), rediscovery of which by CLAUSIUS in 1857 started a new

---

[9] J. J. WATERSTON, Props. I and II of "Note on the physical constitution of gaseous fluids and a theory of heat", at the end of *Thoughts on the Mental Functions*, Edinburgh, 1843 = pages 183–206 of WATERSTON's *Papers*. Of course, nobody could be blamed for not noticing something published so obscurely, but the bashaws of science cannot have it *both* ways, advising the youngsters with original thoughts both to keep out of major journals *and* to gain their initial recognition by publishing in pages nobody reads.

will continually impinge against each other by reason of their *vis insita* alone, and will suffer no loss of collective momentum. The surface of impact is supposed to be spherical, and may either be the actual surface of the solid molecule, or a surface of powerful molecular repulsion surrounding it.

I. The distance traversed by a molecule, after imping-ing on one and before encountering another, is inversely as the density of the medium.

If all the molecules were arranged equidistant, the number at any distance from a central point will be as the square of that distance, but the intercepting effects of each of that num-ber is as the inverse of the square of the distance ; hence the intercepting effects are uniform, and accumulate as the distance increases. Let $d =$ diameter of molecular surface of impact ; then *as* the distance $(x\,d)$, beyond which a molecule cannot go without impinging on another, *is to* the mean distance of the molecules $(n\,d)$, *so is* the number of circles (n^2) (whose com-mon diameter is equal to that of the molecular surface of impact) that can be inscribed in an equilateral triangle, the side of which is equal to the mean distance of the molecules *to* unity. Hence, $x\,d \doteqdot n^3 d$, and $x \doteqdot n^3 \doteqdot$ to the inverse of the number of molecules contained in a given volume. [Upon the accuracy of this reasoning depends almost all that follows.]

II. The density of the medium being constant, the impinging distance will vary reciprocally as the square of the diameter of the molecular surface of impact.

For $n\,d$ being constant, $n \doteqdot \frac{1}{d}$, and $x\,d \doteqdot n^3 d \doteqdot \frac{1}{d^2}$.

III. The impinging distance being constant, the den-sity will vary directly as the diameter of the molecular surface of impact.

For $x\,d$ being constant, $x = \frac{1}{d} = n^3$, hence $d = \frac{1}{n^3} =$ den-sity.

IV. Let there be two elastic parallel immovable planes, and between them, in a common perpendicular, let two molecules that have unequal velocities continu-ally impinge, then shall the sum of the squares of their velocities be always equal, and the velocities will con-

Figure 27. A page from WATERSTON's *Thoughts on the Mental Functions*, 1843, explain-ing his concept of mean free path.

period in the kinetic theory. To CLAUSIUS is due also the first explicit assignment of a probability in physics, if on the vaguest of grounds. In 1860, only fifteen years after the disgrace of WATERSTON, appeared the first of the four great memoirs by MAXWELL which raised the whole kinetic theory to a new level of abstraction, rigor, and applicability and created the science of statistical mechanics.

For his first paper on the kinetic theory, MAXWELL, a cannier Scot, though nineteen earlier papers on other subjects had given him a high reputation chose an informal journal[10] in which to make "the hypothesis that... minute parts are in rapid motion" the "subject of rational curiosity". In 1859, when he first turned to the kinetic theory, he informed STOKES of his discoveries. Two years before that, STOKES as Secretary of the Royal Society had offered to sponsor him for election, but MAXWELL had preferred to wait. It seems that STOKES also encouraged him to submit something to the Royal Society's publications. From the second of MAXWELL's references to the kinetic theory in the published correspondence it seems that STOKES, warning him against "speculations about gases", had not regarded the Royal Society as likely to receive such a research with favor. Perhaps in response to STOKES's advice, in those years MAXWELL submitted to the Royal Society only experimental findings. After having received two notable honors from the Royal Society, MAXWELL decided to stand for election. That was in 1860, the year in which his first paper on the kinetic theory made its modest appearance.

A body so august as to intimidate MAXWELL from presuming to present himself might have been expected to maintain the highest possible standard. By looking at what the *Philosophical Transactions* did publish in the years around 1860 the reader will easily judge of that.

Six years later, in 1866, and twenty-one years after having scorned WATERSTON's mathematical theory because it rested upon the

---

[10] The *Philosophical Magazine*, which in the first half of the century had published, along with many fine papers, a good deal of plain rot and outright error. Had HERAPATH and WATERSTON chosen it in the first place, their papers might well have been accepted and have appeared in instalments, easy to read and assess. Of course the *Philosophical Transactions* would seem to be the fittest place for orderly and complete exposition of a new branch of physics, but a splendid format can and often still does present *in extenso* what is mere routine, too dull to deserve presentation at all unless beneath the literally betitled name of a big apparatchik of royal science.

In the second sentence of this paper MAXWELL names BERNOULLI, HERAPATH, and JOULE as being the first to explain the gas laws on the basis of rectilinear motions and impacts.

hypothesis of the molecular constitution of gases, the Royal Society printed in its *Philosophical Transactions* a great memoir of MAXWELL starting from just this same hypothesis. With its specific calculations of probabilities and its detailed reckonings of the effects of collisions on an infinite assembly of molecules subject to that very strange and special inverse fifth-power law of repulsion, this paper is vastly more mathematical and vastly more speculative than the rejected essay of WATERSTON. Partly, it was a difference of persons: an established Fellow, not just some nobody out in Bombay. More than this, it was the difference of twenty-one years: The Royal Society was ready to catch up with the Germans.

5. IN PRAISE OF DISORGANIZATION OF SCIENCE

The older historians loved to point a moral from their tales, sometimes going so far as to alter history in order to conform it with their own preconceptions. With the recent rise of history of science as a profession, not only alteration of the facts but also all kinds of moralizing based on history are fallen into disrepute. Not being a historian of profession, I feel myself privileged to remain a reader of history in the simple old way. I am not ashamed to agree with HERODOTOS and MACHIAVELLI and MONTAIGNE and LEIBNIZ in believing that the great use of history is for the lessons it teaches us for our conduct today. The recent work in continuum mechanics has drawn inspiration from fresh, direct study of the classics of mechanics: NEWTON, the BERNOULLIS, EULER, CAUCHY, STOKES, and MAXWELL in particular were consulted for their methods of inquiry, standards of inference, and philosophies of natural science, so as to correct the bad morals grown common and even exalted through the so-called applied mathematics and applied mechanics of the first half of this century. This essay on the early kinetic theory furnishes a lesson of a different sort, a lesson about the organization of science.

We have seen the Royal Society twice in thiry years with maximally pompous humbuggery and humbugging pomposity stifle the truth in favor of the wrong, twice bury a great man in contempt while exalting tame, bustling boobies whose whole lives add nothing to the science passed on into our day. To see how ludicrous was the position of the Royal Society, we need only state it. The Society, or at least its officers, regarded the Society itself as committed to support any paper it published. Consequently, to accept a "speculative" paper was dangerous, even admitting the chance that some such papers turn out to be right. Responsibility for the decisions was delegated to referees. Today, it is hard to find something less important about the kinetic theory of gases than what two anonymous referees who had never touched the subject thought about it in 1816 or 1845. If that thing hard to find can be found, it is whether or not the Royal Society as a body was to commit itself to the kinetic theory in 1816 or 1845. Nevertheless, the Society's position in those two years was so strong as to stop dead two careers in mathematical physics which at their short beginnings gave promise second to few in the century, and of which even the exhumed frusta stand among the monuments of mechanics, while for most of the Fellows of that body from 1800 to 1850 it would be like pulling hens' teeth to find a single paper or experiment worth a moment's notice now. Such is not the picture the layman has of a great academy of science. He is taught that scientific truth is demonstrable, irrespective of persons, and that those who sit in the great academies are the great scientists of the time. Since the stories I have just told do not bear out either of these beliefs, they may be taken as criticism of the Royal Society, but they are not so intended. I could tell of similar happenings in the French Institut, in the Berlin Academy, and in the Academy at Petrograd—in a word, in *every* famous academy down to 1850. (I should not dare to come any closer to today.)

That my moral is of a different kind, may be seen from the fact that no such story can be told of any of our own numerous academies of science in America. These range from local clubs ever eager to recruit those who will pay dues in exchange for a title to a semi-official body of scientists and administrators of big science whose advice is sometimes asked but rarely heeded by the State Department and the President. No-one could justly claim that the American academies in their egalitarian numerosity uphold standards of qualification for membership anything like those of their elite Continental progenitors, yet no-one can justly impute to our more modest and plebeian bodies such fatal mistakes as have been made again and again by the august official Olympuses of Europe. I have good reason to doubt that our academicians be wiser or more kindly or less rancorous. While certainly they would never so much as consider admitting someone openly critical of their ways, even if he were endowed with genius equal to WATERSTON's, they would, I am sure, given the chance, one day have elected HERAPATH, not, of course, for his youthful wild oats sown upon his off-beat, odd-ball kinetic theory, but for the second half of his life, which was devoted to good, sound railway science. The main difference is to be found, not in the persons, but in the importance attached to them. *Science* is independent of passions and titles; *scientists* are not. That the Royal Society twice unanimously considered the kinetic theory of gases to be nonsense from top to bottom, has no more bearing on the soundness of that theory than the true value of π would have been affected, had a certain midwestern legislature passed the bill it indeed seriously considered which would establish a standard of $\frac{4}{3}$ in law. What is, unfortunately, not only ludicrous but also poisonous about European academies is the importance which they attach to themselves and which others allow them to enjoy. Up to the present time, scientists in the United States have been happier. Few of them care enough about our academies to learn who is and who is not a member, and the academies, as yet, do not set themselves up as arbiters over the truths of science and the lives of scientists.

The difference, starting at the top, goes down all through our work. The European assistant who disagrees openly with his professor risks losing all chance of going on with his research, not to mention failure ever to get a decent job. In the United States, a paper is no more esteemed if it appears within covers sealed by an academy or professional society, no less so if it has been rejected by such a body before being published in a private journal, and for the young giant, trampling upon his professors is a more honorable path to fame, promotion, and such modest prosperity as the scientific trade allows than is the fawning filial piety the European professor expects and receives from his disciples as long as he lives. Our academic life presents to the foreigner a lamentable scene of chaos. No-one knows who is on top. If in University 1 Professor A is a demigod, we have only to consult Professor B in University 2 to learn that in *his* department A would not qualify even as an assistant. True, A belongs to six national committees, has a million dollar grant from the Central Spy Bureau, and has published eight successful textbooks, but B, who points to A's textbooks as models of nonsense, has written 216 research papers with twenty-three co-authors and also is consultant for four major corporations, assistant editor of five journals and second vice-president of a professional trade-union. The courses A sets down as minimum requirement for a degree under his ægis are not even given in B's university. Then there is always C, who upholds "good teaching" by letting his students from sight of anything less than fifty years old. Even our administrators, frantic though they are to bring to heel their own faculties, add their part to the exterior disorganization in refusing to allow full "credits" to the transferring student who has drunk his mediocrity in another grove of the Muses. Finally, in their precaution lest one of their hirelings usurp any semblance of rule or become so notable that the university could lose fame by losing him, they blow now hot, now cold upon each local hero in turn, so as to keep their pantheons fluid.

You may think I fantasticate. Just read what DYSON[11] wrote in 1958:

A few months ago Werner Heisenberg and Wolfgang Pauli believed that they had made an essential step forward in the direction of a theory of elementary particles. Pauli happened to be passing through New York, and was prevailed upon to give a lecture explaining the new ideas to an audience which included Niels Bohr. Pauli spoke for an hour, and then there was a general discussion during which he was criticized rather sharply by the younger generation. Finally Bohr was called on to make a speech summing up the argument. "We are all agreed," he said, "that your theory is crazy. The question which divides us is whether it is crazy enough to have a chance of being correct. My own feeling is that it is not crazy enough."

The objection that they are not crazy enough applies to all the attempts which have so far been launched at a radically new theory of elementary particles. It applies especially to crackpots. Most of the crackpot papers which are submitted to *The Physical Review* are rejected, not because it is impossible to understand them, but because it is possible. Those which are impossible to understand are usually published. When the great innovation appears, it will almost certainly be in a muddled, incomplete and confusing form. To the discoverer himself it will be only half-understood; to everybody else it will be a mystery. For any speculation which does not at first glance look crazy, there is no hope.

We need not join DYSON and his co-editors in their extreme position, regarding whatever is incomprehensible as being worth a hearing. At least by the time he was fifty, GOETHE advocated the opposite criterion:

Vergebens werden ungebundne Geister
Nach der Vollendung reiner Höhe streben.
Wer Großes will, muß sich zusammenraffen.

Neither HERAPATH's papers nor WATERSTON's would have seemed crazy to any mathematically literate person of their respective periods who would take the trouble to read them. But their papers were put into the hands of no such person. They were rejected for heresy! Such a posture deserves no place in science, then or now or ever. I am sorry to feel compelled to add that even today there are parts of the

---

[11] F. J. DYSON, "Innovation in physics", *Scientific American* **199** (1958): 74–82.

world where it is maintained by organizations that tyrannize science and bully scientists, and that not all of those are in the East.

It is bad taste to praise one's own land and dispraise the foreign without strong justification. I am doing so because today, the day of massive unanimity, when social democracy triumphantly levels us all in informally uniform, cordially timorous, voluntary servitude, there are pressures from all sides to unify science. The societies and academies begin to demand a uniform format for papers as well as a "style" of polysyllabic baby-talk after Dutch patterns. Officials, accustomed to dealing only with other officials, perplexed in a world where the applications of science hold the fortunes and lives of all in balance, turn to the heads of these bodies and impute to them an authority which in fact is theirs only by default if not arrogation. The government, which three decades ago began to give and accustomed us to receive, now grasps and throttles. The public, ever fearful in ignorance, recognizes the size of the budget as the measure of excellence in science as in welfare and turns the spigots of the common purse toward those readiest to spend, be it to dig to China, to carry syphilis to the moon, or to choke the human mind in a dense dust of digits. Great social activities such as these, like the building of the pyramids, require a central organization and a clear chain of command. While they themselves touch science at most on its margins, they are so gigantic and so poisonous that even marginal contact may be fatal.

Several events in modern physics that DYSON recounts at some length in the article of his just cited join the early history of the kinetic theory of gases in illustrating the principle that in science, the majority is always wrong. This principle is difficult for a social democracy to swallow. We cannot move a step without being reminded that the persons who prefer silence to soft music are outvoted; they must hear the soothing syrup because "most people like it". The great society, the affluent society, sets for itself the humane doctrine of nourishing the weak and the maimed, of subsidizing the incompetent and the foolish; it looks with suspicion and fear on the strong of heart, the thoughtful, the wise. An unsurrendered mind proclaims an antisocial man. "Protest", indeed, is met with high tolerance if not capitulation, so long as it is the stupid mob cry of the unlucky, the by choice and now often also by inheritance unemployed, and the jacobins. What modern society cannot stomach, is a man who thinks. Even if social democracy is now so enlightened as to stop short of deciding a good value for π by plebiscite, it takes refuge in polls of "experts". We all hear of what "doctors say" and what "scientists say". A scientist who lets the laymen who know him find out that he rejects some tenet the newspapers have published as "accepted by science today" will arouse their suspicion that he himself is not quite all right.

All this is of no importance so long as it is only the public that plays the booby. The danger comes when scientists allow *themselves* to be organized, when they begin to *respect and obey* pronouncements on science by academies, universities, societies, and, finally, governments. May that day never come! May our academies remain clubs, may our universities remain disjoint snarls of local red tape and campus-club poolroom politics, may our professional societies remain subscription lists for advertisers! So long as we preserve the disparate, entwined, mutually suspicious inefficiency which rules the circumstances of scientific work in this country today, a HERAPATH or a WATERSTON, should he arise, may eat and work and discover.

Note for the Reprinting

The text printed above is condensed and revised from the first half of Essay VI in my *Essays in the History of Mechanics*, New York, Springer-Verlag, 1968. The acknowledgments printed there should be taken account of by anyone who seeks the sources of the foregoing, too. The remainder of that essay is superseded and greatly extended by my "Early kinetic theories of gases", *Archive for History of Exact Sciences* 15 (1975): 1–66.

35. GENIUS AND THE ESTABLISHMENT AT A POLITE STANDSTILL IN THE MODERN UNIVERSITY: BATEMAN (1976, 1981)

CONTENTS

Since The Johns Hopkins University is most famous as a teacher of teachers, we must not be astonished to learn that many of its eminent scholars left it long before they reached the age of retirement. That the greatest men of Hopkins are no longer there, is such a commonplace as to make those who remain uneasy. Some who left did so because they had come only to take a degree in the first place; others, because they were let go or at least not encouraged to remain. It is fitting that a part of the centennial commemoration should refer to these migrant if not fugitive sons of capricious and miserly Alma. One of them, C. S. PEIRCE, has been celebrated already. Our lectures beginning today are dedicated to the memory of another, HARRY BATEMAN (1882–1946), both a doctor of this university and a teacher here who did not win acceptance.

1. BATEMAN'S EARLY YEARS

HARRY BATEMAN was born in Manchester in 1882, son of a drug-gist and commercial traveller. He was a scholarship boy in grammar school and then at Trinity College, Cambridge. At the age of 21 he was bracketed Senior Wrangler; two years later he was Smith's Prize-man and won a fellowship which allowed him to travel on the Continent, visiting Paris and Göttingen. Perhaps it was then that he began to acquire his few un-English traits. One of these was a thorough knowledge of the literature of mathematics and physics in French, German, and Italian; another was his preference for mathematics that was rigorous by the standards of his day. G. H. HARDY, five years older than BATEMAN and in the end a more famous mathematician, related that in 1908 he had asked a dozen candidates, including several future Senior Wranglers, how to sum the infinite geometric series but had "not received a single answer that was not practically worthless." HARDY recalled that it had been BATEMAN who first told him about uniform convergence. At the same time he deprecated BATEMAN's later willingness to accept arguments in terms of formal operators, lacking in ε–δ rigor. HARDY's teacher was A. E. H. LOVE, a famous elastician and a foreign member of the Lincean Academy, a man of unusual learning and breadth, who advised him that in order to find out what mathematics really meant, he should read JORDAN's *Cours d'Analyse*—a most un-British book. I do not know who was BATEMAN's teacher. BATEMAN himself in the preface (1916) to his *Differential Equations* acknowledges his gratitude to "Dr. Glaisher, who first roused my interest in the subject, Dr. Forsyth and Prof. Whittaker, who developed it by their lectures and writings." He states also that he "partly borrowed" passages from DARBOUX' *Théorie des Surfaces* and found invaluable, "the treatises on analysis by the great French writers Jordan, Picard, Goursat and others . . .." I conjecture he was not close to anyone at Cambridge. Certainly he was not retained there. Working within the finest tradition of British mathematical physics, for pure mathematics he turned to Continental authors, as had GREEN, STOKES, KELVIN, and MAXWELL before him. I possess his collection of the Borel tracts, signed on the flyleaf when he was at Trinity College and by his own hand joined between homely boards. (The California Institute of Technology, to which he bequeathed his books, sold for a few cents apiece those that dupli-cated its holdings. They would have come in handy later, after the students began to steal.) Apparently very short of money, at Cam-bridge BATEMAN taught in a correspondence school and coached for the tripos.

Figure 28. HARRY BATEMAN (1882–1946), after a drawing by S. SEYMOUR THOMAS, by courtesy of the Reverend EDWARD E. HAILWOOD.

BATEMAN went on to the universities at Liverpool and Manchester, neither of which chose to retain him; his first post in the U.S.A. was at Bryn Mawr. In an obituary of him FRANCIS D. MURNAGHAN, an Irishman who for two decades had dominated the activity in mathematics and mathematical physics at Johns Hopkins, wrote[1]

> we can only surmise that he was not particularly successful as a teacher of young ladies or that he did not find the work particularly congenial. In 1912 he received an appointment as Johnston Scholar at Johns Hopkins University where Frank Morley, an old Cambridge mathematician, was head of the department of mathematics. The Johnston Scholarship in those days at Hopkins was a research scholarship, and the holder could give a seminar if he felt like doing so and if there were any students who felt like taking the course.

A later Johnston fellow was to be OSCAR ZARISKI, who came in 1927 and soon thereafter by departing went over to the majority of Hopkins' great mathematicians.

2. BATEMAN AT JOHNS HOPKINS

MURNAGHAN continues,

> In 1912 [Bateman] was thirty years old, had published some 64 papers and had been two years in America . . .. As we look back on the situation we cannot escape the inevitable Why? Here was a man of international reputation, pleasant (if self-effacing) personality, and he had to spend the next five years in a position designed for a young unmarried Ph.D. of promise or for an established scholar on leave-of-absence . . .. When we think of the "odd-jobs" he had to do to eke out a subsistence, the reading of papers for the Weather Bureau, the hot Washington summer at the Bureau of Standards, the teaching at Mount Saint Agnes and then recall that during this period he wrote his book on electrical and optical wave-motion, we can only subscribe to the old Latin tag: *Per aspera ad astra.*

When MURNAGHAN thus expressed his astonishment that BATEMAN had accepted so poor an offer from Hopkins, he was himself

---

[1] F. D. MURNAGHAN, Obituary of BATEMAN, with bibliography, *Bulletin of the American Mathematical Society* **54** (1948): 88–103.

chairman of the mathematics department. The tactfulness for which that important position calls may have restrained him from going on to ask the question that comes to our minds at once on learning the circumstances: Why did Hopkins not retain BATEMAN? The reason could not have been lack of formal qualification, for, perhaps recognizing the grasp of what WILLIAM JAMES had already denounced as "the Ph.D. octopus", in his second year BATEMAN, stooping to a compliance that must have grated upon an M.A. of Cambridge, with his sixty papers behind him sought and received an American doctorate from examiners not one of whom had ever done any research of importance.

MURNAGHAN's account of BATEMAN at Hopkins may provide part of the explanation. He wrote,

> In 1914 I was awarded a Traveling Studentship in Mathematical Physics . . . and was looking about for some place to study. My professor, A. W. Conway, told me that there was a young man, Bateman, at Hopkins and that he thought that I could not do better than study with him. I followed this advice and, looking back over a third of a century, I judge the advice to have been sound. Bateman, a frail slight man of 32, was lecturing on The Absolute Calculus and Electrodynamics (remember that this was 1914 and that four years or more had to elapse before most of us in this country heard of Einstein's General Theory of Relativity). As I recall the situation, six students started the course and by March I was, if my memory is correct, the only student. I do not think that this diminution of the size of his class bothered the lecturer very much, and I have sometimes thought that if the vicissitudes of student life had prevented my attendance, the lecture would have been none-the-less delivered. By common standards he was not a good lecturer. He was too detached, too objective and perhaps too scornful of histrionic effects, and we were too untrained to profit as much as we should have from the instruction he gave us.

3. MATHEMATICS AND PHYSICS AT JOHNS HOPKINS IN BATEMAN'S DAY

The years BATEMAN spent at Hopkins, 1912–1917, are seldom mentioned in accounts of our glorious heritage. We do hear today, nevertheless, that a vital university cannot stagnate but must change with the times, must grow bigger and broader, must progress. Hopkins, having begun at the top, could progress in but one direction. Long past were the days of SYLVESTER and GILDERSLEEVE; long ago PEIRCE had retreated into penurious isolation; ROWLAND had been dead for a decade; REMSEN was soon to retire as President Emeritus.

The mathematicians of academic rank during BATEMAN's time were named COBLE, COHEN, HULBURT, and MORLEY; the physicists, already outnumbering them, were AMES, ANDERSON, BLISS, PFUND, SPARROW, and WOOD. Of these R. W. WOOD is the only man whose work is remembered today. He, and only he among the physicists and mathematicians then at Hopkins, was regularly receiving honors. I have heard that he was not tactful with his colleagues in physics, and that they did not love him. The fourteen big volumes of the *Dictionary of Scientific Biography*, which inclines heavily toward minor figures of the late nineteenth and early twentieth centuries, of course include WOOD but do not mention any of the rest except AMES, who is described as having done research "limited in quantity" and with no great results; long before 1913 he "had given up research and turned to administration." The *Dictionary* gives a page to BATEMAN and mentions no other Hopkins mathematician of the period.

From the annual reports of the president we learn that in 1912–1913 BATEMAN lectured one hour per week on integral equations; for the first half of the year at least, WOOD was supported by Columbia University for work in his small laboratory at his summer home. In 1913–1914 BATEMAN lectured on the theory of potentials; WOOD was away on leave all year at the Sorbonne. In 1914–1915 BATEMAN lectured on the differential equations of mathematical physics. He was appointed lecturer in applied mathematics for one year, and MORLEY took the occasion to praise the Johnston Scholarship "and to express a hope that the number of such foundations may be increased, until every department which is concerned with the eternal verities has one at its disposal when occasion arises." (In fact the reverse occurred: Johnston Fellowships, though still listed in the yearly catalogue, are no more, for at some apparently undeterminable time the funds for them were diverted from mathematics to the unfathomable depths of some general budget.) In 1915–1916 BATEMAN, with load doubled to two hours, lectured on relativity; WOOD was away on leave all year. For 1916–1917, BATEMAN's last year at Hopkins, no report of the president can now be found. In the succeeding year, 1917–1918, the only changes besides BATEMAN's departure are that mathematics had been displaced from the head of the list so as to receive its alphabetical deserts, an undergraduate mathematics club had been formed, and a special course in mathematics for chemists had been instituted. In physics WOOD was again away on leave, thus continuing to exemplify the tradition of the most famous savants associated with the university founded by JOHNS HOPKINS.

Other contrasts between the activities in mathematics and physics may be discerned. The reports for mathematics list graduate courses only and end with a statement that "the usual undergraduate courses were conducted...."; the physics department, more progressive, described the research done by its staff and listed only courses the titles of which suggest they were very elementary. A PARKINSONian note is struck by the record for 1915–1916, which states that the physicists were about to move up to the Homewood campus, where they had been allowed a little space in one of the new engineering buildings provided by the State of Maryland; it closes with the now routine plea for "a large increase in appropriation for apparatus and equipment." The *Dictionary*'s biography of AMES praises him for having kept physics at Hopkins alive by causing his men to do their research in the excellent laboratories of the Bureau of Standards in Washington. Certainly the reports include no evidence of any theoretical work in physics at the level of what was then being done in Britain and on the Continent, with one exception: BATEMAN's single lecture on time and space, given in his first year. Today we might expect that a theoretical physicist active in 1913 would have been interested in relativity; we forget that relativity was then a shocking subject, cultivated only by the elite and the crazed. BATEMAN, an expert in relativity from its start, did not lecture a second time to the physicists at Hopkins. WOOD, who as ROWLAND's successor dominated the physics department, R. B. LINDSAY describes as expressing himself in "physical pictures", which "he felt he (and many others) could understand better than mathematical equations, which he found rather boring." Indeed, "His academic record was undistinguished in the required fields of languages and mathematics", and his graduate study at Johns Hopkins, Chicago, Berlin, and M.I.T. was not stigmatized by a doctorate. Of the students listed as speakers in the reports of the departments of mathematics and physics but one is remembered today, namely MURNAGHAN, who had come to Hopkins only because BATEMAN was there.

During this period BATEMAN was working mainly on electricity and magnetism; part of his results are published in his classic book, *The Mathematical Analysis of Electrical and Optical Wave Motion on the Basis of Maxwell's Equations*, Cambridge University Press, 1915. The reports for 1912 through 1917 list about forty publications by mathematicians and physicists; of them twenty-eight were by WOOD and eight were by BATEMAN, but in fact during that period BATEMAN wrote at least thirty of his papers and two of his books. Perhaps someone had suggested to him the alternative of "publish or perish". If so, BATEMAN was not the last to learn that at Johns Hopkins too much research not only is evidence of poor teaching but also cuts deeply into precious time needed for campus politics.

4. The Johns Hopkins University from its Foundation to Bateman's Day

The Johns Hopkins University, upon which, with its faculty and fellows mainly not even from Maryland, the Baltimoreans had at first bestowed the suspicion due to an outlandish evangelist, by 1912 had been accepted and then enshrined as an object of local veneration.

At its foundation[2] in 1876 the trustees of Johns Hopkins had declared their intention to proceed gradually with a few leaders in the main departments of study, a company of nonresident professors and lecturers, a strong body of adjunct or assistant professorships, and a promising group of associates or holders of fellowships. Of the five research professors chosen for the "few leaders", three were near thirty in age. They had been engaged in the confidence that they were "soon to be the men of scientific and literary renown"; they were promised a free hand in their research, the means of publishing their results, and the company of few students, these to be of exceptional capacity. GILMAN, the founding president, stressed the importance of giving to university professors "only students who were far enough advanced to keep them constantly stimulated."

To this end fellowships were provided. GILMAN desired thereby to attract "men of mark, who show that they are likely to advance the sciences they profess." Of the first twenty-one scholars to enter upon their fellowships, two had doctorates from major German universities, and all but one had a lower degree. The stipend of a fellow was $500 per year. In those days an Irish navvy in Baltimore was paid $1.50 for a twelve-hour day; skilled union labor received from $2 to $3; porterhouse steak sold for 18¢ per pound, live chickens $2.50 per dozen; a standard house cost $2000–$3000; and $5500 bought a luxurious townhouse on St. Paul Street with elevator and carriage house. Thus the young scholar's stipend was the equivalent of at least $20,000 taxfree today (1976). Such was the diversity of wealth at that time that some of the fellows complained of penury. Certainly nowhere else in the country was there any such opportunity for free graduate education unburdened by routine teaching. Harvard had founded a graduate school six years earlier, but as one of its later presidents was to state it "started feebly" and "did not thrive" until

---

[2] My statements of fact regarding the early years of The Johns Hopkins University are quoted or paraphrased mainly from the account of JOHN C. FRENCH, *A History of the University Founded by Johns Hopkins*, Baltimore, The Johns Hopkins University Press, 1946, in lesser part from the later and more detailed book of HUGH HAWKINS, *Pioneer: A History of the Johns Hopkins University, 1874–1889*, Ithaca, Cornell University Press, 1960.

"the example of Johns Hopkins forced our Faculty" to take graduate study seriously. In his history of The Johns Hopkins University JOHN C. FRENCH wrote in 1946

> Though the graduate departments were, as they still are, regarded as of supreme importance, college courses were offered from the first and for two reasons. First of all, the Trustees believed and the President agreed that the University was under strong obligations to the city and the state . . ..
>
> A second consideration was the fact that the University also needed a college as a source of students fitted to do advanced work.

Some years after the opening, GILMAN stated that

> A university cannot thrive unless it is based upon a good collegiate system; and it may rightly encourage or establish a college, if needed, as an important department of its activity.

There is no evidence that The Johns Hopkins University ever throve on the basis of the College, as the undergraduate activity came to be called, nor did the College ever provide a great fraction of the graduate students. Indeed, the standards of the College were too lax to ensure that its products would be fitted for graduate work—even, as a report soon to be quoted suggests, for any "future occupation".

In his address upon the twenty-fifth anniversary of "the Hopkins" in 1901 NICHOLAS MURRAY BUTLER[3] wrote

> It was significant of the university that its early prestige was gained through men, not buildings. The rooms in which the first instruction was given were modest in the extreme. Though comfortable, they were simply apartments in remodeled dwellings. This fact, full of meaning as it was to scholars, helped to hide from Baltimore and from the country the true character of the work which had been begun. Where were the great libraries and laboratories; where the vast piles of brick and stone; where the chapels, the dormitories, and the gymnasiums which popular fancy assumed to be the necessary evidence of the existence of a college or a university?

---

[3] N. M. BUTLER, "Creating a university", *The American Review of Reviews*, January, 1901.

It was America's congenital distrust of learning, the arts, and intellectual activity, which had been remarked by many visitors ever since the founding of the Republic, that made the undergraduate program necessary to "hide from Baltimore and from the country the true character of the work" and to discharge pragmatically the "strong obligations to the city and the state" that the President and the Trustees were made to feel by the aborigines.

Thus entered the Johns Hopkins undergraduate. With him came the infinity of pressing undergraduate problems which the man who devotes himself to learning, be he student or be he teacher, has outgrown, those adolescent and social and psychological problems on which university scholarship founders while university administration flourishes. Within six years a great committee, headed by the president himself, was established to discuss the vociferated needs of the undergraduates; after twenty long and thoroughly deliberated meetings it delivered its message: seven undergraduate curricula. At that time there were only 108 members of this supremely important set of problem children. The result of adopting the curricula, it is claimed, did "secure a positive amount of regulation with a certain amount of freedom" and did "provide a liberal education which should have a tendency toward some future occupation". There is no evidence that this "tendency toward some future occupation" often pointed to graduate work. FRENCH tells us that the College drew

> more than half its students from Baltimore and the immediate vicinity. As numbers increased, the line between collegiate and graduate work, already distinct enough in method, had to be more sharply defined in administration. In 1889 it was found necessary to have a dean of the college faculty . . ..

Plainly the students of the College were not expected to be "far enough advanced to keep [the university professors] constantly stimulated." A largely distinct "collegiate faculty" had been hired; it included "a considerable number of professors engaged chiefly in directing research," and we are told that it "was highly competent. In chemistry, for example, lectures to beginners during their first term were given by Dr. Remsen himself . . .." After learning that, anyone familiar with American universities and their officials will see that it was REMSEN who would be chosen to succeed GILMAN as President. The advantages offered to undergraduates were small classes, association with graduate students and a small part of the research faculty, and use of the university library. With these "went some inconveniences which the college students were not slow to complain about. There was no campus, and no adequate provision could be made for athletic sports."

In 1901, writes FRENCH, the University was given

> a new and ample site in what was then a suburban area . . .. The undergraduates had happy visions of an attractive campus, broad athletic fields, and the "college life" which they fancied their cramped quarters had hitherto denied them; and their enthusiasm stirred misgivings among their elders. Enrollment in the college, still largely local, had been slowly increasing . . ..

While plans for the new site were being argued, some of the professors proposed a new solution to the ever more urgent problems provided by the College.

> They suggested that at Homewood the college should be remodeled with such high standards of admission and achievement that the enrollment would be automatically much reduced. It was their idea that a college so designed would attract only superior students who had ambitions for distinction as scholars, and that those who greatly valued the sports and frivolities of campus life would go elsewhere. They . . . were optimistic enough to suggest that the college which they proposed might soon come to occupy a unique place in our higher education.
>
> The plan was debated with interest by the faculty, but nothing came of it. President Remsen was said to be timid about the danger of an embarrassing anticlimax if the undertaking should after wide publicity result in failure. Moreover, the sense of responsibility to the community was still definite and it could not be denied that the proposed college would meet the needs of only a few Baltimore boys. It is probable, too, that the Trustees, though quite willing to keep the college from running away with the graduate departments, were not blind to the importance in their finances of the tuition paid by the undergraduates whom they already had.
>
> The move to Homewood opened a new outlook for the college and ended forty years of unostentatious but solid achievement at the downtown site. The collegiate students usually numbered less than two hundred, and most of them lived in or near Maryland. Applicants from more distant places were likely to come for special reasons, for no attempt was made to recruit undergraduates from other states; and the college work received little publicity outside the *Register*. Indeed, so much was heard throughout the land of the prowess of the Johns Hopkins doctors of philosophy and of medicine that most persons assumed that the University had no collegiate department.

This unstable situation—a perfectly ordinary, routine college for Baltimore boys maintained on the excuse that it was to provide bachelors who might progress to the nationally renowned graduate school kept largely out of sight of the Baltimoreans, a college whose real purpose was to gather money paid for tuition, to placate the town, and to keep swelling the administration—was doomed to change in conformity with GRESHAM's Law. While in 1912 the school of philosophy had 179 undergraduates and 217 graduate students, even then the President's report concerns little else than the activities for and of undergraduates. In 1917, before the celebration of the first half-century, the number of undergraduates had doubled, the number of graduate students had shrunk to less than one-third of the total enrollment, and progress had developed The Johns Hopkins University—the first real university in the Western Hemisphere—into what to some it may seem today, a small undergraduate college in which some research professors and their few students are tolerated. Such was the Hopkins to which BATEMAN came. In mathematics and theoretical physics, at least, except for BATEMAN's the courses seem to have been much the same as those common at state universities in the midwest and the far west.

5. SYLVESTER AND THE FIRST HOPKINS LINE IN MATHEMATICS AND MATHEMATICAL PHYSICS

We all hear of the great tradition of Hopkins mathematics, beginning with the arrival in 1876 of SYLVESTER, the first professor and star of the first faculty, in which he was the only man older than forty-five. At sixty-one he was boiling with ideas and activity, but he was to return to England in less than seven years. The foremost mark of his influence remaining today, apart from recollection of the unfortunate quarrel with PEIRCE, is *The American Journal of Mathematics*, which GILMAN had "badgered" him to found almost from the moment of his arrival. The first list of fellows included two mathematicians: GEORGE BRUCE HALSTED and THOMAS CRAIG. HALSTED was the first man to obtain a doctorate under SYLVESTER's direction. SYLVESTER recommended that he thereafter study in Germany and wrote him warm recommendations. Although Hopkins gathered its later faculty mainly from its own graduates, HALSTED was not on it. Apparently CRAIG was SYLVESTER's favorite pupil; he remained at Hopkins for the rest of his life, not publishing any research of importance; he became editor of the mathematical journal; later faculty gossip had it that he drank himself to death.

Anyone who looks at the list of courses given in the first year will fail to find there anything to suggest a great school of mathematics or theoretical physics. In the fall semester SYLVESTER lectured to nine students on higher algebra, ROWLAND to seven on thermodynamics. In the spring semester SYLVESTER lectured on spherical harmonics; ROWLAND, on electricity and magnetism; and CRAIG, to eight students, on rational mechanics. Apart from one semester on elasticity taught by STORY, these were the only courses that could have been advanced in any sense. The main burden, carried by CRAIG and STORY, was at the level of a good preparatory school or the first two years of a college. Of course we must recall that in a society of scholars the best teaching is often done outside classrooms.

In 1878 PEIRCE, and a little later ROWLAND, urged that an effort be made to win the "rarely excellent" J. WILLARD GIBBS for a chair of mechanics. GIBBS came to give a series of lectures on rational mechanics during January and February of 1880. He was favorably inclined toward the professorship and privileges offered him shortly thereafter, but the accompanying salary, $3000, was too low to move him. SYLVESTER wrote GILMAN that "no inducements held out to [GIBBS] could be too high", and that at $5000 he would be "dirt cheap", but even in this early period of greatest financial sanguineness no further offer was made. Until that time GIBBS had served Yale as an honorary professor, unpaid. The powers there seem to have been astonished to learn that some other place might find worthy of pay such an eccentric fellow, who just sat and thought and wrote equations, and so with the magnanimity typical of university administrations they offered him a salary of $2000 to remain. Remain he did. E. B. WILSON, a student and worshipper of GIBBS, on 3 September 1953 in a conversation I straightway wrote down told me that

> GIBBS always lectured over the heads of his students and always refused to teach undergraduates at all. He knew his students did not follow him but did not alter his style on that account, having a definite idea of how the subject should be presented. He once told me that in all his years of teaching he had had only six students sufficiently prepared in mathematics and in physics to follow him; these included E. H. MOORE (later of Chicago) and myself.

Such a professor was obviously unsuited as a teacher of Baltimore boys admitted only because their parents were willing to pay the tuition. The mature wisdom of GILMAN in refusing to humor PEIRCE, ROWLAND, and SYLVESTER was in line with the retreat of Hopkins from the lofty aims with which it began.

There was to be no theoretical physics at Hopkins until BATEMAN came, over thirty years later. By then there was no longer a ROWLAND to set value on *theoretical* physics in a physics *department*, no longer a SYLVESTER to regard mathematical *physics* essential to a mathematics *department*. The faculty and program in mathematics had gained in organization both of courses and of persons, but, as HAWKINS writes in his history, "the inspiration of an ecstatic creator and living link with the mathematical past had departed with the stocky, absent-minded Victorian gentleman who was so poor an organizer." SYLVESTER's chair had been given to NEWCOMB, the astronomer, who never taught anything but astronomy. GILMAN's plan had included lecturers not necessarily taking up residence in Baltimore, possibly from the staffs of other universities and from foreign countries, and at first many eminent lecturers, for example GIBBS and Lord KELVIN, had been engaged for brief periods. Such lecturers could have saved mathematics and theoretical physics for a time. On the contrary, by 1886 those who directed Hopkins had lost their early ambition and the courage to take the risks that achievement of it had required; in that year the executive committee declared against any "long courses of public lectures, by persons from a distance."

Two years later a promising young German, BOLZA, came to Johns Hopkins as a reader in mathematics, but after one year he moved on to Clark University and to the University of Chicago at the times of their respective foundations. In some regards imitators of Johns Hopkins, they drove forward vigorously in mathematics and theoretical physics, which Johns Hopkins had effectively abandoned. BOLZA founded a major school of American mathematics centered upon the calculus of variations, but at the University of Chicago, not at Hopkins.

The last, faint glimmer of SYLVESTER's line at Hopkins seems to have been a lecture by CRAIG when EISENHART was a graduate student. That lecture introduced EISENHART to differential geometry, which he chose for his research. He studied DARBOUX's treatises on his own and received his Ph.D. in 1900. Thereupon he went to Princeton, where he took a main part in founding and leading the American school of differential geometry.

The mathematicians at Hopkins after the departure of SYLVESTER and before the arrival of BATEMAN—nearly thirty years—seem to have been research zombies. MORLEY was in charge from 1900 to 1928. The best G. D. BIRKHOFF in his survey of fifty years of American mathematics could find to say of him was "the staunch and kindly remembered British geometer".

HALSTED, who had been SYLVESTER's first pupil, was one of the Americans who took up HILBERT's axiomatic treatment of geometry, destined to have a profound and permanent influence upon American mathematics. WILDER[4] described HALSTED as "one of the most forceful personalities of American mathematics". Upon his return from Germany HALSTED went to Princeton, where he spent fifteen years. His main interests lay in the foundations of geometry and in a futile attempt to introduce some elements of clear and precise thought into the training of undergraduates, who were taught then (and long thereafter, and in many instances still today) from "practical" or "applied" cookbooks which did not so much as suggest that the quality of the ingredients listed by a recipe might influence the success of the dish.

HALSTED's most productive period, 1894–1903, was spent at the University of Texas. While there, he inspired R. L. MOORE, then an undergraduate from the Lone Star boondocks, to prove that one of HILBERT's axioms of geometry was a consequence of the rest. For his graduate work R. L. MOORE went to the University of Chicago, the faculty of which BOLZA had joined in 1893. There R. L. MOORE came under the influence of E. H. MOORE, Chicago's first mathematician, and more particularly of his student VEBLEN. One of the greatest of American mathematicians, G. D. BIRKHOFF, was also a student of E. H. MOORE, just a little later.

After taking his degree at Chicago R. L. MOORE held short appointments including a year at Princeton and nine at the University of Pennsylvania; in 1920 he returned to Austin, where he became a celebrated teacher of outstanding mathematicians. VEBLEN went to Princeton, where he founded schools of topology and mathematical logic; with EISENHART he also founded and led a great school of differential geometry.

As was to be expected of early doctors of The Johns Hopkins University, HALSTED and EISENHART were great teachers. While the line of SYLVESTER was dead at Hopkins, we may regard it as having continued to some extent at Austin and Princeton, at both of which it conflued with the line of the University of Chicago, which combined German scholarship and thoroughness with the stubborn independence, even eccentricity, that American science had adopted from American life in the nineteenth century—the independence shown earlier in physics by HENRY, GIBBS, and ROWLAND and in philosophy by C. S. PEIRCE, the independence that in mathematics was to produce its finest flower in G. D. BIRKHOFF, a son of immigrants, and in S. LEFSCHETZ, an immigrant electrical engineer from France who

---

[4] R. L. WILDER, "The mathematical work of R. L. Moore: its background, nature and influence", *Archive for History of Exact Sciences* **26** (1982): 73–97.

got his mathematical training at Clark University. The predominant influence was not SYLVESTER's but HILBERT's. HALSTED, the two MOORES, EISENHART, VEBLEN, LEFSCHETZ, and their associates and students produced, long before the influx of refugees from HITLER, a characteristically American school of geometry and topology—a school that made HILBERT's program of foundational research and conceptual analysis through explicit axioms flourish more abundantly than it did in the hands of HILBERT's own students in Germany.

6. BATEMAN AND THE SECOND HOPKINS LINE IN MATHEMATICS AND MATHEMATICAL PHYSICS

Hopkins's failure to retain BATEMAN can scarcely have grown from any ill will, for he and his wife maintained ever afterward the warmest relations with the MORLEY family. A more likely reason was the ever-increasing predominance of the undergraduate College for the youth of Baltimore. Matters were to come to a head in 1926, the fiftieth anniversary. President GOODNOW[5] was then to state

> with the ever-increasing number of students who are entering the college courses, the members of the faculty are being so loaded down with detail and routine that their time for research, and for the instruction of small groups of advanced students, is being encroached upon seriously.

GOODNOW was to propose that the first two years of college work be done away with and the bachelor's degree discontinued, and along with it "the social glamor of 'college life' and . . . stupendous athletic contests to attract students or popular fame." By 1926 it was long too late for even this compromise. Nothing was to come of the plan; an undergraduate engineering school pointed toward the needs of the community had been operating for some time, and afterward every sort of professional or semiprofessional training at the lowest level, even a school of "education", a night school, and centers of influence and charity for the neighborhood, were to drown substantial, academic research and the training of research students. "Research professor" was to become a term of denigration at The Johns Hopkins University.

---

[5] According to CHARLES K. EDMUNDS, writing in *The American Review of Reviews* for November, 1926.

BATEMAN's influence on mathematics at Johns Hopkins, despite the modest and temporary positions he held, was greater than he could have expected or is generally perceived; he himself can scarcely have been aware of it. He it was who, unknowingly, attracted MURNAGHAN; MURNAGHAN's work shows BATEMAN's influence again and again; and MURNAGHAN was retained. MURNAGHAN became a capable, devoted, and creative mathematician. Though lacking BATE-MAN's ingenuity, brilliance, and early depth, to some extent he shared BATEMAN's breadth, seeing mathematics and mathematical physics in fruitful union. He had a good eye for talent; he dared challenge the administration; while finally he chose to leave after a quarrel with President BOWMAN, he had meanwhile succeeded in bringing the Hopkins mathematics department back up to a level respectable for the country as a whole. Some of the outstanding mathematicians whom MURNAGHAN engaged (to mention only those already deceased, retired, or otherwise departed)[6] were HARTMAN, VAN KAMPEN, VAN DER WAERDEN, WHYBURN, and WINTNER.

Like BATEMAN and many other professors of former times, MUR-NAGHAN left his books to the university that he had served faithfully and well for most of his life. The Librarian of The Johns Hopkins University Library, deeming them not worth the expense of process-ing, disposed of them (along with legacies from professors eminent in other fields), at first by sale at a few cents per pound to a local booksel-ler and later, after the transpired notice of the sale had burst into scandal, by directing "work-study" students, paid for by the taxpayer, under a pledge of silence to bear them off to the incinerator. The *terminus technicus* of Library Science for this process is "de-accession".

7. BATEMAN AT THE CALIFORNIA INSTITUTE OF TECHNOLOGY

BATEMAN is associated in our recollections with the California Institute of Technology, which employed him from 1917 until his death in 1946. It would be romantic justice if he had been recognized in his Baltimorean obscurity by MILLIKAN, a latter-day GILMAN, but that is untrue. The position that BATEMAN accepted in 1917 was at Throop College of Technology, not to change its name and rank until 1920, a year before MILLIKAN took full possession. CalTech's favorite

---

[6] The *Dictionary of Scientific Biography* includes only the following mathematicians associated with Hopkins as professors or graduates: SYLVESTER, PEIRCE, HALSTED, EISENHART, BATEMAN, and WINTNER—a short list for an institution allegedly devoted to research and allegedly favoring mathematics. Doubtless MURNAGHAN would have been included if he had died in time.

graduate, a man who became a Hollywood director and thus gave CalTech much intellectual prestige and also some property, had been in fact a fraternity boy at the old Throop, from which he received his degree in 1918.

MORLEY had told BATEMAN that Throop was looking for someone, and BATEMAN had written off at once to the astronomer GEORGE ELLERY HALE. HALE advised the president of Throop to offer BATEMAN a professorship; MILLIKAN, who then held a part-time post called Director of Physical Research, concurred; and BATEMAN went to Throop in 1917 as professor of mathematical physics and aeronautical research at a salary of $2000 per year plus $500 for expenses of moving. Perhaps BATEMAN's main charm for Throop and MILLIKAN was his low price. (Forty-one years earlier, GILMAN had made ROWLAND, then twenty-eight years old and with scarcely anything yet published, first professor of physics at Johns Hopkins with a salary of $3000.) The next year Throop recruited another man from Hopkins, L. E. WEAR, who had received his doctorate in the same year as BATEMAN; he was still at CalTech in the 1940s; he did no research, he regarded himself as a good teacher of undergraduates, and he was never promoted to full professorship; he was in charge of undergraduate instruction in mathematics, and some teaching assistants regarded him as a hazard. By 1922 BATEMAN's salary had been increased to $3000, which was the average for professors there and half of the highest salary; in 1925 the figure was $3500, and the relative status the same. In 1930 his salary stood at $6000, halfway between the average and the maximum. A. D. MICHAL, who had been his closest friend, in 1948 gave me the same figure for BATEMAN's final salary. While $6000, nearly if not entirely taxfree, was a good salary for 1930, perhaps of greater value than what any university allows any professor today, by the end of the war ROOSEVELT's "belt-tightening" inflation had reduced it to little more than a beginner's level. In regard to professors, the universities had demonstrated their spotless patriotism by being the only group of employers to comply strictly with ROOSEVELT's noisy and ineffectual request that pay be kept fixed; at the lower academic levels they had forgotten patriotism entirely and followed the labor unions, doubling pay and requiring less qualification, while with clerks, janitors, and the like still less capacity was paid still more, as also for the new class of experts in administration, which was soon to save the universities entirely from the impractical influence of professors.

BATEMAN's early life had taught him thrift, and somehow he learned to invest, for he left an estate of $250,000, no mean achievement for a man of sixty-four who until he was thirty-five had barely subsisted and who thereafter had lived through the Great Depression.

MURNAGHAN wrote of BATEMAN,

> ... I remember well a feeling of amazement, mingled with discouragement, which came over me when I discovered the thoroughness of the man. He already possessed a large carefully indexed card-catalogue on each card of which was written in his minute, but beautifully clear, handwriting an abstract of a paper which he had read. I am told that in later years this card-catalogue crowded him out of his office and almost out of his home.

At some time after his arrival at CalTech BATEMAN's title was changed to professor of mathematics, theoretical physics, and aeronautics. While his three nominal chairs brought him but one salary, they did provide him at least two offices. He kept in those his files, which were not of cards but of slips of letter paper, carefully cut to size from the backs of reports of companies to stockholders. He stored the slips in shoe boxes, with few dividers, for he remembered, more or less, where each slip belonged. He lived just across the street from the campus, and he had a similarly crowded study in his home. He got up at five o'clock in the morning, went to the library when it was empty and quiet, and read the current journals in mathematics, physics, astronomy, and several branches of engineering, abstracting each paper as he read it. In his late fifties he complained that his memory was failing, for up to that time he had remembered the title of each paper and the reference, but then he was beginning to forget the page numbers and sometimes even to miss the year by one or two. He had instant and usually rather accurate recall of the entire literature of the preceding century, from LAPLACE's time down to the 1920s.

MURNAGHAN wrote of BATEMAN's lecturing,

> As time went on the scene changed and he must have changed with it, for I have heard enthusiastic reports of his lectures from students who took courses under him in the late twenties and thirties at the California Institute of Technology.

That may be so, but he did not draw large classes, and none of his courses was recommended for any particular curriculum. They were reputed to be too difficult. He did not use notes, except sometimes one or two of his abstracts of papers. He was small, stooped, and of fragile appearance, though he walked and moved briskly. When he looked at you, his pale blue eyes seemed to pierce and see right through into the distance behind. He usually turned his back to the students while lecturing; he continually bit his lips and shot his cuffs.

He covered the board systematically with immense equations interspersed with carefully written statements in elegant, flowing block capitals, for he had learnt that American students could not read a British gentleman's handwriting. He was mild, affable, and willing to be questioned, but rarely could he descend to the level of the questioner, and so students profited little from his answers, which tended to discourse upon generalization of the result which had not been understood in the first place.

Most of BATEMAN's colleagues admired him. While MILLIKAN considered him "not practical", MILLIKAN's son CLARK chose BATEMAN as his adviser for the doctorate in aeronautics, which he gained with a thesis on biplane wing theory and variational principles of hydrodynamics.

CalTech was friendly to mathematical theories of nature. There was no "social science" and no psychology; the undergraduate program in pure science required basic courses in English composition and reading, history, economics, U.S. Constitution, French, and German and allowed as electives one or two trimesters of academic philosophy, topics in English literature, business law, astronomy, genetics, *etc.* Some, at least, of the physicists appreciated BATEMAN's interest in physics, and he enjoyed evaluating difficult quadratures for them or summing their series of Bessel functions and Legendre polynomials. W. R. SMYTHE, whose required course in electricity and magnetism was designed, as he writes in his recollections, "to weed out weaklings" among the graduate students, always spoke of BATEMAN with awe. No wonder his students, shaken by the rigors of the course SMYTHE himself described as being elementary, did not put their noses through BATEMAN's door. V. KÁRMÁN repeatedly described him as the world's greatest expert on compressible fluids. While, I suppose, he did not have to teach, he regularly did so, although it was a rare student who stayed through a course. In 1940–1941 he lectured four hours a week on partial-differential equations of physics. He stuck to his principle of including in a course almost nothing that could be found in a book. Since he himself had published a tome on this very subject, the topics taken up in this course of his were often recondite indeed, and as a third-year undergraduate enrolling at the beginning of its second trimester I found it harder than I had imagined anything in mathematics could be. I earned a D at best, but BATEMAN gave me a C; for a boy who the year before had broken the record for straight As the experience could not have been sweet, but it taught me the difference between a good ordinary teacher and a great mathematician, and after that I never cared what grade I got in anything. In 1941–1942 BATEMAN lectured five days per week for all three trimesters on methods of mathematical physics; from start to

finish there were just two listeners: C.-C. LIN and I. Each day BATE-
MAN assigned three or four problems and suggested we read five to
ten articles in journals. ROOSEVELT was goading America to enter the
war on the British side, and BATEMAN, rousing to the call to the
colors, took on an additional course in the last two trimesters: in the
second, aerodynamics of compressible fluids, and in the third, potential theory, making nine hours of teaching, three times the usual
amount for a CalTech professor who regularly engaged in research. I
spent roughly twenty hours per week on each course; I tried to read
all of BATEMAN's references that could be found in the Institute's
library, but I did not succeed in doing all his exercises. Some of
BATEMAN's problems were very easy; others required hours of study
and thought. The disquieting thing about BATEMAN was that he
seemed not to discern the difference, so if you got one of the problems
straight off, you always thought there must be something wrong, and
you wasted time in ascertaining that what you had done was really the
problem assigned. Each of the half dozen problems on the final
examination required either long study in the library or the ability to
replace such study by original research on short notice. Here is an
example:

> In studying the phenomenon of mirage Biot used the law $n^2 = a^2 + by$ for the refractive index n while Tait used the law $n^2 = a^2 + b^2 y^2$. Find the rays in each case and indicate the law you prefer.

It was a rare man who solved one examination problem fully. The
notes and problems I wrote in the thirty-five trimester hours of BATE-
MAN's lectures I followed make a stack six inches high and exceed in
bulk all the rest of the notes and problems I compiled in four years at
CalTech.

In the mathematics department in the early 1940s four men left
over from before MILLIKAN's time made up the majority, over twenty
years in grade. One of these was BATEMAN. The other three had
never done any research after the Ph.D., and two of them were not
even competent to teach talented juniors; outstanding in hebetude,
they formed the pædagogic party. The gifted undergraduates went to
lengths to avoid their courses, preferring to hear any student assistant. MILLIKAN had retained for himself the headship of the division
of mathematics, physics, and electrical engineering; he ignored the
mathematics group, whose leader seemed to be E. T. BELL. The only
course BELL taught was abstract algebra; while he did little to excite
the students in that subject, he was admired for his science fiction and
his *Men of Mathematics*. I was shocked when, just a few years later,

WALTER PITTS told me the latter was nothing but a string of Hollywood scenarios; my own subsequent study of the sources has shown me that PITTS was right, and I now find the contents of that still popular book to be little more than rehashes enlivened by nasty gossip and banal or indecent fancy. Opinions differ regarding BELL's status as a mathematician. While other than BATEMAN he is the one member of the group then that the *Dictionary of Scientific Biography* includes, the article about him mentions no specific achievement. The others of note in the mathematics group were A. D. MICHAL, a pioneer in the study of generalized differentials, the group properties of partial-differential equations, and integral invariants, and MORGAN WARD, an excellent teacher whose interests lay in the theory of numbers and lattice theory. MICHAL was active in research, and most of the graduate students worked with him. Although it was not a happy department, in the years 1940/1942 the BATEMANs and MICHALs were often together, and both families were hospitable to the graduate students. MICHAL and WARD were at war with each other, and BELL made no secret of his considering abstract spaces to be "sterile junk". While BATEMAN had a low opinion of BELL, both as a mathematician and as leader of the group, he abstained from disputes. One of my fellow students who was doing research under MICHAL's direction transmitted to me MICHAL's statement, as, I think, MICHAL wished him to, that BATEMAN had no influence in the affairs of the group. I was shocked; I recall reading in INFELD's autobiography how shocked he had been when EINSTEIN told him his influence was insufficient to do anything for anybody at the Institute for Advanced Study. Now I find both cases merely ordinary for scholars in institutions of learning. When my sophomore engineers complained that I worked them too hard, BELL and WARD, and of course also the researchless faculty, took their part. I asked BATEMAN what could be done. With a compassionate expression he described the students in words still ringing in my ears: "Nothing. They want to be fed it with a spoon."

8. BATEMAN'S RESEARCH

BATEMAN's published research ranges over geometry, algebra, analysis, differential and integral equations, electromagnetism, relativity, quantum mechanics, radioactivity, optics, acoustics, seismology, fluid mechanics, special functions, and numerical calculation. A careful, thoughtful obituary by ARTHUR ERDÉLYI[7] fills ten pages but takes

---

[7] A. ERDÉLYI, Obituary of BATEMAN, with bibliography, *Obituary Notices of Fellows of the Royal Society* **5** (1947): 591–618.

up only some aspects of BATEMAN's work. He was not an interdisci-
plinarian, because for him there were no disciplines. He saw the
mathematical sciences as a continuum with no compartments; the
center of his interest was the mathematics of physical problems and
the physical interpretations of mathematics. As BELL[8] wrote of him,

> In pure mathematics, his dominating interest was in the analysis
> that has developed from classical mathematical physics. ... His
> numerous contributions to mathematical physics are marked by a
> vivid, at times almost romantic imagination.

He saw no difference in style or standard between pure mathematics
and applied, between mathematics and theoretical physics, between
physics and its application. From the great range of BATEMAN's interest you might expect his papers to be superficial, but they are by no
means so. Each one displays virtuosity, but the choice of problem is
sometimes more for its difficulty or curiosity than for its importance.
In a period of increasing specialization BATEMAN remained a natural
philosopher. He published in too many fields to be admitted to the
inner circle of cyclic citation in any of them, though he was awarded a
few medals.

BATEMAN was a star but not a leader. For example, he was the first
to publish a survey of the theory of integral equations; appearing in
1910, it presented a field scarcely a decade old and then central to
pure analysis and to some branches of mathematical physics, but of
course BATEMAN's exposition was cast into the shade by HILBERT's
masterpiece, which appeared two years later. The contrast between
the two works is typical of their authors. BATEMAN amassed everything he could find on the subject, and his report provides a full
bibliography. HILBERT included only what he thought important,
only what would fit into his masterly organization of the field; he left
the rest unmentioned and cited hardly anybody. Another encyclopædic work by BATEMAN is his report of 1932 to the National
Research Council. It is commonly referenced as "Dryden, Murnaghan, Bateman's *Report on Hydrodynamics*". DRYDEN and MUR-
NAGHAN provided eighty-eight pages of standard, old, introductory
routine on incompressible ideal fluids; BATEMAN, over 500 pages
containing compact summaries of hundreds of research papers on
viscous fluids, turbulence, and compressible fluids. His part of the
work remains classic as an annotated bibliography of published
research, mainly from the preceding 100 years.

---

[8] E. T. BELL, Obituary of BATEMAN, with bibliography, *Quarterly of Applied Mathematics* **4** (1946/1947): 105–114.

BATEMAN's darting thrusts were not consolidated by an organizing talent. He could see a complicated network of interconnections among the details of mathematics but not the structure of a theory. For example, later, to divide into chapters his famous book on partial-differential equations, he could think of nothing better than to group problems according to the co-ordinate systems in which they were most conveniently expressed. This eccentricity reflected a fatal fault which kept BATEMAN from reaching the first rank among the mathematicians of his day and rendered his books less widely useful than those of his somewhat older contemporary, E. T. WHITTAKER.

The bland, flat style in which BATEMAN lectured appears in his papers, too. For example, in 1906–1907 he introduced and illustrated the method of inverse Laplace transformations for the solution of linear differential and integral equations, but his work was not noticed, it seems, except by INCE. According to a recent history[9] of the subject, BATEMAN gave "the first example we have of the modern Laplace transform". BATEMAN began a treatise on the method, but growing interested in something else, he did not complete it, and the standard expositions of this once very popular field, books written by men whose entire lives were devoted to this one subject, do not mention BATEMAN.

Sometimes in retrospect a connected line of thought can be discerned in BATEMAN's work. From 1915 to 1926 he sought to find models for the atom by special solutions in electromagnetic theory. This approach would not win a Nobel prize in an age that gloried in revolutions and rejected Victoriana, especially since BATEMAN was not one to publish a guess as a solution or to maul the equations into submission to a preconceived answer. He did succeed in exhibiting creation of point charges, spinning electrons, and quantized radiation. One of his most beautiful discoveries is a solution of the wave equation that expresses the explosion of a singular line into a singular cylinder; another is a solution that at a given instant assumes arbitrarily assigned values upon the surface of a sphere but vanishes both inside and out.

While BATEMAN's interest centered upon problems of physics, his outlook was out of date. The difference is expressed in an extract from a letter PAUL EHRENFEST wrote to his wife from Pasadena on 11 January 1924, which Professor MARTIN J. KLEIN has kindly sent me:

---

[9] M. A. B. DEAKIN, "The development of the Laplace transform 1737–1937. II. Poincaré to Doetsch 1880–1937", *Archive for History of Exact Sciences* **26** (1982): 351–381. DEAKIN states that BATEMAN was wrong in attributing the idea to POISSON.

> Also I discussed with Bateman a few of his works in <u>mathematical</u> physics; after a few questions, in half an hour I could hardly see what they <u>deliver</u> and what they do <u>not</u> deliver. He is a dear fellow . . .—can <u>calculate</u> quite wonderfully, <u>'calculational</u> <u>intuition'</u>, but his spirit jumps about helplessly in the volcano of calculated results in his publications,—he has no <u>physical</u> sense. Not a soul understands him, and he has taken a liking to me (and I to him!!!) because I listen to him with <u>interest</u> and grasp all that he says—and sometimes help him to understand himself better.

BATEMAN wrote clearly and precisely, even in handwritten notes; I do not recall ever having seen in them floating phrases set off by repeated exclamation marks or words twice or thrice underlined. The EHRENFESTS' article on statistical mechanics, with its "models" to indicate the kind of thing that ought to come out of statistical mechanics but did not seem to, served to cleave the skeptics from the believers, and the believers' party won. It was published at the time when "physical intuition"—a term then just recently invented—was conquering research in physics. Earlier physicists had faced the basic equations of a mathematical theory with patient, scrupulous industry; they had asked them for answers, which sometimes and at long last they had succeeded in extracting from them. Physicists of the new school used mathematics not as a tool for discovery but rather as something fit to be twisted and mangled in the course of rhetoric aiming to persuade the reader of conclusions which their "intuition" had revealed to them. Hence arose the modern distinction between "mathematical physics" and "theoretical physics".

I hope that some day an expert on classical electromagnetism as a mathematical theory will go through BATEMAN's work, which with modern notations, concepts, and experience will no longer seem complicated, and evaluate it for what it is, not for its failure to be intuitive quantum physics.

9. BATEMAN'S REPUTATION

It is perhaps inevitable that a man's memory be colored by recollections of him in his last years. BATEMAN died in 1946. Those who knew the BATEMAN of the 1930s and 1940s recall him awash in definite integrals, special functions, and clever formal transformations. This impression is reinforced by the volumes written by the Bateman Project, which was directed to arrange and publish the material BATEMAN had left unfinished. The volumes it did produce are valuable and reliable handbooks, but their authors were not in sympathy with BATEMAN's encyclopædic brilliance and made little use of the contents of his shoe boxes. One of those authors, WILHELM MAGNUS, informs me that he found there as many as twenty definite integrals separately filed when in fact they were special instances of a single one in the same file. Indeed, BATEMAN's slips could scarcely have been used by anyone else, since he relied upon his astonishing memory to find his way among them, classified as they were by accidental traits rather than any intrinsic order. Another of the authors, FRANCESCO TRICOMI, later wrote in his autobiography that BATEMAN had been "a man of scant mathematical culture". In fact TRICOMI never made any attempt to understand or use BATEMAN's work; with the egotism common to many mathematicians, he simply wrote out his own treatment of confluent hypergeometric functions, which is far from exhaustive in content or just in citation. Perhaps a heavier example is the judgment of AUREL WINTNER, long the dragon of the Hopkins mathematics department. HUGH DOWKER told me in 1944 or 1945 that WINTNER had once bet he could find an error on a page taken at random from BATEMAN's monumental *Partial Differential Equations of Mathematical Physics*; WINTNER won the bet, finding that a certain statement was in fact false in just one case, a trivial one, which BATEMAN had neglected to remark. Styles of mathematics do change, and today neither TRICOMI nor WINTNER, despite the respect they deservedly earned, would be regarded as much less old-fashioned than BATEMAN. The young mathematicians of Italy called TRICOMI in his last years "the greatest nineteenth-century mathematician the twentieth century has produced"; I have never heard a single word of praise for WINTNER's research from his surviving colleagues. COURANT, who was a specialist in partial-differential equations and a masterful organizer in two senses of the word, wrote of BATEMAN's book, "there is no other work which presents the analytical tools and the results achieved by them equally completely and with as many original contributions . . .." Of course any statement by COURANT has to be weighed against what he fancied he might gain from making it. BATEMAN's book is for virtuosi, profitless to those not already familiar with the field as it stood in the 1920s.

My review of the products of the Bateman Project, published[10] in 1954, includes the following passage:

---

[10] "Review of '*Higher Transcendental Functions,* . . . , based, in part, on notes left by Harry Bateman. By the staff of the Bateman manuscript project . . .', New York, McGraw-Hill, 1953", *Bulletin of the American Mathematical Society* **61** (1954): 576–578. In reprinting this extract I have altered one phrase because as first published it did not convey accurately what I meant to express.

A few mathematicians will be disappointed by this work. To them I can remark that, having known BATEMAN toward the end of his life, I think it unlikely he would have finished either of his planned works no matter how long he had lived. As can be seen from the sequence of books he did finish, in him the [common British distaste for structured (pedantic!) exposition] grew with age to monstrous intensity, and the huge task of organization before actually beginning to write these two works, he was reluctant to face. On the other hand, the staff which has written the work under review has taken up a large responsibility, for to them alone was given any opportunity to make use of the great mass of material left by BATEMAN, an opportunity, as they tell us, they have decided to decline. It seems unfair both to BATEMAN and to the distinguished authors themselves that BATEMAN's name, not theirs, appears on the title page, which is cluttered besides with government gobbledegook.

The analysts of the twenties and thirties turned away from the "classical" approach with its formulæ and explicit calculation, preferring instead to seek generality, method, and idea. To many trained in this "modern" line of thought, works such as BATEMAN planned are a voice from the past, of interest only to "applied mathematicians". Perhaps there was a trace of this view behind the decision to write a handbook instead of a treatise. If, as one often hears, special functions are used only by numerical practitioners who do not have a large XXXAC at their disposal, the compactness and selection of the present work are advantages. We note that on one of the back covers the publisher advertises engineering handbooks. But, while the children of the "classical" analysts tend to refer to them with a shade of patronism, sometimes grand-children are closer than children. With the resurgence of applied *mathematics* (not merely *application* of mathematics), a noticeable trait of the mathematical scene of the fifties, it is possible that exhaustive treatises in analysis may come into the regard that the classical exhaustive treatises on hydrodynamics and elasticity have again and rightly been granted. That the handbook scheme permits a reduction in *size*, to which the introduction several times refers, may not always be a recommendation. It is to be hoped that BATEMAN's notes remain intact for possible future use.

I am told that now no trace of the contents of BATEMAN's shoe boxes can be found at CalTech, though two empty boxes survive in a professor's office. ERDÉLYI left a note indicating that he had weeded BATEMAN's correspondence.

Anyhow it is wrong to regard BATEMAN as being no more than a technician in the minutiæ of classical hydrodynamics and special functions. His early work is mainly in pure analysis, with a sprinkling of geometry. Mr. MAGNUS recalls having seen three different versions of an essay on binomial coefficients, each manuscript starting from the beginning, and all three with unnumbered pages. I recall that he told me in 1949 that had this essay been published in the early 1900s, when it was written, it would have anticipated several discoveries of later persons. Indeed H. W. GOULD[11] cites an unpublished manuscript of 520 pages on binomial coefficients that ERDÉLYI allowed him to study. I am informed that the archives of CalTech still contain one such manuscript and also more or less finished manuscripts on four other subjects. BATEMAN's unpublished work provided GOULD with a particular integral formula that suggested to him a general functional transformation which seems to deserve further study. I cannot help wondering if the shoe boxes contained further treasures that sympathetic heirs might have recognized.

While BATEMAN's late papers suggest primary interest in formal manipulation, he was by no means unaware of the progress of rigor in analysis. Professor ANGUS TAYLOR has given me an example of BATEMAN's erudition in pure mathematics:

> About 1934, while studying the independence of the axioms for a vector space, I decided there might well be a real function f of a real x, other than $f(x) = cx$, such that $f(x+y) = f(x)+f(y)$. I asked Bateman if he knew any literature on this functional equation. He promptly (within a day, or perhaps even right away—I don't recall for sure) gave me the reference to the paper by G. Hamel in which he establishes just what I wanted to know, by means of a "basis" for the reals—i.e., a set of reals, linearly independent over the field of rationals, such that every real number is a finite linear combination, with rational coefficients, of basis elements.

If we were to judge the fields of BATEMAN's papers by their titles, we should have to say that his abiding interest lay in electromagnetic theory, but as that subject lends itself to almost every kind of mathematical thought, BATEMAN's devotion to it reflects also the universality of his interest in the mathematical sciences.

BATEMAN's old problems never died for him, and he kept returning to them. When, in 1941, I asked him for a topic of research, he gave me an offprint of a paper on potential theory he had published

---

[11] H. W. GOULD, "Generalization of an integral formula of Bateman", *Duke Mathematical Journal* **29** (1962): 475–479.

in 1915, saying he had not yet gotten back to follow up some ideas put forward there. This was my first contact with Johns Hopkins, though at that time I did not know I was having it, for I had not heard that there was a mathematics department connected with the famous hospital in Baltimore.

The contemporary who most appreciated BATEMAN's work was also one of the greatest American mathematicians: G. D. BIRKHOFF. In his article "Fifty years of American mathematics" he wrote as follows[12] regarding applied mathematics:

> In default of a better term we use the designation of applied mathematics for that large part of mathematics which seems to be closely connected with physics or some other branch of science. Inasmuch as most of the so-called "pure" mathematics of the present day was at one time "applied," the term is a very vague one. Nevertheless, the field of applied mathematics always will remain of the first order of importance inasmuch as it indicates those directions of mathematical effort to which nature herself has given approval.

Unfortunately, American mathematicians have shown in the last fifty years a disregard for this most authentically justified field of all. It was remarked at the outset that the American tradition was at first of quite the opposite character. Nevertheless today we recall only six Americans who are deeply concerned with applied mathematics in the usual sense, of whom four were brought up in the great British tradition. These are Harry Bateman, Ernest W. Brown (recently deceased), F. D. Murnaghan, H. P. Robertson, J. L. Synge, and R. C. Tolman. Among these men it should be remarked that Brown was the world's foremost lunar theorist, while Tolman is to be regarded as primarily a physical chemist. All six men possessed an extremely broad scientific outlook. The names of Bateman and Tolman will always be mentioned among those who were closest in spirit to the special theory of relativity at the time of its discovery. Furthermore, Bateman has added to classical electromagnetic theory while Tolman has contributed to the relativistic theory of the expanding universe in which he has shown his daring speculative spirit. Robertson has also contributed in the same relativistic direction. Murnaghan and Synge alike have been creatively interested in geometry, dynamics, classical hydrodynamics and elasticity, and relativity.

---

[12] Pages 270–315 of Volume **2** of *Semicentennial Addresses*, American Mathematical Society, 1938 = pages 606–651 of GEORGE DAVID BIRKHOFF, *Collected Mathematical Papers*, Volume **3**, New York, American Mathematical Society, 1950.

After the passage of nearly fifty years we have a different perspective on these men. The type of work BROWN did has become the province of computing machines. TOLMAN was a physical chemist of some mathematical competence who wrote mainly vague and wordy expositions; of him, the *Dictionary of Scientific Biography* records only outstanding qualities and high office, no achievement. ROBERTSON was a physicist who never redeemed his early promise; the *Dictionary* omits him. BATEMAN's greatest contribution was not to special but to general relativity and the foundations of electromagnetism, and it is permanent, as I shall presently explain.

By the time BATEMAN left Hopkins, he had written about half of his life's output in papers and three of his six books and major expositions. It is tempting to suggest that in accord with PARKINSON's Law his best work was finished before he got a decent job. I think that this is true *a fortiori*, because he had done his best work even before he came to Hopkins. The obituaries remark that BATEMAN in 1908–1909 proved the electrodynamic equations to be invariant under conformal transformations of a space of four dimensions, thus recognizing the conformal Lorentz group and providing a step toward general relativity. WHITTAKER, who in his *History of the Theories of Æther and Electricity* aroused colossal antagonism by trying to set the record of relativity straight on the basis of print and record rather than recollection and folklore and professional propaganda, perceived the importance of BATEMAN's analysis and its priority to EINSTEIN's work of 1912–1914 but did not sufficiently grasp its contents. BATEMAN, who always assumed that the total charge was conserved under the transformations he considered, went on to prove the converse, that only under conformal transformations was the form of the electromagnetic equations invariant. Nobody gave any other reason for considering this group or explained why it should have any physical meaning.

The importance of BATEMAN's paper lies not in its specific details but in its general approach. BATEMAN, perhaps influenced by HILBERT's point of view in mathematical physics as a whole, was the first to see that the basic ideas of electromagnetism were equivalent to statements regarding integrals of differential forms, statements to which GRASSMANN's calculus of extension on differentiable manifolds, POINCARÉ's theories of STOKESIAN transformations and integral invariants, and LIE's theory of continuous groups could be fruitfully applied. To see now in this great paper of 1910 the arsenal of pure mathematics BATEMAN expressly and unobtrusively brought to bear upon this basic physical problem, besides giving the reader an uncanny sense of recent thought expressed in archaic notations, makes it clear why the Johns Hopkins of 1912–1917 was no place for BATEMAN. First there is BATEMAN's decisive step of freeing electromagnetism from a metric space-time. Though he himself did not say so, clearly there is no empirical basis for any idea of distance, real or imaginary, between events that are not simultaneous. Second, as BATEMAN with characteristic understatement remarked, his integral equations provide "a very concise expression of the electrodynamical equations, which promises to be of considerable importance in future development." They assert, first, that charge and current are always interconvertible and that the total charge-current in each three-dimensional submanifold without boundary is conserved, and, second, that the magnetic flux through each two-dimensional sub-manifold without boundary is conserved.

The future developments had to wait, as had those of GRASS-MANN's calculus, for many years to pass. KOTTLER in 1922 and VAN DANTZIG in 1934 called attention to BATEMAN's results and made passing use of them. In 1958 Mr. TOUPIN recognized their power. He remarked that if total charge were regarded as an axial scalar rather than an absolute one, BATEMAN's electrodynamic field equations would become universal, invariant under all changes of space-time co-ordinates. Conformal invariance is essential only for the MAXWELL–LORENTZ æther relations, which endow the space-time manifold with a special structure dictated by the physics of electromagnetism. At last it becomes concretely conceivable that explanation of all gross phenomena could be based upon electromagnetic ideas, with no prior concept of distance in space-time. TOUPIN writes[13], "[These *Maxwell–Bateman laws*] are *the cornerstones of electromagnetic theory.* The laws of nature embodied in these postulates are perhaps the most lasting achievements of the classical theory of electromagnetism." I have been told that some physicists are now convincing themselves in their own way that charge should be regarded as an axial scalar in space-time. If they do so, they will find ready to hand a systematic, general exposition of electromagnetism on that basis, published by TOUPIN[14] in 1960 in one of those volumes of the *Handbuch der Physik* that physicists usually set aside as being only pure mathematics or engineering.

---

[13] Page 667 of C. TRUESDELL & R. TOUPIN, "The classical field theories", pages 226–793 of FLÜGGE's *Encyclopedia of Physics*, Volume **III**/1, Berlin *etc.*, Springer-Verlag, 1960. As BATEMAN tells his readers, the integral statements can be found in a paper by R. HARGREAVES, "Integral forms and their connexion with physical equations" (1980), *Transactions of the Cambridge Philosophical Society* **21** (1912): 107–122. That paper concerns purely algebraic transformations and does not enter into the physics of electromagnetism or anything else.

[14] TOUPIN's theory of electromagnetism and gravitation is presented systematically, on the basis of his published lectures of 1965 at Bressanone, in Chapters 7 and 8 of C.-C. WANG's *Mathematical Principles of Mechanics and Electromagnetism*, New York & London, Plenum Press, 1979.

I have dwelt upon this paper of BATEMAN's for several reasons. I think it is his best. Its example, with its explicit calls backward half a century and its now proven message forward another half century, demonstrates better than any preaching could that the truly mathematical sciences make one whole, in which pure mathematics and its *bona fide* applications blend together, and true progress, tranquil and contemplative, in its short and often hesitant steps looks not only forward and sidewise but also backward. Finally, it is Mr. TOUPIN who as director of mathematical research at I.B.M. has made our BATEMAN lectures possible.

One of BATEMAN's last papers is called *The Control of an Elastic Fluid.* It was delivered as a Gibbs Lecture of the American Mathematical Society in 1943. That annual lecture, devoted expressly to the applications of mathematics, is often entrusted to a pure mathematician whose closest contact with applications is a few old terms wrested beyond recognition, to a physicist wandering in realms of abstract operators whose domains, codomains, and operational rules float forever in the wastes of intuition, or to an engineer who detests mathematics and wishes to tell the mathematicians why. The choice of 1943, a real mathematician who devoted himself to real applications, was most unusual. BATEMAN's words express the footed assay of a man who stands above his own time, seeing both backward and forward:

> Mathematicians should pause periodically in their own work and peruse the progress in astronomy, biology, chemistry, economics, engineering, and physics to see if recent advances in these fields suggest problems of mathematical interest. One reason why the Gibbs Lectureship was founded was, indeed, to facilitate a fruitful friendliness between mathematicians and other scientists.
>
> The subject of control is now very important and promises to be so in the future. Much has been written about the control of the air, the control of ships, airplanes, balloons, bombs, gliders, robots and torpedoes. The regulation of rotation became important in the early days of the telescope and steam engine. The related problem of stability is important now for electric motors, marine engines, hydraulic turbines and the generating plants for the distribution of gas and electricity for there is generally an economical speed of operation. In radio telegraphy a certain speed may be needed in order to get a desired frequency.
>
> Controls are necessary in the chemical industries and in mining. They are useful in entertainment and were much needed

when arc lights were used for illumination. Fountains which begin to play automatically at sunset are used at exhibitions. Appold's home in London had many automatic devices to interest visitors.

The control of conditions under which observations are made is of great importance to the astronomer, the physicist, and the aeronautical engineer. The designer of an engine plans to regulate the flow, pressure, temperature and composition of his working fluid so that the engine will run smoothly and economically.

The control of combustion may be important not only for economical reasons but also to avoid the production of smoke. On the other hand this production may be desirable sometimes when a smoke screen is needed. In such a case there should be flexibility of control. The subject of control is important also in refrigeration, air conditioning and the preparation of food. Great attention is being paid to human comfort. We are in an era of air conditioning on a large scale and this requires the solution of many problems of control. It is now understood throughout the land that the provision of the proper atmospheric conditions for the comfort of workmen and the performance of good work is even more important than the regulation of the supply of air and fuel to an engine. Precise weather is needed for precision work and for the manufacture of instruments of precision such as gauges. Proper air conditioning is needed for the production of quality fabrics. The proper temperature must be maintained when stained glass windows are being made. In small arms munition works where dry explosives are handled there is inevitably a certain amount of dust and for safety the amount must be regulated. A gas company must regulate the pressure of gas which it distributes and must also regulate the composition so that an escape of gas may be readily detected by the odour of the escaping gas. Controls are needed for the safety of miners and of workmen in many industries. In the purification of drinking water the rate of supply of chlorine must be regulated.

The subject of control is clearly an enormous one and it is well to bear in mind that advances made in one branch of the subject are sometimes useful in another.

The paper is forty-six pages long and has 112 footnotes; one of these contains eleven references to works with titles such as "electrical dust and fume precipitation", published by the Institute of Mining Engineers; other footnotes cite a history of air-conditioning by W. H. CARRIER and an article by HANDLEY PAGE on the HANDLEY-PAGE wing. BATEMAN's references run back 100 years in the most natural way; he writes of POISSON and JACOBI, two of his favorite authors, as if they were still living, and in fact he sometimes takes up a problem just where one of them left it. We are not astonished to find cited HUYGENS's book on the pendulum clock, published in 1658. All references are to points BATEMAN raises. The body of the lecture is thoroughly mathematical and displays among other things a mastery of theories of stability altogether extraordinary for 1943. The contents was a decade ahead of its time in scope and nature, but the style of presentation was already old-fashioned for mathematicians; it was also too abstract for the engineers then. To walk out while a man was speaking was considered impolite in those antiquated days, but the response of the audience allowed no doubt that the lecture had fallen flat. The mathematicians were not interested in air-conditioning or parachutes or helicopters; they were already accustomed to statements in the terminology of abstract algebra and topology, which BATEMAN did not use; and stability theory, which BATEMAN included in almost every course he taught, was long out of fashion and not yet ready to come back in. In rereading BATEMAN's virtuoso lecture I have the impression that if it were to be delivered today, it would be appreciated and largely understood by an audience of research engineers in mechanics, though they would know some theories more recent than those reported in it.

10. The Bateman Lectures at Johns Hopkins, 1976–1979

BATEMAN had devoted his life first to out-of-date physics, then to out-of-date mathematics, and at the end to engineering of the future. The announcement of the first series of lectures in his honor at Johns Hopkins reads as follows:

> The Harry Bateman Lectures in natural philosophy are conceived in BATEMAN's spirit. The [first series is] devoted to a subject which, while it has roots going back three centuries, has flowered in the thirty years since BATEMAN's death: the mathematical theory of finite elastic strain. Though concerning physical phenomena of everyday occurrence, this subject can be approached fruitfully only with the tools of modern pure mathematics . . .. In return, the phenomena of finite elasticity may point the way toward wholly new avenues in the theory of partial differential equations. Many of the particular problems of the subject arose in contexts of engineering, and solutions of those problems have been put to use in the manufacture of polymeric substances and the design of highly elastic structural members.

The first lecturer, STUART ANTMAN, was introduced as

> a man who regards natural philosophy as BATEMAN did, a man for whom pure mathematics and its physical interpretation illuminate each other. By proving that non-linear theories of elastic rods do allow the possibility of necking and shear instability STUART ANTMAN has solved a major problem which had been standing open for two centuries. He is today of the same age as was BATEMAN at Hopkins. Let us not draw the parallel too closely. Mr. ANTMAN is a prime expositor and is the author of the standard reference work on non-linear theories of elastic rods, published four years ago in the *Handbuch der Physik*. Furthermore, he is neither unrecognized nor, as professors go nowadays, impoverished. After his doctorate in aeronautics and engineering mechanics he went to the Courant Institute as a visiting member, soon gained academic rank, and was promoted almost at once. He became a professor of mathematics just six years past his doctorate. He is already everywhere respected as one of the handful of leaders in the thrust to master the partial-differential equations of finite elasticity by creating the mathematics their solution requires. Whether he will leave an estate of millions of dollars, remains for later generations to learn.

> The audience today is a small one, but still too large for the example I had hoped to follow. I was expecting exactly twenty-one. You will recall that when Lord KELVIN, then Sir WILLIAM THOMSON, came to lecture here in 1884, he chose as his subject "Molecular Dynamics and the Wave Theory of Light". Though the title may not sound that way today, to KELVIN physical optics was an application of the theory of elasticity, and the main authors he cited were CAUCHY, POISSON, ST. VENANT, GREEN, and STOKES. KELVIN referred to his hearers as his "Baltimore co-efficients" because their number, twenty-one, was the maximum possible for the independent elasticities of a crystal according to GREEN's theory.

NOTE ON THE BATEMAN LECTURES

The first series of BATEMAN lectures was offered by the Natural Philosophy Group as its (unsolicited) contribution to the centennial of The Johns Hopkins University in 1976. It and the four subsequent series were financed solely through the International Business Machines Corporation's graduate scholarships in mathematics granted to RICHARD JAMES (1975–1978) and CHI-SING MAN (1978–1980).

The sketch printed above is based on the introduction to the first series, read on 15 March 1976. In preparing it for the press I have corrected some factual details and added others made available to me by Professors MARTIN J. KLEIN, WILHELM MAGNUS, KNOX MILL- SAPS, and ANGUS TAYLOR, by the archivists of the California Institute of Technology and The Johns Hopkins University, by Dr. JOHN L. GREENBERG, and by Mr. CHARLES PURCELL; I have profited from some corrections by Professor T. APOSTOL and some suggestions by Professors SAMPSON and WILDER. None of these bears any responsibility for such errors of fact and judgment as remain.

The speakers and subjects of the series were as follows

1976 STUART ANTMAN
Nonlinear Analysis and Nonlinear Elasticity (4 lectures)

GIANFRANCO CAPRIZ
Signorini's Perturbation Method (1 lecture)

1977 CONSTANTINE DAFERMOS
Non-linear Hyperbolic Systems, Continuum Mechanics and Thermodynamics (4 lectures)

1978 JAMES SERRIN
The Concepts of Thermodynamics (4 lectures)

1978 JOHN M. BALL
Constitutive Inequalities and Problems of Existence in Finite Elastostatics (4 lectures)

1979 DAVID OWEN
The Concepts of Accessibility and Restorability in Classical and Modern Thermodynamics (4 lectures)

Requests to the University and to outside sources for funds to continue these lectures fell upon deaf ears.

PART IV

TRAINING

36. THE SCHOLAR'S WORKSHOP AND TOOLS (1970, 1976, 1981)

Just as we may picture an ideal scientist, we may set forth qualities ideal in one who is to trace the development of science. I do not use the terms "history of science", "physics", or "mathematics", since these now denote established professions, while a great scholar, although he must stand on the same ground as his fellows, is taller and sees above their heads.

Before creating a work of art, an artist must have first a workshop and then a good set of tools and the skill, discipline, and taste to use them to good purpose. Only after these have been acquired and arranged can his conception be given form worthy of its inspiration and plan.

The workshop of the scholar in the history of science is the periods in which his authors lived. He should know those periods' ways of life and belief and education, both the common and the eccentric; their political histories; their variety in aspects; their social and economic structures; their architectures, literatures, and arts. He should feel at home in houses of those times, sit easily in their chairs, both figurative and wooden, and discern what was then mostly admired or rejected in painting and sculpture and decoration. He should have read not only the books that carried the intellectual products of his period but also those that were then the fare of young minds as they were taught, such books having been commonly of an earlier time. For example, a reader of GALILEO who does not have at hand the favorite situations in *Orlando Furioso* is at a disadvantage, and the student who does not command, as a minimum, the main episodes of Holy Scripture, classic mythology, and the corpus of golden Latin is glaucomatose in the modes of thought of Western men educated before 1900. Philosophy enters in the same way, as a part of general education, since philosophers have mainly followed at the respectful distance of a century and expatiated upon insights of science after they were already superseded in science itself; although indeed some few scientists were active also in philosophy, mainly they did not so much enter the arena of their own day as react, as bright boys will, against the bilge they had been made to ingest while under the schoolmaster's lash.

In addition, the scholar in the history of science should know the lives of the scientists if they are available or if he can construct them. For such men as HUYGENS and EULER, at least the outlines of their biographies seem indispensable if we are to enter into converse with them, yet we must admit that had all their personalia been destroyed, they would still live in their works, and we should still have to meet them, even though only through their surviving writ, as today we must meet EUCLID and PTOLEMY if we are to meet them at all. Our scholar, while setting just value upon biographies of scientists and accounts of scientific mores and fashions, will not fall into the current puerility of confusing these ancillary social sidelights with science itself. In common fairness and common sense, he will ultimately estimate the scientific work of NEWTON and D'ALEMBERT with the same standards as he uses for those of the unknown JORDANUS and the perhaps composite HERON.

We come now to the tools with which our scholar will stock his workshop. The tools of science are languages, logic, and the faculty of criticism.

Languages are of two kinds: the common tongues of man, and the defined vocabularies specific to science. Not all scholars need the same languages, but the scholar ought to know well such languages as his authors knew well, for it was in those that they spoke, read, and thought.

There is first the matter of mere translation. Translators often understand at most the simpler parts of the works of science they translate, and translators of a later period tend to admix words rooted in ideas they take for granted because they met them in school. That such translators may produce ninety-nine decent sentences out of a hundred, is not enough, for the gross mistranslation of the remaining one percent of the text occurs, fatally, always at the critical passages, the very ones where a scholar ought to pause and ponder each word, and here the translators almost without fail give him short weight, or, even worse, overweight. For example, many readers have been misled, and many more will be misled when they consult the only history of hydraulics in English to find DANIEL BERNOULLI's own statement of his most famous theorem. The details may be read above in Essay 26. The American authors, unable to read Latin, thought the French professional historian DUGAS could be trusted as a translator. On the whole, he could, but if there is such a thing as truth in history of science, disastrous mistranslation of the two most important sentences in a major source cannot be counterbalanced by any number of pages correctly rendered routine from the same work. My friend the late

LADISLAO RETI[1] has provided two even finer instances of gross mistranslation by the editors of LEONARDO DA VINCI's codices, "scholars of outstanding literary merit, but insufficiently versed in scientific and technological matters". The first:

LEONARDO, Ms. I 39v:

albero della folla del pappiro

RAVAISSON-MOLLIEN:

tree of the leaf of the papyrus

RETI:

camshaft of a fulling mill for paper

The second:

LEONARDO, Ms. F 9br:

Il lapis si disfa in vino e in a ceto o in acquavite e poi se può ricongiugnere con colla dolce.

RICHTER:

Chalk dissolves in wine and in vinegar or in aqua fortis and can be recombined with gum.

RETI:

Hematite is dispersed in wine and in vinegar or in strong spirit and can be put together again with glue.

In the days of the vogue for undisciplined, unbounded claims that LEONARDO had discovered nearly everything, this second passage was used to infer that he had invented the technique of pastel.

Sin exists in this vale of tears and cannot be eradicated, but the saintly attempt to eschew it. It may be true that translations, like secondary sources, sometimes have to be consulted, but the forearmed scholar should shut his eyes to both except *in extremis*. While he runs the chance of error or omission, it is better for him to slip a dozen times on his own in good faith than to parrot a single blunder by some preceding historian. Few if any parrots cite their sources, and if they do, it is by rote, their teacher's sources at nearest, not their own. A historian who quotes with acknowledgment is thereby demoted to scholiast; by so cutting his risk, he cuts his rank.

Our scholar must know the changing usages of words and must not by prolepsis inject later meanings into the terms his authors used.

---

[1] Pages 67–68 of Volume III (commentary) of LEONARDO DA VINCI, *The Madrid Codices*, New York *etc.*, McGraw-Hill, 1974.

For example, the English word "mechanical" in NEWTON's day meant "working like a machine, having a machine-like action, acting or performed without exercise of thought or volition". These are definitions given by the *Oxford English Dictionary* and unequivocally supported there by quotations from NEWTON's period; the usage of "mechanical" as being different from "chemical" is only 150 years old, and the separation of "mechanical" from "electrical" is too recent to be listed in that dictionary, published some seventy-five years ago, or even in the *Supplement*, 1976. The forces of chemical bonds and of magnetic attraction or repulsion were in NEWTON's day, by mere convention of language, mechanical forces[2]. Also, well into the nineteenth century, "particle" or "molecule" meant "a little part" or "a little heap", a vague word which sometimes stood for a discrete entity, a prototype of modern molecules, but at least as often indicated merely a vanishingly small element of a plenum. When a word is inclusive, the scholar must learn to infer from the context any special sense it may have and to refrain from giving it one if there is no such context, this latter case being much the commoner. In his use of words in their modern senses a certified, indoctored historian of science today may be guilty of that very "present-mindedness" with which he cantingly reproaches those unenlightened and outdated folk who pronounce old scientists to have been "right" or "wrong" in terms of the doctrines today accepted. Is it not possible that failure to maintain the ordinary standards and methods of textual criticism is the main source of the journalism about "Newtonians" and "anti-Newtonians"? May not this division be no more than a fancy that

---

[2] To Dr. JON DORLING, who disagrees with me here, I am obliged for some interesting quotations. As he has remarked, CHAMBERS' *Cyclopedia* refers "mechanical powers" to "the six simple machines", namely, "balance, lever, wheel, pully, wedge, and screw". This usage seems to me ridiculously archaic for the *practice* of NEWTON's day, especially since the six superannuated relics of classical antiquity just listed could not possibly be wrested into an explanation of several situations NEWTON does treat in the *Principia*, which he clearly does regard as a book on mechanics: efflux of water from a vessel, the speed of advance of waves, the propagation of sound, the internal friction of fluids. Thus, I should say, CHAMBERS on this point was simply uninformed. Dr. DORLING does cite one author of the eighteenth century, THOMAS JOHNSON, who uses "mechanical" in a strictly CARTESIAN sense, but I do not see that a single author's preference for narrow and delimiting definitions, however commendable, shows his own special usage to be universal or even common in the language of his period. JOHNSON is certainly not an author whose opinion any expert on mechanics today would regard as important, or whose work any historian would regard as dominant in the period.

Nonetheless, Dr. DORLING's ability to argue the point with me is of greater importance than whether he can bring me over to his opinion, since it illustrates the very quality I here lay down as a necessary one for scholarship, namely the capacity to make an informed judgment of the contemporary meanings of words.

historians read into a past whose language they confuse with their own? Perhaps "scientific revolutions" are distortions, illusions fostered by those who, trained in progressive social history today, exalt revolution wherever it raises its head, to the point that in order to sell a new, particularly disgusting color of lipstick, the manufacturer need only proclaim it "revolutionary" while marking it by a suitably hideous brand.

Our scholar must be able to edit a text at need, though such work should be at most incidental for him. The capacity to edit, like the capacity to calculate, may be required at any time in the course of historical scholarship, and it should be ready on call. A yet unedited text, like a yet unsolved equation, may block the road to understanding, and the scholar cannot wait years or decades for someone else to open it.

The languages of the second kind are those peculiar to science. These range from the concepts, expressed by words common to all Classical or Western languages of the period but by no means common to all men who spoke those languages, which represent ideas in the sciences themselves, to mathematics, in which some but not all of those ideas are subjected to operations as clearly stated as the rules of chess. Just as the rules of chess are obeyed by all the players but do not dictate their moves, so the rules of mathematics do not at all exhaust the subject but merely delimit it. Of course there are many disciplines within mathematics, and our scholar must take care to know all that were familiar to the author he studies. These include, obviously, not only the ones today called "mathematics" by professional mathematicians but also what was regarded as mathematics formerly and hence known to any competent theorist. For example, to MAXWELL mathematical hydrodynamics was mother's milk, and with it he drank in both field concepts and methods of handling them with precise logical arguments. If our scholar wishes to study MAXWELL's creation of electromagnetic theory, he should know the contents of, say, the first edition of LAMB's *Hydrodynamics* forward and backward, just as he must know calculus; otherwise, he will be gravelled at many a point that to MAXWELL needed no explanation at all, or he will attribute to MAXWELL as innovations concepts MAXWELL himself took for granted from his schooling and applied easily to new uses, with no illusion he had invented them. The fact that hydrodynamics is not taught in departments of mathematics or physics in today's computerized education mills is as irrelevant to its central role in the creation of electromagnetism as is the fact that MAXWELL had not had the benefit of a course in Marxism as a basis for Liberal thought. This remark might seem so obvious as to be idle, yet I have seen accepted by departments of the history of science in the most renowned of our universities thesis after thesis which gave evidence on every page of blank and self-righteous ignorance of the parts of the mechanics of continua which were taught regularly to mathematicians and physicists in every decent university throughout the very century whose achievements the juvenile authors of those theses claimed to explain, interpret, and motivate.

Our scholar, of course, is not to speak only to himself. He must be able to translate his authors' words into a language that can be read now. This language, too, is double: both the common speech of our time and also a mathematical dialect that can be understood fairly easily today by an educated layman—a layman in history but reasonably proficient in mathematics. If such a layman sees a short text of ARCHIMEDES translated merely into English, he will find it incomprehensible without days of study. Even a historian so old as HEATH recognized this fact; his translations, in a sense double ones, are next to useless to a working mathematician now because the dialect of mathematicians has changed greatly since HEATH's day, much more than has the English language, and so they must all be done over. In order to translate into a speech comprehensible today, our scholar must know a good deal also of what is now regarded as mathematics. To a historian I need not say how delicate if not dangerous this kind of translation is, but it is one he must undertake, however much he may deplore the need for it. Thus he must strive to emulate both his authors and his best contemporaries in the style of his languages, the common one and the mathematical. History is not isolated in time. It is a bridge between past and present.

With mathematics come, of themselves, logical standards. Thus our scholar will not fall into the trap of confusing symbolism with mathematics. For example, he will recognize nearly all of GIBBS's difficult arguments in words as being fairly strict mathematics though not abbreviated by use of mathematical symbols, and he will equally easily perceive that some long strings of symbols and equations published by LAPLACE and FOURIER and WRONSKI, not to mention THOMAS JEFFERSON JACKSON SEE, range from the unconvincing to the ridiculous and hence are not mathematical at all.

Our scholar must become fluent not only in the mathematics of the period he studies but also in those of preceding ones, for mathematics, as everyone has agreed until the most "revolutionary" historical disclosures of the last few years, is an accumulative science, in which truth, unlike entropy, does not decrease[3]. Thus he will be

---

[3] This simple fact makes it less difficult (for a competent mathematician) to clear the history of mathematics than for a biologist to learn the history of the biologies. *Cf.* ERWIN CHARGAFF, "Preface to a grammar of biology", *Science* **172** (1971): 637–642:

unmoved by those ninnies who say that because the standards of
mathematical rigor have changed in time, rigor itself is merely subjective and hence of no account, as if the change in vocabulary and style
from CICERO to AUGUSTINE to JORDANUS to NEWTON were good
reason to refuse to study Latin at all! The term "rigor" as used by
physicists and historians today seems often to refer to a particular
style of εs, δs, and convergence proofs in vogue about the year 1900,
which, while it played an essential role in the mathematics of that
time, had the socially unfortunate effect of making many natural
scientists hate mathematics blindly then and transmit that hate to
their students ever since. That these proofs have largely disappeared
from mathematics today, is less important than to learn from the
history of mathematics that each level of superior rigor arose, not
from pedantry but from the need to resolve what seemed contradic-
tions or paradoxes in mathematics itself. Thus anyone today competent in mathematics would say at once that KELVIN's work in hydrodynamics was essentially rigorous, since he was not dealing with any
circumstance in which discontinuities necessarily arise, but that the
eminent referees of FOURIER's prize memoir were right in objecting
to it as being unrigorous, because one of its major claims was contradicted by counterexamples known since fifty years before its time[4].

"It is almost impossible to retrace the course of the history of science to an earlier stage,
for not only should we be required to forget much of what we have learned, but much
of what a previous epoch knew or believed to know has simply never been learned by
us. We must remember that the natural sciences are as much a struggle *against* as *for*
facts. Every 30 years, a new growth makes the old forest impassable . . .. To the scientist
nature is as a mirror that breaks every 30 years; and who cares about the broken glass
of past times?" [pages 637, 639].

In mathematics, including geometry and rational mechanics, "fact" means something else than it does to a man looking through a microscope. What CHARGAFF writes
of "the natural sciences" is in a minor way true of the mathematical sciences as well but
is there just a high hurdle to be leapt, not a wall of incomprehension.

[4] In his later work on partial-differential equations EULER had introduced the now
commonly accepted description of a function as being a rule assigning to each real
number x in an interval another real number $f(x)$. He had used fluently and with great
success the pulse function defined on (0, 1) as follows:

$$f_{ab}(x) = \begin{cases} 0 & \text{if } 0 < x < a, \\ b & \text{if } x = a \qquad (0 \leq a \leq 1), \\ 0 & \text{if } a < x < 1, \end{cases}$$

he had discussed finite sums of pulse functions, and he had represented solutions of the
wave equation by superimposing infinitely many pulse functions. The "Fourier" series
of all the infinitely many different pulse functions f_{ab} are the same, independent of
the values of a and b. In particular, they are the same as the "Fourier" series for the
zero function. While this fact is easily understood in terms of *later* theorems about
trigonometric series, it does not fit at all into FOURIER's replacement of functions by
their "Fourier" series. Since, as EULER had remarked with profit, any function on [0, 1]

As in any education, much that is easy to learn must nevertheless be learnt, and so far I have mentioned only the little things. A cabinet-maker must be able to hammer a nail and plane a board, but these two skills fail to suffice for practice of his craft. Were it not that the counterparts of hammering and planing are nowadays often decreed unnecessary for formal certification as expert in the new profession called "history of science", I should not have taken the time to list qualifications which, even so little as two decades ago, were taken for granted.

To gird such graith of literature, history, and mathematics, if it has not driven our scholar back to the complacent routines of professional science or professional humanities, will certainly have given him a good idea of what science is. That is the second-most difficult thing for a student of the development of science to learn. Certainly many professional historians of science confuse science itself with membership in the Brotherly & Protective Order of Loons—alas, a description all too true of professional science today. Certainly the best way to learn what science is, is to become a scientist. Even a mediocre addition to the body of scientific knowledge gives its author an insight into the peculiar kind of thinking called science that is sounder than decades of study as a mere observer are likely to provide. Indeed, scientists are human beings, subject to human failings; the converses, however, are false: Not all human beings are scientists, and ordinary human failings do not exhaust scientists' peculiar failings, nor do ordinary human virtues exhaust the virtues possible for a mathematician. *A man who has never done scientific research can scarcely gain an inkling of how any scientist of any period ever conceived, approached, and solved a scientific problem.* The scholar whose only contact with science is second-hand, especially if his only personal intercourse is with historians and philosophers, is liable to fall into jejuneness if not error more glaring even than the anachronisms and pure fancies of that competent scientist but crudely biased and historically ignorant person, ERNST MACH. For example, the late ALEXANDRE KOYRÉ persuaded himself, and then through his skill in literary pleading persuaded many others, that since GALILEO came to be critical of ARISTOTLE, he must have been a PLATONist who saw all of nature as a poor approximation to ideal laws which had only to be thought, not inferred from refined and designed observation, and that his experiments were rhetorical flourishes, unperformed in fact. With pleading

---

may be regarded as a sum of pulse functions, FOURIER's rule of superposition of series does not hold if the class of functions considered is broad enough *to include all functions used in his own day.*

almost as special as MACH's, who positively described[5] as GALILEO's experiments that GALILEO never claimed to have done, KOYRÉ took GALILEO's failure to claim them as evidence he never did them. While MACH pictured GALILEO as a grubby Victorian materialist who fought the Church by tabulating data, KOYRÉ turned him into a twentieth-century professor of philosophy, not a scientist of any kind. Different as were MACH and KOYRÉ from one another, their views are as anachronistic as their faces would seem before CARAVAGGIO's *Barefoot Madonna* or under the great ceiling of the banquet hall of the Palazzo Farnese, covered by the CARRACCI with the naked gods and goddesses of revived and reconciled antiquity in the very years when GALILEO began to study local motion. Today we may learn from the work of THOMAS B. SETTLE[6] that the experiments on motion along inclined planes, the outcomes of which GALILEO proclaimed with errorless exactitude, are and were practicable and if carried out according to GALILEO's description do yield, to within acceptable error, the specific results GALILEO claimed. Whether GALILEO did the experiments he could have done, is a subtler question, regarding which SEGRE[7] has presented evidence and arguments. A monograph by WINIFRED LOVELL WISAN[8] teaches us that GALILEO evolved his main propositions regarding falling bodies through persistent, repeated attempts at mathematical proof and mathematical generalization, the very reverse of the appeal to inspired affirmation which is now called "physical intuition" by fellows who when in school never did quite get the hang of mathematics. In GALILEO's writings I read no evidence of an approach much different from that used by HUYGENS and MAXWELL, once the student shall have learnt their

---

[5] See Chapter 2 of E. MACH, *Die Mechanik in ihrer Entwicklung historisch-kritisch dargestellt*, 9th edition, Leipzig, Brockhaus, 1933. In § 1.2: "From his extensive observations and experiments on oscillating pendulums...." In § 1.3: "... we cannot doubt that he also put the law of falling bodies to experimental test." *Ibid.*: "... from his assumptions he derives the relation between height of fall and time of fall, and this was tested by experiment." § 1.4: "The connection between t and s can be tested by an experiment, and Galileo carried it out in the way we shall now describe." *Ibid.*: "Thus the conclusion from Galileo's assumption and also, consequently, the assumption itself were confirmed by experiment." Later researches regarding GALILEO's work on inclined planes do not absolve MACH of having reported as fact what was, on the basis of evidence available to him, no more than conjecture if not wishful thinking. MACH's description of GALILEO's discoveries reads like a Hollywood script.

[6] T. B. SETTLE, *Galilean Science, Essays in the Mechanics and Dynamics of the "Discorsi"*, Ph.D. dissertation, Cornell University, 1966.

[7] M. SEGRE, "The role of experiment in Galileo's physics", *Archive for History of Exact Sciences* 23 (1980): 227–252.

[8] W. L. WISAN, "The new science of motion: a study of Galileo's *De motu locali*", *Archive for History of Exact Sciences* 13 (1974): 103–306.

respective languages. The proportion of prejudice, special pleading, logical mistakes, exaggerated claims, bombast, and boasting in the work of GALILEO is high, even for a great scientist, and so if a man who never did any scientific work and probably never had to endure the painful trial of personal contact with a creating theorist fails to recognize these uncommendable traits as the typical reverse of the medal whose obverse bears the profile of a lonely, suspicious, resentful, and arrogant creator of major scientific doctrine, he may be forgiven. The way in which a real scientist makes use of fancy, images, calculation, proof, experience, and experiment is so different from the mental processes of merchants and lawyers and physicians and literary men that it cannot be understood from philosophic preaching or schematic diagrams. No amount of sincerity, let alone high influence, office in the artists' union, and a fist locked upon the public teat can enable a blind man to mix colors for painters. May it not be that the current fad of antithesis between "internal" and "external" history of science is no more than a reflection of modern students' refusal to undergo the social and mental discipline needed in order to face science as science?

Contact with research thinking, even once established, is easily lost. A scholar of the development of science would be well advised to keep his hand in living science, be it by teaching or be it by creative research, so as to reduce his danger of missing the forest by excessive devotion to the dryads within some one or two stunted trees.

In the past the common pattern of growth for a competent historian of science applied to an early, perceptive, and continuing interest the reserve of years sufficient to amass the great knowledge and to sharpen and mature the faculty of criticism, combination of which alone can produce a major historian. For a recent example I may cite the late LADISLAO RETI, almost the only person who ever succeeded in retaining normal standards of reason and critical analysis despite having devoted much of his life to study of the inventions of LEONARDO DA VINCI. This pattern is not obsolete. We see a prime specimen of it in the treatise on experimental solid mechanics by JAMES FREDERICK BELL[9], a man who never before published a word on the history of science but for three decades found that the history of his field and his own present leadership in research on its basic problems illuminated each other mutually. In his book we find the first general presentation and the first critical analysis of the basic experiments of the past 150 years on the mechanical qualities of

---

[9] J. F. BELL, "The experimental foundations of solid mechanics", in FLÜGGE's *Encyclopedia of Physics*, Volume **VIa**/1, edited by C. TRUESDELL, Berlin *etc.*, Springer-Verlag, 1973.

solids. Just as sound research on the history of theory must recreate the actual steps, the mathematics itself and not merely a specious biographical essay on putative influences and philosophic currents, so BELL gives us the results of real experiments, experiments critically repeated, experiments recast, experiments reformed, and experiments invented—a treatment never before attempted in study of the history of any branch of experimental science.

Yet I do not say that engagement in modern research is indispensable to the historian of science. In the work of DEREK T. WHITESIDE[10] we find an example, if a very rare one, of magisterial scholarship in the history of science without the usually prerequisite experience in science itself, and the work of SETTLE and WISAN was done at least formally in university departments divorced from live science. The only generalization here that seems common and exceptionless is that true capacity for research in the history of science is highly exceptional, much more so, I think, than capacity of equal rank in science itself.

Our scholar is now ready to acquire his most costly tool: a critical mind. A historian of any kind must select, even if only by defining his subject of study. A student of the development of science must know what science is; otherwise he will become either an ordinary historian of society—that is, warfare—, a biographer or chronicler, or—alas, nowadays more likely—a sociologer. Not every human activity is science, nor are the distinguishing qualities of science arbitrary. Our scholar must select what is science; within science he must select what is permanent; within what is permanent he must select what is important; within what is important he must select what is true. Of course, what is untrue, unimportant, or transitory has played its part and cannot be overlooked; all this must be known, and some of it needs mention, but the louts and loons ought not be let shout FAUSTUS or TAMBURLAINE off the stage. Neither may our scholar fall into hero-worship; although the historian today is little likely to accept the Golden Legends which the tribes of professional scientists teach their apprentices, he is all too tempted to disregard both the mediæval caution: Beware the man who has read only one book!, and its opposite, the caution of the Preacher: ". . . of making many books there is no end."

Like any good dramatist, a scholar must face his protagonists as equals, questioning them in their own languages, understanding their answers, providing a reply where a reply suits, but otherwise letting them speak for themselves. The language of science is frank if not

---

[10] D. T. WHITESIDE, editor, *The Mathematical Papers of Isaac Newton*, 8 volumes, Cambridge, at the University Press, 1967–1981.

heartless. However much a scientist may dislike to be found wrong, he begs for intense, minute, and critical study, study which is almost certain to detect flaws if not errors. His greatest hope is to be found right in the main, if not by the judgment of his jealous contemporaries, bound by their tribal taboos, then at least (and even better) by the higher tribunal of his peers in later centuries, when even the names of most of the pashas of science in his own time will have been forgotten. To be called, uncritically, right only by the standards of his own day, is the bitterest insult a well-meaning chorus of KOYRÉsters can direct against a mind like NEWTON's[11].

A scholar in the development of science must be able to follow science itself not only in its birth but also in its growth down to his own time. The history of science must be rewritten repeatedly, not so much because new letters or bits of scratchpaper are discovered or because the "values" of history change in accord with the laws of GRESHAM and PARKINSON, as because science itself keeps on growing, and a seed planted a century ago may sprout tomorrow into a great tree, the roots and nourishment of which the historian must trace. Scientists, unlike old soldiers and heads of state and apostles of faiths, do not die. Old works of science are read and reread, not only by antiquarians but also by scientists whose main interest in them, an interest often fruitful, is to learn their content so as to use it and to better it. A classic of science, like a classic of literature, must be read in its author's own words, not the paraphrase of a pædagogue. The work of GIBBS in thermodynamics, after it had lain some ninety years scantly understood and partly misrepresented (though indeed much admired), was read critically by COLEMAN & NOLL in the 1960s and made by them the basis of part of the splendid new science of rational thermomechanics[12]. All historical studies of thermodynamics I have

---

[11] It is no exception here that the criteria of rigor required by mathematicians have been tightened again and again. The woolly-headed have pounced upon this fact as evidence in support of the virtues of vagueness. In this regard the professional mathematicians, for the most part, stand on firmer ground when they speak of the "essential" correctness of an old proof. The historians of science, most of whom have not yet caught up with the rehabilitation of divergent series in the early years of the twentieth century, would do well to read the preface of T. H. HARDY's *Divergent Series*, Oxford, Oxford University Press, 1949. An instructive example of timeless mathematics is furnished by POINCARÉ's proof of his recurrence theorem of 1890, which for strictness today requires only that (Riemannian) "extent" be replaced by "measure" in all its occurrences. A devotee of "changing values" would gleefully pronounce POINCARÉ's proof right only by the standards of its own day; any competent mathematician regards it as essentially right forever.

[12] See the introduction to my book, *Rational Thermodynamics*, New York, McGraw-Hill, 1969, and the "historical introit" of its augmented second edition, New York, Springer-Verlag, in press.

seen ignore the aspects of the works of GIBBS that have proved most fertile in the last decade. Moreover, they accept the mathematical limitations shown by most of the thermodynamicists who preceded GIBBS as evidence that the customary jabber in books on thermodynamics is essential to the concepts rather than merely an unfortunate accident of history. While stubbornly refusing to let the far-seeing lenses of a COLEMAN or a NOLL of 1963 shape their images of what thermodynamics is, for the definition of their subject they cleave uncritically to the physics texts of the 1930s or 1940s which they themselves happened to study when undergraduates, thus committing the new sin, peculiar[13] to our day, of middle-mindedness, a specialized case of ordinary muddle-mindedness. Much earlier GIBBS, and a few others, read the difficult, abstruse book of a Gymnasium-professor from GAUSS's day named GRASSMANN; from the vector analysis GIBBS built by combining with GRASSMANN's simplest concepts some aspects of the shallower but then more popular algebra of HAMILTON's quaternions, half a century later grew the vector spaces of pure mathematicians; some of those mathematicians, tracing references, went back to GRASSMANN, read his work themselves, and from it developed abstract multilinear algebra, which is the basis of the theory of manifolds. To see GRASSMANN only in the dim light of his own day, the day which left him almost in obscurity, and to neglect the century of brilliant life after death he has enjoyed, would be foolish refusal to recognize the object whose history a historian of science claims to write.

The workshop is ready, fitted out with fine tools. My plaint, long as it has been, is only the prologue. What will the scholar now do? That I cannot tell, any more than anyone can teach a student of art, after he has been trained to draw and paint as well as fitted out in a studio with perfect light and the best canvas and brushes and pigments, decorated by a garland of superbly qualified and sympathetically cooperative models, how to create a masterpiece. But we must not follow the path most comfortable for a recognized and remunerated profession: We must not replace the creation of art itself by the manufacture of workshops no living artist is trained to use.

---

[13] For example, anyone trained between, say, 1880 and 1960 is likely to regard the "quasistatic process" and the "thermodynamic state" as fundamental concepts of thermodynamics; such a person will search for them in the early works and will reproach as being insufficient or irrelevant any modern paper which neglects them. In fact neither concept played any part in the considerations of LAPLACE, FOURIER, CARNOT, or KELVIN, or in those of CLAUSIUS until about 1865, and in modern rational thermodynamics both are certainly unnecessary and at most of scant use as ordinarily conceived. In some recent work the term "state" is re-introduced in a sense general enough to include functions of time such as the history of a process. Few indeed are the historians who would be able to come so near to the present day, should they wish to.

Note for the Reprinting

The first draught of this essay was written between supper and an evening's chat at the conference on "The Interplay between Mathematics and Physics in the Nineteenth Century, the Rise of Mathematical Physics", held at the Institute for History of Science, Aarhus University, in August of 1970. That was the first time I had heard a roomful of young historians of science arguing about historiography (the most pressing problem in the history of science!). Fortunately I was not asked to read what I had written. After some revision it was published in *Centaurus* **17** (1973): 1–10. Some parts of it, further rewritten, were included in "The scholar, a species threatened by professions", *Critical Inquiry* **2** (1976): 631–648, reprinted with some corrections in *Speculations in Science and Technology* **3** (1980): 517–532. I have revised the text again for the present printing.

Although an anchorite in the desert may shiver with pride in his isolation, comforting himself by repeating inwardly that so spake NATHAN, he cannot but be reassured if he finds himself to have been instead a BAPTIST. Thus I gladly diminish now that the master, ANDRÉ WEIL, has spoken to the (perhaps Philistine) ears of a general congress of mathematicians in 1978: "History of mathematics, Why and how", pages 434–442 of Volume **3** of WEIL's *Œuvres Scientifiques*, New York *etc.*, Springer-Verlag, 1980.

37. HAS THE PRIVATE UNIVERSITY A FUTURE? (1976)

I have been reading about Tulane University in the history, written by JOHN P. DYER, which Mr. COWIN caused to be sent to me. I learned there that in the years 1905–1912 Tulane conferred sixteen doctorates, of which fifteen were honorary. That makes me a little uneasy. In fact, in a work of mine now so far along toward publication that when your President called by telephone it was already too late to stop the presses, I wrote

> While once the title of "doctor" meant teacher, now the "earned" doctorate is become a formal statement of what the candidate need not know, and its award, like freedom of a guild, makes him formally free to stop learning, while the "honorary" doctorate is most often a certificate that learning has never begun.

That being so, you will, I hope, forgive whatever I say now as being just the sort of thing that uneducated folk do say.

We Americans are a people devoted to progress. I am not sure what progress is, but nearly everyone agrees that it has occurred in abundance. Not far wide of the mark would be the definition: Whatever does happen is progress. That gives us the comforting assurance that progress will continue until abolished by the Last Judgment; nevertheless, while it tells us just what progress has been until now, it does not tell us what progress will be like in the future.

Since the fall of Rome, at least, prophets have done their work ill. In the 1920s they pictured the time we now live in as one in which everybody would have his private airplane, skyscrapers would be a mile high, and life would be prolonged indefinitely. In the 1930s they foresaw progress into a society which would abolish war and unemployment, causing crime to wither away, and the happy, collectively recognized labor-unionist would devote his vastly increased leisure to amateur theatricals, making pottery, and the study of MILTON and BROWNING. The great war of the 1940s, which ended by betraying half of the peoples whose heritage and culture are our own into the adamantine grasp of an Eastern police state more efficient in

oppression than any the West had ever produced in its 1000 years,
while much of the rest of the world was set adrift in grinding, pitiable
poverty subject to thugs and brigands more cruel, bloodthirsty, and
ruthless than any colonial ruler had ever been, was greeted as the
struggle through which America would give the whole world freedom
of speech and worship, freedom from want and fear. In the 1950s the
prophets saw all Americans going to school; the University of California grew at a rate which if continued for half a century would have
matriculated the entire population of the Golden State. These are all
examples of predicting progress as continuance of just what was happening. By the 1960s the readers and advisors had learned how to
calculate not only the first derivative but the second, and even in
simple cases how to integrate; they concluded that the gross national
product would increase exponentially for the "foreseeable" future.
The event proved that the future had not been foreseen.

During these decades the nature of the university changed more
slowly. Before the First World War the "liberal education" was
supreme. It was a simple, common, rudimentary training in the Latin
language, English literature, European and American history, mathematics, and a little natural science. Around 1900 Tulane's College of
Arts and Sciences offered only four courses of study, of which three
were of just this kind, differing merely in the proportions of the
mixture; the fourth was for mechanical artisans and technicians. (In
that archaic age it was fairly secure to presume of a matriculated
student in an American institution that he was already able to read the
English language and even to write it a bit. Social democracy and the
program of making "education" essentially "remedial" from the first
day had yet to be imposed on the nation by the enlightened Liberals,
who dismissed the experience of mankind and the wisdom of preceding sages as being no more than evidence that primitive peoples
lacked the Liberal principles of "social science".) Largely through the
immigration of scholars from Germany and countries overrun by the
Nazis, between the two wars America discovered research, which up
to that time had been encouraged and rewarded little and at that only
by a few snooty private universities: Johns Hopkins first, then Clark,
Chicago, and Columbia, then Harvard and Yale, and finally even
Princeton. The old liberal education gave a youngster a basis of a
culture common to his ancestors and his peers in all the West, a basis
from which he could—though he rarely did—advance in any direction
he pleased. The university of the 1940s and 1950s, which focused upon subsidized, organized research using costly glassware, monstrous
machines, and billions of miles of wire, destroyed the liberal education
and turned every student into a specialist passed or a specialist failed—
in most cases, a specialist passed who had deserved to fail. Research in

the old "academic" disciplines is obviously not for everyone; partly to provide subjects in which new "knowledge" would be easy to get, curricula were so inflated by new "sciences" of every hue that the subjects previously called "academic" withered away, and "esoteric" became a term used to denigrate old-fashioned scholarship. Administration was made an end in itself, indeed a "science", the rules of which are just the same for prisons and universities and fast-food chains; in the words of ARTHUR GORDON WEBSTER even in 1914, it is practised by a gang of "geniuses for accountancy, for hustle and grind", which seeks to "relate academic achievement to the product of the factory and the machine shop", to which more recent times have added a cadre of applied psychologists and welfare workers who minister to and in part provoke the dreadful personal crises common among young folk afflicted by a surfeit of irresponsible leisure.

The second change crept upon us more quietly; faculties hardly saw what was happening, and many students and parents do not see it even now. It was a bloodless coup—a coup of administrations encouraged by and at the same time ever more regulated by the Federal Government. These forces exploited the craze for research, the classic egotism, naivety, and ineptitude of professors, and campus unrest. Researcher, teacher, and student were gradually, gently, and respectfully devalued into unorganized but greatly featherbedded labor for administrations; for government, into entries in a computer program to render the university the minion of the state in return for a fluctuating pension granted as a measure of success in courtship. The research of professors came to be regarded as a source of income for universities, income labelled "reimbursement of overhead". A professor who did not engage in this fund-raising for dear old Siwash received in effect a lower salary and perhaps was not allowed much use of facilities for such research as he chose to do on his own account.

I am not here to bewail facts or to predict the future. I have summed up progress in universities as I have seen it in my lifetime. You who are graduating today may expect things to continue just as they are, but only more so. You may expect to see even more freedom from curricular restraint, a campus enlivened by even more subsidized entertainment, more and more "privileges" and "activities" in a protected environment, less and less of the responsibility, denial, punishment, and misfortune that mark the life of an ordinary adult. You will be wrong. What kind of university your younger brothers and sisters will see, I do not know, except that it will not be a simple extension of what you have seen.

Progress—which, I repeat, I equate with what does happen—rarely pleases all who suffer it. Teachers of my age were once students. Most of them have not in later life experienced more of what they had in their student days, or what was predicted for them, or what they desired to experience. They have not always liked what has happened.

A decade ago the beer industry engaged a wizard to determine America's taste. The many kinds of beer once made in this country had by then been reduced to two, one very dry and flavorless, the other just a little sweeter and thicker. The wizard concluded that 49% of Americans favored the richer, sweeter beer; 51%, the dryer. The losing beer quickly vanished, and now all of us must drink alike, if under different labels. My connections with the Central Espionage Agency are insufficient to determine whether the universities employed the same wizard. If I may judge by the result, they have followed the example of the beer industry, for in courses of study and regulations and attitudes one American university would now scarcely be distinguished from any other by the blindfold test. Once almost infinitely various, they are now copies of each other except in bulk and labelling. They are even alike in that each must boast how different it is from the rest.

It is no wonder that prophets have forecast the end of the private university. Harvard, indeed, can for a long while afford to do a good job of imitating Berkeley at higher cost, but few if any other private institutions could keep up. Inflation devalues endowments; the private wealth, whether actual or potential, without which there can be no new endowment, our rapacious governments are systematically destroying through the progress of taxation. Arithmetic shows that to continue much longer the private status of most of our presently private institutions will be impossible. Nevertheless, their administrations and faculties still campaign for money to build more buildings, to buy fancier equipment for entertainment and research, and to seduce more and more youths, maidens, and even senior citizens. Arithmetic shows that as a whole they cannot succeed. Most must face, in a few years, the alternative: Close, or be annexed by the state.

Arithmetic does not show, on the contrary, that any one institution has only these two alternatives. The sea fowl cannot change the course of the great wave on which it rests, but it may fly off the breaker that will dash its fellows onto the rocks. Some private institutions may find ways to survive. Most of these will have to abandon one or more of the four dominant educational principles of the 1960s:

(1) Big is beautiful.

(2) Keep up with the Joneses.

(3) Whatever a student asks for, he should be given free or even paid to receive.

(4) Research is a state of beatitude.

As for the first, retrenchment is just as normal an experience in life as is expansion. Few are the investors who get only richer. Some schools may see the wisdom of becoming smaller by choice, and in so doing choose what they will retain; perhaps they will strengthen that while abandoning most of whatever else they presently do. Some may find it to their advantage to move to smaller, plainer, and cheaper quarters; those that do not may come to rue their having forgotten a documented corollary of PARKINSON's Law: "A perfection of planned layout is achieved only by institutions on the point of collapse."

For an example of what a small, excellent school may do, and of the hazards it faces in a world doomed to ever uglier bigness, we may refer to DYER's history of Tulane. In regard to the program in architecture in 1912, the year its first full-time professor was engaged, he writes

> The quality of instruction was very high, if one may judge it from the records made by the alumni in professional life or as students in other schools. No better example of this can be found than in the 1912 examinations for entrance into the Paris Ecole des Beaux Arts. In this year, 185 non-Frenchmen took the examinations, and only eight passed. Two of the three Americans in the successful group had been students in the new architecture program at Tulane. In spite of the quality of the work, however, it was touch and go for several years as to whether the department would survive or not. Enrollment was small, student fees were insignificant, and the budget of the College of Technology was inadequate to absorb the deficits. For a period of time, consideration was given to abolishing the department completely, but local groups interested in seeing it continue came to the rescue financially.

Today, too, one full-time professor and four or five part-time associates can make a great school. Much has changed since 1912, but nobody, so far as I know, has suggested that men have grown wiser or brighter, and in the world of the intellect one man still makes the measure of a man's work.

A danger in becoming smaller is that universities may continue to keep up with the Joneses; if so, each will cut out just what the others do, and the result will be continued multiplication; moreover, a narrower prospect of study for students as a whole would drive more and more of them to the state institutions, and the private schools would be worse off than ever. In order to survive, the small school must state and achieve a purpose. The only purpose that would be new and different for it would be to excel in scholarship. The idea has been

noised before but never tried without ruinous compromise. In 1935 a
faculty group at Tulane signed the following statement:

> Tulane should not and cannot compete . . . in mass education,
> which should remain the obligation of the State, but should pro-
> duce a select and higher type of scholarship to furnish . . . intellec-
> tual leadership. . . . Because of their freedom and independence,
> endowed universities are the better fitted to perform such tasks.

In order to excel now, the private school must focus its effort. What
several others already do well, it would be wiser not to attempt. Diversity and excellence foster one another. Both together, spread over all
the remaining private institutions, can strengthen them all.

The problem of choice for the student in a specialized private
school is much easier. He would not have gone there in the first place,
particularly in view of the high cost in money, had he not been
strongly inclined in its direction. Should he decide that he had chosen
the wrong line altogether, he probably would wish to leave. If, as our
administrators blandly tell us, the university carries each student at a
loss, the departure of a dissatisfied student should save money for the
college. The private university can no longer afford to sustain the
student who does not know if higher education is for him, or who
matriculates so as to find himself and decide whether to major in
anthropology, journalism, or modern dance. Human mistakes, not
only the mistakes of business enterprises, must be recognized and
written off as losses.

The directors of a small, channelled, impecunious, private university may come to see that if students are adults, as now they are
certified to be—adults who can vote for measures that give them
money by taxing it away from those who earn it—they ought to provide their own entertainment, as other adults do: at their own
expense, in using the facilities the civil community affords, and at
times left free by their obligations. It is true that most young people
would not like a school without frills, but the few who know what they
wish to learn could more easily find it; they might consider the high
cost well spent if no part of it were wasted on athletics, social events,
and "cultural" programs. A great university needs only a few students. Its problem is to find the right ones.

The value of research is well known. Two aspects of research are
now often forgotten: Unsuccessful research is worth little, and most
research is unsuccessful. Training for research is not enough; it must
be training for successful research. Few professors, and even fewer
students, are qualified for successful research, and no purpose is
served by coddling them into pretense that they are. Perhaps a

small, sober, private institution, recognizing this fact, may see three different viable lines of instruction. To describe these I shall use terms appropriate to engineering, but the distinctions apply to all fields of knowledge:

(1) *To train fine technicians,* people who learn a particular kind of engineering practice thoroughly and securely. Such students should be taught primarily by professors who have had enough contact with research to sense its value, but whose primary interest is in teaching. Such teachers should be encouraged to keep abreast of the important discoveries rather than try to do research themselves. The heavy equipment typical of most engineering schools today will soon be beyond the reach of any private institution's budget. Small loss, for use of it is better learned in industry or government laboratories, and this aspect of the technician's training could well be supplied by a program in which study in the classroom alternates with work in the field.

(2) *To train leaders in the professions of engineering.* These students should be taught to see not only how engineering is practised but where it is going. Their training should include at least two branches of engineering as well as a solid basis in writing and reading English, at least one foreign language, and European history, as well, of course, as the two or three pure sciences basic to their branch. Their professors should be men having some experience in research, broad knowledge of basic sciences as well as their own branches of engineering, and some experience in practice. Since leadership is a mental activity, not a manual one, extensive routine work in testing and design would be unnecessary. For a future leader of his profession, a brief experience on a production line or in an industrial laboratory may do more to foster grasp of reality than many semesters of taking measurements in the coddling environment of a school.

(3) *To train research scientists.* As this is the aspect of university life that has been most emphasized in recent years, I need not describe it. I hope I need not explain how important it is that some persons of engineering bent take up research as their vocation. If man is to learn more about the physics, chemistry, and biology of everyday life, it will be from research by engineers, for the professions of pure science, turning away from all that we can see and touch, devote themselves to the artificial worlds they create in their costly laboratories and frenzied imaginations—the horribly dangerous, the invisibly small, the unattainably swift, the times before man existed and after he shall have ceased to exist, the airless reaches of outer space and the interiors of cells that cannot fairly be regarded as either dead or alive. Students

drawn to research should be scrupulously selected and given special instruction aimed toward it. Their professors should be, from the beginning classes, the finest creative investigators available. It would be as foolish to deny to the brilliant freshman instruction by the leading expert as to compel that expert, in what was formerly (and often still is) discharge of moral debt to Academic Society, to shiver, tremble, and gasp for air in what WEBSTER in 1914 called "that cold bath to enthusiasm that the dreary driving of unwilling pupils frequently is to the aspiring scholar." If research is to be done at all, it should be recognized by the university as the main function of those few who can do it well. The recent explosion of research has provoked most universities in their frenzied lust to swell enrollments in graduate programs, titillate undergraduates, and provide excuses for larger and larger administrations, to cut research itself down to the routines of the mediocre, bestowing the once deserved title "doctor"—that is, worthy to teach—upon any drudge who will endure a few years of meanly paid, often demeaning servitude to some laborious and dreary project for government or industry. The brilliant students are those least valued and most neglected in universities today.

It is unlikely that any one small, private college could provide programs of all these three kinds well. I can see easily one college for technicians, one for practical leaders, one for research. Each, by itself, should be cheap and easy to administer by a small admiralty. It is proved abundantly already that no seamen need correspond to the post of admiral; it has not been proved that two admirals are better than one.

Returning to the small private institution of any kind, I state four conditions that seem to me necessary to its survival:

(1) *It must be truly private.* Today we suffer regulations and restrictions upon personal liberty which our grandparents would have rejected as tyranny. Statism, whatever its cast and label, grows everywhere. We cannot fight it head on. Most men are doomed to become indistinguishable except by their identification numbers in the state's computers. Not all men are so doomed. *The private university should become the haven of the private man.* The apprehensions of Tulane's principal benefactor, JOSEPHINE LOUISE NEWCOMB[1], in her

---

[1] Mrs. NEWCOMB founded and later endowed Newcomb College for women, co-ordinated with Tulane University. DYER writes, "One of the assurances given her when she founded the college was that Tulane was free from political control." Less than ten years later "she was alarmed over the attempts of the president and the administrators to secure state financial aid by *having Tulane declared a state institution.*" She feared "that attempts would be made to divert the Newcomb funds to the general use of the univer-

wise guidance of the college she founded have proved, a century later, to be all too true. Grateful as we must now be for the support the Federal Government gave and continues to give to research, we cannot fail to perceive its ever tighter grasp upon education. The crippling restrictions of academic freedom and the outright destruction of academic standards imposed by arms of the government have rendered the sugar too dear for the pie. The private university must move toward a position from which it can and will refuse outright all grants and contracts from governments. A welcome consequence could be drastic reduction in the number of administrators needed, which would at once improve teaching and the standards of selection and retention of students; not only that, it would at once and even more strongly throttle the flood of "inevitable" research and open space for investigations of quality and potential value.

(2) *It must reduce its faculty to a number it can afford to pay, from its endowment and regular private income, a decent annual wage.* If Tulane in the years around 1900 could pay its few professors $2500 to $3000 a year, free of income tax, I can see no reason why a university today cannot pay a professor worth his salt the current equivalent of the same. The present scale of salaries reflects the great co-operative fallacy: Three chickens cackling together are smarter than one raccoon.

(3) *It should appeal to and support students from the two most neglected, indeed abused groups of citizens*: the talented, irrespective of race or origin, and those whose parents are neither very rich nor very poor.

(4) *It should withdraw from the public eye.* While state institutions may need to become increasingly slaves to the public, which means slaves to politics and politicians and the rajahs of civil service, somewhere among us there should be place for detachment and, let us hope, wisdom. Professed, direct public service is dangerous. He who seeks public praise runs the risk of incurring public blame. One outburst of public blame can do a university more harm than a dozen public statements of praise can remedy. The common citizen will tell you that universities are centers of political and social agitation if not radicalism. There are many persons on campuses as well as outside them who welcome and promote the socially committed university. It is possible that some private institutions specialized in politics or other social sciences can play and win the game of involvement. The small,

---

sity." She even proposed to build a new institution in another state. No sooner was she dead than her fears were substantiated. Indeed, a major theme in the history of Tulane in the early years of this century is its administrators' schemes, finally successful, to demote her college into a fief and appropriate its endowment to whatever purposes listed them.

private school of arts, natural sciences, or engineering should leave the social sciences to those who can afford the risks. Applied social science is used more and more by states to foster the programs that the rulers choose to impose upon their subjects. It is unlikely that any state will much longer tolerate in the social sciences the freedom of thought and inquiry and opinion without which no true center of learning may subsist.

In regard to the Tulane of the decades just before and after 1900 DYER wrote,

> It was a period before the passion for Ph.D.'s took over the American college campus, a time before the professor had to publish or perish, a time when he dared get out of his own narrow field and write something within some other discipline. It was a leisurely and noncompulsive . . . life. . . ..

Let me suggest that even in an engineering school, today full of plumbing, tapes, and blinking lights, there is reason to reinstate thinking. Expensive complexes of pipes and wires call for many cheap hands to keep the gadget awhir until it obsolesces. The main product of this frenzy has been frantic calls for more and more of the same. The hardware is more profitable for the suppliers than for the students.

For a small college, not only private but also *independent*, the administration may find the task of raising endowment astonishingly easier than for a standard educational house of resort. To get a good husband, the seeking woman need not appeal to every man on the street; indeed, if her appeal is of that kind, she may find it hard to get a husband at all. It is enough that she find and capture the man she wishes. Most women succeed; some succeed again and again. The administrators could learn from them. Easy virtue goes very well in a public institution designing to serve all comers, but a small university run as a tight ship on a clear course might more readily find donors who, like the informed and wise Mrs. NEWCOMB, were disgusted by the way things were going elsewhere.

List of Works Cited and Quoted

JOHN P. DYER, *Tulane, the Biography of a University 1834–1965*, New York & London, Harper & Row, 1966. The passages quoted or referred to are on pages 84–85, 89, 102, 107, 117, 127–128, 133, 142, 145, 151.

ARTHUR GORDON WEBSTER, address published on pages 60–62 of Volume 3 of *Twenty-fifth Anniversary of Clark University, 1889–1914*, Worcester, Massachusetts, Clark University Press, 1914.

Note

The text above differs only by a few turns of phrase from an address delivered to the engineering commencement at Tulane University upon receipt of an honorary degree, 15 May 1976. The issues refer not only to engineering but to universities in general. The colleges of "liberal arts" are in far worse case today than schools of engineering. It is they, above all, who prostitute education. They degrade "arts" of every kind into mere conversation and propaganda, and they interpret "liberal" as license to be garrulously idle and to remain functionally illiterate in every language. To speak to the point regarding them, far harder words than mine here are called for. Like the universities, they are dominated by the social sciences.

The dangers incident upon having social sciences in an institution of learning are of two kinds:

(1) *Public reaction.* For example, it is well known that most social scientists stand to the left in politics and in their classes propagate their views as being products of "science", and so when public opinion swings to the right, it naturally becomes hostile to institutions that harbor social scientists. While many social scientists equate progress to growth of leftist policies, recent as well as much older experience shows that humanity does not progress steadily according to their definition but sometimes regresses. Another example is provided by the opinionative faddism of the social doctrines. Parents who were taught social science themselves and then see their children taught altogether different doctrines under the same titles have reason not only to doubt the value of their own "educations" but also to be unwilling to pay for their children's having to repeat the same dismaying experience in later life.

(2) *The intellectual and cultural mediocrity if not actual charlatanism of most social scientists.* For this, the greater of the two dangers, I refer the reader to S. ANDRESKI's *Social Sciences as Sorcery*, New York, St. Martin's Press, 1972, especially the chapters "Manipulation through description", "Censorship through mass production", "Ideology underneath terminology", "The Law of lighter weights rising to the top", "Gresham's and Parkinson's Laws combined", and "The Barbarian assault on the corrupted citadels of learning". On pages 201–202 ANDRESKI describes a difference between the natural and the social scientists:

> A mediocre natural scientist, albeit unable to think of anything new, or even to keep fully abreast of current progress, remains nonetheless a repository of useful (even if limited and perhaps superficial) knowledge, whereas a mediocre social scientist, unable to distinguish between worthwhile ideas and the half-truths and inanities which flourish in his controversial field, will be an easy victim of deluded mystics and charlatans and will act as an agent of mental pollution. This difference explains why the vast expansion of educational institutions has had beneficial effects on the level of technical skills while helping to turn the humanistic studies into a massive pollution of the mind.

On page 203:

> While demanding at least as much work and ability for a proper understanding as the natural sciences, the social sciences (with the very

partial exception of economics) differ from them by having no natural threshold of acceptability. A physicist or a chemist may hold cruder views on politics, aesthetics or ethics than a shop assistant, but this is not what he is paid for. His status and salary are justified by his knowledge of chemical reactions, of the structure of the atom or whatever his speciality may be . . . and in such matters there is little room for bluffing. No amount of plausible talking and posturing will make a bridge stand if it has been incompetently designed; while ignorant dabbling with chemicals will soon lead to a fatal explosion. In contrast, nothing will immediately blow up or fall down in consequence of a politologist's or economist's inanity; while the harm caused by his ignorance or dishonesty may not materialize until years later, and will in any case be debatable and difficult to blame on a particular man. Related to this is the fact that, as the criteria of excellence are so dubious, it is impossible for a layman seeking advice to find who the real experts are. Neither a degree, nor a university chair, nor membership of a famous society or institution constitutes a warranty that a given social scientist deserves being taken seriously, because in the competition for these honours knowledge and integrity often matter less than skill at intrigue and self-advertisement. It is not surprising, therefore, that—far from being particularly good in sociology or political science—the wealthiest American universities contain an unusually large proportion of phonies who bask in the collective glory deservedly won by their colleagues in the exact disciplines.

On pages 207–208:

The absence of minimal standards offers unlimited scope for numerical expansion, which in the exact sciences is constrained by the scarcity of talent. This is the chief reason why (like the arts) the social sciences have been allowed to expand so much, because educational bureaucracy has a vested interest in boundlessly increasing the number of inmates of its establishments, regardless of whether they learn anything; and in fostering one of the grossest superstitions of our times (bolstered up by the golden calf of pseudo-quantification) which equates the progress of education with an increase in the number of individuals kept within the walls of educational institutions. In reality (and especially in the case of the United States) one could say that never have so many stayed in school so long to learn so little.

The last sentences of his book deserve to be engraved on bronze tablets set over the portals of all institutions of real learning (if such there be):

Apart from the consequences of almost everything becoming a part of the entertainment industry and being affected by the methods of high pressure salesmanship and advertising, another unexpected influence has begun to operate in the fields we are discussing. It seems that since they have become an established occupation, the social sciences have begun to attract the type of mind which in the olden days would have taken up dogmatic theology or preaching. This has been an unfortunate change, because the old theology and mysticism (regardless of which denomination) were linked to a moral code, whereas the new cults enjoin no firm rules of conduct, adherence to which was the price for an opportunity to satisfy a desire for the kind of admiration normally bestowed upon the licensed interpreters of the Holy Writ.

Instead of entertaining visions of a final victory of reason over magic and ignorance, we have to reconcile ourselves to the fact that the norms and ideals which permit the advancement of knowledge have to be defended in every generation against new enemies, who reappear like the heads of the Hydra as soon as others are decapitated, and who employ ever-new labels, catchwords and slogans to play on the perennial weaknesses of mankind. Whatever happens in the instrumental exact sciences, we can be sure that in matters where intellectual and moral considerations mesh, the struggle between the forces of light and the forces of darkness will never end.

The pioneers of rationalism inveighed against the traditional dogmas, ridiculed popular superstitions, campaigned against priests and sorcerers, and castigated them for fostering and preying upon the ignorance of the masses—hoping that a final victory of science would banish for ever the evils of unreason and organized deception. Little did they suspect that a Trojan Horse would appear in the camp of enlightenment, full of streamlined sorcerers clad in the latest paraphernalia of science.

PART V

PHILOSOPHY?

38. IS THERE A PHILOSOPHY OF SCIENCE? (1973)

A Scientist who writes on philosophy faces conflicts of conscience from which he will seldom extricate himself whole and unscathed; the open horizon and depth of philosophical thoughts are not easily reconciled with that objective clarity and determinacy for which he has been trained in the school of science.

H. WEYL, *Philosophy of Mathematics and Natural Science**

I felt that the philosophers moving in the realm of infinity without the precautions and experiences of the mathematicians were like ships in a dense fog in a sea full of dangerous rocks, yet blissfully unaware of the dangers.

M. BORN, *My Life***

MEDAWAR's *Induction and In ition in Scientific Thought* is an extraordinary book. First of all, it is well printed in black ink with a plain, legible type on good white paper, with running heads and page numbers where they belong, decent margins, paragraphs properly indented, a centered and dignified title page, and a sober yet clearly stamped binding. Such bookmaking is nearly extinct. It deserves a laurel wreath.

Scientists scarcely ever read anything about the philosophy of science, and with good reason. A philosophy of science expressed by someone who is a sound enough scientist to know the object of which he speaks is rare. Rarer is a creative scientist competent enough in philosophy, history, and language to write something worth reading on the philosophy of science. Such a rarity is MEDAWAR.

---

* Princeton, Princeton University Press, 1949, page v.
** New York, Scribner's, 1978, page 74.

Here he presents three lectures, carefully thought out and worked out, brilliantly if simply presented with a witty good humor which softens but does not lighten their smashing rebuttal of a popular philosophy of science which many practising scientists uphold today, especially in the life sciences, and which is preached by one of the senior schools of philosophy. This philosophy is called "inductivism". The acute, morbid variant known as "operationalism" has been described by MARIO BUNGE as "not a possible interpretation of theoretical physics but its antithesis, for no theory is possible without *theoretical concepts*, i.e. constructs overreaching experience (and thus making the explanation of experience possible)[1]."

MEDAWAR's Lecture 1, "The Problem Stated", begins by contrasting two seemingly diverse layman's concepts of a scientist: "a discoverer, an innovator, an adventurer into the domain of what is not yet known or not yet understood", and "a questioner of received beliefs". As an example of questioning MEDAWAR adduces GALTON's statistical inquiry into the efficacy of prayer, and to show that such critical tasks are not yet all done he suggests that one day someone might dare to fix a similarly cold and unbiased skepsis upon the efficacy of psychoanalytic treatment.

> If such a thing were done, might it not show that the therapeutic pretensions of psychoanalysis were not borne out by what it actually achieved? It was perhaps a premonition of what the results of such an enquiry might be that has led modern psychoanalysts to dismiss as somewhat vulgar the idea that the chief purpose of psychoanalytic treatment is to effect a cure. No: its

---

[1] MARIO BUNGE, *Foundations of Physics*, Springer Tracts in Natural Philosophy, Volume **10**, Berlin *etc.*, Springer-Verlag, 1967. See page 27.

Indeed, before operationalism was born it had been anticipated by PIERRE DUHEM, the master of us all in the philosophy of science. See § VII of Chapter VI of Part II of his *La Théorie Physique, son Objet et sa Structure*, Paris, Chevalier & Rivière, 1906. (There is also a translation of the second edition, *The Aim and Structure of Physical Theory*, Princeton, Princeton University Press, 1954.) All the subsequent quotations of DUHEM's writings are from this chapter unless otherwise noted:

> *In the course of its development*, a physical theory is free to choose whatever way it pleases, provided logical contradiction be avoided; in particular, it is free to take no account of experimental facts.
> It is not the same *when the theory gains its full development*. When the logical structure has reached its full height, then we must compare the whole of its mathematical propositions, obtained as conclusions from these long deductions, with the whole of the facts of experience . . . .

It is an error "to allow only 'reasoning about *realizable operations*' or 'to introduce only *quantities accessible to experiment*'." (DUHEM here refers to and quotes G. ROBIN.)

purpose is rather to give the patient a new and deeper understanding of himself and of the nature of his relationship to his fellow man. So interpreted, psychoanalysis is best thought of as a secular substitute for prayer. Like prayer, it is conducted in the form of a duologue, and like prayer (if prayer is to bring comfort and refreshment) it requires an act of personal surrender, though in this case to a professional and stipendiary god.

Nevertheless,

> The exposure and castigation of error does not propel science forward.... To prove that pigs cannot fly is not to devise a machine that does so.

How have scientists achieved the understanding that makes scientific discovery possible?

> What methods of enquiry apply with equal efficacy to atoms and stars and genes? What *is* "The Scientific Method"? What goes on in the head when scientific discoveries are made?

Here is the place for a philosophy or methodology of science, but if its purpose is

> to prescribe or expound a system of enquiry or even a code of practice for scientific behavior, then scientists seem to be able to get on very well without it. Most scientists receive no tuition in scientific method, but those who have been instructed perform no better as scientists than those who have not. Of what other branch of learning can it be said that it gives its proficients no advantage; that it need not be taught or, if taught, need not be learned?

Indeed, most philosophers of science are small men as scientists or not scientists at all, but perhaps the real scientists can speak prose, like M. JOURDAIN, without knowing they do so.

> Ask a scientist what he conceives the scientific method to be, and he will adopt an expression that is at once solemn and shifty-eyed: solemn, because he feels he ought to declare an opinion; shifty-eyed, because he is wondering how to conceal the fact that he has no opinion to declare. If taunted he would probably mumble something about "Induction" and "Establishing the Laws of Nature," but if anyone working in a laboratory professed to be trying to establish Laws of Nature by induction we should begin to think he was overdue for leave.

You must admit that this adds up to an extraordinary state of affairs. Science, broadly considered, is incomparably the most successful enterprise human beings have ever engaged upon; yet the methodology that has presumably made it so, when propounded by learned laymen, is not attended to by scientists, and when propounded by scientists is a misrepresentation of what they do. ...

One way out of this dilemma is to argue that scientific methodology is understood intuitively by scientists and needs to be propounded only for the benefit of other people. Nearly all scientists are loud in deploring the utterly unscientific way in which everyone else carries on—politicians, educationalists, administrators, sociologists—and it is upon *them* that they urge the adoption of the scientific method, whatever it may be. ...

... Many modern methodological texts have therefore a strong orientation towards the social and behavioral sciences, as if sociologists and social anthropologists were backward because (poor things) they had not been properly brought up in the manners and usages of polite science. While I respect this evangelistic mission, I am not in sympathy with it. The 'backwardness' of sociology (as in the nineteenth century of biology) has little now to do with a failure to use authenticated methods of scientific research in trying to solve its manifold problems. It is due above all else to the sheer complexity of those problems. I very much doubt whether a methodology based on the intellectual practices of physicists and biologists (supposing that methodology to be sound) would be of any great use to sociologists. On the contrary, the influence of inductivism, the subject of my next lecture, has in the main been mischievous.

Despite the failure of scientific methodology in all that is claimed or hoped for it, MEDAWAR thinks it has a place. Its agenda should be:

1. The problem of *validation*: of the grounds upon which general statements may be judged true or false or merely probable, and the methods by which we may quantify their degree of imprecision ... ,

but in fact scientists fear error much more in the generation of scientific knowledge than about what experiments "really" mean.

2. *Reducibility, emergence*. If we choose to see a hierarchical structure in Nature—if societies are composed of individuals, individuals of cells, and cells in their turn of molecules, then it

makes sense to ask whether we may not 'interpret' sociology in terms of the biology of individuals or 'reduce' biology to physics and chemistry.

But this problem seems hopeless:

> Each tier of the natural hierarchy makes use of notions peculiar to itself. The ideas of democracy, credit, crime or political constitution are no part of biology, nor shall we expect to find in physics the concepts of memory, infection, sexuality, or fear. No sensible usage can bring the foreign exchange deficit into the biology syllabus... or nest-building into the syllabus of physics. In each plane or tier of the hierarchy new notions or ideas seem to emerge that are inexplicable in the language or with the conceptual resources of the tier below. But if... we cannot "interpret" sociology in terms of biology or biology in terms of physics, how is it that so many of the triumphs of modern science seem to be founded upon a repudiation of the doctrine of irreducibility? ...
>
> 3. *Causality.* The problems raised by the notion of necessary connexion, and the discussion of its actual and proper use.

Here MEDAWAR gives an example from genetics, in which a broad assertion of causality, easy to grasp but in fact wrong, was refined in four stages so as to become almost incomprehensible (and certainly, to him who is not a geneticist, so much vaguer as to be empty) but, presumably, better biology.

Lecture 2 is called "Mainly about Induction", but a more specific title would be, "Induction is Bilge". With the self-abnegation which is the Briton's most seductive quality, MEDAWAR begins by blaming the whole thing on the "English-speaking world". First, induction

> cannot be a logically rigorous process.... No process of reasoning whatsoever can, with logical certainty, enlarge the empirical content of the statements out of which it issues. If it could indeed do so then all scientific research could be carried out in a recumbent posture, with the eyes half closed.

It would seem that MEDAWAR here refers to the notoriously bad habits of mathematicians, late risers, often undressed, always half asleep if not sleepwalking.

But perhaps induction is only a "style" of reasoning. If so, its shortcomings are many and grievous.

> § 1. At the very heart of induction lies this innocent-sounding belief: that the thought which leads to scientific discovery or to the

propounding of a new scientific theory is logically account-able . . . . Even if they are not apparent at the time (because they have been short-circuited or speeded up), a retrospective analysis can reveal the processes of reasoning and the logically motivated actions which conduct the scientist toward what he believes to be the truth. There is a grammar of science, and the language of science can be parsed.

This is quite different from saying that *given* some belief or opinion of would-be natural law, no matter what its origin (whether by research, by revelation, or in a dream), *then* its acceptability can be tested by procedures that involve the use of logic. In the inductive view, it is the process of *getting* an idea or formulating a general proposition that can be logically reasoned out. It follows that, in the inductive scheme, discovery and justification form an integral act of thought. . . .

This concept of the inductive process must have arisen out of a misleading formal analogy with *de*duction. In deductive reason-ing, e.g. in Euclid, we discover or uncover a theorem by reasoning which, if we have carried it out correctly, guarantees the theorem to be true if the axioms or premises are true. Our ability to deduce Pythagoras' Theorem from Euclid's axioms—i.e. to dis-cover Pythagoras' Theorem in Euclid's axioms—is in itself our justification for believing it to be valid. In a purely formal sense, therefore, discovery and justification are the same process in deductive logic . . . .

But in fact even the theorems of mathematics, at least mainly,

> entered the mind by processes of the kind vaguely called intuitive; deduction or logical derivation came later, to justify or falsify what was in the first place an "inspiration" or an intuitive belief.

Indeed,

> Deductivism in mathematical literature and inductivism in scientific papers are simply the postures we choose to be seen in when the curtain goes up and the public sees us. The theatrical illusion is shattered if we ask what goes on behind the scenes. In real life discovery and justification are almost always different processes, and a sound methodology must make it clear that they are so.

§ 2. Inductive theory insists on the primacy of Facts: of propo-sitions that put on record the simple and uncomplicated evidence of the senses. Karl Pearson was a great believer in facts:

> The classification of facts and the formation of absolute judgements upon the basis of this classification . . . essentially sum up the aim and the method of modern science.

> The classification of facts, the recognition of their sequence and relative significance, is the function of science.

Pearson felt that the study of facts was conducive not only to good science but to right-mindedness in general:

> Modern Science, as training the mind to an exact and impartial analysis of facts, is an education specially fitted to promote sound citizenship.

It may therefore seem downright subversive to question the primitive authenticity of facts or to cast doubt upon evidence in the form in which it is delivered to us by the senses; it is worse still to ask, as Whewell did, how often "facts" can be stripped of a mask of interpretation and theory. It is very un-English, to be sure, for to put such a question is to challenge the greatest philosophic tradition of the English-speaking world, the tradition of philosophical empiricism which we inherit from John Locke. Nothing enters the mind except by way of the senses (its fundamental principle goes); and though the senses may sometimes be clouded, though we may sometimes be the victims of deception and illusion, yet if we can only get at it in its primitive simplicity, the evidence of the senses is the foundation of all knowledge . . . . There is an *essential* trustworthiness about the evidence of the senses, and therefore about the simple observational statements which put that evidence on record.

It won't do, of course. No one now seriously believes that the mind is a clean slate upon which the senses inscribe their record of the world around us. . . . "Everything that reaches consciousness is utterly and completely adjusted, simplified, schematized, interpreted," said Nietzsche . . . . Innocent, unbiased observation is a myth: "experience is itself a species of knowledge which involves understanding," said Kant. What we take to be the evidence of the senses must itself be the subject of critical scrutiny. Even the fundamental principle of empiricism is open to question, for not all knowledge can be traced back to an origin in the senses. We inherit some kinds of information. A bird's song is in some sense the transcription of a chromosomal tape recording, and the same goes for the entire repertoire of all that can properly be called "instinctual" behavior.

§ 3. Although inductive exercises often begin with an injunction to assemble all the "relevant" information (relevant to what?), inductive theory provides no *formal* incentive for making one observation rather than another. Why indeed do we not count and classify the pebbles in a gravel pit . . .?

§ 4. [In the science of real life t]heories are repaired more often than they are refuted, and a methodology of rectification . . . is something we shall expect to find in any satisfactory formal account of scientific reasoning. Sometimes theories merely fade away: no one now believes in the doctrine of "Protoplasm," but no one to my knowledge has ever refuted it. More often they are merely assimilated into wider theories in which they rank as special cases . . ..

§ 5. . . . Inductivism . . . fails altogether to explain how it comes about that the very same processes of thought which lead us toward the truth lead us so very much more often into error.

Methodologists who have no personal experience of scientific research have been gravely handicapped by their failure to realize that nearly all scientific research leads nowhere—or, if it does lead somewhere, then not in the direction it started off with. . . .

Why do scientists hold or come to formulate erroneous opinions? That, surely, is a central problem of methodology. . . .

§ 6. What are we to make of *luck* in our methodology of science? In the inductive view, luck strikes me as completely inexplicable; it can arise only from the gratuitous obtrusion of something utterly unexpected upon the senses; it is like winning a prize in a lottery in which we did not buy a ticket. . . .

§ 7. . . . Classical inductive theory reveals no clear grasp of the *critical* function of experimentation. . . .

Experiments are of at least four kinds:

(i) Inductive or Baconian experiments . . . ("I wonder what would happen if . . .").

Such experimentation "is not a critical procedure"; it merely enlarges experience and nourishes the senses.

Sciences which remain at a Baconian level . . . amount to little more than academic play. . . .

(ii) Deductive or Kantian experiments . . . ("let's see what happens if we take a different view").

These account for mathematics, *e.g.* the discovery of non-Euclidean geometries.

(iii) Critical or Galilean experiments: actions carried out to test a hypothesis or preconceived opinion by examining the logical consequences of holding it. Galilean experiments discriminate between possibilities. . . .

(iv) Demonstrative or Aristotelian experiments, intended to illustrate a preconceived truth and convince people of its validity. . . . Thomas Sprat took a very poor view of experimentation in this style—". . . a most venomous thing in the making of sciences; for whoever has fix'd on his Cause, before he has experimented, can hardly avoid fitting his Experiment to his own Cause . . . rather than the Cause to the truth of the Experiment it self."

So much for the "ruinous shortcomings" of inductivism.

Lecture 3, called "Mainly about Intuition", presents the "hypothetico-deductive scheme" nowadays usually associated with the name of KARL POPPER.

According to this second view, science, in its forward motion is not *logically* propelled. Scientific reasoning is an exploratory dialogue that can always be resolved into two voices or two episodes of thought, imaginative and critical, which alternate and interact. In the imaginative episode we form an opinion, take a view, make an informed guess, which might explain the phenomena under investigation. The generative act is the formation of a hypothesis: "we must entertain some hypothesis," said Peirce, "or else forgo all further knowledge," for hypothetical reasoning "is the only kind of argument which starts a new idea." The process by which we come to formulate a hypothesis is not illogical but non-logical, i.e. outside logic. But once we have formed an opinion we can expose it to criticism, usually by experimentation; this episode lies within and makes use of logic, for it is an empirical testing of the logical consequences of our beliefs. "If our hypothesis is sound," we say, "if we have taken the right view, then it follows that . . ."—and we then take steps to find out whether what follows logically is indeed the case. If [it is,] then we are justified in "extending a certain confidence to the hypothesis" (Peirce again). If not, there must be something wrong, perhaps so wrong as to oblige us to abandon our hypothesis altogether. . . .

[Thus] "falsifiability" marks the distinction between, on the one hand, statements that belong in science and to the world of common sense, and on the other hand statements which, though they belong to some other world of discourse, are not to be dismissed contemptuously as nonsense. Metaphysics is a compost that can nourish the growth of scientific ideas. But if we accept falsifiability as a line of demarcation, we obviously cannot accept into science any system of thought (for example, psychoanalysis) which contains a built-in antidote to disbelief: to discredit psychoanalysis is

an aberration of thought which calls for psychoanalytical treatment. (The critic cannot win against such a contention—but he
does not have to compete.)

MEDAWAR concludes that the "hypothetico-deductive scheme"
measures up to "the specifications of a good methodology" under the
following heads:

1. Discovery is distinguished from justification.

2. The action that is distinctively scientific comes not from "the
apprehension of 'facts' but from an imaginative preconception of
what might be true."

3. A "special incentive" is provided for our observations by
confining them to "those that have a bearing on the hypothesis under
investigation."

4. "Continual rectification . . . of hypotheses" is allowed for.

5. Scientific error becomes "an ordinary part of human fallibility:
we simply guess wrong . . . ."

6. ". . . The lucky accident fulfills a prior expectation . . . ."

7. Due weight is laid upon "the critical purposes of experimentation . . . ."

Turning devil's advocate at the end, MEDAWAR lists some shortcomings of the "hypothetico-deductive" method:
A. In principle there is no limit to idle hypotheses.

To exchange Whewell's system for Mill's is, on the face of it, to
trade in an infinitude of irrelevant facts for an infinitude of inane
hypotheses.

There must be, outside the method as stated, "some internal censorship . . . not . . . wholly logical . . . ."
B. ". . . we may yet be fallible in our imputation of fallibility." Our
observations may be faulty, misunderstood, ill conceived.
C. The "hypothetico-deductive" process is not distinctively
scientific, being "merely a scientific context for a much more general
stratagem".
D. The "generative act" in scientific inquiry lies outside the
scheme.

Scientists are usually too proud or too shy to speak about creativity and "creative imagination"; they feel it to be incompatible

with their conception of themselves as "men of facts" and rigorous inductive judgments. The role of creativity has always been acknowledged by inventors, because inventors are often simple unpretentious people who do not give themselves airs, whose education has not been dignified by courses on scientific method. Inventors speak unaffectedly about brain waves and inspirations: and what, after all, is a mechanical invention if not a solid hypothesis, the literal embodiment of a belief or opinion of which mechanical working is the test?

MEDAWAR finishes his apology by bringing what he calls "intuition" back into the picture of scientific discovery:

(a) *Deductive intuition*, which allows us to skip logical steps yet reach the logical conclusion.

(b) *Inductive intuition*, which hits the right hypothesis on which to base logical deductions.

(c) *Wit*, which is the instant apprehension of analogy.

(d) *Insight*, which invents an experiment that will serve as a searching test of a hypothesis.

Finally,

> The scientific method is a potentiation of common sense, exercised with a specially firm determination not to persist in error if any exertion of hand or mind can deliver us . . . . The purpose of scientific enquiry is not to compile an inventory of factual information, nor to build up a totalitarian world picture of natural Laws in which every event that is not compulsory is forbidden. We should think of it rather as a logically articulated structure of justifiable beliefs about nature. It begins as a story about a Possible World—a story which we invent and criticize and modify as we go along, so that it ends by being, as nearly as we can make it, a story about real life.

The quotations abstract MEDAWAR's case and illustrate the flair, the brief and easy elegance of language, and the polite yet biting irony with which he presents it. His booklet should be read and weighed by many who, I fear, will not read it: the working scientists, whose rote preconceptions it will offend, and the philosophers, who will be repelled by its sound base in workaday science and its plain English.

Much of its contents, I think, should stand uncontested. Although I must to my shame admit that I first heard POPPER's name only recently, I have always held to what he calls the "hypothetico-deductive scheme", not from any illusion I had invented anything so obvious, but because reading of the sources had convinced me that all the major mathematicians and physicists had used it, at least until quantum mechanics turned the philosophy of science into a stamping ground for the war dances of the tribe of obscurantists. Also, it seemed familiar from the writings of POINCARÉ and DUHEM. While many readers sorely need to see inductivism (and operationalism) demolished as MEDAWAR demolishes it, others will regard Lecture 2 as a brilliant proof that after all, it is not the stork who brings the babies.

However much the "hypothetico-deductive" description of the act of creation in science may shock hills of scientific ants who have been taught to chant that they are hardshell empiricists, it is nearly empty as a philosophy of science because it covers nearly all reasonable human action (MEDAWAR's objection C). I have seen carpenters and plumbers use it and use it consciously; they seem not to allow the possibility of any other approach to their problems; and, without knowing any philosophical name or theory, they scorn BACONian empiricism as being just stupid; yet in problems a scientist would consider scientific, they do not so much as grasp the questions, let alone see how to approach them even in the crudest terms. I think science has more to it than just plain common sense.

While inductivism accounts, if falsely, for far too much, the "hypothetico-deductive" scheme accounts for too little, merely *allowing* that to occur in science which really does occur there. Almost all of MEDAWAR's accusations of failure against inductivism can be turned around to make instances of jejunity for the "hypothetico-deductive scheme". For example, if inductivism fails to account for the frequent defeat of scientific inquiry, the "hypothetico-deductive scheme" fails to account for its occasional success, except, perhaps, by a "lucky" chance.

Indeed, luck in science finds a place in the "hypothetico-deductive scheme"—too big a place. We cannot win by a "theory" of odds which tells us merely that sometimes some people are lucky. Why does luck in science strike the successful scientist often, the unsuccessful one not at all? That it does not strike the unscientific is clear enough, since fortune favors the prepared mind, but some of the unsuccessful, particularly the old high priests of the scientific trades, are better prepared, or so they ought to be, than the brilliant, half-trained youngsters who make many of the great discoveries. Counter to BACON's naive yet wily contention that the method of inquiry he promulgated, which in his day had been little tried if at all, should place "all wits and understandings nearly on a level," MEDAWAR might have adduced the record of Nobel prize winners, despite all the recrimination, superabundantly just though it is on the brazen scale of lasting merit, against the caprice and faction of the popularity polls to which the waxen awards lead and resort. But were the EINSTEINs and BOHRs merely the luckiest of professional physicists? Can we teach the young (or old) scientist to be luckier? Indeed, inductivism is not the grammar of science, but what is?

Also, neither inductivism nor the "hypothetico-deductive scheme" seems to offer anything at all, let alone anything helpful, on the second of MEDAWAR's central philosophical problems, that of reducibility, yet examples of successful and enlightening proofs of reducibility and irreducibility occur again and again in mathematics. Perhaps the trouble here lies in the implicit faith of natural scientists and philosophers that reducibility, if possible, is unique. While today the beginner in mathematics has to master the concept of isomorphism, I have seen no trace of it in the literature on method and application, where it should find its most helpful use. Even rather intelligent theoretical physicists often proclaim that experiments *prove* matter to be discrete. Many opinions about nature which physicists now hold untrue were once inferred by logic just as unsound and believed just as fanatically by the profession, only to be destroyed later by the discovery of new facts and the creation of new theories[2]. With such examples of misinterpretation of fact and logic set before him by the most successful of the natural sciences[3], what can a biologist be expected to think?

---

[2] While many physicists of the last century interpreted FOUCAULT's determination that light moves slower in water than in air as a proof that light can be nothing but a wave motion of a plenum, DUHEM rejected the inference as wholly illogical even though he found the statement congenial to his own view of physics as a whole. This was in 1906, before the corpuscular nature of matter had become orthodoxy, and of course long before any logically clean theory of light quanta had been proposed.

[3] The flourishing new branch of mathematics called "model theory" seems not yet to have turned its attention to the problems that distinguish the modelling of physical science from the modelling of one mathematical structure by another. In a paper called "Logics appropriate to empirical theories", presented at a conference in 1963, P. SUPPES remarked upon the relatively superficial and extemporaneous level at which studies of the foundations of natural science remain even today:

It is sobering [to consider] the scientific contrast between the majority of papers read here and the standard sources in the methodology and philosophy of science. Yet it is encouraging, because the hope is engendered that many of the methods, and perhaps above all, the intellectual standards of these papers, will extend themselves in a natural way to logical investigations of the empirical sciences. The logical and philosophical foundations of physics, for example, seem to be at about the stage where the foundations of mathematics were during most of the nineteenth century. Nearly any physicist and a large number of philosophers are

I confess also to being dismayed by MEDAWAR's resort to "intu-ition". The only kinds of avowed intuition I meet in real life are called "feminine" and "physical"; the former I should not dare to describe, but the latter seems often to be an excuse for sloppiness, employed by those who would teach in a vague and confusing way something mathematicians have understood simply and well for half a century or more. Both kinds are used as clubs to stun reason. MEDAWAR means something else, but it is not clear that naming four different kinds of intuition brings us any closer to understanding scientific discovery, let alone to effecting it. A beautiful woman, indeed, may be beautiful in four different ways, yet the listing of these is unlikely to help the standard aspirant to Venus' girdle even make any better use of such meager charms as "luck" has given her.

MEDAWAR is very modest about his own competence in mathe-matics, but I think a little boldness here might have helped him. His discreet remarks, nearly all in footnotes and set off by disclaimers, reveal more grasp of mathematical methods and aims than one encounters commonly in a natural scientist or unmathematical phil-osopher. Now mathematics has two enormous advantages for the student of scientific method and philosophy, and both of them are obvious and indisputable.

First, mathematics, in the sense in which the word is used today, has an abundantly documented history extending over 2000 years, while most of the natural sciences are bound in space to Europe and America and in time to the industrial revolution and its fearful after-math of holy wars, widespread suppression, tyranny, and the extinc-tion of whole races and cultures, imposed by "revolutions", whether acute and bloody or chronic and asphyxiating, under the banner of that phantasmagoric materialist idol, "progress". Mathematics, on the other hand, has been done and done well by Hellenic aristocrats, Chinese sages, Arab astrologers, schoolmen of the Middle Ages, Renaissance upstarts, Baroque swashbucklers, squires and parsons and lords of the Enlightenment, as well as, more recently, by pro-fessors hired to lecture to shabby auditoriums in the midst of indus-trial slums and by obedient place-holders in the anthills of socialistico-democratic bureaucracy—not only under governments and beliefs

---

prepared to deliver at a moment's notice a lecture on the foundations of quantum mechanics. The situation is far different with respect to the foundations of mathe-matics. With an ever-increasing volume of deep and rigorous results, mathe-maticians unacquainted with the literature are not prone to deliver casually-put-together *obiter dicta* on logic and related topics.

The volume is called *The Theory of Models*, edited by J. W. ADDISON, L. HENKIN & A. TARSKI, Amsterdam, North-Holland Publishing Co., 1965.

regarded as "enlightened" by some dominant school of "progressive" thought today but also and equally in "reactionary" ones. Thus the *empirical* basis for philosophy and method, namely, examples from different periods, circumstances, and cultures, is greater beyond compare for mathematics than for any other science. Perhaps that is the very reason that philosophers tend to exclude mathematics from the sciences by definition.

I think the only really grave defect in MEDAWAR's book is its oblivion of history. To a biologist the history of biology before, say, 1900, may seem like the history of ignorance and error, but the history of mathematics records an almost unremitting increase not only in the details but also in the wide aspects of truth and knowledge, not only in the quarrying of much good stone but also in the creation of new or enlarged styles of sound and beautiful building. Thus mathematicians can and do read lessons in the past. For a recent example, one may look up the entry "Morals, Bad" in the index of my *Essays in the History of Mechanics.*

Lest this essay seem (to those who do not weigh it) no more than a hymn to the Queen of the Sciences, I remark that MEDAWAR does not face the real subtlety of experiments, which of necessity are projected, described, and interpreted by means of words, or, if you prefer, those operations of the mind we denote by the terms "concepts" and "logic": abstract symbols and rules of operation with them. DUHEM wrote, *"Experimental check of a theory does not have in physics the same logical simplicity as in physiology."* In physics

> there can be no question of dropping at the door of the laboratory the theory we wish to test, for without it we cannot calibrate a single instrument or interpret a single reading. . . . In the mind of a physicist who experiments two apparatus are constantly present: one is the concrete apparatus, the apparatus of glass or metal which he manipulates; the other is the schematic and abstract apparatus with which the theory replaces the concrete apparatus, and about which the physicist reasons; the two ideas are indissolubly linked in his comprehension; each necessarily calls up the other; the physicist can no more conceive the concrete apparatus and not associate with it the notion of the schematic apparatus than the Frenchman can conceive an idea and not associate with it the French word that expresses it[4].

---

[4] *Cf.* BUNGE, page 77 of *Foundations of Physics*: [While]

> the experimental physicist . . . may think of himself as an untheoretical operator, . . . [he] is a man full of ideas . . . , with a flair for finding objectifiers of unobservables and skill in designing material counterparts of theoretical models,

Hence no experiment ever tests a hypothesis as such, but rather the image cast by that hypothesis against the background of an unquestioned structure of concepts, methods, and even standards of taste. By careful trial, test, or analysis we may determine the pitch of roof most secure against December wind, but if our house is built of mud and founded upon a summer sand flat in the river of time, the superbly designed and fashioned telescope in our attic will serve us scantly for charting the heavens during the next March floods.

Since DUHEM's day, the concepts and methods of the biological sciences have become subtler and more evolved, as MEDAWAR's book shows. The "schematic apparatus" is now as important in histology as it is in chemistry, and the direct, wide-eyed question put straight to nature can no longer distinguish physiology from physics. Indeed such questioning, which once belonged to physics, too, and is served forth *ad nauseam* in BROWNE's *Vulgar Errors* and the early numbers of the *Philosophical Transactions of the Royal Society*, is not a distinctor of one science from another but merely a mark of infancy in any science where it appears.

While DUHEM may have attempted to specify what an experiment in physics can really show, he seems to have succeeded only in stating what it can not show:

> *An experiment in physics can never condemn an isolated hypothesis but only a whole theoretical assembly.* .... A physicist sets himself the task of showing a certain proposition to be incorrect. In order to deduce from this proposition that a certain phenomenon should occur, and in order to set up the experiment that should show whether this phenomenon does or does not occur, in order to interpret the results of the experiment and establish that the phenomenon predicted has not in fact occurred, he does not limit himself to use of the disputed proposition. Rather, he employs in addition a whole assembly of theories, admitted by him unquestioned. The prediction of the phenomenon, failure of which to occur should cut off the debate, does not flow from the disputed proposition taken by itself, but from the disputed proposition joined to all this assembly of theories. If the predicted phenomenon does not occur, it is not the disputed proposition alone that is at fault, but the whole theoretical armature which the

---

computing in terms of orders of magnitude, putting aside irrelevant details, and making plausible inferences. Without theories his manipulations would be gadgeteering or even white [grey?] magic and his data would not count as evidence for or against some idea . . .. "[E]vidence" is a relational not an absolute item . . . . In advanced science there is no theory-free observation.

physicist has used. All the experiment teaches us is that among all the propositions which have served to predict the phenomenon and to establish its failure to occur, there is at least one error, but where that error lies, the experiment does not tell us. Does the physicist declare that this error is contained in the proposition he wished to refute and not elsewhere? Then he assumes implicitly that all the other propositions he has used are true. His conclusion is worth no more nor less than this confidence of his . . . . Thus we are come far indeed from the experimental method as persons who are strangers to the way it works like to conceive it.

Hence for DUHEM

[a] crucial experiment in physics is impossible . . . . Contradiction by experiment, unlike the reduction to absurdity used by mathematicians, cannot transform a physical hypothesis into an incontestable truth. For that, we should have to list completely the various hypotheses to which a given group of phenomena could give rise. However, the physicist is never sure to have exhausted all the possible assumptions. The truth of a physical theory is not decided by heads or tails.

For a physical theory, the only experimental test which is not illogical consists in comparing *the whole system of that physical theory with all the experimental laws and in judging whether the latter are satisfactorily represented by the former.*

Having convinced ourselves by experiment that something is wrong with a theory, what can we do? While the layman or the philosopher might suggest we throw the whole theory away and in a grand flash of genius make a new one from the bottom up, as GALILEO is popularly credited for having abolished ARISTOTLE by the drop of two balls, no such thing, and nothing even roughly like such a thing, has ever happened in science. What should we do? Exactly what we do do, namely, discard, or more often weaken, some particular hypothesis, while leaving the rest as they were. Which of all our many hypotheses do we fell? The choice, says DUHEM, is left to "the sagacity of the physicist". Here blows up the blizzard of unreason. DUHEM advises us now to consult "reasons which reason does not recognize, reasons which speak to the sense of quality and not to the sense of geometry (à l'esprit de finesse et non à l'esprit géométrique), . . . , that which we justly call *common sense* (le bon sens)." Thus while MEDAWAR calls for stout British intuition graced by plucky luck, DUHEM before him had relied upon refined Gallic sagacity and good sense. Indeed these qualities are admirable, as are dogged Teutonic perseverance

and elegant Italian flair, which might help here too, but they have
been merely named, not defined. The choice of hypotheses, particularly in the special case of rectification of hypotheses, is the central
problem of the theorist. Our guides have led us to this problem and
stated it, but they have not taken the smallest step toward solving it.

If our philosophers have served the theorist no more than Timon's
feast, what can they offer the experimentist? Certainly there have
been and can still be experiments of the kind MEDAWAR[5] calls "critical
or Galilean". Indeed, these tell us much less than the folklore of
science (called "the scientific method" in the beginning stages of much
of the "life sciences" as they are taught today) would have an innocent
believe. Nevertheless, some experiments do test two or more alternative hypotheses while leaving the remaining hypothetical basis
unchanged. Such experiments are more enlightening than others. A
philosophy of experimental science should teach us how to conceive
and project experiments of this kind. MEDAWAR offers no suggestion
here, perhaps because he does not analyse what an experiment is.

The projection of an experiment, especially a decisive one, is itself
neither an experiment nor a fact: It is a deed of the mind, more like to
the creation of a theory than to the reading of a dial or the interpretation of such a reading. If this deed of the mind is to refer to nature, it
must reflect the regularity, the precision, the diuturnity we recognize
in the processes of nature. Deeds of the mind delimited by regular
and precise controls are called "mathematics". Not all of them, by any
means, are yet developed and incorporated in the formal structures
cultivated by professional mathematicians and described in books full
of written symbols. The realm of mathematics, like the realm of
thought itself and unlike the surface of the earth, is unlimited. No
matter how much of it men have explored, after any finite number of
human lifetimes more of mathematics necessarily remains uncharted
than the part which has been tamed or civilized.

MEDAWAR considers mathematics to be distinct in kind from all
other sciences. So did DUHEM, who wrote,

> Physics does not progress in the way geometry does. While
> geometry grows by continual addition of new theorems, proved
> once and for all, physics is a symbolic picture which continual
> retouches extend and unify. The *whole* of these produce a picture
> which is more and more a likeness of the *whole* of the experi-
> mental facts, although any one detail in this picture, cut off and
> isolated from the rest, loses all meaning and represents nothing at
> all.

---

[5] DUHEM did not pause to discuss these, perhaps because his subject of study was
theory.

Here, I think is one of the few places where we know today a little more about science than DUHEM did[6]. Naturally enough, he saw mathematics as it had been taught to him in the 1870s, a vast and ancient tree whose trunk seemed everlasting, whose boughs grew only longer, and whose verdure and floretry increased year after year. He did not know that while the outside was green and healthy, the trunk was rotten. Since his time, many of the boughs have had to be cut off and freshly rooted. Nowadays mathematics seems much more like a formal garden of the high renaissance[7]. The principles of horticulture and most of the plants are the same as they were; it is they that are eternal in their renewals; but the garden, lovingly tended by a band of experts, is frequently dug up and rearranged in new patterns, patterns created by the gardeners, sometimes to find ordered places for plants newly imported or discovered. Physicists are wont to complain that when they seek some mathematical bananas, they find that someone has moved the banana tree. [That is why they, especially those upon whom devolves the task of Nestors of their mystery, invariably demand that the hapless mathematicians who are ordered to pluck and serve forth bananas to the abecedarians destined to carry the torch of physics for the morrow shall lead them through the plat of yesteryear (preserved in hallowed textbooks fifty or more years old).]

---

[6] DUHEM is easier to refute than BUNGE, who distinguishes between "formal truth", "factual truth", and "partial truth" (*Foundations of Physics*, pages 28–29). It seems to me that BUNGE's approach to science is tinged by an occupational hazard of the trade of physics: the belief that if we could only understand quantum mechanics, we could understand *everything*. SUPPES, in the essay cited above, prefers a three-valued logic for the construction of mathematical models of natural phenomena, but again it seems to be quantum physics that he regards as the hero-villain of science. I wonder if we might not find, on the contrary, that if we could once get a clear picture of the relation between classical physics and observation and the process of creating new "classical" theories for extended ranges of phenomena, we might find in the end that quantum mechanics was not so mysterious or so all-important or so unavoidable as the physicists frantically and ceaselessly chant. Perhaps MEDAWAR, with his suggestion that we "quantify [the] degree of imprecision" of theories that are neither so good as we wish nor altogether bad comes nearer to the mark here. Rather than merely state that some truths are partial or "uncertain", we might introduce degrees of [factual] truth— instead of truth values, whether two or three, a truth measure something like a probability. In such terms we might create a mathematical model of the relation between theory and experiment.

[7] I recall once having been harangued by a prized and honors-laden guru of the Anglo-Indian way of "applied" science and "applied mathematics" to the effect that all this modern abstraction could easily become "too high baroque". That was in 1952. The distinction between "baroque" and "renaissance" was and is rather subtle for the "applied" but seems to be minor, since it is the "high" that is reprehensible to them.

The second fault in DUHEM's contrast between mathematics and physics reflects his own commitment to a now extinct particular school of physics, the energeticists, men who were mainly more interested in tacking the parts of physics together than in probing their logical and empirical foundations, their inner relations to each other, their reducibility or irreducibility. To my mind, this passage is the only one in DUHEM's great book that has aged in the sixty-five years since it was written. Indeed, physical theory is a picture, but DUHEM seems to have thought only one picture could serve. His simile suggests to us a single vast canvas, over which crawls the whole army of physicists, each working with his own tiny brush and pigments specialized to the needs of his own corner. Indeed, the theorist is an artist who depicts nature, but there is more than one picture in the gallery. One man may have many true portraits. There are many media, and for each there are many styles. A picture is a likeness, not a duplicate. The medium by force, and the artist by choice, omits much if not most of the subject. An engraving omits color, a painting omits one dimension, a statue omits breath, and all three omit flesh and blood. One engraving can indeed furnish a better likeness than another, but improvement is seldom gained by tinting what was designed in black and white, and to make a statue spurt blood when punctured would not serve any purpose for which a statue is wished.

Here MEDAWAR with his "tiers of the natural hierarchy" comes closer to the mark, though he does not stay near it. HELMHOLTZ[8] seems to have been the first to state that a theory can be no more than a *mathematical model*[9] for nature.

---

[8] See § 1 of his *Vorlesungen über Theoretische Physik*, Volume 2, *Dynamik continuirlich verbreiteter Massen* (lectures of 1894), Leipzig, 1902. HELMHOLTZ spoke of the difference between analytical dynamics and continuum mechanics, a difference that most modern philosophers of science, if I may judge by their voluminous publications, seem not yet to have caught up with. The former he described as "a picture which is only an abstraction which simplifies the considerations, and whose consequences are fit to represent the facts with sufficient completeness and brevity in certain domains of phenomena." Continuum mechanics, "the opposite limiting case", is also "only an abstraction, a picture, and indeed one which corresponds to the perceptions of our senses...." Moreover, "as the designation 'abstraction' or 'picture' indicated, it is not claimed ... that the concept of continuously distributed masses ... corresponds completely to the structures occurring in nature."

[9] The term "model" is not used here in the technical sense favored by physicists and historians and philosophers, who, antiquated as their views seem to anyone active in research on the foundations of mechanics today, allow only models within the theory of dynamical systems composed of mass-points and rigid bodies subject to pairwise equilibrated central forces and rigid constraints, perhaps with an occasional linear spring or dashpot.

A model represents one *aspect* of nature[10]. As the late WILLIAM
FELLER wrote,[11]

> In applications of geometry and mechanics theoretical concepts
> are identified with certain physical objects, but the method is
> flexible and varies from occasion to occasion so that no general
> rules can be given. The concept of a rigid body is useful and
> essential to mechanics, and yet no physical objects meet the
> specifications. Only experience teaches us which bodies can, with
> a satisfactory approximation, be treated as rigid. Rubber is usually
> given as a typical example of a non-rigid body, but in discussing
> the motion of automobiles most textbooks treat the wheels,
> including rubber tires, as rigid bodies. This is an example of how
> theoretical models are chosen and varied according to con-
> venience or needs. Depending on our purposes, we feel free to
> disregard atomic theories and treat the sun as a tremendous ball
> of continuous matter or, on another occasion, as a single mass
> point. We must always remember that mathematics deals with
> abstract models and that different models can describe the same
> empirical situation with various degrees of approximation and
> simplicity. The manner in which mathematical theories are
> applied does not depend on preconceived ideas and is not a mat-
> ter of logic; it is a purposeful technique which depends on, and
> changes with, experience.

Also

> ... like all mathematics, the theory of probability builds theoreti-
> cal models which are applied in many and variable ways. The
> technique of applications can be understood only after the theory.
> The intuition develops with the theory.

A better model may indeed represent that one aspect more closely;
it may represent in addition a different aspect, but it need not.
Different models are used for different ends, and commonly a second

---

[10] *Cf.* § 4, "Mathematics and its physical interpretation", in "The Classical Field
Theories" by C. TRUESDELL & R. TOUPIN in Volume **III**/1 of FLÜGGE's *Encyclopedia
of Physics*, Berlin *etc.*, Springer-Verlag, 1960. See Essay 2, above.

[11] W. FELLER, *An Introduction to Probability Theory and its Applications*, New York,
Wiley, 1950; see §§ 1–3. This passage was condensed, to its loss, in the second edition,
1957. I take this occasion to mention with gratitude that it was a series of lectures on
probability by FELLER in 1947 to the Theoretical Mechanics Subdivision, U.S. Naval
Ordnance Laboratory, White Oak, Maryland, that first made this matter clear to me.

aspect of nature calls for a model different in kind, not merely a "retouching" or refinement of a model already in use for some other aspect. A perfect theory of genetics, if we had one, might indeed incorporate or rest upon physics, but it might equally well leave physics altogether out of account. Even in pure physics itself the idea that all the phenomena follow from and hence are "explained" by a single, final set of "laws", is merely a popular illusion. Nearly fifty years ago G. D. BIRKHOFF stated[12],

> no theories appear to be fundamental in physics—it is merely that some are more fundamental than others in certain directions. Although we have a vague feeling of the unity of the physical universe, and are in possession of beautiful mathematical abstractions which account for numerous phenomena, nevertheless we have just begun to discover what is going on.

Much has been discovered since BIRKHOFF's day, but his statement still stands.

MEDAWAR's "reducibility" reflects only a part of the problem of interconnection of models. Equally important is the problem of constructibility, for a model of some refinement may need to be explained and understood in terms of simpler models. Here, again, is where we have recourse to mathematics, for mathematics not only unfolds the implications of logical models but shows how to create and interrelate them. On both these counts MEDAWAR goes astray. Formally, indeed, "no process of reasoning ... can enlarge the empirical content of the statements out of which it issues," but are we to conclude that there is no "empirical content" in calculating the future motions of the planets because they are already "contained in" or "implied by" the laws known to govern them? Does it make more than strictly juridical sense to say that the calculations enabling astronauts to correct their course so as not to become corpses lost or absorbed in space did not "enlarge the empirical content" of the routine initial data and the long-known general laws of mechanics and rules of mathematics fed into the thoughtless maws of the computers? A PARMENIDES might contend that the mass, position, orientation, linear and rotational velocities, and tensor of inertia of a weightless object in the vacuous heavens far beyond human sight had, until the mathematicians NEWTON and EULER had created and developed their theories, no empirical content whatever. According

---

[12] G. D. BIRKHOFF, "Newton's philosophy of gravitation with special reference to modern relativity ideas", pages 51–64 of *Sir Isaac Newton, 1727–1927, A Bicentenary Evaluation of his Work*, Baltimore, Williams and Wilkins, 1928.

to BUNGE[13] "... factual science presupposes and contains certain formal theories which it does not question and cannot subject to doubt because facts are irrelevant to pure ideas." I am not sure what BUNGE means here, but I think that, conversely, without ideas there would be no facts.

If, as MEDAWAR writes, "It is indeed true that deduction owes its existence to the infirmity of our powers of reasoning: it cannot bring us news of the world, but ... it can bring us awareness," is this not true also of the senses? The regular heavenly motions, most of us are taught to believe, have been always, since the Seven Days of Creation, such as they are today, yet the imperfect senses of man were not trained to apprehend them well until the industrial disaster ("revolution") had put him in a position to spend on the construction of built-in obsolescence human efforts of a magnitude that once could be spared only for fanes, dedicated to eternity. After all, have the millenia and, I presume, millions of indefatigable astronomers done anything more than "bring us awareness" of what our theretofore imperfect powers of sense let pass by unheeded? If, on the other hand, as MEDAWAR seems to believe, observation and measurement have created new empirical facts, the same may be said with equal justice of the results of purely logical inquiry. When the minds of men, who, I should think, make up a part of the "world", for millenia have fed upon and developed the numbers 1, 2, 3, ..., so intensely that now they cannot even think of experiment, not to mention trade, in any but numerical terms, did the discovery and proof that the fraction of primes in the first n integers was asymptotically $(\log n)^{-1}$ as $n \to \infty$

---

[13] *Scientific Research* 1, New York, Springer-Verlag, 1969, page 23. I am not sure I understand BUNGE's distinction, two pages earlier, between "formal" and "factual" sciences:

> The diversity of the sciences is apparent as long as their objects and techniques are focused on; they vanish as soon as the underlying general method is disclosed.
> The first and most remarkable difference among the various sciences is the one between the *formal sciences* and the *factual sciences*, i.e. between those dealing with ideas and those dealing with facts. Logic and mathematics are formal sciences: they refer to nothing in reality. ... Physics and psychology are among the factual sciences ....

In that case I do not see what BUNGE means by saying that factual science "contains" certain formal theories.

If I may be allowed a conjecture supported mainly by inexperience, I timidly suggest that the "empirical" or "factual" quality of a science is quantitative, not qualitative. Those sciences in which the proportion of unstated and hence unformalized theoretical axioms is greater, and in which accordingly, what is by some asserted to be "true" may be "false" to others and is insecure for all, are the ones commonly called "empirical" or "factual".

really add no more than a bit of "awareness"? Here, I think, DES-
CARTES saw the issue better: In explaining his third rule toward
guidance of the genius, he described

> *deduction*, through which we understand everything that is
> unavoidably inferred from certain other things surely known. ...
> [M]any things, though not by themselves evident, are known with
> certainty, but only because they may be deduced from true and
> known principles by a continuous and nowise interrupted move-
> ment of cogitation which contemplates every single thing per-
> spicuously.

In the deductive process, more is at hazard than first to conjecture
and then to prove that A → B. Before even guessing that A → B, the
deducer must formulate A. To feign a definition is even so original an
act, and potentially even so idle or so fruitful, as to project an
experiment.

The second great advantage of mathematics for the analyst of
scientific method is that mathematicians have been very frank, even
obscene, in confessing the intimacies of their thought-lives. (Here I
must note with regret that MEDAWAR repeats the old saw about
mathematicians' taking pains to ensure that their "intuitions" be concealed.) As MAXWELL wrote[14], mathematicians are "guided by that
instinct which teaches them to store up for others the irrepressible
secretions of their own minds ...." If, indeed, NEWTON and GAUSS in
their celebrated reticence stopped barely short of the silence of a
Trappist, EULER laid bare again and again the facts and thoughts that
led him to the conjectures he attempted, sometimes short of success,
to prove, and CAUCHY in later life published the secretions of his
mind week by week, just as they spurted out. In addition to the
revelations of POINCARÉ we have such papers as HADAMARD's "Comment je n'ai pas découvert la relativité" (Congress of Philosophy,
Naples, 1924) and his book, *Psychology of Invention in the Mathematical
Field*, Princeton, 1945. DESCARTES' in his *Rules toward Guidance of the
Genius* tells us, more clearly than in his *Discourse on the Method of
Reasoning Well*, exactly how to discipline ourselves so as to gain a
position from which we might be able to discover something worth
knowing. Not only that: Today, in the face of intensive specialization
and heavy artillery, many a research paper in pure mathematics starts
by revealing a heuristic basis on which to conjecture the main results
and even to sketch their proofs.

---

[14] J. C. MAXWELL, "On the dynamical evidence of the molecular constitution of
bodies", *Nature* 11 (1875): 352–359, 374–377.

Indeed, there is a major book, concrete but not requiring of the reader that he be a professional mathematician, on the way mathematicians conjecture, discover, and prove: G. PóLYA's *Mathematics and Plausible Reasoning*, 2 volumes, Princeton University Press, 1954. This is not a philosopher's book but rather a book on method as revealed by the words and deeds of masters. It is a book less concerned with naming and classifying outlooks or the nature of knowledge than with teaching us how to discover truth for ourselves. It is a lesson book, the lessons being taken from history, for[15] "the true school of scientific method is study of the masters."

PóLYA's lesson book presents the very opposite of the kind of philosophy of science on which MEDAWAR justly levels the reproach that "it gives its proficients no advantage". Indeed, most philosophers of science today seem to disregard what every carpenter knows: You must learn how to do something yourself before you can rightly explain how it is done, let alone teach someone else to do it. (Of course, the philosophers are not alone here: The educationists for decades have taught our children that common sense is an unfortunate aberration of the uneducated, that the accumulated wisdom of millenia of human experience is outmoded superstition, and that gabble about how something has been done or might be done is an improvement on all forms of physical and mental toil.) PóLYA's book will stand, along with those of DESCARTES, POINCARÉ, DUHEM, and WEYL, as long as persons remain who have completed apprenticeship in the specific and concrete knowledge and skills which alone form a sufficient basis for understanding science. These books will die only when science itself is replaced by a routine of some sociobureaucratic faith.

PóLYA denies the strict difference between mathematics and natural science upheld by DUHEM and MEDAWAR. On page v of

---

[15] The quotation is from my lecture to a congress of mechanics in 1962, reprinted in my *Essays in the History of Mechanics*, New York, Springer, 1968.

The approach of PóLYA is not so restricted as the one DUHEM advocated and called "the historical method" (*op. cit.*, § 6 of Chapter 7), for it is not intended as "the best means, indeed the only means of giving to those who study physics a just idea and clear view of the very complex and living organization of this science." Indeed, the "organization" of mathematics has changed again and again, and failure to see change of this kind was one of DUHEM's limitations as a historian and philosopher of science. Those learned in the history of mathematics as a whole have rarely been more than barely competent, if that much, in the mathematics of their own days, and all too often even their historical work suffers from failure to see where mathematics is, or even was, going. PóLYA does not trace the history of any branch or period of mathematics; rather, he uses the history of mathematics as a quarry for examples of method, for it is method above all that the masters teach us.

Volume 1 PÓLYA writes of mathematics that "no other sub-ject . . . affords a comparable opportunity to learn plausible reason-ing." While MEDAWAR eliminates mathematical induction as "a special usage", PÓLYA entitles his first volume "Induction and Analogy in Mathematics," and on page 111 he writes "Has mathematical induc-tion anything to do with induction? Yes, it has, and we consider it here for this reason and not only for its name," and he makes good this point, like all his others, by solid, specific examples from the *practice* of mathematics. No less a master than EULER himself, referring to those most mathematical of entities, the pure numbers, wrote[16]

> It will seem not a little a paradox that even in the part of mathe-matics usually called pure I have set much store by observations, which commonly seem to have a place only in external objects striking our senses. Since indeed the numbers in themselves must be referred to the intellect alone, we may scarcely see how observations and, as it were, experiments be worth anything toward exploring their nature. Nevertheless here it has been shown by very solid reasons that most properties of numbers which we have by now come to know were first revealed to us by observation alone, and this for the most part long before we had confirmed their truth by strict demonstrations. . . . Nevertheless, examples are not lacking in which induction alone has cast us into error.

Of course PÓLYA, like EULER before him, requires in the end a strict logical proof; for him induction suggests the definitions to be laid down, the hypotheses to be feigned, and the propositions to be deduced, but the scientific act is completed only by deduction. Thus he regards "the inductive attitude" as the basic character of science: "This attitude aims at adapting our beliefs to our experience as efficiently as possible." I suggest that they who wish, after having purged their minds of the puerilities of inductiv*ism*, to learn how induction goes, and what part deduction plays in ripening its fruit, should read PÓLYA's book[17]. The main difference between mathe-matical sciences and verbal doctrines in regard to indûction lies in the

---

[16] In the editorial summary (draughted by himself) of his "Specimen de usu obser-vationum in mathesi pura" (1753), *Novi commentarii academiae scientiarum Petropolitanae* **6** (1756/1757): 19–21 (1761) = LEONHARDI EULERI *Opera omnia* (I) **2**, 459–460.

[17] The roles of induction and intuition and proof in mathematical thought, past and present, are illustrated in the Socratic dialogue on the EULER polyhedron formula by I. LAKATOS, "Proofs and refutations", *British Journal of the Philosophy of Science* **14** (1963/1964), repaginated as a monograph; [second edition, prepared for the press by J. WORRALL & E. ZAHAR, Cambridge *etc.*, Cambridge University Press, 1976; correc-ted reprint, 1979. This book's brilliant logical analysis is marred by its author's ignor-

higher standard of mathematics, in which induction is not only the first step toward precise formulation but also the first step toward precise proof. The difference can easily be exaggerated, since even for a purely empirical study the final product is not merely a table of numbers but some kind of general summary, some statement capable of mathematical formulation as an axiom to be accepted or a theorem to be proved, even though the experimentist commonly stops short of the last step.

I contend that deduction, although indeed it is ideally possible without induction, rarely if ever succeeds without it, and that induction without deduction is empty if not impossible. Both deduction and induction owe their values jointly to the strength of our powers of reasoning, not to their weakness!

---

ance of mathematics in the eighteenth century and his irresistible yearning for revolutions.

[LAKATOS in a footnote on page 79 of the first edition, page 74 of the second, finds what he considers the only flaw in PÓLYA's ideas: "he never questioned that science is inductive, and because of his correct vision of deep analogy between scientific and mathematical heuristic he was led to think that mathematics is also inductive." In the sections titled "Induction as the basis of the method of proofs and refutations" and "Deductive guessing versus naive guessing" (pages 73–82 of the first edition, pages 68–76 of the second) LAKATOS argues that mathematics grows from deductive guessing rather than induction from examples.

[Here is the place to remark on two errors LAKATOS makes regarding EULER's paper cited in the preceding footnote. First he thinks author and editor of this paper are different persons and states that PÓLYA was wrong in confounding them (footnote on page 11 of the first edition), although both Professor PÓLYA and I explained to him in 1965 or earlier that EULER wrote the editorial summaries of all mathematical papers in the *Novi commentarii* at this time, the same error occurs in the second edition (footnote on page 9). LAKATOS was correct in doubting that EULER would have referred to himself as "most illustrious" but naive in presuming there was no clerk at the Academy to insert the normal graces. Second, he accepts PÓLYA's translation of "observationes et quasi experimenta" as "observations and *quasi-experiments*" (LAKATOS' italics). I cannot find "quasi experimenta" anywhere in the text of the memoir to which the summary refers. It would have been foreign to EULER's ways to introduce a new term without defining or at least explaining it. While "quasi-" as a prefix occurs in English from 1643 onward, I can find no such usage of the common Latin word "quasi" by Classical authors or in the Latin scientific literature of the eighteenth century. In the translation I have given in the text above "observationes et quasi experimenta" is rendered as "observations and, as it were, experiments", "experiment" being used here in the sense of "trial".

[In regard to method the first sentence in the second paragraph of EULER's text would seem to support PÓLYA's view:

Indeed, Fermat must be regarded as having inferred these remarkable properties of numbers by induction, long before he learned how to prove them.

The sense of the passage shows that EULER's "quasi experimenta" and "per inductionem" meant one and the same thing to him: "by trial".]

How can the result of a purely logical process serve as evidence for induction? By the mere fact that we know it. What you do not know, however true it may be, you cannot use, even though it may be known to others. ARCHIMEDES, in books of his now lost, may well have proved theorems not yet rediscovered by anyone. The indisputable truth of such theorems serves us nothing and is presently empty; such truth is as ideal and unreal as are the other truths known presently only to YAHWEH but from us still withheld, although both kinds, tomorrow, may spring to instant life, one by the luck and skill of a paleographer whose keen eye seizes and penetrates a palimpsest, the other by a thunderbolt of JOVE. A yet undeduced or now forgotten consequence of the axioms is as much unknown as a yet uninduced datum of an experiment not yet conceived. That the axioms of a deductive system contain *in posse* all their consequences, is as true and as idle a fact as that all the appearances of nature exist already and merely await the experimentists who can and will someday observe them. (Some experimentists, it seems, while boasting their puritan empiricism, would have us think they created, rather than merely observed, the data of their experiments. Alas, this may even be true!) A datum of experiment and a proved theorem of mathematical physics, though obtained by different courses of thought and action, are often put to use in just the same way: They are known, kept in mind, and ready to hand as stones in the foundation of new discovery, be it experimental or theoretical. In their use for such foundation, the differences of their origins may be of little moment.

Now mathematics cannot possibly fall outside the domain of a philosopher who lists as one of the four main types of experiments the KANTian ones ("let's see what happens if we take a different view"), for these are the bread and meat of mathematical research. Even less could MEDAWAR exclude mathematics on the basis of intuition, for all four kinds he describes are in the standard repertory of any mathematician, the "experiments" mentioned under "insight" being KANTian ones.

Indeed, I think that factual analysis of daily mathematical experience could dispel the mystery that MEDAWAR leaves enshrouding intuition and inspiration by adding to the mere names and abstract qualities concrete examples of success. I think that fruitful intuition[18] ("feminine" and "physical" apart) is a name for a process

---

[18] "Intuition" has come a long way. In explaining the third of his rules toward guidance of the genius, DESCARTES wrote

> By *intuition* I understand, not the uncertain faith in the senses nor the false judgment of contriving imagination, but rather the easy and discerned concept of

which, however vaguely apprehended it now be, is systematic, rational, and hence a fit subject not only for analysis but also for husbandry. PÓLYA's book could justly have been entitled "The Anatomy of Intuition", but anatomy is no more than a prerequisite for physiology. To dissect and depict the corpse does not by itself disclose the functions of the healthy body, let alone teach us to cure the sick or raise the dead. The physiology of intuition could be a fit object for the extension of the new mathematical theory of models.

While MEDAWAR reproaches inductivism with failure to explain reducibility and emergence, he does not indicate how the "hypothetico-deductive scheme" can do so. In fact, reducibility is another standard problem for mathematics[19], so much so that when one pure mathematician asks another for "applications", he usually is satisfied by a specimen showing that the new theory somehow

---

a pure and attentive mind, such that no doubt can remain henceforth about what we understand, which is born from the light of reason alone . . ..

The reference to *reason alone* suggests the very opposite of physicists' and engineers' tribal faith in their own unreasoned "intuition", the origin of which I have endeavored in vain to trace. In *Isis* **50** (1959), 480, I published the following query: "The term 'physical intuition' is frequently on the lips of physicists nowadays and appears occasionally in their writings. When did this notion arise, how did it gain currency, and what does it signify in regard to scientific method?" No answer was ever published or sent me. The only one of the several senses of "intuition" defined in the *Oxford English Dictionary* that comes near to that of current physics I have discussed in the footnote on page 212 of my *Essays in the History of Mechanics*.

[The *Dictionary*'s supplement of 1976 gives much attention to "intuitionism" in mathematics but does not mention physical intuition. The easiest way to it is suggested by the definition in the *Dictionary* itself for Sense 4, appropriate to Scholastic Philosophy: "The spiritual perception or immediate knowledge, ascribed to angelic and spiritual beings, with whom vision and knowledge are identical." We should simply replace "angelic and spiritual beings" by "physicists".

[The earliest instance of "physical intuition" that I have found occurs in a passage by J. LARMOR, 1906, quoted above on page 228.]

[19] Unfortunately the examples philosophers use as a basis for their generalizations are all too often abortive. *E.g.* when E. NAGEL in his book, *The Structure of Science*, New York *etc.*, Harcourt Brace & World, 1961, Chapter 11, chooses to discuss "the reduction of thermodynamics to statistical mechanics", he so grossly glosses over the conceptual and mathematical problem, in fact totally confounding at least two different problems, as to burlesque the result itself (or, better, the class of results presently proved or conjectured) and so confuses any but the most expert in the subject.

Reducibility in the natural sciences is only rarely a matter of showing that Theory 1 is included as a special case of Theory 2. So far as I know, in classical physics the only case of this kind is afforded by the inclusion of optics in electromagnetic theory, and even here there are some subtle points. More usually Theory 1 is shown to be in some sense a limit case of some special case of Theory 2 provided additional assumptions be

includes or illuminates some old ones. The compartments of the physical sciences do not confine mathematics. If, as MEDAWAR says, we
shall not expect to find in physics the concept of memory, we certainly
do find extensive theories of memory in mathematical structures, and
perhaps such names as "credit" and "punishment" and "democracy"
and "tyranny" can find logical models more easily than cellular
ones.

MEDAWAR, following POPPER, has enthroned falsifiability as the
discriminator of science from other doctrines and arts. Falsifiability of
what? Of a "prediction". Now a prediction in a true science, be it
biology or physics or mathematics or any other, is a conditional statement: "If A is true, then B follows," and it is B that must be falsifiable.
Conditional statements, whatever the empirical and inductive content
or justice of their protases, are in themselves either true or false. If
the statement "A implies B" is wrong in logic, the falsification of B by
experiment is no more than a BACONian pebble in a gravel pit, gotten
up in whorish trappings to seduce us to give it a respected place as
a truth of science. Thus the deductive aspect of natural science,
however much it may be masked sometimes by leaving simple or
commonly received premises unstated, is just as much a discriminator as is falsifiability itself. In the practice of science, all but the
most trivial cases of deductive reasoning are codified and organized in
terms of mathematics. The one common feature of all the true sciences is the substructure of mathematics upon which they rest. In the
study of methods in mathematics and its application lies the only
viable philosophy of science as a whole.

Mathematics has retained from its venerable antiquity a character
that once belonged also to the natural sciences as well but now is
unique to it. The new natural philosophy of the seventeenth century,
abrogating authority no matter how august, told man to try each new
or old truth for himself, to put his questions, once he had thought
them through in the light of his own experience and purified them by
his own reason, to nature herself. A curious experiment was to be
confirmed, as soon as news of it arrived, by independent trial. Thus
the individual man, not the Politburo of Organized Scientific
Authority, was the measure of science, and if John Smith could not

---

made. Such is the reduction of classical thermostatics to mechanics: No *single* conservative dynamical system in a finite region can illustrate the laws of thermostatics, but those
laws do follow from consideration, at a fixed time, of an *infinite sequence* of ever more
numerous systems of *like composition*, provided we make some *further and extra-
mechanical assumptions* regarding improbability.

Rational science cannot—it *dares* not—accept pronouncements of some Supreme
Soviet or Pope of Science as a substitute for unbiased trial or recreation of logical proof.

himself confirm a conclusion of ISAAC NEWTON, let alone the latest catamite of the Nobel Prize Committee, he should reject it. In the natural sciences today no such tests are feasible. The freshman is taught to prate by rote as the very foundations of science a heap of dogma about invisible entities and their intercourse, the existence of which has been inferred by the potentates of a mighty, tax-subsidized bureaucracy in command of mountains of the most expensive plumbing the world has ever seen. All the beginner can do with this dogma is believe it, memorize it, and use it[20]. He cannot possibly recreate it for himself, let alone search and test it. At least six years of intensive indoctrination are needed before the modern student of natural science can begin to trace the subtle interrelations of measurement, calculation by colossal machines, application of classical theory, and speculative pseudo-mathematics that have led to the bald assertions that made up most of his freshman textbook. While the product of all this may be and doubtless is sound science, the student's path to it is pure antiscience: blind acquiescence in the dogma of some faith, one that while lacking a personal deity and ethics is as abject as any religion of the so-called Dark Ages. Though the aim is different, the process of learning is much the same as for Communism, Naziism, Democracy, or other revealed truths.

Mathematics is still taught in the old way. While mathematicians no longer try to provide their students with total knowledge of all the mathematics considered important today, or with any *Weltanschauung* of mathematics, from the first day of the first course they strive to show how mathematics is done. Often encountering strong resistance from the students, most of whom would rather learn by rote, and unremitting vilification from the local votaries of physics, chemistry, and engineering, the teacher of mathematics insists upon clear statement and logical proof. The student who cannot by himself, by his own mind unaided by authority or computing machines or television pictures of impressive complexes of pipes and wires connected in flow

---

[20] DUHEM saw and bewailed the beginning of this kind of instruction. He inveighed against an "ideal inductive" method of teaching physics, based not on real experiments but imagined ones, "experiments that not only have not been done but could not be . . .", and even "absurd experiments [that] claim to establish a proposition which it is contradictory to regard as a statement of an experimental fact", by use of which so as to avoid founding an experimental science "on a postulate one founds it instead on a pun." To escape this "monstrous jumble of petitiones principi and vicious circles, . . . , endless defiances of logic . . . ," DUHEM recommended the historical method. Unfortunately the greater subtlety of natural science and the higher critical standards of the history of science today combine to make the historical method impractical in ordinary instruction, though it is supremely enlightening for those who have the mental apparatus and the devotion sufficient to follow it.

diagrams enlivened by talking mice and ducks, provide logical steps by which the prime number theorem is proved from the properties of the real number system, has not learnt that theorem. If he does learn it, his logical steps need not be the same as those in the book, or as those any predecessor has used. The beginner in mathematics may not get far, but however far it be, it is by real mathematics, the same act as that which the greatest mathematicians have performed and do perform to make their discoveries. *Mathematics is the one science that is still a science even for the beginner.*

The hindrance here is that while mathematics is used by every scientist, even by the experimental biologist, it is commonly used unwittingly and ill if not wrongly, and almost always with distaste. Mathematics is a language which, while it can be learnt by any intelligent and studious person, in fact nowadays is treated like hieroglyphics by all but the professional mathematicians[21]. Why should anyone not going to be a professional mathematician learn mathematics? Indeed—and, I fear, in the midst of the triumph of social democracy the comparison may not seem as doltish as it is, since the concept of "dolt" has been devalued—why should anyone learn English who is not going to be by profession an author, or, better, an "editor", in that language? It is as if the only persons who learnt to read Latin were those who spoke it as natives—that is, nobody—and all studies of HORACE in relation to the literature of the world were made by Hottentots who reproached him for having deliberately concealed his meaning by use of an arcane language they could decipher only in part and by conjecture.

Note for the Reprinting

The book reviewed is PETER BRIAN MEDAWAR's *Induction and Intuition in Scientific Thought*, Philadelphia, American Philosophical Society, 1969, 62 pages.

My essay review of it first appeared in *Centaurus* **17** (1973): 142–172. Here and there I have shortened or extended the original text a bit. My few substantial additions are set apart by square brackets.

---

[21] *Cf.* P. R. HALMOS, "Mathematics as a creative art", *American Scientist* **36** (1968): 375–389.

39. SUPPESIAN STEWS (1980/1981)

In Memoriam G<small>EORGI</small> H<small>AMEL</small>

C<small>ONTENTS</small>

0. Suppesians

I. Stegmüller's Concept of a "Theory of Mathematical Physics"

II. The Suppesians' Only Application: "Classical Particle Mechanics", Their "Paradigm"

III. Noll's Axioms for Systems of Forces and Dynamics

0. Suppesians

> The ancients said,
> To be virtuous without instruction,
> Is this not sagacity?
> To be virtuous after instruction,
> Is this not nobility?
> To be virtueless even after instruction,
> Is this not stupidity?
>
> (?) WU CH'ÊNG-ÊN, *The Journey to the West**

1. THE SUPPESIAN STYLE

The subtitle of WOLFGANG STEGMÜLLER's pamphlet *The Structuralist View of Theories* is "A Possible Analogue of the Bourbaki Programme in Physical Science". That raised my hopes. I thought to find

---

* Translated by ANTHONY C. YU, Chicago & London, University of Chicago Press, Volume 2, 1978, pages 359–360.

something constructive: a sketch of a set of general assumptions and statement of some key-theories through creative axiomatization in HILBERT's style, enough to cover at least in principle some part of some physical science.

Alas, I had to face a pile of authorities big enough for a mediæval scholiast but uglier: "Sneed's conception of theoreticity", "a new 'Kuhn-reconstruction' so to speak", "a 'Super-Super-Montague'", "'Kuhn Sneedified'", "'Sneedification of Suppes'", *etc. etc.* A jewel is *"Feyerabend's 'Kuhn Sneedified' ought to be replaced by 'Sneedification of Suppes'."* The title of § 3 is "The Force of *T*-theoreticity and the Ramsey-View-Emended. Non-Statement View$_2$ and Non-Statement View$_{2,5}$."

Such jargon (or thinking?) serves as its own parody. It outrages science and plain English alike. A reader unaccustomed to journalese in works offered as philosophy may find queasy the political labels *"too liberal"*, *"too conservative"*. Then there are the "revolutions", without which philosophers and historians of science might waste away for lack of fodder. BALZER's formal appendix, dense with symbols, provides nine theorems having no apparent bearing on the text or connection with the words of the authorities adduced there; perhaps it is meant only as a gesture of respect for the contributions made to axiomatics by great logicians of the past.

Nevertheless, as a reviewer I must try to decipher and analyse STEGMÜLLER's views and arguments. In doing so I will not use the mathematical symbols with which he punctuates his text, for they seem to be mere boiler plate.

2. THE SUPPESIANS

STEGMÜLLER tells us he belongs to a school he calls "structuralist", but the more easily pronounced "SUPPESian", with the ictus on the "pes", is more specific, for the members of that school cluster about PATRICK SUPPES. STEGMÜLLER distinguishes (page 4)

two different trends in the philosophy of science, both of which endeavour to obtain systematic rational reconstructions. For the sake of vividness I call the procedure advocated by the first school the *Carnap approach* and the method recommended by the second school the *Suppes approach*. In both cases the first step of rational reconstruction with respect to a particular physical theory consists in an axiomatization which is intended to lay bare the mathematical structure of the theory in question; but there is a fundamental difference in the way this task is performed.

According to the Carnap approach the theory is to be axiomatized within a formal language. It was Carnap's firm conviction that only formal languages can provide the suitable tools to achieve the desired precision. . . .

The Suppes approach is altogether different. Like Bourbaki, and unlike Carnap, Suppes uses only *informal logic* and *informal set theory* for the purpose of axiomatization. . . .

It is now more than twenty years since Suppes advanced the claim that philosophers of science should use *set-theoretical* instead of *metamathematical* methods. With this claim he began to lay the foundations of what I shall call the *structuralist view.*

The "Carnap or formal language approach" STEGMÜLLER calls *statement view*₁; the "Suppes approach" or "structuralist approach" he calls "non-st. v.₁".

I have not studied the works of CARNAP. Some who have done so deny that he proposed or endorsed the approach to which STEGMÜLLER attaches his name. Rather, they say, he deemed it valuable to make plain that a theory already sufficiently rigorous in the ordinary mathematical sense was capable of being rendered logically formal. He did not claim it desirable to express an entire theory in formal language. He advocated use of formal logic in regard to existing informal theories not as an end in itself but to clear up questionable points, mainly regarding foundations, in ordinary mathematical treatments.

I. Stegmüller's Concept of a "Theory of Mathematical Physics"

3. STEGMÜLLER'S MAIN THESIS

Here is STEGMÜLLER's first and apparently only main thesis (page 5): It is not possible for human beings to achieve the aim he attributes to CARNAP, no matter how desirable it might be to do so. Formal set theory is too cumbrous for application, and so those who follow the approach he imputes to CARNAP (page 5) "are forced to use simple, fictitious examples instead of instances from real science." STEGMÜLLER admits that this contention by itself does not drive us into the camp of the earlier SUPPESians, who (page 7) "confined themselves basically to rendering precise the *purely mathematical* aspect of physical theories."

4. "General" and "Special" Philosophies of Science

STEGMÜLLER distinguishes (page 42) between *"general philosophy of science"* and *"special philosophy of science"*. The former provides "sweeping remarks on all of science". STEGMÜLLER regards (page 43) search for "definitions, giving necessary and sufficient conditions, within general philosophy of science" as being "possibly *the* fundamental error of most present philosophers of science". He describes himself (page 42) as having now in "a drastic change of my philosophical attitude" moved closer to SUPPES' position in asserting that only necessary conditions can be given. These necessary conditions are to be sought from (page 44) *"special philosophy of science*, which deals with particular theories".

SNEED's achievement here, says STEGMÜLLER (page 13), is "an essentially *semantic supplementation* of the method of informal axiomatization. ... Following the SUPPES approach *without* 'Sneedification' would ... entitle us to nothing more than the claim that the mathematical structures of physical theories are *integrated into* the Bourbaki programme." The next sentence is garbled. I guess that STEGMÜLLER intended something like this: Use of an informal theory of models in SNEED's way "justifies talk of an *analogue to* or an *extension of* the Bourbaki programme in which *real physical theories* and not only their mathematical skeletons are studied."

In plain English, we should provide systems of axioms for special sciences, and in doing so we should employ only informal set theory, much as BOURBAKI does. In 1900 HILBERT[1] took a position in this matter:

[Problem] 6. MATHEMATICAL TREATMENT OF THE AXIOMS OF PHYSICS.

The investigations on the foundations of geometry suggest the problem: *To treat in the same manner, by means of axioms, those physical sciences in which mathematics already plays an important part; in the first rank are the theory of probabilities and mechanics. ...*

If geometry is to serve as a model for the treatment of physical axioms, we shall try first by a small number of axioms to include as large a class as possible of physical phenomena, and then by adjoining new axioms one after another to arrive at the more special theories. ... The mathematician will have also to take account not only of those theories coming near to reality, but also, as in geometry, of all logically possible theories.

---

[1] HILBERT [1901].

5. Stegmüller's Basic Question, and His "First Approximation" to an Answer

STEGMÜLLER raises (page 7) "the fundamental question of what distinguishes a *theory of mathematical physics* from a mere *mathematical theory*." For SUPPES himself (page 9), "*this role is played by theories of fundamental measurement*." As these theories are usually ignored by physicists, SUPPES' writings have attracted no attention (writes STEG-MÜLLER) from scientists engaged in studies of the foundations as they conceive them. For STEGMÜLLER, explaining the view of SNEED[2], distinction between "*a theory of mathematical physics*" and "*a mere mathematical theory*" is to be sought in the kinds of mathematical statements, without reference to a theory of measurement. To a given mathematical axiomatization ("a set-theoretic predicate" (page 12)) STEG-MÜLLER adjoins "the set of possible models" of the predicate and SNEED's "class of ... intended applications" (page 11). We may (page 12), "in a first approximation, identify a physical theory" with these aggregates.

6. Application: Every Mathematical Theory is "in a First Approximation" a "Theory of Mathematical Physics"

Let us test STEGMÜLLER's approximation by applying it to Euclidean geometry. For "the set-theoretic predicate" we take the abstract axioms of an n-dimensional Euclidean point-space as presented in a good book for undergraduates today. For "the set of possible models" we take the set of all equivalent systems of axioms. A serious example of an equivalent set would be one presented in terms of rectangular Cartesian co-ordinates. A specimen is provided in the appendix to EISENHART's freshman text[3]; $n = 2$, and the abstract axioms are HILBERT's. Of course there are infinitely many abstract models: simple ones, in which one axiom is replaced by a different statement to which it has been proved equivalent, the other axioms being granted, and complicated ones, in which no axiom is the same as one of the original ones, *etc. etc.* For "the open set of 'really' intended applications" (page 12) there are many mathematical examples. We may give n a fixed value such as 1, 2, 3, or 4, and indeed whole treatises are devoted to each of these applications. We can also limit the class of figures allowed, so obtaining the geometry of spheres, the geometry of polyhedra, *etc.* STEGMÜLLER's only example of a "para-digmatic subset" of "the open set of 'really' intended applications" is

---

[2] SNEED [1971].
[3] EISENHART [1939].

one for "classical particle mechanics" that he claims (page 12) NEWTON[4] "gave": "the solar system; various subsystems . . . like earth–moon or Jupiter and its moons; pendulum movements; free-falling bodies near the earth's surface; the tides." He does not tell the reader whether these words refer to the dirty earth, the flaming sun, and the tides of polluted water or to the mathematical models that NEWTON by use of his mechanics was able to formulate as representing the real bodies of the several kinds. In either interpretation, all these "subsystems" of mechanics refer to bodies of specified shapes such as spheres or spheroids and to the locations of such shapes. Thus every one is also a member of "the open set of 'really' intended applications" of Euclidean geometry. NEWTON spoke to that effect, though in other words, in the preface to his *Principia*, and HILBERT referred to mechanics as an extension of geometry. An extension of a structure contains all the elements of the structure extended.

Therefore Euclidean geometry satisfies STEGMÜLLER's approximate criterion for "a physical theory". So, I think does any mathematical structure, unless STEGMÜLLER means by "applications" only structures intended to represent physical systems. But then to make sense he would have to define "physical".

We are left with only the former possibility, which makes every mathematical theory become "in a first approximation" a theory of mathematical physics if we specify a few members of the "open" set of its applications.

We are back again at HILBERT's program: ". . . geometry is to serve as a model for the treatment of physical axioms . . .." HILBERT here uses "model" in one of its old and still popular senses. Had he been condemned to express himself in the jargon of philosophers of science in 1981, he would have been required to write "paradigm" instead.

I, for one, am happy with the idea that a mathematical theory of physics is as a theory no different in principle from any other mathematical theory. The difference comes only in the class of applications the theorist has in mind. A physical theory should apply to some aspect of physical experience and perhaps also of physical experiment, that is all. "Physical" here is used in its ordinary, conversational sense. I do not have to try to define it, because I do not claim to provide a philosophy of science. It is STEGMÜLLER who has set himself that task. In publishing his fifth book on the philosophy of science he must do more for the reader than merely translate into jargon an opinion today common among mathematicians. Common knowledge

---

[4] NEWTON [1687].

may need to be refined, cleared, or even rejected, but nothing is gained by rewording it darkly.

He who reads further in this review will learn that it does not point at the edifice the SUPPESIANS continually design, alter, and redesign as an armature for theories of physics but rather at their competence as structural engineers. The body they model and remodel again and again[5] upon their successive versions of the skeleton of theoretical science is something they call "classical particle mechanics". This review weighs their words and postures regarding the mechanics of mass-points and other aspects of classical mechanics. A reader wishing to view SNEED's scheme of a theory of mathematical physics will find it sketched in the appendix.

II. The Suppesians' Only Application: "Classical Particle Mechanics", Their "Paradigm"

7. NEWTON'S "PARTICLE THEORY" OF THE TIDES, 1687: SUPPESIAN FICTION

STEGMÜLLER starts with NEWTON himself, whose *Principia*[4] he regards as being a book on particle mechanics; in his list, quoted above, of the "paradigmatic examples" of particle mechanics he attributes to NEWTON we find *the tides*. Apparently he got this idea from MOULINES, who wrote[6], ". . . tides were regarded by NEWTON and his collaborators as firm applications of particle mechanics; the theory-element they developed in order to account for tides was the so-called 'equilibrium theory of tides'." Perhaps MOULINES will someday disclose the names of NEWTON's "collaborators". The passage about the tides, very short, is *Principia* Book III, Proposition XXIV, Theorem XX in the first edition, Theorem XIX in subsequent edi-

---

[5] Their literature includes descriptions of special relativity and an axiomatization of "simple equilibrium thermodynamics" by MOULINES [1975]. Even the title of MOULINES's paper suggests that he confounds the thermodynamics of processes with GIBBSian thermostatics. Alternative sets of axioms for the classical example of the former, in which the systems considered are homogeneous bodies of some pure, frictionless fluid which does not undergo a change of phase, are presented and analysed by TRUESDELL & BHARATHA [1977]. PITTERI [1982] extends their theory to bodies described by any finite number of substate variables and susceptible only of reversible processes.

Note added in proof. BALZER & MÜHLHOLZER [1982] have constructed a SUPPESian axiomatization of the classical mechanics of impact.

[6] MOULINES [1979, pages 425, 432].

tions. It is qualitative, drawing only upon the general idea of gravitation. Apart from some references to observations, the only quantitive statements refer to "three hours" before or after certain events, but NEWTON does not reveal where he got that figure. He lets the reader infer that his theory squares with the three-day lag of the tides at Plymouth and Bristol, but he does not explain how. Neither the word "particle" nor any mathematics occurs anywhere. Anyone who has come closer than mere gossip to the theory of the tides will know that this "firm application" has been celebrated since the day it was published as one of the shakiest of the many ill-anchored passages in the *Principia*. BROUGHAM & ROUTH[7] made a brilliant and sympathetic attempt to provide reasoning that NEWTON might have used to reach his conclusions. AIRY[8] in his famous article "Tides and Waves" wrote

> The theory of Newton is rather a collection of hints for a theory than anything else.

Also

> Newton treated the general explanation of the Tides as a matter of Wave-theory entirely, (though not without errors,) . . . .

AIRY does not mention any "particle". But for MOULINES[6]

> . . . Laplace dismissed tides from the domain of particle mechanics and treated them rather as an application of hydrodynamics by means of his 'waves theory'. This development took place because only a portion of tidal phenomena could be considered a firm application of particle equilibrium theory; the rest proved to be recalcitrant to this theory-element.

Also

> A negative aspect of Laplace's work was his dismissal of the tides from the domain of applications of particle mechanics.

Wondering where MOULINES got the idea that NEWTON had a "particle equilibrium theory", I looked up the one and only "historical source" that he cites here. It is a modern exposition of theories of floods and tides[9]; "particle" does not appear in it either in word or by

---

[7] BROUGHAM & ROUTH [1855, Chapter VIII].

[8] AIRY [1845, §§ 16–19].

[9] DEFANT [1957, § 7].

implication; NEWTON is mentioned just twice, once as a progenitor of the "equilibrium theory", dismissed as valueless, and once as the author of a theory that was to be replaced by LAPLACE's.

8. PARTICLES IN NEWTON's *PRINCIPIA*, 1687: SUPPESIAN RHETORIC

Those who have studied NEWTON's *Principia* will have noticed that what the SUPPESIans call a "particle" is neither defined nor described there. The Latin *particula* means "a small part". Throughout NEWTON's writings and afterward, even down to the middle of the nineteenth century, the word "particula" and its derivates in modern languages were used in that radical and unspecific sense, which allows as instances a geometrical point, a grain of a granular body, an atom in a philosophical or physical sense, a mass-bearing geometric point, a small but finite increment, an element of integration, *etc.* The entry for "particle" in *The Oxford English Dictionary* gives under Sense 2 quotations ranging from 1398 to 1901 in illustration of usages, including some that the *Dictionary*'s definition omits, one of which is from EMERSON's *Fluxions*, 1743: "To find the Motion of any Particle of the String as suppose of X the middle Point". Because the mass of the particle X is 0, that particle would not move at all if it were subject to the laws of SUPPESIan "classical particle mechanics". The problem EMERSON treats belongs essentially in the group that NEWTON attacks in *Principia*, Book II, Section VIII; in it NEWTON in the first edition uses the word "particula" exactly once (Proposition XLVIII, Theorem XXXVII, in later editions Proposition XLVII); there he refers, as does EMERSON in the *Dictionary*'s quotation, to a differential element in a plenum. In the second edition NEWTON inserted in Proposition L, Problem XII, the famous passage about "the crassitude of the solid particles of the air". Those particles have specified, positive diameters. Particles of that kind make up the "rare medium" discussed in Section VII. They, too, have finite size.

In the celebrated Sections XII and XIII of Book I, "The attractive force of spherical bodies", and "the attractive forces of bodies which are not spherical", the word "corpusculum" is used frequently, always meaning "little body". In the proofs of some of the theorems of these sections, brilliantly conceived and brilliantly demonstrated, the words "corpusculum" and "particula" both appear. In his usage NEWTON distinguishes between them with almost perfect consistency. "Corpuscle" could, indeed, refer to a punctual mass, but in my opinion, which rests in part on NEWTON's diagrams and in part on my being able to read the language commonly used in the mechanics of the seventeenth century, he meant by it a sphere of positive diameter, small enough to fit the kinematical conditions considered. "Particle", on the other hand, seems almost invariably to refer to an element of mass in an integration, for which NEWTON has prepared the reader in the scholium following Proposition LXXIII, Theorem XXXIII. An especially clear example is provided by the "equal particles of any body" considered in Proposition LXXXVIII, Theorem XLV. The *Oxford English Dictionary* gives no definition or instance of the English word "corpuscle" as referring to a punctiform body.

Section XIV of NEWTON's Book I, celebrated for its difficulty, concerns "corpora minima", very small bodies. These tiny bodies traverse media. They serve for a theory of the reflection and diffraction of light, regarded as a corpuscular emanation. In Proposition XCVI, Theorem L, the word "corpus" is used both for a very small bit of light and for an infinitely broad and very thin sheet of penetrable material; in the proof NEWTON lets the thickness of each sheet diminish while the number of sheets increases "so that the action of attraction or impulse . . . becomes continuous". The analysis is too sketchy to justify any definite delimitation of the very small bodies; the two sources drawn upon are Galileian parabolic trajectories and the rule for composition of forces, neither of which necessarily refers to punctual masses.

It is my opinion that NEWTON never once, anywhere in his *Principia*, used the words "particula" or "corpusculum" to mean a SUPPESian particle. The concept *mass-point*, a punctiform body of positive mass, is often attributed to LAGRANGE; in fact it was introduced and defined by EULER[10] in 1736.

Very well, forget the word. Turn to the deed. Do NEWTON's axioms pertain only to mass-points? Is NEWTON's mechanics a mechanics of mass-points and no more?

Far from it. The whole of Book II treats media consisting of infinitely many points, the mass of each of which is 0. The quotation from EMERSON is a typical example: The NEWTONian body there is a differential element. Book II is widely known as a major source for hydrostatics and hydrodynamics. BROUGHAM & ROUTH[11] describe this book briefly and write "that it would have conferred lasting renown upon anyone but himself, had it been the only work of another man, is certain." The second half of their book presents and analyses NEWTON's attempts at field theories of hydrostatics and hydrodynamics. Thus NEWTON himself cannot have regarded his mechanics as being restricted to bodies of the kind later called mass-points.

---

[10] EULER [1736, § 96].
[11] BROUGHAM & ROUTH [1855, pages 162–165].

The axioms NEWTON put at the head of his book concern "bodies".
His famous comment on Law I refers to a spinning top. The points
making up a top are infinite in number; the mass of each is 0. One of
NEWTON's most celebrated discoveries is the theorem stating that
each of two homogeneously layered spheres attracts the other as if all
its mass were concentrated in a corpuscle at its center (Book I, Proposition LXXVI, Theorem XXXVI). This theorem, too, is inaccessible to
the mechanics of mass-points, for each point of the solid sphere has
mass 0. NEWTON's proof rests on use of the two preceding propositions, the former of which concerns a "corpuscle" outside a sphere,
and both of which refer to the "attractive particles" that make up the
sphere. Only if the "particles" are interpreted as differential elements
of mass, do NEWTON's arguments there make sense. Without the
theorem, NEWTON's "paradigmatic" treatment of sun and earth would
be nonsense, for while those two bodies are nearly spherical, any yokel
leaning on his hoe at sundown can see with his own eyes that they
occupy regions too big for human beings to regard as points.
NEWTON's figure for Book I, Proposition LXIV, Problem XL, shows
spheres of different diameters, S[ol] being naturally much larger than
T[erra] and L[una].

According to STEGMÜLLER (page 11), "The best way to discover the
application of a physical theory . . . [is to look] at the author's
examples . . .."

9. THE SUPPESIANS' APPEAL TO THE HISTORY OF MECHANICS

It is well known that the historical notices in textbooks of physics
mainly transmit and embroider old fictions having scant basis in fact.
No reasonable person will blame the physicists, for their notices are
not serious, being meant only as ornaments in the propaganda of
faith; both the authors and their students enjoy chanting creeds and
would not like to see their prophets desanctified by accurate accounts
of how discoveries were made, often for reasons contrary to modern
dogma and by methods infracting the modern physicist's code of
decency.

On the contrary, the SUPPESians reject both the faith and its rules.
We cannot condone their statements about history as being no more
than witnesses to their good standing with the physicists' union.

MOULINES in his paper on "the example of Newtonian mechanics" includes a section[12] called "The evolution of Newtonian

---

[12] MOULINES [1979, § 3]. While MOULINES seems to be regarded by the SUP-
PESians as their authority for the history of mechanics, he states that his facile gen-

particle mechanics during its first hundred years":

> This account relies on relevant historiographical literature, which has been listed in the "Bibliography" at the end. So, concrete historical statements made in the present reconstruction derive from the assumption that the descriptions and interpretations of the historians of science mentioned are, on the whole, accurate. In some particularly significant cases, I explicitly mention the particular historiographical work from which the data have been taken.

Alas, he chooses to cite a book and an article[13] of mine! To my grief I infer that my writing there is so bad as to allow MOULINES to divide the evolution of his "theory nets" into four "periods", in each of which only "particle mechanics" was of any importance. I hoped sufficient the evidence I presented to make my readers conclude that no part of mechanics developed in isolation from the rest, and that at no time before, say, the days of HAMILTON and JACOBI did the mechanics of mass-points play the leading role. Although in the eighteenth century the mechanics of continuous media remained special, limited for the most part to rigid motions, three-dimensional hydrodynamics, and one-dimensional hydraulics and theories of flexible lines and elastic bands, nevertheless it was in these theories, not in the mechanics of mass-points, that the main issues of the foundations arose and were given their first, still brilliant analyses. I refer especially to the concepts of rigid motion and change of frame, the principle of rotational momentum, and the concepts of torque and shear force. Even with all that omitted, MOULINES' second and third "periods" occurred at the same time. The explanation:

> First, notice that from a purely chronological point of view, the different periods distinguished partially overlap. This gives support to the general statement that the purely topological ordering relation $\preceq$ is to be preferred to the numerical $\leq$ for the representation of historical issues.

MOULINES writes of "particle collision" as one of the "direct empirical applications of ... the fundamental laws ..." and cites NEWTON's "Corollaries III, IV, VI". The famous scholium following

---

eralities derive only from his scan of secondary sources. He claims that the French géomètres "made great efforts to solve difficult many-body problems in celestial mechanics".

[13] TRUESDELL [1968] [1976].

the first six corollaries to the "Axioms, or Laws of Motion" refers to "spherical bodies", not mass-points. The second figure there suggests that the experiment (a real experiment!) employed pendulum bobs whose diameters were almost as great as a third of the distances from their centers to their points of suspension. From the results of this experiment on large bodies NEWTON concludes, "And thus the Third Law, so far as it regards percussion and reflection, is proved by a theory exactly agreeing with experience."

According to STEGMÜLLER[14]

> [W]hat is really interesting or important about various disciplines is their evolution and development, *i.e.*, the *dynamic* aspect which completely eludes logical analysis. In order to research this aspect, the logical method must be replaced by the *historical method*. Thus, it is no accident that the four outstanding critics of traditional philosophy of science . . . command an extensive knowledge of the history of science and endeavor to support their criticism of traditional notions with *historical arguments*.

One of the four "outstanding critics" STEGMÜLLER lists is the late N. R. HANSON, unforgettable for his dauntless imagination of how things surely must have gone. MOULINES & SNEED[15] refer to "NEWTON's own formulation of classical particle mechanics".

10. THE "CLASSICAL PARTICLE MECHANICS" OF McKINSEY, SUGAR, & SUPPES, 1953

In his "Mathematical Problems" of 1900 HILBERT[1] wrote

> Indeed, research on the foundations of a science has an especial charm, and to examine those foundations will always be one of the foremost tasks of the man of science. "The goal we must always keep in sight", said Weierstrass once, "is to strive to reach a secure judgment regarding the foundations of a science. . . . But to penetrate the sciences at all, study of special problems is certainly indispensable." In fact a satisfactory treatment of the foundations of a science prerequires deep understanding of its special theories; only that architect is qualified to dispose securely the foundations of a structure who himself knows in all detail the functions it is to discharge.

---

[14] STEGMÜLLER [1976, page 3]. Much of the material in this book derives from STEGMÜLLER's earlier work [1970].

[15] MOULINES & SNEED [1979, page 60].

STEGMÜLLER agrees (page 6): In

> the Suppes approach . . . , the investigation . . . is done by a person
> who is an expert in the field in question . . . ,"

A physical theory (page 16) is to be "*axiomatized in a precise way.*"
Citing with approval a passage by SNEED, STEGMÜLLER continues:
"the only presupposition made . . . is that the mathematical structure
of the theory meets the Bourbaki standard (or, for that matter, the
Suppes standard)." Later (page 85) STEGMÜLLER recalls "the high
standard of technical *and* philosophical precision required by the
Suppes–Sneed approach." MOULINES & SNEED go even further[16]:

> Philosophy of science in [Suppes'] style requires that *its* prac-
> titioners know as much, if not more, about the specific sciences
> they treat as the practitioners of these sciences themselves.

The reader may interpret the ambiguous "as" to mean either "as do"
or "as about"; the panegyric tone of the passage suggests the former.
Indeed, MOULINES & SNEED continue,

> In the work of Suppes, the philosophy of physics approaches the
> standards of rigor and precision that have become commonplace
> in the philosophy of mathematics.

Also

> . . . his philosophical work consists mainly . . . of painstaking, de-
> tailed and highly technical analysis of concrete, specific scientific
> concepts and theories.

The sole example that STEGMÜLLER and the other SUPPESIANS call
upon to illustrate this standard is the axiomatization of "classical
particle mechanics" in 1953 by MCKINSEY, SUGAR, & SUPPES[17].
STEGMÜLLER writes (page 4),

> . . . I shall presuppose that [Suppes] and his co-workers succeeded
> to a large extent in their endeavour. Besides, I assume as known
> that their axiomatizations are much superior to earlier attempts as
> far as clarity and precision [are] concerned.

---

[16] MOULINES & SNEED [1979, page 60].
[17] MCKINSEY, SUGAR, & SUPPES [1953].

Thus, for the SUPPESians, that axiomatization meets the BOURBAKI standard, and SUPPES is an expert on classical particle mechanics.

Here are the axioms:

A system $\Gamma = \langle P, T, m, s, f \rangle$ which satisfies Axioms P1–P6 is called an *n-dimensional system of particle mechanics* (or sometimes, when there is no danger of ambiguity, simply a *system of particle mechanics*).

Kinematical Axioms

Axiom P1. P is a non-empty, finite set.
Axiom P2. T is an interval of real numbers.
Axiom P3. If p is in P and t is in T, then $s(p, t)$ is an n-dimensional vector such that $d^2/dt^2 \, s(p, t)$ exists.

Dynamical Axioms

Axiom P4. If p is in P, then $m(p)$ is a positive real number.
Axiom P5. If p is in P and t is in T, then $f(p, t, 1), f(p, t, 2), \ldots, f(p, t, i), \ldots$ are n-dimensional vectors such that the series $\sum_{i=1}^{\infty} f(p, t, i)$ is absolutely convergent.
Axiom P6. If p is in P and t is in T, then

$$m(p)\frac{d^2}{dt^2}s(p, t) = \sum_{i=1}^{\infty} f(p, t, i).$$

Although STEGMÜLLER[18] calls them "Newton's version of particle mechanics as formulated by McKinsey *et al.*", anyone familiar with the practice and history of mechanics will see at once that these axioms are not broad enough to cover all the problems on falling bodies treated by GALILEO, or all the problems solved by NEWTON in Book I of the *Principia*, or all the problems OSGOOD includes in his undergraduate textbook.[19] Can STEGMÜLLER find in BOURBAKI a set of axioms for X insufficient to cover the elementary principles of X as presented in an ordinary course on X?

McKINSEY, SUGAR, & SUPPES regarded NEWTON's First Law as "a trivial consequence of our axioms". Next they discussed the possibility of adjoining further axioms to express NEWTON's Third Law and statements differentiating internal from external forces. They proved eight simple theorems, some of them standard, some regarding subsystems and unions of systems, and one to the effect that any system can be embedded in a system that satisfies their interpretation of

---

[18] STEGMÜLLER [1976, § 6.1].
[19] OSGOOD [1937].

NEWTON's Third Law. Finally they showed the primitive notions m, s, and f to be independent. In a succeeding paper[20] they discussed properties preserved or destroyed by transformations of systems of particles into systems of particles.

Before analysing these paradigmatic axioms of a theory of mathematical physics I describe the circumstances leading to the publication of the paper by McKINSEY, SUGAR, & SUPPES and its sequel.

11. CIRCUMSTANCES OF THE PUBLICATION OF McKINSEY, SUGAR, & SUPPES' AXIOMS

On 29 July 1952, shortly after McKINSEY had submitted the manuscripts of both papers to the *Journal of Rational Mechanics and Analysis*, I wrote as follows to him:

> The objective of your paper is most commendable, but I do not think it is achieved. For the reasons detailed below, I should be inclined to return the paper unconditionally. Since, however, I have myself the intention ultimately to write very differently on the subject, it may be that my own opposed views make it more difficult to appreciate the value of your paper; moreover, this *Journal* is very likely the only place where such an article can be published in this country; therefore, I should like to give you the opportunity to discuss the matter, and if you still feel the work should appear in print, I will publish it with a note as follows: "The communicator is not in agreement with the view of classical mechanics expressed in this paper. He believes, however, that it should be made available to the student of rational mechanics." . . .
>
> Basically, I do not feel that particle or rigid body mechanics is worth the trouble. . . . You do not mention either the basic paper of Hamel[21] in the *Annalen* for 1908, where axioms for genuine mechanics are given, or the recent rather messy maccaronic treat-

---

[20] McKINSEY & SUPPES [1953].

[21] HAMEL [1908]. For other early work see "Additional Bibliography P: Principles of Mechanics" at the end of TRUESDELL & TOUPIN [1960]. There is no article on HAMEL in the *Dictionary of Scientific Biography*. More remarkable yet, the editors of that compendious work chose to devote a column apiece to men like AMES and E. T. BELL though the authors of their biographies could not find a single specific contribution to science, even ephemeral or tentative, that might deserve mention. The *Dictionary* gives less than two columns to ST. VENANT and omits J. H. MICHELL and PIOLA, all three of them great elasticians.

ment in Brelot's papers and book[22]. Of the latter treatment, as of
yours, I feel that the basic problems are buried under rather
superficial and unnecessary paraphernalia and terminology of
modern mathematics. You refer several times to rigid body
mechanics as if it were the only alternative. Actually, a true system
of axioms of mechanics must deal with finite deformable bodies;
rigid bodies are included as a special case, and the discipline of
mass-point mechanics emerges in its true light as a rather degenerate and crude first approximation.

Why should a particle be subject only to a denumerable number of forces? When a particle is attracted by a solid sphere, is it
acted on by the non-denumerable number of forces exerted by
the volume elements of the sphere, or by the single force
effectively concentrated at the center of the sphere—or by two
forces, one exerted by the left half, the other by the right half of
the sphere? Isn't the matter of composition of forces part of the
mechanism of setting up a special problem, rather than of the
axiomatic structure of mechanics? *Cf.* your footnote 8. Moreover,
your including the composition of forces leads to your strange
Axiom P5. The word "absolutely convergent" rings very oddly in
an axiomatic system. Absolute convergence, or even convergence,
is a concept introduced in mathematics so as to be able to get
sufficient conditions for certain mathematical results. It does not
correspond to any physical idea. ...

Your taking P6 as an axiom rather than a definition does not
seem to differ from the traditional hazy way of doing things (*cf.*
[pages 270–271 of the paper in print]). ... [T]here is no meat in
Newton's second law when it is regarded as a *mathematical* axiom.
Mathematically it is only a definition; its real content, like that of
the problem of time measurement (page [256 of the paper in
print]), is "epistemological and experimental". When Hamel
states that if we regard $f = m\ddot{s}$ as a definition "then Mechanics
would not be a natural science", he clearly refers to the *interpretation, (natural* science), not the formal system. Hamel himself, if I
recall correctly, somewhere speaks of force as a concept derived
from our muscular feelings. In a somewhat more sophisticated
way, we could say that the *physical* axiom of Newton's second law
may be stated as follows: "there exists a Euclidean frame in which
the observer has *a priori* knowledge of both the mass and the
acceleration of observed bodies, without having first to determine
the motions themselves." This, I believe, is the real content of
Newton's second law.

---

[22] BRELOT [1943] [1944] [1945].

Your axiom P3 seems a step backward from Hamel. Your suggestion [page 259 of the paper in print] that a single problem be treated "by considering two realizations" seems artificial and unnecessary. Moreover, properly general axioms of mechanics can and should include impacts, which should certainly not be left, as you suggest on page [259 of the paper in print], to "a different branch of mechanics." ... I do not understand the philosophical point on page [263 of the paper in print]. It seems that the usual center of mass theorem for continuum mechanics, the purely kinematical proposition that if p be the total linear momentum, M the mass, and x the coordinate of the center of mass of any body, then

$$p = M\dot{x}$$

is devoid of subjunctives and conditionals. From this we have at once $\dot{p} = M\ddot{x}$. Your remark on page [263 of the paper in print] about the finite set is most mystifying, since the above theorem holds under the weakest sort of measure assumptions. The important thing about the theorem really is that the corresponding proposition on moments,

$$h = r \times M\ddot{x},$$

is *not* true in general. ...

The remarks on concatenation and Theorem 8 seem to me to be the only contributions made by the paper. ...

While you cannot in such a paper give a complete historical account, nevertheless the relevant things should be cited. To omit the name of Euler in footnote 2 is serious, since much of what Lagrange put into the *Mécanique Analytique* was taken from Euler's papers. Also, as far as mass-point mechanics is concerned, you make no mention of the possibility of defining a "system" in totality, without using the separate forces—the method of D'Alembert, Lagrange, and Hamilton. The [book] by R. Dugas[23] ..., while not very accurate, is quite interesting on these points. The originals are easily available in Jouguet's "*Lectures . . .*"[24].

Finally, it seems to me that that axiomatization if successful must do more than put down in black and white what everyone knew before. Hilbert's axiomatization of geometry really got to the heart of the matter and showed which axioms are necessary in order to prove the usual theorems. Your axiom system is close to

[23] DUGAS [1950].
[24] JOUGUET [1908] [1909].

the unformalized traditional one, and to me it seems to share the traditional inadequacy. You do not show what major results in mechanics would fail to follow if these axioms were modified or if some were omitted. The system is moreover partly too general, since it is n-dimensional and devoid of geometry, but partly not general enough, since it does not include representations for the mechanical behavior of the commonest objects—a glass of water or a beam.

Finally, I should remark that Professor HAMEL, the only living person who has made a significant contribution to the axiomatic and conceptual foundations of mechanics, is a member of our Editorial Board. If you still feel that your paper deserves publication, you might wish to send it to him.

On the same day I wrote as follows to HAMEL:

I inclose a copy of a letter to McKinsey in regard to a paper on the axioms of mechanics, in case he should decide to send the paper to you. It is possible I was too harsh with it.

McKINSEY did not enter into communication with HAMEL.

On 30 September McKINSEY sent me manuscripts slightly revised and replied to my criticisms. He was correct in pointing out that some were ill taken and some reflected misunderstandings on my part; in the foregoing quotations of parts of my letter I have omitted those passages. To the remaining points I had raised he replied as follows:

Regarding your general remark to the effect that classical particle mechanics is not "worth the trouble", we feel that even though—as you point out—this discipline is not the "true" system of mechanics, this kind of an approximation has shown itself sufficiently useful (in celestial mechanics, for example, and in statistical mechanics) to warr[a]nt axiomatization. . . .

In regard to [your remark on our axiom P3], it seems to us that it is not desirable to use Hamel's generalization to piecewise twice-differentiable functions—especially in view of the recent result of Gale (*American Mathematical Monthly*, 1952, pages 291–295) on the indeterminacy of impact mechanics. When impacts are permitted, the analogue of our Theorem 2 (first paper) can no longer be proved. . . . [W]e hope that you will find it possible to publish these papers, in accordance with the suggestion in the first paragraph of your letter. None of us has any objection to your adding the suggested note ("The communicator is not in agreement . . ."), but unless some technical errors should be found, we should prefer not to make substantial changes.

On 6 October I replied to McKINSEY:

> I still feel such doubts about your papers that I have sent them to Professor Hamel. My reason for doing so is twofold. First, if he approves of them, we can publish them under his name as communicator, without qualification. Second, if he does not, he may be able to express his objections in a form more compelling to you than mine appear to have been.
>
> Since I am sending the MSS by air mail, requesting their return via air mail, and since the next date for sending material to our printer is December 1, no delay should result.
>
> My promise to communicate the paper, with a qualifying footnote, of course still holds. However, I believe that both of us will feel more comfortable if we can get Hamel's opinion, whichever way it goes.

At the same time I wrote to HAMEL:

> I send herewith the MSS . . . by J. C. C. McKinsey *et al.* My own opinion of the first paper is contained in a letter I sent you some time ago, but McKinsey persists in wishing to publish.
>
> You will recall that according to our refereeing system the opinion of the Member of the Board is transmitted without secrecy to the author of the paper. However, the Board Member is under no obligation whatever, and in case you do not wish to express an opinion, there is no need to do so. If you really feel the papers to be worth publishing, I should feel much more secure if your name, rather than mine, appeared as communicator, since my competence in the foundations of mechanics is by no means equal to my interest.

HAMEL replied eight days later. I print here in English translation his entire reply, apart from salutations:

> As editor of the *Journal of Rational Mechanics and Analysis* you ask me to evaluate two papers by Mr. McKinsey and his collaborators. I must recommend against publishing them in their present forms.
>
> What these gentlemen claim, to have set classical mechanics upon a more rigorous foundation, I cannot regard as justified at all. Their considerations in this regard offer nothing new; in fact they represent a considerable step backward insofar as they restrict the subject to the most meager mechanics of points and leave the concept of force in vague generality. Nevertheless, in

the concept of force lies the chief difficulty in the whole of mechanics. Moreover, a scarcely lesser difficulty lies in the inclusion of continua, which require careful definition of two pairs of concepts: applied force and force of reaction, interior force and external force. Not a trace of that from our authors. How foreign to them is the whole matter, is shown by their polemic Footnote 3. The axiom there faulted as being unnecessary is fundamental for the construction of the entire theory of mechanics, namely, for the definition of the concepts just named.

Probably of some value is Theorem 8 of the first paper, which makes every system a subsystem of a Newtonian system. Nevertheless, I think the presentation should be made more transparent; the basic idea should be expressed clearly and not choked by formulæ. Moreover, our authors should make clear how their proof differs from Hertz's, which probably Footnote 11 means to cite, though it gives no specific reference. Perhaps this matter would do for a short note. Also the transformation theorems in the second paper seem to deserve attention. I regret that here [in Landshut], far from all libraries, I cannot determine to what extent the results are new. Also unfortunately, I no longer have the book of Hertz.

I have no objections to your letting the authors know the contents of this letter.

Here is "their polemic Footnote 3" as printed:

[3] Thus, although HAMEL's formulation of mechanics is rightly regarded as one of the clearest existing treatments of the subject, we find in HAMEL[25] the following strange axiom: "The forces dt are determined by their "*causes*", that is, by variables which represent the geometrical and physical state of the surrounding matter. This dependence is single-valued and in general continuous and differentiable." One does not see how this axiom could intervene in the proofs of theorems, or in the solution of problems.

On 20 October I wrote to HAMEL:

I haven't quite decided what to do about publication. Our *Journal* is, I believe, the only one in this country, either in mathematics or in physics, which would publish any serious paper on the axioms of mechanics. If the work is published, it might stimulate discussion and awaken interest in a subject which is hardly

---

[25] HAMEL [1927, page 3].

realized to exist here. However, my personal opinion of the paper coincides entirely with yours, and the *Journal* will certainly have nothing to be proud of in this work.

The next day I wrote as follows to McKINSEY:

> Herewith I send a copy of Hamel's evaluation of your papers on the axioms of mechanics. It is my honest opinion that this evaluation would be shared by virtually all students of the foundations of mechanics. Thus I think you would do yoursel[f] and your co-authors only harm by publishing.
>
> As soon as the MSS are received back, I shall return them to you for your further consideration.

McKINSEY's reply is dated 27 October:

> Thank you so much for your letter of October 21ˢᵗ, with the enclosed report of Professor Hamel on the two papers by Sugar, Suppes, and me. I should like to tell you that we have not been very much impressed by Hamel's criticisms, and that we still want to publish these papers in their present form.
>
> In the first place, with regard to Hamel's criticism that our treatment is restricted to "the most meager mechanics of points", it does not appear reasonable to us to object to a scientific paper on the ground that it has not accomplished something which the authors were not even trying to accomplish: one does not criticize a paper on linear differential equations for not also covering non-linear differential equations. We are of the opinion, moreover, that, as a preliminary to any adequate treatment of the mechanics of extended bodies, it is desirable (or perhaps even necessary) to present classical particle mechanics in a clear and precise form. In addition, such a presentation would be useful for an analysis of relativistic particle mechanics—and of quantum particle mechanics, both in classical and relativistic form.
>
> I must admit that we find Hamel's strictures on our treatment of force difficult to comprehend. He has perhaps overlooked our painstaking and detailed discussion, on pages [260–262 of the paper in print] of the notions of internal and external forces.
>
> So far as regards Hamel's remarks about our Footnote 3, we are still of the opinion that the "axiom" of Hamel which we quote there is strange and useless. In this "axiom", as in Hamel's whole axiomatic system, it is not clear what are the primitive notions for which an implicit definition is being given. It is not explained, for

example, whether the terms "variable", "physical", "state", and "matter" in the axiom in question are to be regarded as: (1) primitive notions to which properties are being assigned by the axioms; or (2) terms of everyday language, which are to be given their usual meanings. And either interpretation of Hamel's usage makes it difficult to maintain that he has given anything like an adequate axiomatization of mechanics: for, if the first interpretation is made, then the axioms are clearly not sufficient to establish the kind of theorems one wants to prove in mechanics; and, if the second interpretation is made, then Hamel is not axiomatizing physics at all, but taking it as intuitively given.

I should like to add that I have showed [sic] Hamel's report to some friends of mine out here whom I consider good mathematicians, and especially qualified as judges of methodological and foundation problems; and they insist that Hamel's report should not induce us to abandon the idea of publishing these papers. Thus, Alfred Tarski feels very strongly that we should not worry about Hamel's objection that the papers are not concerned with anything more complicated than classical particle mechanics—and says that, so far as regards the axiom of Hamel mentioned in our Footnote 3, not merely does it fail to satisfy modern standards of precision, but it would not have been considered sufficiently clear to be taken as an axiom even in the time of Euclid. He also points out that Hamel himself refers to the two main theorems of the two papers as being of value. A. P. Morse has stated that he is in substantial agreement with Tarski's views in this regard.

Professor Tarski, who read both papers carefully some time ago, has also authorized me to say that, although he does not feel competent to decide on the worth of the papers from the point of view of physics, he considers them logically and methodologically flawless, and mathematically interesting.

In closing, I should like to repeat that we still wish to publish these papers in substantially their present form. However, in the enclosed copies, following the suggestion of Professor Hamel, we have added some remarks in connection with Theorem 8 to make clearer the intuitive character of the proof. Moreover, in addition to correcting some typographical errors, we have simplified the formulation of Theorem 2 of the second paper.

In view of your letter of October 6, 1952, we assume that there will be no further delay in publishing this work. We do not object to your adding to the papers your suggested footnote to the effect that the communicator does not agree with the authors' views. Or perhaps you might prefer to write a note for publication in the same issue with our first paper, explaining your objections in a more detailed way.

My reply, dated 3 November:

> I have received your letter of the 27th and your slightly revised MSS. I must take exception to your second paragraph, since it is indeed perfectly reasonable to criticize a scientific paper for not trying to accomplish enough—solutions to unimportant or uninteresting problems, even if completely rigorous and fully achieving their authors' expressed aim, are not usually welcomed by editors. I will publish the papers, with sincere and great misgivings as to the wisdom of doing so.

In that letter I inclosed the text of the footnote I proposed to add. On 9 November McKINSEY closed the correspondence as follows:

> Thank you so much for your letter of November 3. We have no objection at all to your adding the suggested footnote to our papers.

Here is the footnote as printed:

> The communicator is in complete disagreement with the view of classical mechanics expressed in this article. He agrees, however, that strict axiomatization of general mechanics—not merely the degenerate and conceptually insignificant special case of particle mechanics—is urgently required. While he does not believe the present work achieves any progress whatever toward the precision of the concept of force, which always has been and remains still the central conceptual problem, indeed the only one not essentially trivial, in the foundations of classical mechanics, he hopes that publication of this paper may arouse the interest of students of mechanics and logic alike, thus perhaps leading eventually to a proper solution of this outstanding but neglected problem.

III. Noll's Axioms for Systems of Forces and Dynamics

12. PRICKS OF CONSCIENCE, IRREPRESSIBLE IMPERTINENCE

As HAMEL had written to me in 1952, ". . . in the concept of force lies the chief difficulty in the whole of mechanics." I had forwarded a copy of his letter to MCKINSEY (§ 11). Twenty-two years later, Professor SUPPES wrote as follows[26] in the proceedings of a symposium in *pure mathematics* of the American Mathematical Society:

3. Forces in classical physics. It is now 20 years since McKinsey and I began working on the axiomatic foundations of physics. I remember well the argument we had with the editor of the *Journal of rational mechanics*, the irrepressible Clifford Truesdell, who was unhappy with the analysis of force we provided in our original axiomatization. Truesdell felt that our characterization of forces as ordered triples of real numbers satisfying the obvious law of addition for such vectors did not provide any deep conceptual insight into the physical nature of forces. In this judgment I think he was right, but it does not mean that our axiomatization was wrong. The first problem in that early work was to get a set of axioms that were sufficient to characterize what is ordinarily regarded in physics as classical particle mechanics and to characterize it in a way that was logically and mathematically acceptable. Roughly speaking, this means that axioms were given that were mathematically self-contained without any assumptions about the mathematical nature of the objects being considered left implicit, as is often the case in the kind of axioms stated by physicists.

What is missing and what is needed in the analysis of forces is a kind of explicit analysis in terms of elementary primitives. Clearly the primitive notions that McKinsey, Sugar and I used were complicated and already put into the axioms a substantial mathematical apparatus. The axioms were not simple in the way that primitive concepts and axioms of geometry are simple. What is needed is an analysis of the concept of force in the style of Tarski's classic article, *What is elementary geometry?*

MOULINES & SNEED[27] were quick to provide a chorus:

... [T]he ... axiomatization [of McKinsey, Sugar, & Suppes] can only be taken as the first step to clarifying the internal structure of classical particle mechanics. But it should be clear that we consider this accomplishment an absolutely essential first step.

We take care to avoid identifying ourselves with philosophers who tend to minimize the importance of such a first step to clarifying the internal structure of a physical theory. They sup-

---

[26] SUPPES [1974, § 3].
[27] MOULINES & SNEED [1979, pages 74–75].

pose the 'real' task of the philosopher of science is only to clarify the *meaning* of scientific concepts and theories. A good example for this kind of view is the quite negative commentary Truesdell (the communicator) made about the ... axiomatization [of McKinsey, Sugar, & Suppes] when it was published . . .. Truesdell was troubled by the fact that the proposed axiomatization ... did not handle the 'conceptual problem' of the *meaning* of force, while restricting itself to the 'trivial' task of laying down a couple of axioms. Truesdell wanted a semantical clarification of force which, of course, cannot be supplied when the aim is solely to reconstruct the formal structure of the theory.

Not content with imagining that I was one of those philosophers who search for epistemological "meaning", they set my views side-by-side with the operationalism of SIMON[28].

Had it not been for these remarks in print, I should have continued to repress comment, as I did do for a quarter century, on the circumstances that led me to communicate to the major journal for rational mechanics a paper that I expected competent students to find, as indeed abundantly they have, jejune in concept, trivial in mathematics, and philosophically off the mark, as well as insufficient "to characterize what is ordinarily regarded in physics as classical mechanics". My obscure expression must have misled the SUPPESIans, for I have always despised operationalism in all its variants, and I have never sought "semantical clarification" of anything. It is not I but the recent SUPPESIans like STEGMÜLLER (page 13) who call for "semantic supplementation". In my opinion the phenomena of nature come first, and a mathematical theory is set up just so as to model physical objects in a fairly well identified category. Thus an informal semantics is at hand before, during, and after the creation of the theory, and to me that informal semantics has always seemed good enough, as shown by the successes of the great theories. It is the mathematical structure that I insist must be explicit, clear, and reasonably general; it must meet the standards of a branch of pure mathematics, and I do not confuse standards with a notation or a format or a degree of detail. Recently[29] I have explained what "conceptual analysis" means to me. A system of mathematical axioms for forces in the style of HILBERT or BOURBAKI and fruitful for application in modern research on mechanics as a whole—something like abstract measure theory developed to yield a theory of integration useful in analysis—was what I desired in 1952. It has since been provided, but not by the SUPPESIans.

---

[28] SIMON [1947] [1954] [1959] [1970].
[29] TRUESDELL [1978].

13. Suppesian Reaction to Noll's Axioms for Forces, I. Citesmanship and Fear of Mathematics

Noll's axioms for systems of forces, published from 1959 onward, are well known. They are easily available in Noll's[30] *Selected Papers*. I have printed brief descriptions[31] of their most elementary aspects in layman's terms. I have made them the basis of a presentation of mechanics as a whole in a textbook[32] for a senior undergraduate or beginning graduate course which has been published in French (1973), in Russian (1975), and in English (1977). The sales of this book suggest that it is not scarce in any country that fosters the mathematical sciences.

The only statement in regard to Noll's axioms that I can find in the Suppesian literature is a remark by Suppes[33] in 1974:

> The outstanding work of Walter Noll should also be mentioned, but his axioms on the concept of forces acting on bodies require a highly sophisticated and developed mathematical framework. We are still left with the problem of giving, even for highly simplified situations, an elementary axiomatization of the concept of force. Providing such an axiomatization seems to me one of the more interesting problems that remains open in the axiomatic foundations of classical mechanics.

Suppes refers only to Noll's first paper[34] on the axioms, which was written for and presented at a symposium in 1957 at Berkeley, the proceedings of which were edited by Henkin, Suppes, & Tarski. Although, consequently, Noll is not unknown in Suppesian circles, and although his works are easy to find, neither Suppes nor any other Suppesian, so far as I can learn, has ever so much as cited any of the later papers on classical mechanics[35] by Noll and his associates at Carnegie–Mellon University or any of my expositions in terms drawn largely from common speech.

The first paper of Noll, which Suppes sets aside as mathematically too "sophisticated" and "developed", does refer to "mapping", "homeomorphism", "measure", "absolutely continuous", "Borel subset". In mathematics courses today these concepts are made familiar

---

[30] That work is cited in connection with Noll [1958] in the list of references at the end of this review.

[31] Truesdell [1964] [1966] [1968, pages 268–271].

[32] Truesdell [1973] [1975] [1977].

[33] Suppes [1974, § 3].

[34] Noll [1959], hereinafter referred to as "Noll's first paper".

[35] Suppes [1979] cites Noll [1964] but only in regard to special relativity.

to junior and senior undergraduates. Yet in the passage quoted in § 12 from the same paper of 1974 SUPPES claims that he and his collaborators put into their paradigmatic system of axioms "substantial mathematical apparatus". By that I suppose he means n-dimensional vectors, the term "Cartesian product", functions, first and second derivatives, absolutely convergent series, and such signs as $\langle \; \rangle$, $\{ \; \}$, $\oplus$, $\in$, for everything else they used can be found in a typical high-school algebra text of the 1950s. The SUPPESians also decorate their writings with strings of the symbols used in works on mathematical logic; for the most part, these lines merely repeat preceding verbal definitions and assertions.

But it is not a matter of public relations alone. On page 20 of the booklet under review STEGMÜLLER writes "...we can speak about forces *only in the context of particular force laws.*" I do not know what STEGMÜLLER means by "think". It may be true that we can *measure* forces only in the context of special constitutive relations, but even the dullest of philosophers will not confuse measurement with conception and thought. As for thought—which includes the precise reasoning of mathematics—STEGMÜLLER might just as well have written "we can speak about distance only in *the context of particular distance functions*", or "we can speak about polygons *only in the context of particular polygons.*"

14. SUPPESIAN REACTION TO EXISTING FORMAL TREATMENTS:
FEAR OF REAL WORK IN FORMAL LOGIC

The citesmanship just described brings me back to pages 5 and 6 of STEGMÜLLER's booklet, where he explains why he rejects the program of axiomatization using formalized set theory, which he imputes to CARNAP. After telling us that had BOURBAKI followed such a program, he would still be at work on his first volume, STEGMÜLLER continues as follows:

As far as I know, there exist extraordinarily few articles which deal with real physical theories within a formalized language. One of them is a paper by Richard Montague[36]. As is generally acknowledged, Montague was an extraordinary logician, whose intellectual abilities and technical skill very few present philosophers of science, if any at all, can cope with. Comparing this article of Montague's with an analogous one of the Suppes approach, you will quickly recognize an essential difference. You

---

[36] MONTAGUE [1962].

will recognize it at least under the presupposition that with the latter the investigation to be carried out is done by a person who is an expert in the field in question and who is, in addition, sufficiently familiar with informal set theory, but who is not a prodigy of Montague's kind. *It is the difference between a few years of work and a few weeks* (or perhaps *afternoons*) *of work.*

He concludes that only "a Super-Super-Montague" could do the job. In STEGMÜLLER's footnote on page 5, citing MONTAGUE's work, we read

> Another author who ought to be mentioned here is Aldo Bressan; vid., e.g., his applications of modal (type) logic to physical systems . . . ,

and here STEGMÜLLER cites BRESSAN's book[37] on modal calculus, which contains a chapter concerning among other things the foundations of classical mechanics. While many penetrating observations on the concepts of mechanics and the views of MACH, KIRCHHOFF, POINCARÉ, and SIGNORINI may be found in that book, an axiomatization of mechanics there is not. So far as I can learn, this one citation by STEGMÜLLER is the only mention of BRESSAN in all the SUPPESian literature.

In dismissing BRESSAN's work by reference to a book that concerns mechanics only incidentally, STEGMÜLLER leads his readers to ignore BRESSAN's major effort, a long monograph[38] which contains an axiomatic development of the mechanics of mass-points using formalized set theory. Citesman that he is, STEGMÜLLER leaves himself an escape by inserting "e.g.", "for example". But the book he does cite is not an example to the point; unlike the SUPPESians, BRESSAN does not print the same thing over and over again in slightly different expression; the only "example" of his work on axioms of mechanics is his single, mature treatise.

Now the SUPPESians, one and all, cannot fail to know of this treatise, for BRESSAN in his acknowledgment states that he began it under the sponsorship of SUPPES at Stanford in 1959! He also thanks SUPPES and CARNAP "for useful discussions and suggestions directly concerning the present work", and a similar statement is printed on page 103 of the book that STEGMÜLLER does cite.

---

[37] BRESSAN [1972]. See Part II (page 61 ff.), "Some useful concepts definable in the modal language ML$^\nu$"; applications to questions concerning foundations of classical mechanics and everyday life".

[38] BRESSAN [1962].

ALDO BRESSAN is a professor of rational mechanics at Padua; he has made outstanding contributions to various parts of mechanics and electromagnetism, including classical and relativistic theories of deformable continua; he is a member of the Accademia dei Lincei, recognized as one of the leading mathematical physicists of Italy today; and his work on modal logic has been praised highly by reviewers and others whose interests lie in formal logic. How could STEGMÜLLER in 1979 dismiss what he calls CARNAP's program as being unachievable when BRESSAN's extensive formal presentation had then been standing in print for seventeen years?

When STEGMÜLLER in giving as his reason (page 7) for endorsing "the Suppes approach" and rejecting formal language writes "the emphasis lies on my claim of the *nonexistence of a Super-Super-Montague* in present philosophy of science", is he not telling his readers obliquely that BRESSAN (along with everyone else) is unequal to the task of a "Super-Super-Montague"? If so, is deliberate refusal to cite BRESSAN's long and serious tract, full of explicit reasoning, the product of years of hard work, a decent way to express such an opinion in print? Does not STEGMÜLLER's claim that nobody alive is capable of doing what BRESSAN seems to have done, oblige him to reveal wherein lies whatever deficiency he thinks he has found in BRESSAN's treatment?

Perhaps I am not reading STEGMÜLLER closely enough. He refers to "present philosophers of science" as being unequal to the requirements he claims CARNAP imposed. Perhaps BRESSAN, a logician trained in mathematical physics, and NOLL, a mathematician trained in mechanical engineering—both of them regularly lecturing on rational mechanics and other fields of mathematics and mathematical physics—are professionally excluded from the category "philosophers of science". Neither NOLL nor BRESSAN proposed axioms for the sake of axiomatics. NOLL faced the need for a precisely stated mechanics general enough to allow dissipative continua with properties intermediate between solids and fluids and with response dependent upon their past experience. BRESSAN, in contrast, tells us[39] that in studying the informal axioms of mass-point mechanics proposed long ago by PAINLEVÉ[40] he found them insufficient to deliver a theorem that PAINLEVÉ himself had asserted. He noted also the defects arising from want of an existence axiom such as EUCLID's Postulate 1 or HILBERT's Axiom 1 in geometry. Both BRESSAN and NOLL came to

---

[39] BRESSAN [1962, § 10].
[40] PAINLEVÉ [1922].

formulate their axioms not for philosophical explanation but as mathematicians facing mathematical gaps.

That brings us back to MONTAGUE. Few philosophers are in a position like BRESSAN's, who grew up in the competitive and informed domain of Italian *meccanica razionale*, or like NOLL's, who matured in the rapidly expanding field of nonlinear continuum mechanics with its numerous specific problems both mathematical and experimental. A philosopher usually cannot create the object for which formal axioms are to be provided. The well known logical constructions of the natural numbers and the continuum were provided after no doubt remained as to what properties those collections should have. Evidently MONTAGUE thought that such was the case also for the classical mechanics of systems of particles. He wrote[36] in 1962 as if he then regarded McKINSEY, SUGAR, & SUPPES as being what the SUPPESians claim for them now, masters of mechanics, just as PEANO and DEDEKIND had been masters of analysis. MONTAGUE took the ill-fated axioms of McKINSEY, SUGAR, & SUPPES as providing an object worthy of treatment by formalized set theory. Even this task he found too difficult, and he had to rest content with a system the motions of which were limited to a line. Nevertheless, MONTAGUE's attenuated projection of a starved mechanics onto a single bone deserves respect: It is a sincere and laborious effort. So much the more must BRESSAN's work on the three-dimensional theory be respected, until and unless some nullifying error be found in it.

STEGMÜLLER regards the "intellectual abilities and technical skill" of MONTAGUE as something to "cope with". It may be unfair to attribute to an author as vague as STEGMÜLLER any specific meaning in any passage, but, not being a seer or a psychologist, I have to take his words at their dictionary meanings. The *Oxford English Dictionary* lists six distinct verbs "to cope". Only the second of these gives rise to coping "with" something other than a person. The meaning in that usage is "To contend with, face, encounter (dangers, difficulties, *etc.*)". Thus STEGMÜLLER's words mean that "few present philosophers of science, if any" are a match for someone of MONTAGUE's capacities. Is philosophy of science, then, a *contest*? Few mathematicians can contend with the capacities of a HILBERT or a NOLL, but that is not taken as an argument against HILBERT's axioms or NOLL's! We study what HILBERT and NOLL have written, not so as to cope with their work but to learn from it. The SUPPESians, in contrast, find excuses not only for not trying to use formal logic but also for not studying what has been done with it. STEGMÜLLER writes in the passage quoted at the beginning of this section, "*It is the difference between a few years of work and a few weeks* (or perhaps *afternoons*) *of work.*"

Agreed.

15. SUPPESIAN REACTION TO NOLL'S AXIOMS FOR FORCES, II. FEAR OF MECHANICS

In his paper of 1959 NOLL specifies properties of contact forces and body forces, and much of what SUPPES called its "highly sophisticated and developed mathematical framework" refers to contact forces. Contact forces are central to continuum mechanics. In mass-point mechanics they are absent.

The SUPPESians from the days of McKINSEY until now have striven to bar that grisly monster, their bugaboo, continuum mechanics. While McKINSEY let it be known that he regarded his team as having solved HILBERT's Sixth Problem, he expressly disregarded HILBERT's requirement for its solution: "... we shall try first ... to include as large a class as possible of physical phenomena, and then by adjoining new axioms to arrive gradually at the more special theories." No, said McKINSEY (quoted above in § 11), let us begin with the most special. There let us stay, said SNEED[41] twenty years later:

> Just exactly what is classical particle mechanics? What is its mathematical structure like? What are some paradigm examples of the claims this theory makes about the way the world is? In short, just what theory is it that is to serve as our example of a theory of mathematical physics?
>
> Standard treatments of classical mechanics[42] ... sub-divide the discipline into four parts: particle mechanics, rigid body mechanics, the mechanics of deformable bodies, and the mechanics of liquids and gases. Each of these sub-divisions is supposed to deal with the motion of a different kind of thing .... In addition, there are apparently significant differences in the mathematical apparatus used in each of the sub-divisions. Functions like torque and moment of inertia appear in the exposition of rigid body mechanics and not in particle mechanics. The basic partial differential equations of hydrodynamics contain different functions and have a different mathematical form from those of either rigid body mechanics or particle mechanics. These facts, together with the view of theories of mathematical physics that has been sketched, suggest that it might be fruitful to regard each of these sub-divisions of classical mechanics as a separate physical theory— each with its own mathematical formalism and range of applications.

---

[41] SNEED [1971, pages 110–111].

[42] [SNEED's footnote] E.g. JOOS [1934].

　　　Of course, the sub-divisions of classical mechanics are related
in some way. There is something about them that makes it intuitively reasonable to say that they are all "parts" of the same
theory. If we take the view that each of these sub-divisions is a
distinct theory of mathematical physics, in our sense of "theory",
then we are obliged to explain the relation between these theories.
The intuitive nature of this relation is apparent in most expositions of classical mechanics. Particle mechanics is usually ex-
pounded first, and then the concepts of the other sub-divisions
explained in terms of the concepts of particle mechanics. This
suggests that particle mechanics is, in some sense, more basic than
the other sub-divisions, or that the other sub-divisions can be
"reduced" to particle mechanics. Just how this intuitive notion of
one theory's being reduced to another can be made precise is one
of the topics of the next chapter. For the present, it is sufficient to
note that we shall restrict our attention to classical particle
mechanics, laying aside the question of precisely in what sense, if
any, the rest of the classical mechanics is reducible to it.

Having decided that particle mechanics is "more basic" than other
parts of mechanics, SNEED exhumes the axiomatization of McKINSEY,
SUGAR, & SUPPES, devotes a long chapter to it, and apart from some
remarks on rigid bodies, of which more later, leaves all the rest to
future "reduction". Here is MOULINES' echo[12]:

　　　The evolution of Newtonian *particle* mechanics was undoubt-
edly connected with the evolution of other mechanical theories
(hydrodynamics, acoustics, rigid body mechanics). . . . I think the
methodologically advisable procedure is to try to isolate different
theory-evolutions (i.e. evolutions of different theories) and reconstruct the logical structure of each of them . . ..

MOULINES has decided that particle mechanics and continuum
mechanics are "different theories". It is "methodologically advisable"
to abandon HILBERT's program. Who was the "reader and advisor"
who so advised him, MOULINES does not disclose.
　　　In all their thirty years of tergiversation, not one of these philosophers has published evidence that he understands even the
simplest elements of mechanical acoustics, hydrostatics, hydrodynamics, or elasticity as those subjects were taught to every Cambridge
undergraduate in mathematics from 1850 to, say, 1925, or as they
were required of doctorands in mathematics or physics in the United
States from, say, 1876 into the 1930s. I do not refer to anything
"advanced" or "specialized", just to the contents of WEBSTER's

Dynamics, which was the standard American textbook for many years and was used in my time at CalTech by ZWICKY in his basic course, a formal requirement for candidacy in physics. As a school, these philosophers of science have remained content for thirty years with a single testing ground for their thousands of pages of printed claims: a mechanics too narrow to cover even the elements of gravitation as provided by NEWTON himself. Beyond SUPPES' reference[33] to NOLL's first paper,[34] the only SUPPESian mention of modern mechanics that I have been able to find is by SNEED[43]: "various examples of axiomatizations—. . . for example, . . . continuum mechanics by Truesdell . . .". SNEED flatters me by admitting as an "axiomatization" a talk at a philosophy meeting in which I tried to help philosophers get a conversational notion of the concepts employed by people who worked in continuum mechanics then and a sketch of some then current ways to relate and apply those concepts.

SNEED has faith in "reduction" of all mechanics to mass-point mechanics. As evidence for it he discourses at length[44] upon the example provided by rigid-body mechanics. BUNGE disposed of this matter in his review[45] of an earlier book by STEGMÜLLER[46]:

And the sole example of reduction of one theory to another, discussed in this book, is also taken from Sneed, who in turn took it from E. W. Adams [*The axiomatic method. With special reference to geometry and physics* (Proc. Internat. Sympos., Univ. California, Berkeley, Calif., 1957–1958), pages 250–265, Studies in Logic and Found. Math., North-Holland, Amsterdam, 1959 . . .]. Unfortunately, this example too is inappropriate, because Adams did not really succeed in reducing rigid body mechanics to particle mechanics. In fact a continuous body cannot be built out of classical point particles; and even a rigid system of point particles calls for hypotheses on inter-atomic forces that go beyond classical particle mechanics.

All this is old, old. M. v. LAUE[47] in 1919 wrote as follows:

The usual way to reduce the mechanics of continua to the mechanics of mass-points through a visual image (anschauliches

---

[43] SNEED [1971, page 8]. I can find no SUPPESian reference to the paper of NOLL [1967] presented in the same philosophical seminar.

[44] SNEED [1971, pages 216–248].

[45] BUNGE [1978]. Possibly SNEED [1971, page 247] reaches a conclusion of this kind when he writes "provided we are willing to accept certain plausible assumptions about the constraints in the cores . . . ," *etc*. For his meaning of "constraint", see below, § 17.

[46] STEGMÜLLER [1976, § 9].

[47] M. v. LAUE [1919, § 26].

Verfahren) encounters serious logical objections even in
Newtonian mechanics, while the converse limiting process from
the continuum to the mass-point is altogether unobjectionable.

HAMEL had made the same point in more detail, recalling that FELIX
KLEIN had remarked upon it in his lectures. According to HAMEL[48],
". . . what is in practice regarded as the mechanics of mass-points is
nothing else than the theorem of the center of gravity." For evidence
he could have adduced the way NEWTON himself had treated the
mechanics of a planetary system modelled as a collection of homogeneously layered spheres. That would not help the SUPPESians,
for the theory of the attractions of solid bodies is continuum
mechanics and hence excluded by their paradigm. At one point[49]
McKINSEY, SUGAR, & SUPPES write "we have in mind interpreting the
elements of P as the centers of mass of rigid bodies", such as a bullet
passing through a torus; in their view of mechanics such an interpretation would have to be assumed rather than proved tenable, for their
structure is too meager to permit us even to formulate any statement
about rigid bodies.

BUNGE concludes as follows the review quoted above: "the new
philosophy of science advanced in this work is at least as remote from
living science as any of the rival views criticized in it." So is the pamphlet under review here. True. The SUPPESians are consistent: For thirty
years they have honored their resolve to dismiss as being monstrous-
ly advanced and specialized and not "basic" much of NEWTON's
mechanics of 1687.

16. APPLICATION OF NOLL's AXIOMS OF FORCES TO THE MECHANICS OF MASS-POINTS. THE POSITION OF "NEWTON's THIRD LAW"

NOLL's first paper on the axioms of mechanics is phrased in terms
of space-filling bodies. Perhaps presuming that all of NOLL's later
work likewise excluded mass-points, the SUPPESian citesmen have
bestowed silence upon it. Anybody who understands mechanics will
see upon reading NOLL's first paper that the mechanics of mass-
points can be treated along the same general lines and with less
apparatus. NOLL in his Bressanone lectures[50] of 1965 presented
axioms essentially the same but cleared of restriction to continua.
There the set of all bodies $\mathscr{B}$, $\mathscr{C}$, etc., called a *material universe*, is simply

---

[48] HAMEL [1908, page 351].

[49] McKINSEY, SUGAR, & SUPPES [1953, page 260].

[50] NOLL [1966], hereinafter referred to as NOLL's "Bressanone Lectures".

a Boolean algebra. $\mathscr{B}$ and $\mathscr{C}$ are *separate* if their meet is the null body. The axioms of forces concern a vector-valued function $\mathbf{f}$ of pairs of separate bodies: $\mathbf{f}(\mathscr{B}, \mathscr{C})$ is the force exerted by $\mathscr{C}$ on $\mathscr{B}$. The mass $M(\mathscr{B})$ of $\mathscr{B}$ is the value of an abstract nonnegative measure defined on the material universe. NOLL's method and results follow HILBERT's prescription: by "a small number of axioms to include as large a class as possible of physical phenomena". Special choices of material universe correspond to different domains of mechanics. Again, this is HILBERT's prescription: "by adjoining new axioms" we arrive at "the more special theories". NOLL's first and simplest example is "the Newtonian mechanics of particle systems", for which the material universe is the set of all subsets of a given finite set. The mathematics needed to develop the theory of this universe is not only simple but straightforward, suitable for an undergraduate course.

NOLL's lectures were published in a volume put on sale in 1966; a corrected manuscript was widely circulated as a report. While some imperfections may be found in the text, they are removed in NOLL's later work, "Lectures on the foundations of continuum mechanics and thermodynamics", printed[51] in 1973. There can be no doubt that the treatment there provides *a solution of HILBERT's Sixth Problem* in reference to classical mechanics. For the details of "classical particle mechanics" treated on this basis the SUPPESians might have looked at Chapter I of my textbook[52] for beginners, also published in 1973; there the entire structure of "classical particle mechanics" is derived in detail as a special theory comprised by NOLL's axioms; the treatment occupies about seven pages.

One year later SUPPES wrote[53], "We are still left with the problem of giving ... an elementary axiomatization of the concept of force."

McKINSEY, SUGAR, & SUPPES gave no example of how to use their paradigm for any purpose in the practice or development of mechanics, and so far as I can learn, no-one else has done so. While the SUPPESians content themselves with admiring it for its perfection, specialists in mechanics have given it no heed, perhaps because it leaves out of account too much of the meat of the mechanics of NEWTON, EULER, LAGRANGE, and CAUCHY—or, for that matter, of the contents of any current textbook of mechanics for engineers. For example, McKINSEY, SUGAR, & SUPPES, noting that "There is no precise agreement about what the assumptions of classical mechanics are", list[54] among "axioms which might have been assumed but were

---

[51] NOLL [1973]. There he makes gravitational and inertial masses possibly independent constitutive measures on the measurable subsets of $\mathscr{B}$.

[52] TRUESDELL [1973] [1975] [1977].

[53] SUPPES [1974, page 468].

[54] McKINSEY, SUGAR, & SUPPES [1953, pages 260–264].

not" some statements related to "Newton's Third Law". The assumptions McKINSEY and his team refused to adopt can be expressed in the terms commonly used in mechanics:

(1) the mutual forces are pairwise equilibrated,

(2) the mutual forces are central,

(3) the resultant torque exerted by the mutual forces of any set of mass-points on itself is null.

A classical theorem of POISSON asserts that

$$(1) \& (2) \Rightarrow (3).$$

This argument is presented in most textbooks by physicists and is called the "theorem of rotational momentum". Physicists state "Newton's Third Law" sometimes as (1), sometimes as (1) and (2), sometimes as something else. The basic logical question is, are the assumptions (1) and (2) necessary for (3)? The SUPPESians could not and cannot today even state the question, since angular momentum is something regarded by them as appropriate only to rigid bodies, systems about which, since they regard them not "basic", they refuse to learn anything. SNEED puts it thus[55]: "Functions like torque and moment of inertia appear in the exposition of rigid body mechanics and not in particle mechanics." Further on in the textbook[42] that he cites for his information about classical mechanics he could have found a section[56] called "Angular momentum of a system of particles"; it presents the theorem of POISSON, phrased as follows: "For a system of particles in which the forces between any two particles are [pairwise equilibrated and] in the direction of the line joining these particles, the rate of change of the total angular momentum is equal to the sum of the moments of the applied forces." Indeed, the concept "moment of inertia" was created by HUYGENS and JAMES BERNOULLI from contemplation of a model which is equivalent to a finite system of mass-points.[57] Perhaps the SUPPESians need to be told that "torque" in this context is another word for "the sum of the moments of the . . . forces". HAMEL[58] saw the point. Like EULER and CAUCHY, whose work on the foundations of mechanics the SUPPESians give no evidence of knowing, HAMEL laid down a principle of rotational momentum as a basic assumption, but unfortunately he confused it and limited its range by

---

[55] SNEED [1971, page 111].

[56] JOOS [1934, § VI.2].

[57] Cf. DUGAS [1950, § 5 of Chapter V of Book II and § 1 of Chapter IV of Book III].

[58] HAMEL [1927, page 358].

expressing it as "Boltzmann's axiom", which refers to the rate at which the external torque on a body vanishes when the volume tends to null. Nothing about systems of mass-points can be inferred from this axiom. NOLL, on the other hand, has always followed the lead of EULER and CAUCHY in taking the *balance of rotational momentum* either as an axiom or as something that can be deduced from an apparently simpler assertion.

In Footnote 3 of his paper of 1959 NOLL wrote,

> Various statements, mostly quite vague, pass under the title "principle of action and reaction" in the literature. All of these statements, when made precise, are provable theorems in the theory presented here.

An early proof of a statement of this kind, taken from an unpublished report of 1957 by NOLL, was published[59] in 1960. To explain the full result for the mechanics of mass-points, I first introduce some terms. The force $\mathbf{f}(\mathscr{B}, \mathscr{B}^e)$ exerted on $\mathscr{B}$ by its exterior $\mathscr{B}^e$ is the *resultant force* on $\mathscr{B}$. Note that in applications of the general theory the inertial forces contribute to $\mathbf{f}$. A system of forces is *balanced* if $\mathbf{f}(\mathscr{B}, \mathscr{B}^e) = 0$ for every $\mathscr{B}$. A system of forces is *pairwise equilibrated* if $\mathbf{f}(\mathscr{B}, \mathscr{C}) = -\mathbf{f}(\mathscr{C}, \mathscr{B})$ for every pair of separate bodies $\mathscr{B}$ and $\mathscr{C}$. A theorem NOLL derived directly from the axioms of forces[60] states that *a balanced system of forces is pairwise equilibrated*. We easily apply this theorem to the dynamics of mass-points, taking into account the fact that the inertial forces are extrinsic. We conclude that *if the system of forces is balanced, the mutual forces are pairwise equilibrated*. A further simple argument due to NOLL[61] shows that for a system of mass-points

$$(1) \,\&\, (3) \Rightarrow (2).$$

It follows that *in the mechanics of mass-points subject to a balanced system of forces, the system of torques is balanced* if *and* only if *the mutual forces are central*. In rough terms, EULER's Principles of Balance of Linear Momentum and Balance of Rotational Momentum are together

---

[59] TRUESDELL & TOUPIN [1960, § 196A].

[60] NOLL's proof is published in his Bressanone Lectures of 1966; Theorem II of his paper of 1959 had presented a major instance. A proof as a corollary of a more general relation due to GURTIN & WILLIAMS is given by TRUESDELL [1977, § I.5].

[61] NOLL's proof was first published by TRUESDELL & TOUPIN [1960, § 196A]; they cited for it a passage in NOLL's unpublished report listed by them as "[1957, *11*]". BASSET [1894] had remarked upon a weaker statement: (1) and either (2) or (3) for the whole system (not necessarily for subsystems) imply the theorem of rotational momentum for the whole system.

equivalent, in their application to the dynamics of mass-points, to the following:

(1) The system of forces is balanced.

(2) The mutual forces are pairwise equilibrated and central (the strong "Third Law").

Here by "EULER's principles" I mean the usual statements that both the system of forces and the system of torques are balanced, it being understood that "force" incorporates inertial as well as applied force.

Looking back now at the paper of McKINSEY, SUGAR, & SUPPES, we find that they call a system of mass-points "ultra-classical" if only pairwise equilibrated central forces act upon it. They state that "it is easy to construct two ultra-classical systems whose concatenation is not ultra-classical . . ."; indeed, concatenation lays no condition whatever on the forces exerted by the particles of the first system on the particles of the second one. The Principle of Rotational Momentum, which (3) expresses, refers to *any* set of mass-points and in particular to the set consisting in just two mass-points. In the common understanding of NEWTONian mechanics the join of any set of bodies is a body, and the laws of mechanics apply to *all* bodies. McKINSEY and his team in their axioms omitted any concept of "body" and referred only to the particles taken one at a time. Because they did not define a body as a set of particles, the SUPPEsians would find it awkward to state the Principle of Rotational Momentum. As we have just seen, it requires all systems of particles to be "ultra-classical". In particular, it requires that if the concatenation of two systems is to satisfy the laws of mechanics, the new mutual forces, like those of the original two systems, must be central and pairwise equilibrated. The SUPPEsians refuse to adopt the common and natural idea that a given body exerts at a given place and time one and only one force on another body; they refuse also to recognize the clear and natural distinction[62] of mutual forces from extrinsic forces. Their strange Axioms P5 and P6, in which the forces contributing to the resultant force on a particle are restricted only by what their sum should be, reflect the same omission.

It might be instructive if some day some SUPPEsian would explain ATWOOD's machine by application of their paradigmatic axioms.

In defense of the SUPPEsians' claim that those axioms enjoy "mathematical rigor" and provide a treatment "much superior to earlier attempts" in "clarity and precision", I remark that McKINSEY, SUGAR, & SUPPES listed the primitive concepts and specified the

---

[62] HAMEL [1908, end of § 12]. A formal, axiomatic distinction between extrinsic forces and mutual forces in the mechanics of mass-points is given in the book of TRUESDELL [1977, pages 20–21].

domains and codomains of the functions occurring. Other than that, in their treatment I find nothing beyond the ordinary except their omissions.

17. FRAMING, CHANGES OF FRAME, FRAME-INDIFFERENCE, AND NOLL'S AXIOM OF DYNAMICS

The concept "frame of reference" has been central in mechanics ever since the early years of the eighteenth century, when the BER-NOULLIS, CLAIRAUT, and EULER attacked problems with moving constraints. An example of such a problem today is provided by a laboratory in a capsule spinning in space. If the travellers within start to play catch, they will see their cast balls describe arcs whose projections onto the floor (presumed flat) are different from the straight lines which they would be, were that floor quietly horizontal on earth. "Frame of reference" is one of the several fundamental concepts of mechanics that never appear on SUPPESian pages.

In their paper McKINSEY, SUGAR, & SUPPES used rectangular Cartesian co-ordinates. They wrote[63] that "the primitive s", which is an n-tuple of real numbers and is called the position vector of a particle, "fixes the choice of co-ordinate system"; the reader will know that even if $n = 2$, the co-ordinates of one point do not determine a stationary pair of orthogonal axes until a unit of length and one angle are specified, and what about moving axes? McKINSEY, SUGAR, & SUPPES refer once to

> the set of all admissible (inertial) coordinate systems; that is to say, roughly speaking, the class of all coordinate systems with respect to which the particles in question satisfy Newton's Second Law.

Since what they call "Newton's Second Law" is one of their axioms, the reader may infer that their system of mechanics is restricted to a special class of co-ordinate systems, but McKINSEY and his team do not give him any means of determining what that class is, let alone how to determine whether some particular co-ordinate system, such as that defined by the walls of his laboratory, belongs to it. They specify how vectors are added and subtracted but not how they transform under change of co-ordinates. The reader who knows kinematics already may see for himself how accelerations transform, since they are defined quantities, but what about forces? Forces are primitive for the SUPPESians. If there is a transformation law for them, the axiomatizer has to specify it.

---

[63] McKINSEY, SUGAR, & SUPPES [1953, page 257].

Thus the SUPPESians' paradigmatic axioms and the surrounding descriptive material leave the reader unable to locate anything, unable to use frames in which the "Second Law" does not hold, and unable to consider questions of invariance when the frame of reference is changed.

In the succeeding paper[63] by MCKINSEY & SUPPES the transformations of systems of particles apparently presume a single, fixed co-ordinate system. Special instances of the results of that paper could be interpreted in terms of change of framing with respect to which the motion of a given system is described, but only if an axiom on the transformation of forces were given. The authors state no such axiom.

This obscure mess the SUPPESians exalt as clarifying the work of NEWTON and meeting the standards of HILBERT and BOURBAKI!

Part of the SUPPESians' confusion results from their failure to distinguish a frame of reference from a co-ordinate system. Physicists often do likewise, but their basic understanding of the subject saves them from error and vacuity. In fact co-ordinate systems only complicate the matter. Had MCKINSEY, SUGAR, & SUPPES read a hundred pages further in JOOS's textbook, they would have found[64] an excellent if somewhat special as well as informal discussion of space and time in Newtonian mechanics, inertial frames, Galilean transformation, and accelerated frames of reference, with no reliance on any co-ordinate system. Perhaps the title of the chapter containing these matters, namely "Relativistic Dynamics", frightened the SUPPESians away. In a later paper MCKINSEY & SUPPES[64a] introduce a "Galilean carrier" which "[i]ntuitively . . . corresponds to a transformation of a system of particle mechanics from one inertial frame to another . . .." They do not introduce frames as such, do not cite JOOS's book or any other, do not lay down an axiom regarding the transformation of force; insofar as their paper concerns this matter, they merely repeat in their words the statement that a Galilean transformation preserves the dynamical equations. SNEED[65] makes all questions of invariance seem profound and mysterious; he refers only to Galilean transformations and gives no evidence of mastering the difference between the role of inertia and the role of material properties, which do not and should not enjoy the same invariance.

NOLL in his presentation of mechanics made these matters explicit and clear; his ideas are presented abstractly in my textbook[66]. Mr.

---

[64] JOOS [1934, Chapter X].

[64a] MCKINSEY & SUPPES [1955].

[65] SNEED [1971, pages 149–150].

[66] TRUESDELL [1977, Chapter I].

NOLL himself now expresses them informally much as follows. A *frame of reference* is a Euclidean space. A Euclidean space $\mathscr{E}$ is here understood as defined by its geometric properties; it must not be confused, as unfortunately it all too often is, with a Cartesian space, the points of which are n-tuples of real numbers. In this context a body (above, § 16) is taken as being a set $\mathscr{B}$ endowed with a structure by prescription of a suitable class of *placements*, which are invertible mappings of $\mathscr{B}$ onto open subsets of frames of reference. A *motion* of $\mathscr{B}$ is defined by mappings χ of the elements X of $\mathscr{B}$ onto points of $\mathscr{E}$ at each time t in some given real interval $\mathscr{I}$:

$$X \mapsto \chi(X, t), \qquad X \in \mathscr{B}, \quad t \in \mathscr{I}.$$

Equivalently, $\chi : \mathscr{B} \times \mathscr{I} \to \mathscr{E}$. We say that $\chi(x, t)$ is the *place* occupied by the body-point X at the time t in the motion χ with respect to the frame $\mathscr{E}$. The restriction $\chi(\cdot, t_0)$ of χ to the fixed time t_0 is a placement of $\mathscr{B}$ in the frame $\mathscr{E}$.

A frame of reference represents a background against which the motion of bodies can be observed. A *change of framing* is a time-family, $t \in \mathscr{I}$, of distance-preserving invertible mappings $\Phi(t): \mathscr{E} \to \mathscr{E}^*$ from one frame of reference, $\mathscr{E}$, to another, $\mathscr{E}^*$. It represents the relative motion of two backgrounds, each of which may serve for observation of motions of bodies. If $\mathbf{x}$ is the place at time t of some body-point in the frame $\mathscr{E}$, then $\mathbf{x}^* := \Phi_t(\mathbf{x})$ is the place at time t of that same body-point in the frame $\mathscr{E}^*$. If χ is the motion of a body $\mathscr{B}$ relative to the frame $\mathscr{E}$, then the motion χ^* of the same body relative to the frame $\mathscr{E}^*$ is given by $\chi^*(X, t) = \Phi_t(\chi(X, t))$ for all $X \in \mathscr{B}$ and all $t \in \mathscr{I}$.

A theorem of geometry states that one can associate with a change of framing Φ_t, $t \in \mathscr{I}$, a family of orthogonal tensors $\mathbf{Q}_t$, $t \in \mathscr{I}$, each of which maps the translation space $\mathscr{V}$ of $\mathscr{E}$ onto the translation space $\mathscr{V}^*$ of $\mathscr{E}^*$ in such a way that

$$\Phi_t(\mathbf{x}) = \Phi_t(\mathbf{y}) + \mathbf{Q}_t(\mathbf{x} - \mathbf{y})$$

for all $\mathbf{x}$ and $\mathbf{y}$ in $\mathscr{E}$ and for all t in $\mathscr{I}$.

A change of framing induces definite rules of transformation for *defined* quantities. For *primitive* quantities, such rules are not induced; they must be *specified by axioms*. For example, vectors $\mathbf{w}$ and $\mathbf{w}^*$ associated with the motions χ and χ^* of $\mathscr{B}$ relative to the frames $\mathscr{E}$ and $\mathscr{E}^*$ may or may not satisfy the transformation

$$\mathbf{w}^* = \mathbf{Q}_t \mathbf{w}.$$

If they do, they are *frame-indifferent*.

In general, defined vectors turn out *not* to be frame-indifferent. For example, the definitions of velocity and acceleration do not mention any special frame and hence apply in all frames; in contrast, the vectors obtained by applying those definitions are not frame-in-different. We may say that a frame-indifferent vector is "the same arrow" in all framings. It cannot be transformed to null by change of framing. Velocity and acceleration are not frame-indifferent, because an observer may (in principle) employ a background that moves right along with any body-point X he pleases, so making X seem to him to be at rest, with null velocity and null acceleration.

In dynamics not only the motions of bodies must be considered but also, for each time t, a *system of forces* $\mathbf{f}_t$ acting upon pairs of bodies. The values of $\mathbf{f}_t$ lie in the translation space $\mathcal{V}$ of a frame of reference.

NOLL's final axiom for forces is as follows: *The systems of forces are frame-indifferent.* That is, if $\mathbf{f}_t$ and $\mathbf{f}_t^*$ are the force-systems at time t for the frames of reference $\mathscr{E}$ and $\mathscr{E}^*$, respectively, then

$$\mathbf{f}_t^* = \mathbf{Q}_t \mathbf{f}_t,$$

$\mathbf{Q}_t$ being obtained from the change of framing as described above. By rotating the background an observer can make the force $\mathbf{f}_t(\mathscr{B}, \mathscr{C})$ seem to rotate, but he cannot change its magnitude. In particular, the primitive concept "force exerted by $\mathscr{C}$ on $\mathscr{B}$" is not something that can be transformed away by choice of background.

We are now ready to state NOLL's **Axiom of Dynamics**, dating from 1959 but first published[67] in 1963:

> For every assignment of forces to pairs of bodies, the working of the system of forces acting on each body is frame-indifferent, no matter what be the motion.

"Working" is the rate at which the system of forces does work. The rules of transformation of all quantities entering the axiom have been either defined or posited. Thus the assertion of the axiom can be rendered explicit. NOLL proved that necessary and sufficient conditions for his axiom to hold were the following:

(1) The resultant force on every body vanishes, and

(2) The resultant torque on every body vanishes.

Thus NOLL's Axiom of Dynamics is *equivalent* to EULER's *Laws of Motion* with the role of inertia not yet rendered explicit. A corollary of NOLL's Theorem is the general **principle of action and reaction**: *The forces and torques exerted by separate bodies on each other are pairwise equili-brated.*

---

[67] NOLL [1963].

We are now ready to state axioms of inertia such as to render precise the ideas first expressed by NEWTON in his Laws of Motion. We divide the universe into two parts: the set Σ of bodies upon pairs of which we are prepared to specify a system of forces, and the exterior Σ^e about which we know nothing. For example, Σ could be the bodies in a laboratory, the bodies on earth, the bodies of the solar system, or the bodies within a domain on whose boundary lie the "fixed stars". The axioms of inertia will yield somewhat different laws of mechanics in these different choices of Σ. Following NEWTON and EULER, we lay down the following

> **First Axiom of Inertia.** *There is a framing such that if over an open interval of time the center of mass of $\mathscr{B}$ moves in a straight line at constant speed, then in that interval* $\mathbf{f}(\mathscr{B}, \Sigma^e) = \mathbf{0}$, *and conversely.*

The frame here posited to exist is called an *inertial frame*. All members of the class of inertial frames are obtained from any given one by "Galilean" transformations, and conversely.

> **Second Axiom of Inertia.** *In an inertial frame the applied force on a body equals the product of the mass of the body times the acceleration of its center of mass.*

While these traditional statements are superficially similar to the axioms of MCKINSEY, SUGAR, & SUPPES, conceptually they are different, as may be inferred by those authors' statement[68] "we have not assumed Newton's First Law; but this is a trivial consequence of our axioms . . .". It is a bold jack indeed who dares set aside an axiom of NEWTON as being a *trivial* consequence of another. Indeed, in a frame where the Second Law holds, the First Law is a trivial consequence of it. But how is such a frame to be determined? A frame cannot be defined in terms of itself; it must be defined by a statement that applies equally in all frames. The statement that the resultant force upon a body vanishes is of that kind. Conceivably, then, there might be *no* frame in which *all* free bodies of Σ would move in uniform rectilinear motion. The First Axiom of Inertia asserts that one such frame exists. The Second Axiom of Inertia then makes sense, for it refers to the frames whose existence the First Axiom delivers. At the same time, the apparatus of change of framing enables us to convert all statements in an inertial frame into statements in any other frame we choose. Thus NOLL's axioms, like the

---

[68] MCKINSEY, SUGAR, & SUPPES [1953, page 260].

principles of EULER and CAUCHY and unlike the SUPPESians' paradigm, do not require us to use any special frame.

The applications of NOLL's general theory to the dynamics of SUPPESian ·particles are obvious: *Frame-indifference of the working of all systems of forces* is equivalent *to the "Newtonian" equations of motion and the "Third Law" interpreted as the requirement that the mutual forces be both pairwise equilibrated and central.* With this statement we are free of SUPPESian confusion about the "Third Law" and closer to NEWTON.

There is no point in laboring the concept "frame of reference", for it was well (though not universally) understood long before anyone alive today was born.

While the concept of "change of framing" is absent from the system of axioms that the SUPPESians never cease to exalt as the greatest example of their ideas, it is central to the axiomatic approach nowadays used by most students of the foundations of classical mechanics. In the succeeding section we shall consider examples of fruitful use of that approach.

18. THE CREATIVE INFLUENCE OF NOLL's AXIOMS OF FORCES AND NOLL's AXIOM OF DYNAMICS

A system of axioms weighs little if it merely purifies and restates what was known already. HILBERT's axioms of Euclidean geometry immediately had a powerful effect upon researches on geometries of all kinds. NOLL's axioms of forces, after a delay of six or seven years, have had a similar influence upon modern research in mechanics as a whole. Here I comment only upon some profound developments in continuum mechanics.

The concept of mass-point is relatively new in mechanics. The extent of actual bodies is obvious to us. Without that extent, nobody could notice them. Thus the early mechanics, from ARCHIMEDES' time through STEVIN's and GALILEO's and HUYGENS' and NEWTON's, is continuum mechanics, usually rather rough if not crude. A careful and precise development of the concepts of continuum mechanics began hesitantly with JAMES BERNOULLI, attained vigorous maturity with EULER, and was given its classical form by CAUCHY.

CAUCHY took EULER's Laws of Motion as the basic axioms of mechanics. He assumed further that a system of forces acting upon a continuum could be resolved into a system of contact forces and a system of extrinsic forces. His theory of contact forces on three-dimensional bodies could be outlined as follows.

Axiom 1. The contact force $\mathbf{f}_C$ exerted by a body upon a contiguous body is the value of the integral of a field of *traction vectors,* defined upon the two bodies' surface $\mathscr{S}$ of contact, assumed oriented:

$$\mathbf{f}_C = \int_{\mathscr{S}} \mathbf{t}\, dA.$$

Axiom 2 ("CAUCHY's postulate"). At a given point $\mathbf{x}$ on $\mathscr{S}$ the traction vector is determined by the signed unit normal $\mathbf{n}$ of $\mathscr{S}$:

$$\mathbf{t} = \mathbf{f}(\mathbf{x}, \mathbf{n}(\mathbf{x})).$$

From these axioms and EULER's Laws of Motion follows

Cauchy's Fundamental Theorem: If the mappings $\mathbf{x} \mapsto \mathbf{f}(\mathbf{x}, \mathbf{n})$ are continuous for all unit vectors $\mathbf{n}$, then $\mathbf{n} \mapsto \mathbf{f}(\mathbf{x}, \mathbf{n})$ is linear. That is,

$$\mathbf{t} = \mathbf{T}(\mathbf{x})\mathbf{n}.$$

The tensor $\mathbf{T}$ is called the *stress tensor*; it is the protagonist of continuum mechanics.

CAUCHY's concepts were distilled as expressions of the features common to EULER's theories of fluids, flexible membranes, and elastic rods.

HAMEL in his axiomatization followed CAUCHY's program, more or less, but he perceived an improvement. Namely, Axiom 2 can be deduced, subject to assumptions of smoothness, as a consequence of Axiom 1 and EULER's Laws. HAMEL did not state his result[62] in this way, and in regard to analysis his development, in common with most studies of continuum mechanics before 1950, is sloppy, but the essence of a proof is there. NOLL in his first paper included a clear statement (Theorem IV) and provided an elegant, rigorous proof of the theorem HAMEL had roughly formulated and roughly inferred. "Rigor" here is meant in the sense of analysis, specifying the domains, codomains, and smoothness of functions, and with precise treatment of limit processes.

This "HAMEL–NOLL Theorem" provoked a number of NOLL's associates and their students to replace CAUCHY's starting point by more natural and seemingly less restrictive general axioms, from which CAUCHY's assumptions are proved to follow as theorems. The work was done by NOLL himself and by M. E. GURTIN, W. O. WILLIAMS, and their students. The main results, carefully organized and improved by Mr. WILLIAMS, are presented in my textbook[69]. The

---

[69] TRUESDELL [1977, Chapter III].

Figure 29. GEORG HAMEL (1887–1954), after a photograph taken in 1954.

theory rests upon a single

Basic Axiom:

$$\mathbf{f} = \mathbf{f}_B + \mathbf{f}_C.$$

$\mathbf{f}_B$ & $\mathbf{f}_C$ are systems of forces.

$$|\mathbf{f}_B(\mathcal{A}, \mathcal{C})| \leqq KM(\mathcal{A}).$$

$$|\mathbf{f}_C(\mathcal{A}, \mathcal{C})| \leqq KA(\mathcal{S}).$$

Here K is some constant; $M(\mathcal{A})$ is the mass of $\mathcal{A}$; $\mathcal{S}$ is the intersection of the boundaries of the shapes of $\mathcal{A}$ and $\mathcal{C}$; $A(\mathcal{S})$ is the area of $\mathcal{S}$; and the inequalities are to hold for all $\mathcal{A}$ and $\mathcal{C}$ such that $M(\mathcal{A})$ and $A(\mathcal{S})$ are sufficiently small. This axiom posits physically natural properties of the *body force* $\mathbf{f}_B$ and the *contact force* $\mathbf{f}_C$. The former applied to a family of smaller and smaller bodies vanishes ultimately at least as fast as the mass; the latter, as the area of contact. Because of EULER's First Law, $\mathbf{f}_B + \mathbf{f}_C$ is a balanced system of forces. Inertial forces are included in $\mathbf{f}_B$.

Elaboration of NOLL's Axiom and EULER's First Law leads to the following principal results:

Traction Theorem of GURTIN & WILLIAMS: There is an essentially bounded density $\mathbf{t}_{\mathcal{S}}$ such that

$$\mathbf{f}_C(\mathcal{A}, \mathcal{C}) = \int_{\mathcal{S}} \mathbf{t}_{\mathcal{S}} \, dA,$$

and $\mathbf{t}_{\mathcal{S}'} = \mathbf{t}_{\mathcal{S}}$ if $\mathcal{S}'$ is a subsurface of $\mathcal{S}$.

Action-Reaction Theorem of NOLL: While neither $\mathbf{f}_B$ nor $\mathbf{f}_C$, in general, is balanced, both are pairwise equilibrated.

Corollaries: $\mathbf{t}_{-\mathcal{S}} = -\mathbf{t}_{\mathcal{S}}$, and the resultant body force is the value of a volume integral.

HAMEL–NOLL Theorem: CAUCHY's postulate holds on every surface in the interior of the shape of a body.

The Traction Theorem is rather deep; it is not at all easy to prove. The succeeding steps are fairly straightforward except for proof of the HAMEL–NOLL Theorem. After that, CAUCHY's Fundamental Theorem follows as in the older treatments, on the usual assumption that $\mathbf{x} \mapsto \mathbf{f}(\mathbf{x}, \mathbf{n})$ is continuous for all unit vectors $\mathbf{n}$.

From the day NOLL's first general presentation[70] was published, his way of looking at mechanics has had a great and lasting influence on

---

[70] NOLL [1958].

modern research. The reason is not far to seek. NOLL's concepts, axioms, and definitions are clear; his approach is simple and efficient. NOLL's viewpoint renders precise and refounds the mechanics of EULER and CAUCHY—the mechanics that is taught to students of engineering today. While NOLL himself concentrated on single three-dimensional bodies without internal structure, his ideas can be generalized to mixtures and to structured materials; his approach can be applied and has been to the theory of energy and heat, and in some degree also to electromagnetism. NOLL's Axiom of Dynamics is particularly useful for generalization to more complicated kinds of mechanics because it enables anyone who knows how to define the working of a system of forces to read off the explicit forms that the Laws of Motion take for bodies subject to that system. Hundreds of research papers using NOLL's approach—formally or informally, as subject and taste suggest—have been published in the last two decades. Among the important authors, in roughly the order of their first publications in this vein, are B. D. COLEMAN, V. MIZEL, M. E. GURTIN, C.-C. WANG, W. O. WILLIAMS, S. ANTMAN, W. A. DAY, D. OWEN, G. DE LAPENHA, D. E. CARLSON, C.-L. MARTINS, P. PODIO-GUIDUGLI, G. CAPRIZ, G. DEL PIERO, M. ŠILHAVÝ, M. FABRIZIO, J. F. OSBORN. Many of the essential papers have appeared in the *Archive for Rational Mechanics and Analysis*; others may be found in various journals of mechanics, mathematics, and engineering the world over. The SUPPESians give no sign of having looked at any of this work. If they should become aware of it, I wonder if they would hold to the criticism of MOULINES & SNEED[71]:

> These axiomatizations fail, to differing degrees depending upon the specific example, to meet modern standards of logical rigor. Primitive concepts and axioms are sometimes not clearly identified; questions of independence of primitives and axioms are not carefully raised; the epistemological status of the axioms is often fuzzy; "physical intuition" is sometimes employed as an inference rule in obtaining theorems.

Of course in the literature related to NOLL's ideas no appeal will be found to "physical intuition" as a rule of inference. Indeed, independence of the axioms and primitives is rarely regarded as important enough to be taken up explicitly, and only the mathematical structure and its physical interpretation, not epistemology, are developed, but in return there is something altogether absent in the SUPPESians'

---

[71] MOULINES & SNEED [1979, page 65].

work: *great pains to make all axioms physically reasonable and to derive from them interesting and significant general theorems and applications.* It is live mathematical science, growing from the desire to solve specific problems regarding nature.

There has been some desultory argument among the SUPPESians and others as to whether the velocity should be assumed to be differentiable or only piecewise differentiable. A question of this kind is not worth much interest until a concrete problem involving it has to be solved. ANTMAN & OSBORN[72] undertook to determine whether or not EULER's Laws of Motion were equivalent to the principle of virtual work. The former require accelerations as well as velocities; the latter, only velocities; and the NEWTONian concept of impulse ("impulsive force"), in terms of which the mechanics of impact is described, deals with increments of momentum rather than rates of change of momentum. The principle of virtual work involves boundary conditions as well as interior conditions; to yield statements restricting impulse, such conditions are adjoined to EULER's Laws when they are integrated with respect to time. EULER's Laws involve torque; accordingly, the virtual work of the torques must be inserted into the principle of virtual work. ANTMAN & OSBORN prove that with these respective additions, the two formulations of the laws of mechanics become equivalent under very weak conditions of smoothness. The mathematics they bring to bear is not elementary.

NOLL's approach to mechanics is the basis of courses taught in several universities in several countries. It is neither secret nor arcane.

When in 1952 HAMEL in his evaluation of the paper of McKINSEY, SUGAR, & SUPPES wrote "in the concept of force lies the chief difficulty in the whole of mechanics" (above, § 11), he was surely referring to force in its two-fold aspect:

1. What are forces?

2. How are forces determined?

The answer desired is not epistemological or semantic. It is mathematical. Forces are undefined objects like points and lines in geometry. Their properties are mathematically stated. Using those properties, mathematicians can prove theorems about forces. In NOLL's axioms for forces we have seen an answer to the first question. Now we turn to the second.

---

[72] ANTMAN & OSBORN [1979].

IV. Noll's Axioms for Constitutive Relations

19. THE SUPPESIANS' DISCOVERY THAT MECHANICS RESTS ALSO ON "SPECIAL LAWS"

In their paradigmatic paper of 1953 McKINSEY and his team wrote[73] "In dealing with an empirical situation..., each particle is ordinarily subjected to a number of different forces." Even apart from the syntax, which suggests it is the particle that must deal with the situation, my correspondence with McKINSEY in 1953 made me doubt whether he and his collaborators would have been capable of setting up without aid of a crib the equation of motion of the simple pendulum.

The passage of nearly twenty years saw the smugness exuding from McKINSEY's heirs begin to shrivel. In 1971 SNEED[74] criticized his masters' work:

> ... it fails to provide an adequate means of accounting for the use that is made of measured, or calculated, values of theoretical terms. There appears to be no way of providing a rationale for the practice of exploiting values of theoretical functions obtained in one application of the theory to draw conclusions about other applications of the theory.

Since theories are totally useless for physics unless they interconnect applications, SNEED's remarks here seem to dismiss the axioms of McKINSEY, SUGAR, & SUPPES as being empty. He proposed to save the wonderful edifice that had sufficed the SUPPESians for so long by introducing[75]

> the claim that the entire array of theoretical functions ... also satisfy certain other *constraints*. These constraints will require that certain relations hold among the values of theoretical functions employed in different applications of the theory.

More than 100 pages later[76] he tells us more about these constraints:

> The intuitive idea of constraints is this. Not all possible sets of theoretical functions may be used to extend the set of intended

---

[73] McKINSEY, SUGAR, & SUPPES [1953, page 259].

[74] SNEED [1971, page 65].

[75] SNEED [1971, page 66].

[76] SNEED [1971, page 170].

applications of the theory to models for the predicate characterizing the basic mathematical structure. Some are ruled out by constraints on these functions.

He gives two examples[77] of constraints: Mass is "an intrinsic property
of a particle that remains the same at whatever time and in whatever
environment we happen to find the particle", and "at any given time,
the same forces are acting on a particle, in whatever situation we
consider the particle to be. . . ." He does not tell what he means by "the
same forces", and his illustration seem to be wrong in mechanics
unless we consider only forces whose magnitudes are functions of
mutual distances and by "the same force" understand "the same
force-function". STEGMÜLLER[78] explains constraints as being "*cross-
connections between the different intended applications*" of a theory. He
goes on to state the difference between a "constraint" and a "special
law":

> With the notion of a constraint. . . one can recognize an ambiguity
> in the expression "*natural law.*" . . . *But the demands made by con-
> straints could also be thought of as natural laws.* . . . [A] "natural law"
> has *a completely different logical status* than e.g., special force laws.
> While the latter *hold in certain intended applications* of a theory *but
> not in others*, . . . the former does not *represent a feature of some one
> intended application but rather a certain kind of connection obtaining
> between all intended applications*. Therefore such "natural laws"
> must be characterized by means of the notion of constraint. On
> the other hand, natural laws such as special force laws (e.g. the law
> of gravity, Hooke's law, etc.), can . . . be introduced *by restrictions
> on the basic predicate characterizing the mathematical structure of the
> theory*.

The first appearance of the "special laws" in SUPPEsian philosophy
seems to have been a passing remark by SNEED[79]:

> There are also other means of determining force values, more
> practically relevant than those we have mentioned. For example,
> we suppose that a body near the earth suspended from a coil
> spring is acted upon by a "Hooke's Law" force and a constant
> force, equal to the weight of the body. We then determine the
> spring constant by observing the path of the body. . . .

---

[77] SNEED [1971, pages 123, 125].

[78] STEGMÜLLER [1976, pages 73, 77].

[79] SNEED [1971, page 145].

To produce a full-blown logical reconstruction of classical particle mechanics, it would be necessary to examine, in detail, all of the various possibilities for mass and force determinations, or at least all the practically significant ones. We shall not do this here.

STEGMÜLLER[80] rose to the challenge:

> ... in the various particular applications of one single theory ... *various special laws* may hold. .... [T]he Newtonian theory offers a good illustration. The solar system is a special application of classical particle mechanics. Newton postulated that in this *application* each "particle" exerts a force on every other one. This force attracts, works along the straight line between the two particles, and is inversely proportional to the square of the distance. ...
> ... [T]he adequate treatment of special laws consists in the introduction of *restrictions on the basic predicate*.

Also[81]

> Besides the *general* constraints ... , *additional constraints* may appear involving just *those theoretical functions occurring in special laws holding only for certain applications*.

MOULINES & SNEED were not to be left behind[82]:

> [T]hough the ... axiomatization [of MCKINSEY, SUGAR, & SUPPES] provides us with the means of precisely stating various special force laws (e.g. Hooke's law or the law of gravitation) the structure of an *array* of such laws is not exhibited. Very roughly speaking, how the applications of classical particle mechanics in which *different* special force laws are stipulated all "hang together" to comprise the "theory" is not revealed. Even less apparent is how one deals with the development of classical particle mechanics over time as new special force laws are discovered and applied to new applications. Roughly, one wants to say that one-and-the-same "theory" is developing over time.

I do not understand what the verb "comprise" is to mean here; surely the authors do not think that the special laws include or comprehend the general theory; I can only guess that what they mean to say they

---

[80] STEGMÜLLER [1976, page 86].
[81] STEGMÜLLER [1976, page 90].
[82] MOULINES & SNEED [1979, page 74].

do not see is how the "special force laws" are comprised by the theory, but as this particular inclusion offers no more difficulty than any other, I can get no sense at all from what they write.

The "special laws" cause the Suppesians a lot of trouble. On page 12 of his pamphlet Stegmüller writes

> The *special laws* form a particular and difficult problem for informal semantics. The axiomatization of classical particle mechanics given [by McKinsey, Sugar, & Suppes] leaves open the possibility of stating special laws. But it is not indicated how the various applications of classical particle mechanics, for which special laws are stipulated, are related to each other and how they "hang together". *A fortiori*, it does not follow from the account given there how to reconstruct the *development* of classical particle mechanics over time as new special laws and new applications are discovered.

The title of § 2 is "Empiricism Liberalized, Informal Semantics, and the Extended Bourbaki Programme ('Sneedification')", and on page 11 Stegmüller refers to "*referential semantics*" and "*informal semantics* of set-theoretically axiomatized physical theories". While in 1979 Moulines & Sneed[27] with condescension dismissed my "negative commentary" of 1953 as reflecting merely a desire for "semantical clarification", which it did not, earlier in the very same paper[83] they had written

> ...one can formulate the aims of reconstructing empirical theories as:
> —clarifying the *internal structure* of an empirical theory;
> —providing a *semantics* for an empirical theory (clarifying its relation to something "outside" itself).

Also

> It would be difficult to see why a set-theoretic axiomatization of an empirical theory would be of any interest in the absence of *some* way to provide it with a semantics. A fair assessment of the Suppes Program requires us to remember that it *has* a semantic part.

While in one passage[27] they imply that the aim of McKinsey and his team was "solely to reconstruct the formal structure of the theory", here they go so far as to praise them for "dealing with ... the epis-

---

[83] Moulines & Sneed [1979, pages 62, 63].

temological status of 'mass' and 'force' in classical mechanics". In regard to semantics and epistemology, it seems that what is vaunted if JUPPITER takes it up is worth no more than a sneer if attributed to a *bos* from the hoi polloi of active science. Having license from on high, STEGMÜLLER attacks with gusto the problem of "special laws" as one of "informal semantics", something, as we shall see in the succeeding section, that it need not be at all. Further on (page 12) STEGMÜLLER describes the "solution given to this problem" by SNEED and himself in his preceding book as "clumsy" and promises to provide a better one.

Now we are ready to find out what STEGMÜLLER means by his further "non-statement views". On page 22 of his pamphlet he defines "*non-statement view₂*" as "the thesis according to which the empirical claims made by physicists must be reconstructed as Ramsey–Sneed-sentences". And what is a "Ramsey–Sneed-sentence"? To find out, we must have recourse to a passage in STEGMÜLLER's preceding book[84]. STEGMÜLLER now (page 22) writes of a "*modified* Ramsey-sentence because, in it, all the additional semantic aspects mentioned in § 2 come into play . . .." In endorsing for the third time SNEED's two examples (plus "etc.") of "requiring the force functions to take special forms in certain applications" STEGMÜLLER[84a] puts at the head of the list an unhappy addition of his own: Newton's Third Law as the statement that the mutual forces of mass-points are pairwise equilibrated and central. While McKINSEY and his co-authors in 1953 had been just in remarking that "Newton's Third Law" meant different things to different people, NOLL in 1959 cleared the whole matter by his proof, mentioned above in § 16, that for systems of mass-points the particular statement to be chosen by STEGMÜLLER in 1976 as an example of a "special law" was in fact a *general* law, equivalent to the principle of rotational momentum—please, no semantics!—but the SUPPESians do not follow the mathematical literature regarding mechanics.

"Non-statement view₂,₅" occurs in the title of § 3 of STEGMÜLLER's pamphlet and is described on page 24. This view somehow allows for the fact that "if the Ramsey–Sneed sentence of a theory turns out to be false at any particular time, one cannot just pick out one of these laws and make *it* responsible for the failure. . . . [E]mpirical claims are to be interpreted *holistically* as single comprehensive claims." Are

---

[84] STEGMÜLLER [1976, §§ 4.2–4.3]. First "*x* is an enrichment of *y*" if, roughly speaking, *x* is a possible model and *y* is a partial possible model obtained from *x* by erasing the *x*-theoretical terms from *x*. Second, let "is an *S*" be "a set-theoretic predicate", that is, a set of axioms characterizing a concept *S* such as "group" or "system of particle mechanics". Then the Ramsey sentence of a theory is "there exists an *x* such that *x* is an enrichment of the partial possible model *a* of *S* and *x* is an *S*."

[84a] STEGMÜLLER [1976, § 6.4].

the SUPPESians rediscovering in their own jargon the simplest parts of DUHEM's[85] philosophy?

> To try to separate each of the hypotheses of theoretical physics from the other assumptions on which this science rests so as to submit it by itself to test by experiment, is to hunt a chimæra, because realization and interpretation of any experiment whatever imply adherence to a whole set of theoretical propositions.

> For a physical theory, the only experimental test that is not illogical is to compare *the whole system of that physical theory* with all the experimental laws and to judge whether the latter are satisfactorily represented by the former.

STEGMÜLLER refers us to a passage on page 53 that is to make all clear by an example:

> For the sake of illustration, let us consider the fundamental law in the basic core of classical particle mechanics in the Newtonian formulation, i.e., Newton's second law. . . . "Is it an empirical law or an *a priori* truth?" Numerous answers have been given, from "It is an elementary analytic truth, namely a mere definition of force" at one end of the spectrum to "It is an empirically falsifiable hypothesis" at the other. I maintain all of these answers are wrong. We can say even more: this whole discussion concerning the epistemological status of the second law is just nonsense. The sterile dispute has its roots in the misleading way of stating this law *as an isolated and 'self-contained' universal sentence.*

> We get a very different picture if we look at the matter from the structuralist point of view. For we must then ask another question, namely: "What particular claim, for which this law is responsible, must be contained in *all* empirical claims which can be formulated with the help of classical particle mechanics *CPM*?" Roughly speaking, the answer amounts to this:

> *For all intended applications of CPM, forming at any historical time t no clearly defined class but a largely open(!) set I_t, we can find two CPM-theoretical functions f (force) and m (mass) standing in a particular relation to the second derivative of a third, CPM-non-theoretical function s (position) such that the function f will in most (!) applications satisfy certain (!) special laws and both functions f and m will create cross-connections between certain (!) applications through certain (!) constraints.*

> *This statement is undoubtedly not 'analytic' but it is . . . sufficiently empty to withstand any possible refutation.*

---

[85] DUHEM [1906, end of § V of Chapter VI of Part II].

Here "CPM" denotes whatever it is that obeys the axioms of McKinsey, Sugar, & Suppes. The passage just quoted is not the first, dazzled outburst of a reader of 1687, awed by the power of Newton's grasp and method; it states the considered judgment of a voluble professor, into whose brain in 1978, ten years or more after he was certified "expert" in philosophy of science, the realities of Newton's *Principia*, which exceeded the capacities of his teachers and god-fathers in Suppesism, begin to penetrate.

§ 7 announces in its title "Non-Statement View₃", but I can find no statement of that non-statement. On page 45 we read

> By *statement view₃* I shall understand the position according to which philosophical talk about theories, their achievements and their drawbacks *within general philosophy of science* concerns these theories as sentences or classes of propositions. It would be too easy to argue in favor of this attitude as follows: "After all, this is the view accepted by almost all philosophers of science and by most scientists."

Stegmüller then claims that this view "can be tenable only in combination with some kind of philosophy of the as if." After a few pronouncements intended to support this conclusion, he lists seven "disadvantages and dangers" connected with statement-view₃. His nearest approach to a statement of "non-statement view₃" is on page 49: "the non-statement view₃ only mirrors, on the level of general philosophy of science, the position of the non-statement view₁ on the special level . . . ." The term "*non-st.v.₃*" occurs twice more on the same page. Although Stegmüller seems to favor his non-statement view₃, he nowhere explains it any more clearly. On page 87, his next-to-last, he mentions it again and claims for it

> greater flexibility, . . . which manifests itself in its ability to allow more and better differentiations, facilitating our understanding of the systematic-static as well as the historical-dynamic aspects,

and that it allows

> unforced transitions to *pragmatizations* . . . to be more easily performed and [opens] the door . . . to an urgently needed systematic pragmatics.

I hesitate to attribute anything specific to what Stegmüller writes here. Wishing to end this review by a conjecture intended to be as favorable as possible toward the Suppesians, I recall Stegmüller's remarks[78] about "connections obtaining between all intended applications" as well as MOULINES & SNEED's call[82] for "the structure of an *array* of [special] laws" and wish to deal "with the development of ... mechanics over time as new special force laws are discovered and applied to new applications." They lead me to guess that now the SUPPESians may be calling for axiomatic structures such as to include

(1) The "general laws" of a physical theory

(2) The "general constraints" that all "special laws" of that theory must satisfy

(3) The "special laws" which are "additional constraints".

Like BOURBAKI and other mathematicians who do not devote themselves to formal logic, in constructing axiomatic systems of this kind we are to use informal set theory. The desire to represent physical phenomena serves to motivate what "laws" and what "constraints", whether "general" or "special", are to be selected; as is clear from my remarks in § 6, otherwise the outline is the same as is common in theories of pure mathematics. Take, for example, an extract from geometry:

(1) Basic axioms

(2) Requirements for defining figures (*e.g.*, as equivalence classes under appropriate groups of transformations of appropriate sets of geometric objects)

(3) Definitions of particular figures (triangles, circles, *etc.*)

If my guess is right, some of the SUPPESians in their testudineous exercitations have reached the doorstep of the mechanics of NEWTON, EULER, and CAUCHY and now see a glimmering of what HAMEL attempted to express in 1908. We recall the axiom in HAMEL's treatment of mass-points in 1927, which the SUPPESians scorned in 1953 (above, § 10):

> The forces are determined through their *causes*, that is, through variables that represent the geometrical and physical state of the surrounding matter.

Of this MCKINSEY and his team wrote:

> One does not see how this axiom could intervene in the proofs of theorems, or in the solutions of problems.

The axiom means neither more nor less than that the forces are delivered by "special laws", as has long been plain from the applications included in NEWTON's *Principia*. Only by invoking a special law can we ever solve a typical special problem; otherwise we do not have the structure upon which to set up such a problem. True, HAMEL's "axiom" is too vague because he does not delimit what he calls "the geometrical and physical state of the surrounding matter" and does not give a precise meaning to "determined", but the SUPPESians are just now beginning to see that "special laws" like "HOOKE's law" make an essential part of the science of mechanics.

Unregarded by the SUPPESians, an axiomatic program regarding the "special laws" of continuum mechanics has been followed by mathematicians and engineers for over twenty years. Their work provides the subject of the next section, in which we shall see also that NOLL's theory of constitutive relations has a simple, indeed obvious counterpart for the mechanics of systems of mass-points.

20. NOLL's AXIOMS FOR CONSTITUTIVE RELATIONS

Physicists and philosophers today refer to "laws of force" in the classical mechanics of mass-points. NEWTON never applied the name "law" to them; even universal gravitation he put forward as a "hypothesis". The term "law" for specializing relations is mightily unfortunate, especially since most such "laws" are regarded now as being "approximate", honored by nature only in the breach by greater or lesser amounts; modern studies of the foundations of classical physics use *constitutive relation*[86] to denote specializing hypotheses intended to model ideally the response of natural substances, and I shall follow that usage henceforth in this review. The history of the name is included in the sketch of the history of constitutive relations I published recently[87].

In Book I of the *Principia*, which is the only part based fairly systematically on the Axioms, NEWTON's use of constitutive relations set the pattern for theories of discrete systems.

Most early suggestions concerning "particles" satisfied the strong form of "NEWTON's Third Law" automatically. Most made the forces depend upon a *pre-assigned list* of independent variables, typically the differences of positions and the differences of velocities. Some simple properties of invariance were thoughtlessly satisfied. In modern

---

[86] See the article "constitutive" in *A Supplement to the Oxford English Dictionary*, Volume 1, Oxford, Clarendon Press, 1972.

[87] TRUESDELL [1980].

terms, the forces were values of *frame-indifferent* functions of positions and velocities. The early authors on the mechanics of systems of mass-points did not remark on these facts.

While a single mass-point has a permanent mass and exerts mutual forces on other mass-points, the typical "particle" of a continuum is subject not only to extrinsic and perhaps also mutual forces but also to contact forces exerted upon its boundary.

Even for the very simple and usually one-dimensional continua considered in the eighteenth century the problems connected with inferring or inventing constitutive relations are more difficult than those for mass-points. First, such relations characterize a material rather than a body. Second, kinematic constraints provide another kind of constitutive relation. For example, a one-dimensional material may be inextensible, and a three-dimensional material may be incompressible. Constitutive relations for materials were first formulated by JAMES BERNOULLI, who wrote in 1705 that "HOOKE's law" and other such proposals for springs should not be conceived as relations between forces and elongations of bodies one-by-one but instead as relations between the stress and the strain in all bodies of given substance. The history of constitutive relations in continuum mechanics from their earliest antecedents through 1788 is included in my prefaces[88] to the works of EULER on hydraulics, hydrodynamics, flexible lines, flexible membranes, and elastic bars.

When three-dimensional theories of continua were formulated by CAUCHY and others, greater thought and more acute mathematical analysis were required to select the right variables and to impose restrictions that seemed plausible as idealizations of experience with real materials. These restrictions are *requirements of invariance*. Unlike the examples of "general constraints" adduced by SNEED[77], they are not trivial. The book by NOLL & me[89] is essentially a treatise on three-dimensional constitutive relations and their recent applications; it includes historical sections and full references up to 1965, from which an interested and understanding reader can learn for himself how the theory of constitutive relations in mechanics developed.

NOLL proposed axioms for constitutive relations in a paper[90] published in 1958. He alluded to this work in his paper of 1959, delivered to a SUPPESian meeting in 1957. We based our book[89] upon those axioms. In my textbook for beginners[91] I state the first two as follows:

---

[88] TRUESDELL [1954] [1956] [1960].

[89] TRUESDELL & NOLL [1965].

[90] NOLL [1958]. Subsequently NOLL [1972] formulated a more inclusive and abstract theory.

[91] TRUESDELL [1977].

N1. ***Principle of Determinism.*** The stress at the place occupied by the body-point X of the body $\mathscr{B}$ at the time t is determined by the history of the motion of $\mathscr{B}$ up to the time t.

N2. ***Principle of Local Action.*** The motion of body-points at a finite distance from X in some shape of $\mathscr{B}$ may be disregarded in calculating the stress at X.

The third may be stated thus:

N3. ***Principle of Material Frame-Indifference.*** A mapping of the set of histories of motions into the set of stress fields provides a constitutive relation if and only if it is frame-indifferent.

We may explain these in SUPPESian terms as follows. The Principle of Determinism specifies the domain and codomain of "special laws"; in particular, past and present experiences determine present contact forces per unit area. The Principle of Local Action and the Principle of Material Frame-Indifference are "general constraints". The former principle states that the contact forces are not affected by finitely distant elements of the body; the latter, that contact actions are the same for all observers. A determination of the stress through a mapping of deformation histories according to N1 is not a constitutive relation ("special law", "special constraint") unless it satisfies Axioms N2 and N3.

In the practice of continuum mechanics all constitutive relations are made to satisfy the Balance of Rotational Momentum automatically whenever the Balance of Linear Momentum is satisfied. As CAUCHY proved, this requirement is met by making the stress tensor symmetric; that is, by making the codomain of constitutive mappings the set of fields whose values are symmetric tensors. Likewise, in practice Axiom N2, the Principle of Local Action, is satisfied automatically by limiting the domains of the constitutive mappings. Typically, they are subsets of the set of histories of gradients of transplacements of $\mathscr{B}$ at X.

Axiom N3, the Principle of Material Frame-Indifference, is used to restrict any putative class of constitutive mappings. Some quantities derived from the motion must be absent, others can appear only in certain combinations, *etc.* The details in classical examples may be found in Chapter IV of my textbook[32]. The principle itself has a long history in special instances, beginning with HOOKE's intended applications of his "theory" of spring to his "new sort of *Philosophical-Scales*". Fairly general statements of it if restricted to proper-orthogonal changes of framing are due to OLDROYD (1950) and NOLL (1955); the latter called it then "the principle of isotropy of space".

The axioms stated here refer only to materials devoid of internal constraints such as incompressibility. Axioms for constrained materials may be found in my textbook.

The constitutive axioms of NOLL do not apply directly to SUPPESian mass-points. In particular, they do not include mutual forces of the type exemplified by gravitation. This lack reflects simply the practical origin of NOLL's analysis of the basic concepts: Continuum mechanics was then and is still the only portion of classical mechanics that is under active development, and the only mutual force of gravitational type that has ever been of importance in it is gravitation itself, which in classical mechanics offers no conceptual problems not already solved.

Turning to systems of mass-points, I venture to suggest for them axioms of NOLL's type:

MP1. *Determinism.* The applied force upon each mass-point at the time t is determined by the history of the motions of all the mass-points of the system up to the time t.

MP3. *Material Frame-Indifference.* A mapping of the set of histories of motions of the members of the system into the set of forces applied to those members provides a constitutive relation if and only if it is frame-indifferent.

The applied force, to which Axiom MP1 refers, is the resultant force less the inertial force. There is no counterpart of Axiom N2 because in mass-point mechanics the contact forces are naught. Axioms MP1 and MP3 refer only to systems without constraints in the sense of analytical dynamics; they apply also to a holonomic system after it has been reduced by eliminating the constraints, supposed invertible.

The classical kinds of "force laws" in analytical dynamics depend upon the history of the motion of the system only through the present positions and velocities of the members of the basic set of mass-points. The reader will note that if the velocities exist, they are equal to the left-derivatives of the motions and so are determinable from the histories of the motions. Thus Axiom MP1 is trivially satisfied.

A typical extrinsic force on a mass-point is that of uniform gravity: a force of constant magnitude in the direction of the perpendicular from the position of the mass-point to a fixed plane in an inertial frame. A vector in such a direction is frame-indifferent. A somewhat more general statement is given in Part 2 of Exercise I.14.2 of my textbook[91].

Typical mutual forces in analytical dynamics are taken as time-independent functions of the positions of the members of the basic set of mass-points. For a system of this kind that is conservative and allows no nonzero self-forces, we may interpret the consequences of Axiom MP3 by using the result of the first part of Exercise I.14.2 of my textbook. The outcome is as follows[92]:

(1) The mutual forces must be central, and their magnitudes are functions of their mutual distances only.

(2) If the system of mutual forces is conservative and balanced, the Principle of Rotational Momentum is equivalent to the Principle of Material Frame-Indifference.

The second statement is parallel to a celebrated theorem of NOLL[93] on hyperelastic materials. It illustrates the fact that for sufficiently special systems, basic principles that are in general independent may fall into dependence upon one another.

NOLL's axioms have attracted much attention. In any domain of research as active as the basic thermomechanics of continua has been in the past thirty-five years many differences of opinion and practice are to be expected. Some of the most creative men prefer an informal presentation, refraining from use of the formal "axiom" and "theorem". Nevertheless they do state their assumptions openly, and they do demonstrate clear, explicit results. It is fair to say that NOLL's views and organization have dominated much of the field and have influenced nearly all of it.

Coming back to the pamphlet under review, I remind the reader that STEGMÜLLER (if I understand him) in culmination of some thousand pages of print on the philosophy of science has finally called for axioms delimiting (1) general laws, (2) such requirements as the variables entering the general laws and also all constitutive relations must satisfy, and (3) such special constitutive relations as deserve study. He must surely know that much of the literature on mechanics since its beginnings has been devoted to (3). Practically every mathematical paper on a particular problem in mechanics states its assumptions, more or less explicitly. On the other hand, it may be that STEGMÜLLER wishes here a semantic theory of what makes a special constitutive relation valuable or a methodology of constructing constitutive relations. For the mechanics of mass-points such an investigation would be necessarily historical and hence hampered by the scantiness of remains that describe methods and principles explicitly.

---

[92] NOLL on pages 38–40 of his unpublished report listed as "[1957, *11*]" in the bibliography of TRUESDELL & TOUPIN [1960]. In their § 196A TRUESDELL & TOUPIN reproduce NOLL's proof. The theorem emerges with full clarity in the development of TRUESDELL [1977, § I.8 and Exercise I.13.1].

[93] NOLL [1955, part of Theorem 1 on page 42].

In contrast, most of the creative students of continuum mechanics in this century have been pretty frank about the extent of their appeal to or reliance upon physical experiment and how they weigh the results of intellectual and mathematical speculation. To understand what they write, STEGMÜLLER would have to learn much more about mechanics than any SUPPESian so far has thought worthwhile.

In his review of STEGMÜLLER's preceding book BUNGE[43] wrote, as I have already mentioned,

> .... the new philosophy of science advanced in this work is at least as remote from living science as any of the rival views criticized in it.

Not only that, living scientists working in living science have long ago achieved what the SUPPESians are only now beginning to sense need for.

0_{bis}. Suppesians

21. THE SUPPESIAN SUM

Had the SUPPESians provided treatments of several particular sciences in their style, for example optics and electromagnetic theory and general relativity, I should never have taken the pains to analyse their work. A glance shows any serious student of classical mechanics today what babes in its wood the SUPPESians are. "Well", I should have thought, "... maybe they aren't much interested in it, and why should they be? Doubtless they are much better at optics or electromagnetic theory or general relativity. Good enough." No, it is *they* who, again and again, make classical mechanics their main and usually their only victim in applying their ideas. As a student of mechanics I do not flinch at the heavy charge of ambiguity and fuzziness MOULINES & SNEED[94] lay upon practising scientists:

> ... it is not always clear how different theories are related to each other *and* to their applications. ... Even in the physical sciences— usually regarded as paradigms of rigor—practicing scientists rarely express themselves professionally in ways that provide clear answers to questions like this. Yet, presumably, these are things

---

[94] MOULINES & SNEED [1979, pages 61–62].

one must know to answer even the most naive questions about an empirical theory: what does the theory say about the world?— what are the reasons for believing what it says is true?

Such ambiguity and fuzziness may be quite harmless in the literature of ongoing empirical sciences. Most philosophers of science charitably assume that practicing scientists could—if they would just take the trouble—ultimately clear up these questions. Their working hypothesis is that empirical science is not irredeemably confused. The task of the reconstructive philosopher of science is to be clear where "ordinary people" are not—to make coherent sense out of what practicing empirical scientists say and do professionally.

In the foregoing pages the reader will have learned from the SUP-PESians' writings, quoted at length, a good deal about their aims, methods, and products concerning classical mechanics:

(1) They manufacture the history of mechanics to suit their own ends (§§ 7, 9).

(2) They feign the contents of works they cite (§ 8).

(3) They disregard the mathematical literature on mechanics, and they contemptuously reject such objections as are made to them by mathematicians who study the foundations of mechanics (§§ 11, 16).

(4) While they claim to maintain "the Bourbaki standard" and to use "substantial mathematical apparatus", they are deathly scared of mathematics (§§ 10, 13).

(5) They manufacture the views of others (§ 12).

(6) The system of axioms they produced in 1953 and have exalted ever since as a paradigm has a central flaw (§ 16) and is insufficient to cover even the problems treated in Book I of NEWTON's *Principia* (§§ 10, 11, 16).

(7) In asserting that what they call CARNAP's program of providing axioms using formalized set theory is unattainable, they deliberately, repeatedly ignore a formal axiomatization of the classical mechanics of punctual systems published many years ago. They are terrified of more than "an afternoon's work" in formal logic (§ 14).

(8) In finding reasons to dismiss all of mechanics except a schoolboy's toe-taste of the "more basic" theory of mass-points, for thirty years they have feared to dive into the depths of NEWTON's *Principia* and have told each other that[41] "each of these sub-divisions [of mechanics is] a separate theory". While they claim to be masters of mechanics, they are deathly scared of it (§ 15).

(9) While they now call for axioms of forces and constitutive relations, they are silent about the axioms for them that mathematicians have formulated, analysed, and repeatedly applied during the last twenty years (§§ 15, 18, 19).

What of all this? MCKINSEY's last two letters to me in 1952 left me astounded at what then seemed arrogance that deserved no reply. STEGMÜLLER's books and other writings of the younger SUPPESIans correct this impression. Blatherskites who tie themselves into such knots in public are not arrogant. I began to wonder if they were all just charlatans. This suspicion was fostered by the SUPPESIans' practice of fortifying their self-confidence by citing and quoting each other constantly and almost exclusively, an example of what SUSSMAN & ZAHLER[95] in demolishing Applied Catastrophe Theory call "the redundant accumulation of supportive statements":

> Write a paper stating an unsupported theory, and this will not cause the theory to be believed. Write a second paper in which you refer to the theory of the first as "well established", and the acceptance level will increase. Let a colleague of yours write a paper referring to your deep work and the level will rise still more. Multiply all this by two hundred, and you obtain something like Catastrophe Theory. By the time the whole thing reaches the average reader, it will be through articles in which the theory is taken for granted. The reader who wishes to pursue the matter further will be referred to more articles in which the same is done. Few will follow the thread all the way. Those few who do will require such an intellectual effort... that when they reach the end and realize that the thread is... tied... only to itself, it will be hard for them to accept the truth, and to acknowledge that their effort has been vain.

ANDRESKI writes in his delightful *Social Sciences as Sorcery*[96],

> one of the most common ploys is a tacit exchange of praise.... [I]n a field infested by charlatans it... commonly occurs as an unprincipled collusion which enables the partners to circumvent the customary taboo on boasting.

---

[95] SUSSMAN & ZAHLER [1978, pages 207–208].

[96] ANDRESKI [1972, page 49]. *Cf.* SNEED's preface and "updated bibliography" in which "I have included all the relevant literature known to me" in the second edition of his book [1971] in 1979.

On the contrary, the exhausting "intellectual effort" I have put out in trying to penetrate the SUPPESian miasma has disabused me of this idea, too, for charlatans claim to do what has not been done before or cannot be done at all; they will not expose themselves to the ridicule deserved by a club of grown men who, earnestly and in voluminous print, drawing only upon each other's works, struggle to learn in their own way what has been widely known for centuries. As BOILEAU[97] put it,

> No fool too foolish but acquires
> A greater fool who him admires.

The more I read, the greater weight I came to lay upon the evidence the SUPPESians' mode of expressing themselves provides. Such pompousness, undisciplined rambling, verbosity, and frenetic vagueness is ordinary in belles-lettres today; nothing more can be expected of common writers. Even schoolchildren talk that way. Modern "English" patterns itself upon government gobbledygook, the pretentious and asinine drivel of mass media, and the spontaneous excrement of untrained mouths and minds; it designs to evade responsibility, to conceal the absence of knowledge, to stir the proletariat in perpetual alternation of social disquiet and social smugness, and to fill the occasional vacancy of the circumambient air. But the SUPPESians are not common writers. They aim[94] to correct our "ambiguity and fuzziness" by making "coherent sense out of what practicing empirical scientists say and do professionally." They will *clarify* the internal structure of scientific theories and provide *semantics* to relate them with the real world. Yet in their writings there is hardly a single straight sentence in natural words put together with unambiguous syntax. The many specimens I have quoted show sufficiently how tedious and tiresome a task it is to elutriate, enodate, and enucleate their foggy writ.

And let us fill in the blanks in their boasting. STEGMÜLLER (pages 6, 16, 20) and MOULINES & SNEED[16] tell us that SUPPES is the equal of ANDRÉ WEIL in erecting "mathematical structures" and that SUPPES knows at least as much of the "concrete, specific scientific concepts and theories" of mechanics as does WALTER NOLL. I do not think MOULINES & SNEED are joking. Only sincere and thoroughly uninformed enthusiasts could be so foolhardy. Few people today would regard as valid science or valid philosophy the augury and haruspication in which the Romans long placed some faith, but it would be

---

[97] BOILEAU [1674, last line of Chant Premier].

wrong to conclude, as some social historians and promoters have done, on no basis but prejudice, that all *augures* and *haruspices* were rogues. Of sorcerers in the Middle Ages ORESME[98] wrote

> ... the first root of the magical art is the lying persuasion of that which is false. The magician himself is fooled by this and sometimes he deludes others.

Even today, many fools are honest fools.

RAYMOND KLIBANSKY, my teacher in philosophy, said to me once in 1937 or 1938, "the tragedy of the ignorant is that they do not know they are ignorant." In thus expressing what was at least for me a new variant of SOCRATES' defense, he made me see that those who know whereof they are ignorant may have fates sadder than the ignorants':

> Understanding is a wellspring of life unto him that hath it: but the instruction of fools is folly.

Proverbs XVI, 8

---

[98] CLAGETT [1968, pages 336–339].

APPENDIX: SNEED'S ARMATURE FOR A THEORY OF
MATHEMATICAL PHYSICS

Fairness requires me to outline here the edifice currently favored
by the SUPPESians.

STEGMÜLLER[99] phrases as follows DUHEM's[85] fundamental proposition, DUHEM's own statement of which has been quoted above in
§ 19: ". . . each theory is indeed confronted with experience . . . but it
is not *particular* statements which can be subjected to empirical test; it
is always, and only, the *theory as a whole*." Nevertheless, in the practice
of science the theory as a whole is never rejected. If a theory is taken
seriously at all, it directs our modes of thought about the phenomena,
and in thoughtfully re-assessing them we use mainly those same modes
of thought to correct the rejected theory[100]. Somehow the theorist
chooses to regard as suspect one or two of the assumptions, and it is to
test these that he replaces them by others, next both contrasting the
differing deductions and confirming the like deductions, and at last
comparing both the differences and the common features with the
phenomena to the extent that they be known.

To render explicit the distinction between the axioms the theorist
decides to leave untouched and those he might regard as suspect and
hence eligible for replacement by others in his attempt to adjust a

---

[99] STEGMÜLLER [1976, § 17.1].

[100] The early ideas about the planets accustomed mathematicians to represent
their motions by circles about a common center, the center of the earth. PTOLEMY's
system kept circles as the main motions but corrected the details by adding epicycles.
COPERNICUS's system dropped the epicycles and put the centers of motions at the
center of the sun. KEPLER's system changed the circles into ellipses of small eccentricity
with the sun at one focus. The NEWTONian system leaves the representation
unchanged in its major kinematical features but corrects it by appeal to a new dynamical conception of how to determine motions in general. While our understanding of all
motions and in particular of the planetary motions is thereby enormously enriched, our
depiction of the planetary system is changed but little. The planets and the sun move in
nearly elliptical orbits about the center of mass of the entire system. General relativity,
while requiring a basically new way to relate motions to forces, leaves the representation of the planetary systems unmodified except for a minute change in the orbit of
Mercury. The proponents of something new (even if small) regarding what was before
generally accepted must advertise the greatness of their proposed change and keep
silent about the unchanged. Their opponents do the same, with the opposite conclusions. Thus the social and political favor shown to revolution for revolution's sake in
societies dominated by Liberalism/Socialism is not the only basis for calling scientific
discoveries, sometimes great but more often minute, "revolutions". The persons involved in those discoveries, whether for or against them, are rewarded for making
them seem much more original and different than a sober appraisal would allow.

theory, philosophers introduced the idea of *theoretical term*. According to STEGMÜLLER[101], "that which one already 'understands' can, *because it is perfectly understood*, be reckoned as part of the basic language; on the other hand, that which is not fully understood must (for the present) be distinguished as theoretical." It is statements in which "theoretical terms" appear that are eligible for suspicion. A term may be theoretical with respect to one theory, nontheoretical with respect to another. For example (page 21), a distance function is theoretical with respect to physical geometry, nontheoretical with respect to mechanics.

STEGMÜLLER[101] attributes to A. PUTNAM the following reproach: "A theoretical term, properly so called, is one which comes from a scientific *theory* (and the almost untouched problem in thirty years of writing about 'theoretical terms' is what is *really* distinctive about such terms)". This statement STEGMÜLLER calls "Putnam's challenge".

SNEED's answer[102] to PUTNAM's challenge is to make the appellation "theoretical" subservient to a particular theory T. "Those quantities are labeled T-theoretical whose values can be calculated only by 'recourse to a successful application of T'." If T is "classical particle mechanics", then "*force* and *mass* are T-theoretical concepts whereas the position function is T-nontheoretical . . . ." In his pamphlet STEGMÜLLER writes (pages 17–18) "In order to perform an empirical test of an empirical claim containing the T-theoretical quantity f, we have to measure values of the function f. But all known measuring procedures (or, if you like, all known theories of measurement of f-values) presuppose the validity of this very theory T." MOULINES & SNEED give an example from "classical particle mechanics": ". . . all the methods . . . for measuring forces . . . require us to assume at least that Newton's second law holds and commonly that some other special force law holds as well. . . . [T]he values we actually assign to forces depend on how we *use* laws involving forces."

STEGMÜLLER regards it as his mission to explain the "epochal work of Sneed". SNEED's book[103], he writes[104], "is so difficult that I fear the number of those who really absorb his ideas cannot be very great."

---

[101] STEGMÜLLER [1976, §§ 1.2–1.3].

[102] STEGMÜLLER [1976, pages 15, 49], following SNEED [1971, pages 33, 34]. More generally *cf.* STEGMÜLLER [1976, §§ 3.1–3.2] [1979, §§ 2–3].

[103] SNEED [1971].

[104] STEGMÜLLER [1976, page x]. *Cf.* also STEGMÜLLER's reference (page 15 of the pamphlet under review) to a passage of SNEED [1976] "which most readers find incomprehensible."

SNEED[105] defines a *theory of mathematical physics* as an ordered pair $\langle H, I \rangle$, in which H is a "core for a theory of mathematical physics" and I is "the characteristic set of all intended applications". A *core* is a list $\langle M_0, N_0, r, M, C \rangle$, in which N_0 is "the set of partial possible models", "the set of all possible applications". Thus $I \subseteq N_0$. To explain the terms further, it is easiest to refer to STEGMÜLLER's only illustration, "classical particle mechanics". For it SNEED's entities are the following:

M "Classical particle mechanics" as defined by the axioms of McKINSEY, SUGAR, & SUPPES, supplemented by the "constraints" C.

M_0 M with "NEWTON's Second Law" excised, though other statements about the "theoretical terms" mass and force remain. (STEGMÜLLER calls this entity M_p.)

N_0 The kinematics of systems of particles. (STEGMÜLLER calls this entity M_{pp}.)

C The "constraints" (see § 19 above).

r The operator that removes the "theoretical terms" from M.

M is "the set of models", "the *fundamental mathematical structure*", or "the *fundamental law of a theory*". M_0 is "the set of possible models".

I cannot see that SNEED's distinctions correspond to anything that theorists of mechanics do or ever have done. Theorists only rarely test the concept of "theoretical term" by appeal to some intended application. It is mainly the "special laws" that they test and often reject. The only example to the contrary that I have ever heard of refers to the advance of the perihelion of Mercury. In that regard physicists preferred to alter slightly classical mechanics as a whole rather than modify the "special law" of gravitation so as to fit the data. Their reasons for doing so could not be explained in terms of mechanics alone. Rather, dissatisfaction with some mismatches of mechanics with electromagnetism caused them to bend the former into conformity with the latter and then *seek out* instances in which the usually minute adjustment so made might have a discernible effect. *Cf.* § 7 of Essay 41, below.

SNEED's armature, like most other ideas of most philosophers of science, is not proposed as something that scientists would ever use. SNEED's objective is to explain to philosophers the nature of a

---

[105] SNEED [1971, Chapter VII: Identity, Equivalence, and Reduction], STEGMÜLLER [1976, Chapter 7: What is a Physical Theory?]. SNEED presents this armature in general terms after his chapter on classical particle mechanics. The realization above is based on STEGMÜLLER's presentation.

scientific theory. Thus whether the explanation corresponds to anything that scientists really do or think, is altogether of no consequence.

List of Works Cited

1674 N. BOILEAU-DESPRÉAUX, *L'Art Poétique*, reprinted in all editions of his works.

1687 I. NEWTON, *Philosophiae naturalis principia mathematica*, London, Streater. 3rd edition (1726) with variant readings, assembled and edited by A. KOYRÉ & I. B. COHEN, Cambridge, Massachusetts, Harvard University Press, 1972.

1736 L. EULER, *Mechanica sive scientia motus analytice exposita*, St. Petersburg, Volume 1 = LEONHARDI EULERI *Opera omnia* (II) 1.

1845 G. B. AIRY, "Tides and Waves", in *Encyclopædia Metropolitana*.

1855 HENRY Lord BROUGHAM & E. J. ROUTH, *Analytical View of Sir Isaac Newton's Principia*, London, Longman *etc.*

1894 A. B. BASSET, "The foundations of dynamics", *Nature* **49**: 529–530.

1901 D. HILBERT, "Mathematische Probleme", *Archiv für Mathematik und Physik* **1**: 44–63, 213–217 = pages 290–329 of Volume **3** of HILBERT's *Gesammelte Abhandlungen*, Berlin, Springer-Verlag, 1935, reprinted 1970.

1906 P. DUHEM, *La Théorie Physique, son Objet et sa Structure*, Paris, Chevalier & Rivière.

1908 G. HAMEL, "Über die Grundlagen der Mechanik", *Mathematische Annalen* **66**: 350–397.

E. JOUGUET, *Lectures de Mécanique*, Volume **1**, Paris, Gauthier–Villars.

1909 E. JOUGUET, Volume **2** of the preceding.

1919 M. V. LAUE, *Die Relativitätstheorie*, Volume **1**, 3rd edition, Braunschweig, Vieweg.

1922 P. PAINLEVÉ, *Les Axiomes de la Mécanique*, Paris, Gauthier–Villars.

1927 G. HAMEL, "Die Axiome der Mechanik", pages 1–42 of Volume **5** of GEIGER & SCHEEL's *Handbuch der Physik*, Berlin, Springer-Verlag.

1934 G. JOOS, *Theoretical Physics*, New York, Stechert.

1937 W. F. OSGOOD, *Mechanics*, New York, Macmillan.

1939 L. P. EISENHART, *Coordinate Geometry*, Boston *etc.*, Ginn.

1943 M. BRELOT, "Sur les principes mathématiques de la mécanique classique", *Annales de l'Université de Grenoble*, Section Sciences-Médicine **19**, 24 pages. (Offprints are repaginated.)

1944 M. BRELOT, "Sur quelques points de mécanique rationnelle", *ibid.*, **20**, 37 pages.

1945 M. BRELOT, *Les principes mathématiques de la mécanique classique*, Grenoble & Paris, Arthaud.

1947 H. SIMON, "Axioms of classical mechanics", *Philosophical Magazine* (7) **38**: 888–905.

1950 R. DUGAS, *Histoire de la Mecánique*, Neuchâtel, Editions du Griffon.

1953 J. C. C. MCKINSEY, A. C. SUGAR, & P. SUPPES, "Axiomatic foundations of classical particle mechanics", *Journal of Rational Mechanics and Analysis* **2**: 253–272.

J. C. C. MCKINSEY & P. SUPPES, "Transformations of systems of classical particle mechanics", *ibid.* **2**: 273–289.

1954 H. SIMON, "The axiomatization of classical mechanics", *Philosophy of Science* **21**: 340–343.

C. TRUESDELL, "Editor's Introduction: Rational fluid mechanics, 1687–1765", pages VII–CXXV of LEONHARDI EULERI *Opera omnia* (II) **12**, Zürich, Orell Füssli Verlag.

1955 J. C. C. MCKINSEY & P. SUPPES, "On the notion of invariance in classical mechanics", *The British Journal for the Philosophy of Science* **5**: 290–302.

W. NOLL, "On the continuity of the fluid and solid states", *Journal of Rational Mechanics and Analysis* **4**: 3–91 = pages 65–81 of *Continuum Mechanics II. The Rational Mechanics of Materials*, edited by C. TRUESDELL, New York *etc.*, Gordon & Breach, 1965.

1956 C. TRUESDELL, "Editor's Introduction: I. The first three sections of Euler's treatise on fluid mechanics (1766). II. The theory of aerial sound, 1687–1788. III. Rational fluid mechanics, 1765–1788", pages VII–CXVII of LEONHARDI EULERI *Opera omnia* (II) **13**, Zürich, Orell Füssli Verlag.

1957 A. DEFANT, "Flutwellen und Gezeiten", pages 846–927 of Volume **XLVIII** of FLÜGGE's *Encyclopedia of Physics*, Berlin *etc.*, Springer-Verlag.

1958 W. NOLL, "A mathematical theory of the mechanical behavior of continuous media", *Archive for Rational Mechanics and Analysis* **2**: 197–226 = pages 1–30 of W. NOLL, *The Foundations of Mechanics and Thermodynamics (Selected Papers)*, New York *etc.*, Springer-Verlag, 1974. Hereinafter this volume is denoted by "NOLL's *Foundations*".

1959 W. NOLL, "The foundations of classical mechanics in the light of recent advances in continuum mechanics", pages 266–281 of *The Axiomatic Method, with Special Reference to Geometry and Physics* (Symposium at Berkeley, 1957), edited by L. HENKIN, P. SUPPES, & A. TARSKI, Amsterdam, North-Holland Publishing Co. = pages 32–47 of NOLL's *Foundations*. [This paper was composed and delivered in public before the manuscript of NOLL's paper published in 1958 was received by the editor.]

H. SIMON, "Definable terms and primitives in axiom systems", pages 433–453 of *The Axiomatic Method . . .*, cited in the preceding reference.

1960 C. TRUESDELL, *The Rational Mechanics of Elastic or Flexible Bodies, 1638–1788*, LEONHARDI EULERI *Opera omnia* (II) **11**₂, Zürich, Orell Füssli Verlag.

C. TRUESDELL & R. TOUPIN, "The classical field theories", pages 226–793 of Volume **III**/1 of FLÜGGE's *Encyclopedia of Physics*, Berlin *etc.*, Springer-Verlag.

1962 A. BRESSAN, "Metodo di assiomatizzazione in senso stretto della meccanica classica. Applicazione di esso ad alcuni problemi di assiomatizzazione non ancora completamente risolti", *Rendiconti del Seminario Matematico della Università di Padova* **32**: 55–212.

R. MONTAGUE, "Deterministic theories", pages 325–370 of *Decisions, Values, and Groups*, 2 volumes, Oxford = Chapter 11 of *Formal Philosophy, Selected Papers of Richard Montague*, edited by R. H. THOMASON, New Haven & London, Yale University Press, 1974.

1963 W. NOLL, "La mécanique classique, basée sur un axiome d'objectivité", pages 47–56 of *La Méthode Axiomatique dans les Mécaniques Classiques et Nouvelles* (Colloque international, Paris, 1959), Paris, Gauthier-Villars = pages 135–144 of NOLL's *Foundations*.

1964 W. NOLL, "Euclidean geometry and Minkowskian chronometry", *American Mathematical Monthly* **71**: 129–144 = pages 183–198 of NOLL's *Foundations*.

C. TRUESDELL, "The modern spirit in applied mathematics", *I.C.S.U. Review of World Science* **6**: 195–206.

1965 C. TRUESDELL & W. NOLL, *The Non-linear Field Theories of Mechanics* = Volume **III**/3 of FLÜGGE's *Encyclopedia of Physics*, Berlin *etc.*, Springer-Verlag.

1966 W. NOLL, "The foundations of mechanics", pages 159–200 of *Non-Linear Continuum Theories* (C.I.M.E. Lectures of 1965), edited by G. GRIOLI & C. TRUESDELL, Rome, Cremonesi.

C. TRUESDELL, "Method and taste in natural philosophy", Lecture 6 of *Six Lectures on Modern Natural Philosophy*, Berlin *etc.*, Springer-Verlag.

1967 W. NOLL, "Space-time structures in classical mechanics", pages 28–34 of *Delaware Seminar in the Foundations of Physics*, edited by M. BUNGE, Berlin *etc.*, Springer-Verlag = pages 204–210 of NOLL's *Foundations*.

1968 M. CLAGETT, editor, *Nicole Oresme and the Medieval Geometry of Qualities and Motions . . .*, Madison *etc.*, University of Wisconsin Press.

C. TRUESDELL, *Essays in the History of Mechanics*, New York *etc.*, Springer-Verlag.

1970 H. SIMON, "The axiomatization of physical theories", *Philosophy of Science* **37**: 16–26.

W. STEGMÜLLER, *Probleme und Resultate der Wissenschaftstheorie und der Analytischen Philosophie*, Volume **II**/1: *Begriffsformen, Wissenschaftssprache, empirische Signifikanz und theoretische Begriffe*, Berlin *etc.*, Springer-Verlag.

1971   J. D. SNEED, *The Logical Structure of Mathematical Physics*, Dordrecht, Reidel, 1971. 2nd edition with "updated bibliography", same publisher, 1979.

1972   S. ANDRESKI, *Social Sciences as Sorcery*, New York, St. Martin's Press.

A. BRESSAN, *A General Interpreted Modal Calculus*, New Haven & London, Yale University Press.

W. NOLL, "A new mathematical theory of simple materials", *Archive for Rational Mechanics and Analysis* **48**: 1–50 = pages 243–292 of NOLL's *Foundations*.

1973   W. NOLL, "Lectures on the foundations of continuum mechanics and thermodynamics", *Archive for Rational Mechanics and Analysis* **52**: 62–92 = pages 294–324 of NOLL's *Foundations*.

C. TRUESDELL, *Introduction à la Mécanique Rationnelle des Milieux Continus*, Paris, Masson.

1974   P. SUPPES, "The axiomatic method in the empirical sciences", pages 465–479 of *Proceedings of the Tarski Symposium*, edited by L. HENKIN *etc.*, Proceedings of Symposia in Pure Mathematics, Volume **25**, Providence, American Mathematical Society.

1975   C. U. MOULINES, "A logical reconstruction of simple equilibrium thermodynamics", *Erkenntnis* **9**: 101–130.

C. TRUESDELL, *Первоначальный курс рациональной механики сплошных сред*, translation by Р. В. Гольдштейн & В. М. Ентов of an augmented and revised text of TRUESDELL [1973], edited by П. А. Жилин & А. И. Лурье, Moscow, Мир.

1976   J. D. SNEED, "Philosophical problems in the empirical science of science: a formal approach", *Erkenntnis* **10**: 115–146.

W. STEGMÜLLER, *The Structure and Dynamics of Theories*, New York *etc.*, Springer-Verlag, 1976.

C. TRUESDELL, "History of classical mechanics", *Die Naturwissenschaften* **63**: 53–62, 119–130.

1977   C. TRUESDELL, *A First Course in Rational Continuum Mechanics*, Part 1, *Fundamental Concepts*, New York *etc.*, Academic Press.

C. TRUESDELL & S. BHARATHA, *Concepts and Logic of Classical Thermodynamics as a Theory of Heat Engines, Rigorously Constructed upon the Foundation Laid by S. Carnot and F. Reech*, New York *etc.*, Springer-Verlag.

1978   M. BUNGE, review of STEGMÜLLER [1976], *Mathematical Reviews* **55**: 333, no. 2480.

H. J. SUSSMAN & R. S. ZAHLER, "Catastrophe theory as applied to the social and biological sciences: a critique", *Synthèse* **37**: 117–216.

C. TRUESDELL, Address upon receipt of a Birkhoff Prize, *The Mathematical Intelligencer* **1**: 99–103, 193, reprinted above as Essay 11 in this volume.

1979   S. S. ANTMAN & J. E. OSBORN, "The principle of virtual work and integral laws of motion", *Archive for Rational Mechanics and Analysis* **69**: 231–262.

C. U. MOULINES, "Theory-nets and the evolution of theories: The example of Newtonian mechanics", *Synthèse* **41**: 417–439.

C. U. MOULINES & J. D. SNEED, "Suppes' philosophy of physics", pages 59–91 of *Patrick Suppes*, edited by R. J. BOGDAN, Dordrecht *etc.*, Reidel.

W. STEGMÜLLER, "A combined approach to the dynamics of theories. How to improve historical interpretations of theory change by applying set theoretical structures", pages 151–186 of *The Structure and Development of Science*, edited by G. RADNITZKY & GUNNAR ANDERSSON, Dordrecht *etc.*, Reidel.

P. SUPPES, "Self-profile", pages 3–56 of *Patrick Suppes*, cited next above.

1980   C. TRUESDELL, "Sketch for a history of constitutive relations", pages 1–27 of Volume **1** of *Proceedings of the 8th International Congress on Rheology*, edited by G. ASTARITA & G. MARRUCCI, New York, Plenum.

1982   W. BALZER & F. MÜHLHOLZER, "Klassische Stoßmechanik", *Zeitschrift für allgemeine Wissenschaftstheorie* **13**: 23–29.

M. PITTERI, "Classical thermodynamics of homogeneous systems based on Carnot's General Axiom", *Archive for Rational Mechanics and Analysis* **80**: 333–385.

PART VI
DIRGE

40. THE SCHOLAR: A SPECIES THREATENED BY PROFESSIONS (1972)

"There are more ways than one of mourning", said Monkey. "Mere bellowing with dry eyes is no good. Nor is it any better just to squeeze out a few tears. What counts is a good hearty howling, with tears as well. That's what is wanted for a real, miserable mourning." "I'll give you a specimen", said Pigsy. He then from somewhere or other produced a piece of paper which he twisted into a paper-spill and thrust up his nostrils. This soon set him snivelling and his eyes running, and when he began to howl he kept up such a din that anyone would have thought he had indeed lost his dearest relative. The effect was so mournful that Tripitaka too soon began to weep bitterly.

(?) WU CH'ÊNG-ÊN, *The Journey to the West**

Progress cannot be reversed; what it has killed, we cannot restore to life. Professionalism, like pollution, is here to stay. Nonetheless, the fact that professionalism and pollution are facts does not force us to welcome and implement them. Indeed, there are those who would accelerate "progress", their effective definition of which is what is going to happen willwe nillwe. I wonder why progressive thinkers do not, since it is inevitable we shall all die one day, advocate present universal suicide.

Preferring to cling to the remains of life rather than renounce it, preferring to strive for light so long as I can see a glimmer, I have specified the qualities ideal in one who is to search and trace the development of scientific concepts and achievements. [See *The Scholar's Workshop and Tools*, Essay 36 in this volume.]

---

* Page 196 of the selections translated by ARTHUR WALEY and published under the title *Monkey*, London, George Allen & Unwin, 1942.

That a historian of science should be both a scientist and a historian, a few decades ago was a truth so obvious as to seem a platitude. With the general devaluation of all truth into a sequence of fads called "values", this particular truth has withered along with the rest. The sciences themselves have changed in what is sometimes described as an "explosion of knowledge". An explosion it has been indeed, but the dinner table is scarcely a fit place to go into detail about what it was that exploded. Mathematicians, who speak seriously only to other mathematicians, are eager to shield their science from the profane— more accurately, their sciences, since there are now so many semipermeable compartments of mathematics that a mathematical lecture which can be followed in detail by more than twenty professional mathematicians is almost as rare as a mathematical lecture which can be understood by physicists or engineers, and to hold a research conference it is necessary to collect from the corners of the earth the handful of persons who are competent in its subject. The mathematicians' disdain for contact with the natural sciences—indeed, it is forbidden by union rules—is matched by the physicists' disdain for logic. A research paper by a physicist is often no more than a chant of beliefs common to his hogan, the members of which rock back and forth in applause of each repetition of tribal lore[1]. Even I. I. RABI[2], an apostle of "progress" to be achieved through more and more

---

[1] As CHARGAFF remarks on page 641 of "Preface to a grammar of biology", *Science* **172** (1971); 637–642,

> Science has been perverted by public opinion to a sort of Hollywood and has begun to adapt itself to this brutal standard. The noise, enormous even before, has increased with the restriction of available funds. . . . That in our days such pygmies throw such giant shadows only shows how late in the day it has become.

[2] I. I. RABI, *Science: The Center of Culture*, New York and Cleveland, 1970, page 92. Alas, the danger for science is far deeper. On this matter CHARGAFF on page 637 of the work cited in Footnote 1 aptly quotes PEACOCK, who over a century ago wrote

> Science is one thing and wisdom is another. Science is an edged tool with which men play like children and cut their own fingers. If you look at the results which science has brought in its train, you will find them to consist almost wholly in elements of mischief. . . . The day would fail, if I should attempt to enumerate the evils which science has inflicted on mankind. I almost think it is the ultimate destiny of science to exterminate the human race.

To this CHARGAFF adds on page 638 that he is

> convinced that the attempts to improve or outsmart nature have almost brought about its disappearance; just as the all too frequent performance of intelligence tests is more likely to make the testers more stupid than the tested more intelligent. That the end sanctifies the means has for more than a hundred years been the credo of the sciences; in actual fact, it is the means that have diabolized the end.

physics, has seen the danger:

> Science itself is badly in need of integration and unification. The tendency is more and more the other way.... [Physics] breaks up in separate specialties.... As the number of physicists increases, each specialty becomes more self-sustaining and self-contained. Such Balkanization carries physics, and, indeed, every science further and further away from natural philosophy, which, intellectually, is the meaning and goal of science.

Just as the mathematicians spit upon any reference to nature because it is "not mathematics", the physicists are ready with the words "not physics" or "not fundamental" or "only an application" or "merely mathematics" for any research on physical problems that does not follow their party line. Physics is not presented to the student as a gallery of beautiful and enlightening pictures of nature but as a slender, tortuous, and muddy path to blindered research in a few already suffocating popular specialities whose peculiar support from the populace rests only on a suspicion that despite the physicists' protests of innocent love of pure truth for truth's sake, they may become "useful" in developing weapons to destroy the human race before it succeeds in committing suicide spontaneously. The beginner in mathematics is not shown an organized workshop of powerful tools of the mind with manifold application to every aspect of life and understanding; rather, he is indoctrinated in the cult of purity and led as quickly as possible into the particular plotlets of research hoed and watered by the local department. History, too, has progressed; once a chronicle of kings and battles illustrating every human virtue or vice, then a sequence of manifestos about personless peoples and social movements and conflicts of commerce, now it has transformed itself into a culture of self-fecundating research cells, so that the most important thing to teach the freshman is the correct line of historiography, history itself being incidental if not altogether dispensable. Better a blank savage than a heretic, better a heathen than an apostate! The world of "learning" has become a federation of hives of frightened bees, who so as to maintain their little waxen prisms sting all strangers, be they lions or mice or only bees from the next-door office. It is now an administrator's world. Unlike other captains of finance, the chiefs of academe seek only to mendicate and waste money, not to make it.

Once the history of our culture was our common heritage, our pride and our lesson book for conduct both private and public. SAVILE in his preface to his translation of TACITUS in 1591 wrote that "there is no learning so proper for the direction of the life of man as History". 1591 was a long time ago, but old words are not necessarily wrong words: Stones fell in 1591 just as now they fall, and GALILEO's rules have not lost their truth. Scientific research was then a vocation pursued by few, the fruits of which were gathered and enjoyed by many. The history of science, in the words of LEIBNIZ, served "not only to give each his own and to incite others to seek like glories, but also to prosper the art of invention by disclosing method through illustrious examples." For me, as long as I have tried to do research, the one and only school of method has been study, study, and study of the masters.

Today these simple truths are as obsolete currency as gold coin. Unlike such coin, I doubt that they will lend themselves to speculative profit. The beginner in history of science must be taught first of all what will make him, if he completes apprenticeship, different from and independent of historians and scientists alike. Mathematics cannot be defined now except as that which mathematicians do; for physics, we substitute the word "physicists", and soon the history of science will be defined as that which historians of science do and will likewise live a PARKINSONian life, independent equally of science and of history. Just as books on political history are written now to be read by political historians alone, and works on mathematics to be read by none but professional mathematicians, soon we can expect that books on the history of science will be meaningless except to historians-of-science, dumb to scientists and to historians, serving only to produce more and more historians-of-science who are paid, if they can get jobs, to do nothing but indoctrinate more historians-of-science. A meeting in which historians-of-science and historians-of-technology come together will soon be impossible, for the two professions will have created separate jargons and shibboleths such as to defy communication except through specially trained "interdisciplinarians" or, finally, reciprocal computer codes adjusted to the prejudices and wallets of the purchasers.

How has all this come about? Through the professionalization of research. Professionalism is now an end in itself, so that "not professional" is becoming a general insult. To hurl it, you need not even name the union to which the object of the insult has failed to pay his initiation fee.

In an age when everyone is regarded as belonging to some profession or other, we may have forgotten the purposes which alone may justify the existence of such a thing. A profession is organized to protect a consuming public from imposture and incompetence, and to protect its professors from losing their custom to unqualified or unscrupulous outsiders. The goldsmiths' guild assured the buyer of the purity of the metal and a minimum standard of craftmanship, while keeping prices secure from competition by those who would defraud gulls with base metal and botched work, or who would gain a master's profit unprepaid by the tedium and penury of long apprenticeship and day labor. In its abuse it maintained a monopoly of mediocre routine, discouraged efficiency, and stymied talent. The physicians' union requires that in order to practise, even a charlatan must have a licence, and it assures a member in good standing that if he makes a mistake, he may bury it. In its abuse it makes costly mysteries of much that should be cheap and open, and it nourishes an army of parasitic labor and a gouging industry. Such are the professions of purveyors to the public. If they offer us little to admire, we must admit nevertheless that they, like government, while evils indeed, are necessary evils, not unadulterated by some value. One thing is certain: Among goldsmiths and physicians we are not to expect to encounter great discoverers and deep thinkers. To heal our loved ones, we do not seek a physician who indulges in speculative research, and accordingly the training of a medical man seems designed to stamp out from the very first day such temptation toward original thought as the aspiring Æsculapean might have.

If we seek to justify the professions of science on the same basis, we find altogether lacking the first requisite, namely, a public demand for their services. Who among us, in adversity or health, seeks the aid of a nuclear physicist or an algebraic geometer? In an age which has reinstated astrology as decent practice, who needs to be protected against deceit from a man who claims to know homology theory but in fact confuses his Betti groups, or from a pretender to expertise in quantum mechanics who in fact chooses the wrong pseudo-Hermitian operator and even miscalculates his eigenvalues?

Although science lacks altogether the justification of professions, namely, public need, professions of science have been formed that they may usufruct the abuses which are condoned as ineradicable from the true professions, those whose services, bad or good, are publicly indispensable. Research, like prolongation of the lives of the senile at home or of the infants in centers of overpopulation abroad, has become an end in itself, with no thought of what object research might serve. By social command turning every science teacher into a science-making machine, we forget the reason why research ought to be done in the first place. Research is not, in itself, a state of beatitude; research aims to discover something worth knowing. With admirable liberalism, the social university declares that every question any employee might ask is by definition a fit object of academic research; stoutly defending its members against attacks from the outside, which with good reason grows every day more hostile, it frees them from all intellectual discipline; it brings the outside inside by abolishing the distinction between academic learning and any other activity that

makes no money. A study of love-making in Volkswagens is not only
a perfectly good field of academic research; it is even fed from Uncle
Sam's official trough of ambrosia for "social science". Already the
Metropolitan Museum of Art houses a collection of chewing-gum
wrappers, with a special curator whose duty it is to catalogue them
scrupulously for the benefit of scholars—presumably, other specialists
in chewing-gum wrappers[3]. While once the title of "doctor" meant
"teacher", now the "earned" doctorate has become a formal statement
of what the candidate need not know, and its award, like freedom of
a guild, makes him formally free to stop learning, while the "honorary"
doctorate is most often a certificate that learning has never begun.
Although every cow college is now become a great university, staffed
solely by "leaders", we look in vain for those who follow them. Foundations and governments strain their purses to produce billions of pages
of "research" papers, papers most of which are never read once by
anyone, not even, if their slipshod preparation be taken in evidence,
by their own authors.

Such is the outcome of professionalism of fields which lack the just
motives of professions. The work has expanded to fill the posts allot-
ted. Although in politics, warfare, and administration, such is the
ordinary law, forever honored more than any other, in those fields,
which are inherently dissipative if not outright noxious to man, there
is nothing positive to annul.

It is a different matter in scholarly work. There the main function
of the professions now is to redefine their disciplines downward to
suit a complacent and numerous mediocrity. They vilify as "intellectual snobbery" and "ivory towers" the withdrawal and the spiritual
independence which make a life of scholarship possible. They
suffocate scholars in a crowd of overweening clerks and greasy
brokers who, while giving each other the titles of learning, like har-
pies beshit the banquet of scholarship; driving away the scholars, they
join their lackeys in the places of honor and pronounce the befouled
refuse haute cuisine. Dispute, which in the past served as a major
avenue, if often a painful one, to reach the truth, is replaced by secret
infights among grinning rival factions seeking only to ride herd on

---

[3] In my little essay, "Method and taste in natural philosophy" (Lecture 6 of *Six
Lectures on Modern Natural Philosophy*, New York *etc.*, Springer-Verlag, 1966), thinking
to reach the utmost burlesque of modern "culture" and "scholarship", I wrote "An
exhaustive, scrupulously catalogued collection of chewing-gum wrappers will soon
seem at home in the National Gallery of Art." A correspondent then sent me a description of The Burdick Collection of Trade and Souvenir Cards and other Paper
Americana, which had been donated three years before to The Metropolitan Museum
of Art, New York. As in SWIFT's *Voyage to Laputa*, the truth is more bizarre than any
fantasy and serves as its own parody.

their colleagues. Originality is dangerous; it may be heresy; before jumping on the bandwagon, the common guildsman should wait the nod of the hogen-mogens to be sure it joins the right parade.

I write these words, not as one who would exalt one profession above another, but as an outsider, no part of any of them, a person who is best described by the term "idiot"[4]. As such, I have felt free to assume the voice of Ecclesiastes so far, but now it is time to finish with an expression of the optimism required by good citizenship in the democratic state. We must buckle our seat belts.

The history of science is different in kind from science itself and from ordinary history. The material of the history of science is compact: Being history, it necessarily concerns the past, and because in the past science was a tiny and select vocation, not the factory job it is today, there is little to be read. What little there is, includes the highest intellectual achievement of our culture as well as a part of its finest artistic creation. It calls for deep study, not the shallow flood of research that begets research that begets more and more research to burst libraries and out-digit computers. The example set today by the professions of scientists and historians is the worst that historians of science could choose to follow. Indeed, the history of science needs to be cleared and established. Thereafter, it ought to be learned. Although only a handful of persons could ever acquire the eccentric conjunction of skills and knowledge necessary in him who would do sound research in the history of science, there are many who can and should learn the results of that research. History of science should be studied and learned by every scientist, every historian, every person who seeks any intellectual footing in the Western culture. The great need of history of science today is for teachers.

Recently VICTOR F. WEISSKOPF[5] has described as a "destructive element" within the community of science.

> "the low esteem in which clear and understandable presentation is held. ... In music, the interpretive artist is highly esteemed. An effective rendering of a Beethoven sonata is considered as a greater intellectual feat than the composing of a minor piece. We can learn something here: Perhaps a lucid and impressive presentation of some aspect of modern science is worth more than a piece of so-called "original" research of the type found in many Ph.D. theses . . . .

---

[4] ἰδιώτης, a private person, a man holding no office, a layman, an ordinary fellow.

[5] VICTOR F. WEISSKOPF, "The significance of science", *Science* **176** (1972): 138–146; see page 145.

The word "interpretive", though etymologically just, too easily seems to bedwarf what is true giant's work. In their recreation of baroque music, especially the peerless masterpieces of BACH, from the 1950s onward NICHOLAS HARNONCOURT and GUSTAV LEONHARDT and MARTIN SKOWRONECK and their associates and students had first to discover heavenly sounds unheard and styles lost for two centuries, then to learn to execute them upon instruments long mute and to make new ones capable of the old songs, and finally to teach others to do so. In this, their work of supreme scholarship, a perfect collaboration of soul, mind, and hand, SKOWRONECK, LEONHARDT, and HARNONCOURT are true "amateurs", men of talent and education far above the tawdry professional standard of our time, men who work alone, affable and unfettered. They put to shame both the dazed and drugged merchants of "original" cacophany in symphony halls and their counterblanks, the professed scholars of musicology, to whom audible music is as strange and irrelevant as is live science to the historians. Rather than heed the professionals, who would embalm old works in coffins of which only themselves have the keys, I would imitate those musicians, true priests of the muses, who have rescued and made sing again poor forlorn vocal frames lying cracked and crumbling in the shafts of the necropolises called "museums", those grim seats of the Ministry of Love of Art.

ERWIN CHARGAFF[6] writes,

> We posit intelligence where we deny it. We humanize things, but we reify man. I am afraid our sciences have not escaped the process of alienation, of dehumanization, so characteristic of our time. The attempt to describe life in its generalized contours leads to an automatization before which everything—the leap of the cat or the Goldberg variations—appear[s] equally incomprehensible.

WEISSKOPF[7] contends that

> The teaching of science must return to the emphasis on the unity and universality of science, and should become broader than the mere attempt to produce expert craftsmen in a specialized trade. Surely, we must train competent experts, but we also must bring fields together and show the connections between different fields of science.

Even RABI demands that the sciences be "taught more humanistically". If this is true in the sciences themselves, which are by their

---

[6] CHARGAFF, page 642 of the work cited in Footnote 1.
[7] WEISSKOPF, pages 144–145 of the work cited in Footnote 5.

nature accumulative, how much more true is it in the history of science, which is by nature closed! History of science should be taught to laymen, not merely imparted as a trade skill of fledgling historians-of-science. If the results of research are both important in themselves and decently presented—two conditions which usually fail to be met—they can be learnt and taught by any qualified scientist and even by some ordinary historians. Here MORRIS KLINE, himself the author of one of the few general histories of mathematics that is much more than a collection of elementary odds and ends in a capricious selection of special topics, has set the example for years by the doc-torands in mathematical education he has guided. These men have written theses which follow the development of some one branch of mathematics factually for half a century or more; in part research, in part guides to the classic literature, full of concrete mathematics and free of isms and philosophic or sociological or historiographic slants, they are so lucidly written that any college teacher of mathematics can read them with profit and pleasure, for himself and for his students forever after.

Teachers, even good teachers, are easily integrated into a social democracy. Teaching plainly satisfies the two requisites of a pro-fession: public demand and the need for protection. Professionalism is easy for teachers, and they comfortably accept regulation, that inevitable concomitant of professions. Their unions call themselves just that and secure exclusive rights to collective bargaining.

If the history of science should be primarily the concern of teachers, necessarily unionized in law as well as spirit, what will hap-pen to the few scholars, the very few, who can do old-fashioned research? If professionalism of the fields labelled "learned" continues to grow at its present pace, the great scholar, the man who while standing upon the same ground as his fellows is taller and sees above their heads, may soon find himself proscribed. The first stage of proscription may simply forbid him entrance to the university. Indeed, in the university once upon a time scholarship and teaching naturally went hand-in-hand, each assisting the other. The univer-sity's function was even more to select than to teach. What counted was not that the university usually failed to make the student into a scholar, but that it once in a while succeeded. But today, what can the scholar be but an exile in that alternating play school, rabbit warren, and psychiatric clinic?

In our age of frantic affluence, when the assembly lines of the socialized military-industrial complex rain down synthetic manna alike upon the idle and the industrious, the dolt and the genius, reduction of scholarship to a hobby will not extinguish it. A janitor or a salesclerk with unfettered and unrewarded leisure may be in better case to do scholarly research than is a university professor, whose time is occupied in impressing his colleagues and rivals, in swelling himself by swelling his department with admiring and docile boobies, in recruiting as apprentices some of the undergraduate girls from philosophy or art history or ballet or pre-nursing, in squabbling on committees whose task is to fashion a curriculum that appeals to the "inner city", and in scuffling to a deanship. Science, above all, would benefit from less "support" by society; perhaps even experimenters might rise from their prostration before mountains of costly plumbing, of paper defaced by billions of computed ciphers; they might even start to try to think.

However, aberrations from the normal, which are notorious in any society, in a social democracy become evidence of treachery. They betoken paranoia. While we are not yet compelled by law to have in every room a screen showing Big Brother, a man who does not watch an infinity of televised, indistinguishable little brothers at least two hours a day is not quite right. He is out of touch with the times.

Nearly seventy years ago WILLIAM JAMES[8] remarked that soon "bare personality will be a mark of outcast estate . . .." The next step for the scholar is sequestration. In the end, this may not be a bad idea. A madhouse inhabited by scholars would not be much different from what a university once was.

Note for the Reprinting

The text above is essentially the second half of the paper of the same title published in *Critical Inquiry* **2** (1976): 631–648 and reprinted with a few corrections in *Speculations in Science and Technology* **3** (1980): 517–532. The first half of that paper largely repeats material from *The Scholar's Workshop and Tools*. Essay 36 in this volume resulted from combining the two versions just mentioned to form a final and, I hope, better text.

---

[8] WILLIAM JAMES, "The Ph.D. octopus", *Harvard Monthly*, 1903; reprinted in his *Memoirs and Studies*, New York and London, 1924. Then, he wrote (pages 334, 344, 346), we were

> rapidly drifting towards a state of things in which no man of science or letters will be accounted respectable unless some kind of badge or diploma is stamped upon him . . ..
>
> Surely native distinction needs no official stamp, and should disdain to ask for one . . ..
>
> It is indeed odd to see this love of titles—and such titles—growing up in a country of which the recognition of individuality and bare manhood have so long been supposed to be the very soul.

The paper in *Critical Inquiry* was based on an address delivered at the banquet of the History of Science Society and the Society for the History of Technology, Washington, D.C., 29 December 1972. After a dinner that RETI was to remember to the end of his life as the worst he ever ate, my lecture was designed "to replace the missing port and cognac". Fortunately the din of airplanes crossing just above us every fifteen seconds on their descents to the National Airport made my words inaudible. At the time I thought the hall would have been ideal for the debates of the U.S. Congress.

So as to retain definiteness and immediacy, I have not blurred the original focus upon the history of science and technology, trusting that any reader who can understand me at all will be able to turn the same lens upon his own field of learning or pseudolearning.

41. THE COMPUTER: RUIN OF SCIENCE AND THREAT TO MANKIND (1980/1982)

PREFATORY ADMONITION

This essay is designed to be read by an intelligent layman: one who is expert neither in computing nor in mathematics but is competent in some other science such as chemistry or one of the biologies.

While mathematicians and numerical analysts will find the explanations in this essay too obvious to mention, experience over many decades has taught me that much of what is second nature to persons with some training in mathematics is utterly unknown and indeed scarcely believable by those whose schooling did not include serious introduction to the mathematics of infinite processes.

Victrix causa deis placuit, sed victa Catoni.

LUCAN

Unangenehme Seher werden meistens als Narren abgeschrieben.

CHARGAFF

1. Spatial Flight would have been Impossible without Computers

We have seen men shot into parts of space where they became weightless; directed from the ground of this earth, they have themselves moved like planets, have then been ejected from their own newly acquired orbits and given moonweight instead of their former earthweight; after falling upon the moon they have risen from it, reversed their earlier voyage, turned earthward, and finally dropped back upon the old ground of human life and death, the old sea of once boundless wealth and now boundless trouble. Long before the voyage, each step of it was not only imagined but planned in minute detail. Each change of motion was cannily contrived, each span of coasting craftily released and craftily commuted, each error human or mechanical nicely corrected, and thereupon a new course and program calculated and effected. Asking ourselves what has made possible this astonishing effort and astonishing achievement, what resources could be drawn upon that no previous culture had and that even so little as twenty years earlier ours lacked, we easily recognize many. Some of these are political, some social, some financial, some industrial, some technical. All were needed to provide the one most obviously essential tool. That was numerical calculation: swift, copious, directed, accurate. Without great computing machines the entire program of interplanetary flight as well as many activities less flamboyant but equally peculiar to the turn human effort and organization now take had failed, had never even begun.

2. Spatial Flight would have been Impossible without the Classic Equations of Motion

Calculation must not be taken for mathematics. Calculation is a thoughtless routine, like tightening two counter-rotating nuts on an assembly line. Just as a human being had to design both the belt that presents the two nuts and the product that requires them, a human being had to tell the computer, or properly, the computer's team of owners and bosses and hands, which equations to solve. Without those equations, it would have been impossible to conceive what "coefficients" should be determined by programs of experiment, what initial conditions should be ascertained by measurement or assigned by man's will. Without those equations, how could any computer—a brute which can only take common numbers in the order it is told to follow, then add them, then retain or discard the result or enter it upon a selected blank—how could any pile of integrated circuit chips flash out orders continuously directing and adjusting man's voyage to the moon? How could a million digits be put to any use whatever, had the use not commanded the accumulation of those digits in the first place and so provided the key to interpreting them?

I speak of transplanetation not from any love or hate of it but because it is familiar to everyone and easy to dissect in principle. Many factors are indispensable to it, some of them numerical; one is inherently beyond numerics; the mechanics of the motion of the capsule, for without that all the military, political, social, economic, geological, medicinal, chemical, biological, astrophysical (and perhaps astrological) aspects would have failed to exist for want of an object to which to attach themselves. The mechanics of the motion provided the conceptual bones of the undertaking and the central equations which the computers were ordered to solve again and again, thousands or perhaps even millions of times as different initial conditions and different empirical parameters were called for. This is the aspect the mass media never mention. The intellectual proletariat— the masses who believe as divine revelation what "doctors say" and "scientists say" in today's press and forget what those modern priests and astrologers said yesterday—the intellectual proletariat that happily sees billions of dollars it itself paid in taxes spent to stage the greatest television show of all time (until the next nuclear bomb is dropped or biological warfare begins), and which lends to the jingoists of science a credulity beyond a mediæval peasant's before his parish priest, has no idea that this aspect exists.

These central equations are more than 200 years old. They were obtained by men, great mathematicians, who pondered the results of astronomical observations and who put their naked, disciplined minds to the discovery of the simple in the apparent complexity of numbers and numbers upon numbers, numbers seemingly almost casual—ugly numbers.

The differential equations governing the motions of punctual masses and rigid bodies were not enough to determine the conditions of flight through space. Also the physics of gases and heat and radiation, meteorology, the chemistry of fuels, and for manned flight biological and medical experience were needed. It was truly a triumph of science applied. In it was no element of scientific discovery. Without classic science it would have been impossible.

3. CALCULATION WITHOUT CLASSIC STANDARDS IS DANGEROUS. A COMPUTER IS INCAPABLE OF SETTING ITS OWN STANDARDS

In science and engineering today some of NEWTON's and EULER's ideas and discoveries make the very ground we stand upon. Wherever classical physics is applied, when computation goes wrong we can often recognize it; we can say "this outcome is nonsense, because it violates geometry", or "impossible: that contradicts NEWTON's Laws". Frequently in discussing computer codes we hear the statement, "The results do not respect conservation of energy, but that is a computing error, which we can correct by taking a finer mesh if we need greater accuracy." The programmer or his colleague responsible for the physics of the problem has a standard against which he can check, to which he can have recourse in need. This standard is the classics: the conceptual clarity, the logical analysis effected by great men of the past, great thinkers. Today, indeed, there are a few mathematicians—unfortunately very few—who can make sane and sound use of numerical computation an adjunct in the application of mathematics.

In fully modern fields like genetic alteration and the physics of particles with great energies there is no such standard. Research there is based upon semi-empirical guesses and, far more dangerous, uncontrolled numerical calculation. When something goes wrong in the computing, there is no classic foundation to which the student may return. The same mathematical problem usually gets different answers if "solved" on different machines. Hence arises the ugly noun-pair "machine-independence" to denote instances in which disagreement does not follow. Ordinarily only one machine is used, and its spew is accepted forthwith, without giving another machine a chance to fight.

But that is not all. Different codes used to solve the same problem on the same machine often give different answers. If the disagreement is serious, a "benchmark code" can be applied to see which of the combatants comes closer to a true solution known in some simple case because some mathematician has obtained it already without appeal to any machine. It is another instance where the classics are called in to help, but benchmark problems are simple. The fact that a code is accurate for a known and simple instance then breeds confidence, merely emotional, that it will be accurate also for hitherto unsolved problems. The risk of such inference is plain and great. The true solution of the unsolved problem need not be even roughly like the solution of the simple, classical instance. The unsolved problem may well be unsolved just because its solution is in essence different, far more complicated and far more delicate. A field in which such a difference is already familiar—because experiment and human reasoning made it familiar before there were any automatic computers—is fluid mechanics. There the methods used to study flows at low Reynolds number fail altogether for flows at high Reynolds number. Very well. No computer in competent hands will mislead us here. But had the matter first arisen in our day, the day of great computers, all benchmark problems would have referred to flows at low Reynolds number because no true solution was known for any turbulent flow, and no degree of accuracy in application to known solutions would have had any bearing at all on the rightness or wrongness of the code when applied to the far more difficult problems at high Reynolds number. Exploration by computer, no matter how abundant, would have been little likely to discover the fundamental difference. Computers may be used in intelligent application of theories already well understood, but to place confidence in them for exploring unknown domains of science is as dangerous as to suppose that if a small dose of medicine will cure the measles, a large dose of it will cure smallpox.

Then there are the questions of "verification" and "qualification". How can we tell whether the code is really correct for the problem it is set up to solve? Rarely does it so much as touch that problem directly. Instead, the original problem is replaced by a model problem, a problem which can be set and solved in terms of sequences of two marks, say 0 and 1, for the computer can handle nothing else. To see whether the code really can lead to the solution of the model problem—to detect the errors made by the programmer—is something before which the computer stands useless. A human expert on computing, or a team of such experts, is given the task. That is "verification". Next comes "qualification": Does the model problem, the problem cut down to the capacity of the machine, truly represent the problem of natural science it is designed to approximate? A theory is itself a model of nature; the computer is given a model of that model. It is not impossible that the computer can lead us through these detours and pitfalls to natural science, but how can we know whether it does or does not? The very fact that we have appealed to the computer implies that we are floating with no secure bases in human thought: human science and human mathematics.

Even when the original mathematical question is fully and correctly set forth, calculation can introduce disastrous errors. The computer's code replaces the assigned problem by a finite algebraic system, and calculation through such systems tends to smoothe over occurrences that may be catastrophic in the mathematically correct theory. For example, in a situation in hydrodynamics that leads to a shock wave or turbulent flow a computer may deliver smooth answers. I have heard long, inconclusive arguments among experts as to whether a particular, regular, gradual "solution" by computer represented the facts or was merely a result of "numerical diffusion" and hence worthless. Also errors of the opposite kind have occurred. In a notorious calculation directed by FERMI, PASTA, & ULAM, of which I shall say more in § 8, below, the computer was programmed to solve a problem of a type classical in gas dynamics and long known to lead to a catastrophe after a finite time. The proposers, unaware of that, had fixed and very different expectations. CAPRIZ & ONESTO[1] state that

> ... in the first series of [computer] experiments some sort of explosion was observed but was dismissed as being due to instability numerical rather than real—to phenomena in the numerical model rather than in the physical model.

The later computations by the famous directors failed to lead either to the correct "explosion" or—fortunately!—to the conclusions that they desired to get, which in fact were incorrect.

To the double and opposite dangers just specified must be added another listed by BOGGS in a recent, sober, and critical survey[2]:

> Many people erroneously believe that, simply because the computer uses fifteen significant digits, their answers will be accurate to fifteen digits. However, the speed with which some computations can be rendered useless by the cumulative effect of small errors is quite amazing. . . .

---

[1] G. CAPRIZ & N. ONESTO, "L'elaborazione elettronica nelle scienze esatte", pages 83–94 of *Atti della LIII Riunione della SIPS*, Pisa, 1975.

[2] P. T. BOGGS, "Mathematical software: How to sell mathematics", pages 221–229 of *Mathematics Tomorrow*, edited by L. A. STEEN, New York *etc.*, Springer-Verlag, 1981. See pages 224–226 and 228.

[It is a] basic requirement that the computation should continue only as long as meaningful results can be obtained. This implies that estimates must be derived which indicate when the boundaries of the problem domain are exceeded. . . . It is safe to say that, as of today, there are very few problem classes for which such estimates have been derived and implemented in widely available routines. . . .

On the dark side there is the danger, as with all powerful tools, of misuse. Selection of the wrong routine can result in erroneous answers of much poorer performance than necessary. A routine can be applied outside of its range in a way which is impossible to detect. For example, much of classical statistical analysis is based on the assumption of an underlying normal distribution. If this assumption is not warranted, the numbers produced will be meaningless. Unfortunately, there are still those who believe that if numbers are produced by the computer, they must be right.

A critical problem is that managers are sometimes deceived into thinking that with a computer and a good library of programs, they no longer need a mathematician or statistician on their staff. This is a case of over-selling, or failing to provide an honest assessment of the limitations of the product.

With such tools readily available there is a tendency among some to rush to the computer without doing any preliminary analysis or critical thinking about their problem. In some cases, such a practice is institutionalized in the sense that certain computer analyses are required even though the results provide little or no (or even misleading) information. People in this situation often prefer poor codes which always return "answers" to those which warn when problems are present. Of course this is not the fault of mathematical software, but the mere existence of such tools encourages this type of mindless activity.

It is safe to say that for every instance of the sound mathematical guidance that BOGGS recommends in use of computers, in practice there are thousands of the "mindless activity" he deplores.

Failed guesses are usually indistinguishable from faulty numerics. Mountains of digits are become the result of science, not merely its planned and checked application. The answer to failed guesses and failed numerics is more of both. Get a bigger team, spend more money!

The sale of dynamite, cyanide, machine guns, explosive rockets, nerve gas, and thousands of other dangerous things is controlled in an attempt to save mankind from the ubiquitous destruction they might effect if put in the hands of the untrained or unskilled or unscrupulous, not to mention thieves, gangsters, and terrorists. Our lives and fortunes already lie at the mercy of what is promoted as "science" based on calculation by computer. Anybody can buy a computer and use it as he pleases; the merchants of computers urge everyone to do just that. Reactor safety is noisily familiar to everyone. "Computer safety" is a term unheard. Reactors are dangerous in an accident. Computers can be dangerous when they function perfectly.

4. COMPUTERS HAVE HARMED SCIENCE ALREADY

H. R. POST in a brilliant lecture[2a] listed four ways in which the computer has harmed science:

(1) The computer has probably maintained as problems for computation problems which could have been solved meanwhile by mathematics.

(2) "The Computer is certainly the perfect instrument for inevitable research", that is, research which is certain to deliver *some* "results", right or wrong or meaningless, on any problem proposed.

(3) The computer, writes POST, provides

> another example of the modern mania for means instead of ends. Any amount of traffic, but the place of starting and the place of arrival are equally awful because of the traffic. The longest queue of students I have seen was ... to register for a computer programming course. I would have been delighted if any of these students had a real problem in the first place that might warrant the use of the computer or anything else. ... A ... head of a post-graduate college explained in an interview that he had been looking for a central element unifying the many disciplines represented in his college ranging from Literature to Physics. He had found the Computer to constitute the Central Unifying Element.

(4) Use of the computer, says POST,

> substitutes specific knowledge for understanding. You understand a subject when you have grasped its structure, not when you are merely informed of specific numerical results.

[2a] H. R. POST, *Against Ideologies* (Inaugural lecture), Chelsea College, London, 1974.

POST's third count applies an observation of HEIDEGGER[3]:

> In the sciences the subject is not only set by the method; at the same time it is set into the method and remains subordinate to the method. The raging race that sweeps along the sciences today—they themselves have no idea whereto—comes from the increased drive, ever more surrendered to technique, of the method and its possibilities. In the method lies all the power of knowledge. The subject belongs to the method.

HEIDEGGER goes on to say that the way of thought differs from the way of science, but perhaps he has not learnt much of the ways of great mathematicians. True science seeks methods to solve important problems. Once a method has been invented, we may indeed turn it to new applications. In contrast, the addicts of computation, which is only a method, tout it as a panacea, an imperial pill that will cure all diseases, and so no diagnosis is needed before taking it. Any problem will do. No dosage is mentioned. Evidently the bigger the better. Computing will cure what ails you.

Two of POST's accusations, the first and the fourth, need to be enlarged upon. A layman may well ask, "Why would a mathematician's solution be worth waiting for if a computer can solve the problem faster?" Even persons well educated in literature and arts commonly confuse mathematics and numerical calculation. A dear friend of mine is still sure that when I retire to my desk I sit down with delight to add a huge column of figures; we often read in the press that one computer does the work of a thousand mathematicians. Nothing could be more wrong. My wife long ago persuaded me never to attempt the additions and subtractions of entries in the family's checkbook; the thousands of persons whose services the computer replaces, in fact whose potential ultimate capacity the computer renders negligible, are computresses and abacists, much as a gigantic power scoop replaces thousands of coolies with shovels and buckets. Obviously such work is needed in medical analysis and surgery, financial accounting, traffic control, spying by tax officers, totting up bills in stores tended by clerks who after twelve years of democratic schooling still cannot add and never will learn to, some aspects of engineering design, weather prediction, polling illiterate voters, and thousands of other activities essential to the modern state. Herein lies the computer's value to society. Like many other of today's dangerous gadg-

---

[3] M. HEIDEGGER, *Unterwegs zur Sprache*, Pfullingen, Neske, 1959. See page 178. The translation loses some of the subtleties of the German language: "...gestellt... hereingestellt... untergestellt...".

ets, it does work nobody wishes to do or could do even if he wished to. In accord with PARKINSON's Law, work of this kind grows daily more and more colossal, more tremendous, and more indispensable to society. Thus we need more and more computers just to get through tonight and survive until the morrow.

As CHARGAFF writes in his essay *Little ado about much*[4],

> It is not hard to talk people into believing they cannot possibly live without something that a few years earlier they had not the slightest idea existed. In fact the West's entire economy rests on this principle.... Some day this diabolical circle must be broken, or mankind will perish.

5. MATHEMATICS IS THE SCIENCE OF INFINITIES. COMPUTATION IS ESSENTIALLY FINITE

POST's words distinguish a "problem for computation" from a problem "solved . . . by mathematics". Consider a very old example: to find a number x such that $x^2 = 2$. This problem was solved long ago to everyone's satisfaction, though not everybody was satisfied by the same solution. For practical purposes you draw a right triangle with short sides of length 1, then take a ruler and measure the length of the long leg. If you prefer arithmetic, just try numbers. You quickly see that $1 < x < 2$. Next, $1.4 < x < 1.5$. You can go on and on systematically. Multiply out $(1.41)^2$, $(1.42)^2$, . . . $(1.49)^2$; here the first two multiplications suffice to show that $1.41 < x < 1.42$; for each further decimal place at most nine multiplications are needed, and only rarely that many. Soon the multiplications get too big for you, but a computer will come to your rescue! The process of systematic, exhaustive, and exhausting trial illustrates what POST calls "a problem for computation".

A computer could use the very process I have just described, but its bosses would know much faster methods, methods which have been invented over the centuries by mathematicians. These methods involve rapidly convergent infinite series, infinite products, infinitely continued fractions, *etc.* The computer cannot think. Therefore it could neither invent such methods nor demonstrate that they are valid, but mathematicians have proved by strict logic that they are correct and have provided rigorous estimates of error at each stage. A single arithmetic step calculated by one of these methods can supply many correct decimal places. Here is an example of what mathemati-

---

[4] E. CHARGAFF, "Wenig Lärm um Viel", *Scheidewege* **8** (1978): 289–309. See § III.

cal thinking can do: It can guide and check numerics in ways no machine could invent. The mathematical work to which I refer was done long before there were any machines. Using these theorems of mathematics, a programmer can enable his machine to calculate the number $\sqrt{2}$ to hundreds of decimal places with great accuracy and speed.

But how many? Any computer can do before it breaks down only a finite number of calculations. Thus no matter how big and dear the computer is, it can calculate the solution of $x^2 = 2$ only so far. The next decimal place will exceed the capacity of the machine. Mathematicians since the beginnings of mathematics have dealt precisely and successfully with infinitely many quantities. When today's largest computer has calculated all the decimals of $\sqrt{2}$ that it can, any good sophomore student of mathematics will be able to tell you what has to be calculated by the next and bigger machine if it is to get further decimals accurately, and any competent mathematician will know how many accurate decimals the next step will deliver. No machine gives information like that. It comes from logical study of infinite sets of numbers.

Will the computers, bigger and bigger as they spawn themselves, ever get so far that every succeeding decimal they find in their attempt to calculate $\sqrt{2}$ will be 0? That is, will the expression of $\sqrt{2}$ as a decimal fraction ever terminate? No computer can answer this question, blind and brainless brute that it is, for even if it produced a string of 1,000,000 zeros before it stopped, we could still ask it what the next decimal would be, and it would have to remain silent. But Greek mathematicians found the final answer to this question more than 2000 years before the invention of the most primitive digital machine. The answer is no. The solution x of $x^2 = 2$ cannot equal the ratio of any two integers p and q. The equation $p^2 = 2q^2$ has no solution for integers p and q. A statement of this kind cannot be approached by any computer. It is a mathematical statement not amenable to numerics. No matter how many integers a computer could try out, there would always remain bigger ones, millions and billions and trillions of times as large as the computer's maximum entry. In fact nobody needs to know more than a few figures of the approximate value of $\sqrt{2}$; to calculate more, by any means, would be useless; but by every person inclined toward mathematics the fact that $\sqrt{2}$ is an irrational number is cherished as the first step toward comprehending the structure of the set of all real numbers. Computers cannot touch that structure; they cannot handle any irrational number; but the minds of ancient Greek mathematicians succeeded in constructing all of them. You can read about it in EUCLID's *Elements*, Book V.

The mass media have told us that a computer recently solved a famous problem attempted by mathematicians in vain for more than 100 years: To color a map properly (under strict mathematical definitions of "color", "map", and "properly"), do we ever need more than four colors? This problem as it stands involves infinitely many maps, and a computer cannot handle infinitely many anythings. Thus the claim is false. In fact a chain of fine mathematicians stretching over a century, included among them being APPEL & HAKEN, the men who directed the final work by the computer, had by use of regular, traditional, mental mathematics reduced the problem to investigation of *a finite number of instances.* Only this reduction made it possible even to consider appeal to a computer. The finite number was too large for unaided human efforts, so recourse was properly had to a computer to try the cases, one by one. Nonetheless, the algorism for doing so had to be conceived by mathematicians; to this end, theorems demonstrated by a constellation of great men from the past 250 years were called upon.

The press did not mention these heroes of former days and made little of the essential mathematics contributed by APPEL & HAKEN. With fraudulence by omission, doubtless encouraged by the addicts and the merchants of computers, it told the public that the computer solved a problem which had beaten all mathematicians past and present.

No such impression could be gained from anything stated by APPEL & HAKEN. Let them speak for themselves[5]:

> The fundamental reason that the unavoidable set argument worked whereas other approaches to the Four-Color Conjecture did not is that all other approaches need somewhat stronger theoretical tools to make their methods apply. While these might be possible to create, there is no guarantee that they are actually possible; and if they are, there is no obvious way to go about finding them.
>
> On the other hand, many mathematicians have believed that an unavoidable set of reducible configurations might exist, but that a smallest such set was beyond the bounds of reasonable computation. This attitude appears justified when the problem is considered with respect to the tools available prior to 1960. After 1960, with the advent of faster computers, there were still strong reasons to believe that the computations would be infeasibly large,

---

[5] Pages 178–179 of KENNETH APPEL & WOLFGANG HAKEN, "The Four-color problem", in *Mathematics Today*, edited by L. A. STEEN, New York, Vintage Books, 1980, reprinted with their consent.

but there were certainly no theoretical difficulties to overcome other than the choice of a method for obtaining an unavoidable set. Thus by 1970 it became a problem of discovering whether efficient use of known techniques and technical (as opposed to theoretical) improvements would enable one to find an unavoidable set of reducible configurations.

Most mathematicians who were educated prior to the development of fast computers tend not to think of the computer as a routine tool to be used in conjunction with other older and more theoretical tools in advancing mathematical knowledge. Thus they intuitively feel that if an argument contains parts that are not verifiable by hand calculations it is on rather insecure ground. There is a tendency to feel that verification of computer results by independent computer programs is not as certain to be correct as independent hand checking of the proof of theorems proved in the standard way.

This point of view is reasonable for those theorems whose proofs are of moderate length and highly theoretical. When proofs are long and highly computational, it may be argued that even when hand checking is possible, the probability of human error is considerably higher than that of machine error; moreover, if the computations are sufficiently routine, the validity of programs themselves is easier to verify than the correctness of hand computations.

In any event, even if the Four-Color Theorem turns out to have a simpler proof, mathematicians might be well advised to consider more carefully other problems that might have solutions of this new type, requiring computation or analysis of a type not possible for humans alone. There is every reason to believe that there are a large number of such problems. After all, the argument that almost all known proofs are reasonably short can be answered by the argument that if one only employs tools which will yield short proofs that is all one is likely to get.

. . . The example of the Four-Color Theorem may help to clarify the possibilities and the limitations of the methods of pure mathematics and those of computation. It may be that a problem cannot be solved by either of these alone but can be solved by a combination of the two methods.

No-one who reads this balanced estimate will fail to see that the computer here was servant to mathematics and to mathematicians. Let us hope it was a good one. A nasty doubt remains. Did it make just one little mistake? We shall never be sure of that until some mathematician answers the question, which is purely logical, by purely logical means, without intervention of any black boxes. Even so, the computer might have been lucky. It might have made two mistakes which cancelled each other, as mathematicians themselves sometimes inadvertently do. Such mistakes of mathematicians are found by their own checking or by others'. Work of this kind by a computer cannot be checked logically, for appeal to the computer in the first place reflected surrender of logic in favor of routine, routine which was so long that no human being could follow it out even once. The referees of APPEL & HAKEN's paper resorted to an independent computer program to check the correctness of the reducibility calculations, giving some assurance for all parties, but that assurance is only psychological. The doubts mathematicians feel—not only those educated before there were fast computers but many young ones, too—rest on more than an intuitive feeling. That some problems "cannot be solved" by one method or another, is scarcely a valid basis for any conclusion. Our time (not to mention machines' time) being finite, as long as there are mathematicians or computing machines there will be more problems left unsolved than solved by either. We can no more complain because that is so than because we shall not live to see our great grandchildren's grandchildren. As for "a problem that cannot be solved" by traditional mathematics, who can say? Is not the skeleton of the history of mathematics a list of problems that for long could not be solved and then were solved? Much as I respect every statement in the passage quoted from APPEL & HAKEN's paper, I do not think their arguments justify their conclusion. Mathematicians in their search for proof use and always have used many heuristic methods which do not themselves provide a tight proof but may aid in finding one. Among these, computation on fast machines deserves wider use *in mathematicians' hands* than it presently has, but to regard its products even then as being more than guesses toward what can be proved would be foolhardy. Moreover, in the history of mathematics some problems that at first and long afterward seemed to require numerical work have later been solved by other mathematical processes, making no use of numerics. A famous example in this century is provided by the evaluation of "Koebe's constant". It would be wrong to exclude the possibility that one day the Four-Color Theorem may be seen in a new light that delivers a proof making no appeal to the arithmetic of bookkeepers.

Even problems that concern only the integers 1, 2, 3, ... are often beyond the powers of the greatest machine. Take, for example, another old and celebrated problem, to prove what is deceptively called "Fermat's Last Theorem": There are no integers x, y, z such that $x^p + y^p = z^p$ if p is an odd prime number. Many of the great mathematicians for the last 300 years have struggled in vain to prove this simple statement. To this day, they have failed, but their attempts have produced so many wonderful discoveries in number theory that HAROLD M. EDWARDS has been able to write a beautiful textbook in which he develops the subject genetically by telling the story of this one problem. In that book, called *Fermat's Last Theorem*, New York, Springer-Verlag, 1977, on page vi EDWARDS writes ". . . one is in the position of being able to prove Fermat's Last Theorem for virtually any prime within computational range, but one cannot rule out the possibility that the Theorem is *false* for *all* primes beyond some large bound." I know no clearer example of the difference between computing and mathematics. I should add here that computation on great machines has provided valuable information on this problem already and may provide more in the future. For example, it might disprove FERMAT's statement by finding one or more primes for which his equation does have a solution. But computation could never settle the question as EDWARDS has put it. On the contrary, that question may well be solved some day by a mathematician—perhaps, as often in the past, by a young beginner who is just teaching himself mathematics— the lonely realm of thought whence FERMAT and his successors, working with paper, pen, and brain, just a little human brain in a fragile box of bone, have drawn great clarity and beauty for thousands of years.

The layman may not care a whit about all this. He knows that the makers of geographical maps use more than four colors, whether or not they have to; he does not know what a prime number is; he knows that for most practical purposes the inequality $1 < \sqrt{2} < 2$ tells him all he needs to know about $\sqrt{2}$, that the carpenter never needs anything better than $\sqrt{2} = 1.4$, that the machinist rarely goes beyond 1.414 if that far. All that is true. There is no need for the layman to bother with mathematics; arithmetic is more than enough for him, and a cheap little gadget relieves him from even trying to recall what he learned in the subject he hated most in school. Arithmetic will do also for most of the practical engineers and for many kinds of scientists. Wherever there is routine arithmetic to be done, the computer can do it.

It is the scientist concerned with domains which presently make essential use of mathematics or are likely to do so in future who must know better. He must master the tools he is using; otherwise he may hurt himself. This is POST's fourth point: "You understand a subject when you have grasped its structure, not when you are merely informed of specific numerical results." Structure means precise, clear concepts linked by logical inferences. In the United States the justices appointed to the Supreme Court by F. D. ROOSEVELT and their innumerable imitators appointed by later Commanders-in-Chief of various political hues have shown us how a person who knows nothing about a language in its history or the men who spoke it may take old writ and distort it into meanings undreamt of by those who wrote it but politically rewarding for present ends. The latest act for Freedom of Something or Other surely empowers me to imitate those Great Men and go on to abuse classic literature as I please, for my ends. Thus I may wrest to my favor a famous statement by DANTE[6], making use of my freedom to delete two commas and to use one word in a modern sense, unknown in DANTE's day: . . . Don't do theory unless you think you have understood it.

It is not "pure" mathematics alone that the essential finiteness of computation puts beyond the reach of machines. The differential equations that govern the motion of a real or artificial satellite refer to limits, which involve infinitely many numbers. The computer, no matter how much the taxpayer pays for it, cannot calculate any limit. In the sense of the simple calculus every student of engineering is taught in his freshman year—the calculus in terms of which the differential equations governing the motion of a satellite are expressed—the computer can never give us more than discrete approximations. These may be, and often are, good enough, but only a person who understands the structure of the exact problem can decide when an approximation is accurate. "Approximation" makes no sense except in terms of a prior concept of exactness. The test of accuracy lies not in the number of digits a thousand human computresses or the largest acreage of electronic monsters can emit, but in the mind of a man who understands the mathematical problem which the computer is programmed to solve. Here recent developments in mathematics go beyond the capacities of any computer, present or future. In what is called "qualitative analysis" the mathematician demonstrates precise bounds for the effects of variation of the data that must be supplied from experiment or be prescribed by him who controls and directs the process being analysed. Qualitative analysis can prove that the outcome is largely insensitive to change of some parameters, and so those need not be determined or assigned with great accuracy; that small changes of other parameters may give rise to violent alteration of the results, and so these parameters must be measured or assigned precisely. Here the computer is almost helpless. At best it can provide the effects of some particular choices of parameters. It cannot tell us whether those choices are typical; its results are at best points on a graph, not limits within which that graph must lie. For computer graphics to be used safely, the curve sought must

---

[6] *Paradiso* V, 41–42: ". . . non fa scïenza, sanza lo ritenere, avere inteso". SINGLE-TON's translation, following old commentators: ". . . to have heard without retaining makes not knowledge".

have been first proved mathematically to be smooth; if ordered to determine and graph a function that is not smooth, a computer may draw for it a graph of meretricious smoothness.

I regret I cannot make this distinction clear to the total layman; an example reveals it at once to anyone who understands a bit about trigonometric functions. Namely, every beginner is taught how to analyse precisely the behavior of $x \sin(1/x)$ near $x = 0$. Suppose we should ask a computer to solve a problem whose solution—a solution of course unknown to us, because if we knew it we should not ask the computer to get it for us—just happened to involve a function like $x \sin(1/x)$. Such a function has infinitely many maxima and minima in an arbitrarily small neighborhood of $x = 0$. The computer would provide at best a meaningless scatter of points, as dense as its mesh would allow. In fairness to the machine we must say that it would itself announce or otherwise manifest its failure. Failure it would be nevertheless, blank failure in facing a problem of a kind that every beginner learns how to analyse and understand. Data can be and often are far from sufficient to promote understanding; they can even hinder it.

Failure of this kind is not limited to "pure" mathematics. In attempting to estimate the trend to equilibrium according to the kinetic theory of gases three authors[7] in an extensive computation, financed by the taxpayer through three of the biggest Federal foundations and by private industry as well, attempted to verify a conjecture making the entropy a completely monotone function of time. For "a wide range" of times they found that the first thirty derivatives of the entropy had the right sign "with double precision, significant to 33 figures". Thereupon Professor ELLIOTT LIEB[8] by a few lines of elegant, rigorous analysis resting upon a mathematical theorem half a century old proved the conjecture false. He writes that "this case exemplifies the need for great care in using computers to study delicate mathematical properties . . .." Indeed, it exemplifies how misleading computer calculations may be! LIEB reports also that Dr. K. OLAUSSEN, using asymptotic analysis, has estimated that for a computer to reach the correct conclusion it would need to calculate 102 derivatives. In a private communication Professor LIEB has remarked, "Although computers can sometimes give useful hints about properties of solutions to analytic problems, it should be understood (and often is not) that the more delicate is the property under

---

[7] R. M. ZIFF, S. D. MERAJVER, & G. STELL, "Approach to equilibrium of a Boltzmann–Equation solution", *Physical Review Letters* **47** (1981): 1493–1496.

[8] E. H. LIEB, "Comment upon 'Approach to equilibrium of a Boltzmann–Equation solution'", *Physical Review Letters* **48** (1982): 1057.

investigation, the more computer time is needed. In particular, infinitely subtle properties (e.g. the analyticity or complete monotonicity of a function) require infinitely much computer time to verify directly."

If guided by competent mathematicians, computation can offer modest help to mathematical research. It can provide numerical examples which render abstract statements explicit; it can disprove false guesses; it can accumulate material which may suggest conjectures which later somebody may prove true or false. It can do a fine job of plotting graphs, which are easier to assess than are tables of numbers, and can make mathematical statements visual. Beautiful results of a computation of this kind, directed by mathematicians and designed by them to illustrate their completion of the theory of a long-known but perplexing aspect of the convergence of trigonometric series, may be seen in a recent paper by E. & R. E. HEWITT[9].

A computer can even be programmed to deliver exact solutions to problems which can be solved by a finite number of routine substitutions of solutions of problems already solved by mathematicians. For example, it can be taught to replace x^2 by $2x$ or $x^3/3$, leaving x a symbol. It can be made to replace the symbol $\sin x$ by the symbol $\cos x$ without calculating the value of the sine or cosine of x for any particular number x. Thus the parts of literal algebra and calculus that can be reduced to *finite arithmetic routines* can be handled by computers and often should be.

The mathematics of computation is interesting in itself and can be developed abstractly. Computers, like other gadgets, occasion demand for new mathematics; some has been provided already by mathematicians, and more remains to be created.

But how much of the activity surrounding computers is directed by competent mathematicians?

6. COMPUTERS BRING POWER AND THE ABUSES OF POWER. ADVOCATES OF COMPUTING SEEK TO DESTROY MATHEMATICS

Calculation on huge machines is democratic and costs dearly. The small army garrisoned around a big computer is composed of specialists and experts, dear fellows, indoctrinated in computer-worship. Like those engineers and physicists who use computers by habit, most of them never so much as consider the possibility that a problem of theirs might have a mathematical solution.

---

[9] E. HEWITT & R. E. HEWITT, "The Gibbs–Wilbraham phenomenon: An episode in Fourier-analysis", *Archive for History of Exact Sciences* **21** (1979): 129–160.

Well and good, the innocent may say. To each his own. It is not so
easy. For every mathematician who relies on his mind alone, there are
now thousands of navvies on computer gangs. In the United States
today there are more than 500,000 programmers and systems
analysts; these form the officer corps of the computer army; their
number grows swiftly. The American Mathematical Society has about
17,000 members; the number of these whose main activity is research
is hard to determine, but I guess that it would not reach 500. In our
social democracy, numbers bring power. Not only a population but
also wealth pushes computers; computers are incessantly promoted
by advertising; mountains of money are gained by those who design,
manufacture, and sell computers. Wealth, too, brings power. In any
society, power creates abuse. Nobody who knows the computer folk
will expect to find them ready for peaceful co-existence with the tiny
realm of intellectual mathematics—the only source whence informed
criticism of computer-worship may spring. Not only can the computer
addict, benumbed as he is to mathematical ends, perceive only means
of reaching them; not only does he depreciate and deprecate every
means other than computing; but most of all he condemns mathematical research that is not numerical. The tyranny of computers now
threatens to destroy mathematics even as an activity in universities.

Do you think I exaggerate? Rather than adduce instances I quote
in its entirety a manifesto by JAMES C. FRAUENTHAL as issue editor of
SIAM News for April, 1980, under the title *Change in applied mathematics is revolutionary*[10]:

---

[10] Reprinted with permission of *SIAM News* and of JAMES C. FRAUENTHAL.

Many other writers have published similar statements, usually less violent. In *Mathematics Tomorrow*, edited by L. A. STEEN, New York *etc.*, Springer-Verlag, 1981, PAUL
HALMOS in an article on his favorite subject, "Applied mathematics is bad mathematics", writes

> I should guess that in the foreseeable future (as in the present) discrete mathe-
> matics will be an increasingly useful tool in the attempt to understand the world,
> and that analysis will therefore play a proportionally smaller role. That is not to
> say that analysis in general and partial differential equations in particular have had
> their day and are declining in power; but, I am guessing, not only combinatorics
> but also relatively sophisticated number theory and geometry will displace some
> fraction of the many pages that analysis has been occupying in all books on applied
> mathematics.

ANTHONY RALSTON in his article "The decline of calculus—the rise of discrete
mathematics", pages 213–220 of the same volume, quotes WALLACE GIVENS as
follows:

> There is a simple and basic fact about a computer which will, in the decades and
> centuries to come, affect not so much what is known in mathematics as what is
> thought important in it. This is its finiteness.

Shortly after the turn of the century, Niels Bohr and Albert Einstein presented theoretical results which revolutionized physics. This is not news; a quick look through the catalog of my university convinced me that not one of the fifty-five members of the Department of Physics lists classical mechanics as an area of interest. I doubt if a single Ph.D. will be awarded in 1980 by a physics department for the solution of a problem in Newtonian mechanics. Of course Bohr and Einstein and their friends did not instantly solve all the old problems; instead, they created a new set. What happened to all of the old problems which did not simply disappear? They moved into other disciplines called by names like applied mechanics and mechanical engineering.

And what happened to the people who were professors of physics at the turn of the century? What they had been trained to do was out of fashion. Some, no doubt, retrained themselves by learning about quantum mechanics and relativity. Most, I suspect, stayed right where they were and continued to do more or less what they had always done. From an evolutionary point of view, the classicists became extinct in physics departments in a single academic generation (the time from Ph.D. to retirement).

While some of us in a country of medium size and barely two centuries old may hesitate to endorse a declaration of what will happen to the world in "centuries to come", with the sober and factual parts of these statements it would be hard to find reason to disagree.

In the same article RALSTON himself on pages 214–215 takes a position somewhat like FRAUENTHAL's but not so immoderate:

> ...sharp changes should be viewed with great skepticism and should be undertaken only for the most compelling reasons.
>
> Still, I suggest the need for such a revolution [in the teaching of mathematics]. Its cause? The invention and development over the past three decades of the digital computer, perhaps the most important development in science and technology since the invention of the printing press. In any case, it is a development which will not only have profound effects on human life and the social fabric, but which will also—and this is the point here—have a most important influence on the problems on which scientists work and, in particular, on the mathematics they use. (Which is not to say, I emphasize here, that calculus and classical analysis will not continue to enjoy much success; it is only to say that their position of dominance in mathematics and its applications is about to be challenged.)
>
> ...[T]o a considerable degree, the wellsprings of mathematics have always been in the applications of mathematics. Today it is computers generally and computer scientists in particular which generate the need for applications of mathematics in greater volume—and at a much more rapid rate of increase—than does any other area of science or technology. Since the mathematical problems generated by computers and by computer scientists overwhelmingly require discrete rather than continuous mathematical tools, it is hardly surprising that research in discrete mathematics is rapidly increasing.

Although this all seems simple and obvious to us in retrospect, I would guess that the future looked very uncertain in 1920.

As in physics, so in mathematics

What does the revolution in physics have to do with us? Very simply, I believe that we are presently experiencing in mathematics a change which is as dramatic and irreversible as the one which took place in physics earlier this century. The genesis of this change happened some years ago and the effects seem more apparent to me each day. The motivating force: the invention of the computer. The effect: a one time, inevitable change in the field of mathematics. We who consider ourselves to be applied mathematicians must not be so smug as to think that we are either immune to the change, or its only logical beneficiaries.

As I see it, within another academic generation, the mainstream of mathematics will not be analysis, number theory and topology, but rather numerical analysis, operations research and statistics. Already the areas of mathematics which are computation oriented are the most successful in drawing students at all levels and, more important, in drawing funds from university administrators for new faculty members. I am not suggesting that the pure areas of mathematics, or for that matter the classical topics in applied mathematics such as transform methods, partial differential equations and approximation theory, will disappear. Instead, like Newtonian mechanics, they may move permanently from center stage in mathematics departments.

An alternate scenario

There is of course an alternate scenario, though it is no more pleasant for those who enjoy the status quo. Mathematicians can resist the incursion of the computer into their field. They can argue that it is not mathematics to solve a problem using numerical techniques. In fact, at many universities this argument or its equivalent appears to have been made. The result is inevitably that a new department with a name like computer science or mathematical science is created in response to student demand. Then slowly, the mathematics faculty end up doing little more than teaching calculus. The irony is that, as with physics departments where modern physicists teach classical mechanics, computer-oriented mathematicians could offer a more relevant introduction to calculus than many classically trained mathematicians.

Whether it is the name that changes or the focus of the members of the faculty is really irrelevant. What does matter is that by the year 2025 (in my opinion), the vast majority of the mathematicians on university campuses will be either using computers in their work, or studying the fundamental problems which must be solved to advance computer algorithms. In only a few places will there remain centers for research in pure mathematics as we know it today.

This *Bekanntmachung* proclaims the new *tausendjähriges Reich* of science. Not only do computers bring the revolution, displacing 3,000 years of feudal-absolutist-capitalist slavery to naked thought; beyond the computer there is nothing. Computers provide "a one time, inevitable change, ... dramatic and irreversible", the *final solution* of the mathematician question! I confess I cannot see why calculus will be worth anyone's while to learn after the Cultural Revolution shall have reduced all science, discretized, to currying computers; calculus is a theory of limits, and the concept of limit, since it is inherently beyond numerics, must be superseded; but the future serfs who, permanently off "center stage", will "end up doing little more than teaching calculus", will be grateful that the *Sturmtruppen* of the master race, goose-stepping behind the university *Gauleiter*s, permit them to live out their useless days in some silly asylum for private enterprise in the realm of thought. Certainly mathematics done by human minds— before the "one time, inevitable" revolution the modifier "done by human minds" would have been redundant—will find few defenders in an age when everything is decided by some kind of opinion poll, staged and manipulated by mass media.

7. COMPUTING PROMOTES FACTUAL FRAUD. IT HAS HARMED EXPERIMENTAL AND APPLIED SCIENCE IN THE PAST AND IS CONTINUING TO DO SO. BY ITS EMPHASIS ON APPLICATION OF THE ALREADY KNOWN, IT CAN DELAY BASIC DISCOVERY AND THUS REDUCE THE FIELD OF APPLICATION IN THE FUTURE.

Most citizens will feel no regret if creative mathematics disappears. They always hated what little mathematics they met in school. Of the few who liked mathematics, many have no idea that it is possible to discover anything new in it. But mathematics is not the only science that computer addiction can kill. Since computing is advertised by its addicts, by the press, and by the computer merchants as being the oracle of science and society, computer codes are sold or given away gratis to all comers. They can be and have been applied blindly, in disregard of the warnings to users which often are attached to each copy. Fraudulent exploiters can and do promise their clients to solve for a fee any problem posed. He who resorts to a whore may with some confidence expect personal service in fair return for the price or prices he pays; the client of a computer faker is most likely a charlatan himself, who uses his purchase as a tool to help him deceive the public. Fraud is fraud; it can be practised anywhere in any activity; but while a charlatan in medicine is apt to be exposed in time by his victims or their surviving relatives, who can unfrock a computer charlatan?

Many experimental arrangements today feed their data directly into a computer for digesting. Nobody could reverse the process, even if he could disentangle the horrid mess of numbers. The data cannot be recovered; only its interpretation emerges. Scandals in business show us how easy it is for skilled hands to make computers lie. In science now it is even easier to fudge the data as well. If the aim is an explosion of journalism or a Nobel prize, the temptation is great. There are no auditors who must certify the books. The computer not only discourages the attempt to understand before applying, it smoothes the way for factual fraud.

Even in instances when the data can be recovered, computers can and do give contrary interpretations of them. Recently the public, which had been forced to pay billions in taxes for exploring the face of Mars, was shown two conflicting sets of pictures, extracted by "computer-enhanced imaging" from the same dearly bought data by different groups of computer experts. One group found only the grey fog which is the astronomers' usual reward. The other claimed to penetrate the clouds and to discover beneath them a huge face in a desert pocked by pyramids. The work of the second group is either a hoax or not a hoax. If not a hoax, it shows that computer interpretation of data cannot be trusted. If a hoax, it shows that computer experts cannot be trusted. Well and good, the face of Mars, be it but dust or be it a gallery of masterpieces of modern art, is little likely to hurt or help us. The reflections of this affair upon the processing of data by computers are grave. The computer is programmed to remove the "noise" that blurs the image, but criteria to determine what is noise and what is the faint trace of a record of some object must be prescribed for the programmer or by him. The danger to a populace which has confided its welfare to "science" ruled by computers and computer experts is equal, whichever way lie the truth about the diorama of Mars. Just think what mad warriors could do— perhaps now regularly do do—by "computer-enhanced imaging" of spy photographs of the weapons of potential enemies!

Factual fraud in science has reached the public press. In an article called "Fudging data for fun and profit" which appeared in *Time*, December 7, 1981, FREDERICK GOLDEN writes "Findings that were touted only last summer as a fundamental breakthrough in the understanding of carcinogenesis have been branded fraudulent." As he reminds us, "cheating . . . is common to many professions these days . . . ." The earliest meaning of "charlatan" is "a mountebank who sells wonderful drugs"; charlatanism has been the inseparable companion of medicine ever since there have been medicine men (now called "physicians"), and it is no wonder that molecular biology and biochemistry, which are so close to medicine as to be able to gouge into the billions milked annually from the taxpayer by the government and the further billions given by the timorous rich to support the gang warfare euphemistically called "research" in the world of healing, should have learnt what profits mafia science can yield. GOLDEN mentions the frauds discovered at Cornell, Yale, Massachusetts General Hospital, Boston University, the University of California—the tip of the iceberg. It is time for similar scandals in high-energy physics and observational astronomy and every other part of science where there are costly experiment and costly computing which must lead to frequent "breakthroughs" if their funding is to continue. Nothing is more easily forgotten than the "breakthrough" three years ago, for all old accounts have been quitted by the auditors.

"DON'T FEEL BAD ABOUT FALSIFYING THE SOLUTION. I FALSIFIED THE PROBLEM."

Figure 30. Big science as summarized by SIDNEY HARRIS, 1981, reproduced with his permission.

Experiment, yes experiment, is the touchstone of science! GOLDEN observes that "So much is being done in every field that unless an experiment is really important, years may pass before anyone tries to repeat it. Especially at a time when new ideas are at a premium, there is not much profit in doing over someone else's work. Furthermore, repetition is sometimes all but impossible . . .." Indeed. Experiment and advertising are scarcely distinguishable in today's science.

The mania for bigness rules. GOLDEN writes,

> Senior scientists are often so busy scrambling for funds to keep their labs running that they rarely have time to look so closely at what their young whizzes are doing as they would like. What was once a sportsman-like rivalry between researchers has become cutthroat competition. By publishing a paper first, even if some of the data are not quite accurate, a young scientist may beat out a rival for any number of prizes: a tenured post or promotion, a big grant from the Government, an offer from industry . . . and ultimately perhaps the trip to Stockholm.

The foregoing quotations do not mention use of computers. That is so because it is nowadays taken for granted that big science is totally computerized. Computer fraud being the easiest of all kinds of fraud today, anyone who chooses to falsify problems and data will as a matter of course call to his aid the total obfuscation that computed statistics and computed analysis of data can easily be programmed to provide. Indeed, it is unlikely that recent and future frauds have been and will be discovered except through somebody's peaching. In matters such as biological warfare and genetic alteration it could be too late: when it came time to peach, everyone who might be able to do so might be already reduced to functionlessness in mind or body if not actually dead.

For factual fraud there are old precedents. Even in less democratic ages science based upon heaps of data and numerical work has been perilous. PTOLEMY, the Alexandrian astronomer of the second century, made tables of the planetary system which throughout more than 1000 years following were to provide an unshakable bastion for scientific faith and against new doctrines. His work long served as the classic example for comparison of abundant measured data with theory. In the last decade the Royal Astronomical Society has published articles by ROBERT R. NEWTON[11] which show to the satisfaction

---

[11] R. R. NEWTON's work is available also in his book, *The Crime of Claudius Ptolemy*, Baltimore, Johns Hopkins University Press, 1977. NEWTON's arguments are denounced for bias and inconsequence by N. M. SWERDLOW. "Ptolemy on trial", *American Scientist* **48** (1979): 523–553.

of many historians of astronomy that all the observations PTOLEMY claimed to have made himself he in fact fudged to fit his theory. Even the way he went about his fudging has been reconstructed. The late WILLY HARTNER, a profoundly respected and indeed revered historian, claimed to have found similar fudging in the data added by Arab astronomers who later upheld PTOLEMY's system at all costs. Of course, anybody can cheat, at any time and about anything. PTOLEMY's system made factual fraud easy because in its practice it was a numerical scheme. Science is different. As POINCARÉ said, science is not a collection of facts, any more than a heap of stones is a house. Science organizes facts by reason in such a way as to correlate what seems disjoint and to foreshadow facts not yet observed. NEWTON's laws and EINSTEIN's theory of gravitation are not reducible to tables of numbers. No amount of factual fraud could have preserved the NEWTONian planetary system and stopped EINSTEIN's, for they are not numerical. They are mathematical ideas, simple ideas which can be understood structurally first and then applied to instances. The slight correction of planetary orbits that EINSTEIN's theory provides is a minor instance of its value. A small alteration in the NEWTONian scheme could have fitted it to the orbit of Mercury without altering the orbits it delivers for the other planets. That would have been adjusting theory to fit data. Such alteration was proposed and was rejected as being unedifying. EINSTEIN's theory did nothing of that sort. It arose because the NEWTONian view of space-time had proven conceptually inadequate in itself as well as incoherent with electromagnetism. As DIRAC explains[12],

> What makes the theory of relativity so acceptable to physicists ... is its great *mathematical beauty.*

Its formulæ for motions of gravitating bodies emerged as one product of its general revision of basic ideas; the emendment of Mercury's orbit provided not motivation for the change but a test of it. Other relativistic theories of gravitation, for example G. D. BIRKHOFF's, imply the same results as EINSTEIN's in regard to presently possible tests by experiment. NEWTON's theory of the heavens remains today, even in much of cosmology, the basis of our ordinary thought regarding them. The gush of journalism about the advance of the perihelion of Mercury—a tiny and eccentric detail—is no more than an example of the accepted social doctrine that "revolution" is a good thing for everybody. Nobody who does not understand the mathematical theory

---

[12] P. A. M. DIRAC, "The relation between mathematics and physics", *Proceedings of the Royal Society of Edinburgh* **59** (1938/9): 122–129 (1939).

of the electromagnetic field should let the word "relativity" cross his lips except in a question[13].

Against the Scylla of factual fraud stands the Charybdis of the Ptolemaic system itself. If we regard it in its kinetic essence, not in just the particular numerical state PTOLEMY himself decided upon, we find that it contains potentially as many adjustable entries as we like. Inherently the array of deferents, epicycles, and epicycles upon epicycles is a method of interpolation with as many adjustable parameters as the adjuster may wish. In principle it could fit all known planetary observations and be refitted each time a new observation was made[14]. Only the practical limits of numerical calculation in PTOLEMY's day made fudging necessary to get agreement. Only the limits of numerical calculation in KEPLER's day made it impractical to add further epicycles which could have adjusted PTOLEMY's system to agree perfectly with observation for another 1000 years. Had modern machines been available then, KEPLER himself might have formulated his laws nevertheless, but astronomers would not have accepted them. "The old way is more accurate," they would have said: "anyway, our machines are already programmed for it, and we cannot afford the money and delay needed to try a new theory that is, after all, just a theory. Besides, think how many senior epicyclists would be put out of

---

[13] I mean nothing advanced or difficult for any mathematically literate person. In an elegant, limpid textbook for mathematically qualified senior undergraduates, starting from first principles C.-C. WANG presents in less than 200 pages the classical and relativistic theories of electromagnetism and gravitation. I refer to his *Mathematical Principles of Mechanics and Electromagnetism*, Part B, New York and London, Plenum, 1979. Pages 311–314 present and compare the classical and relativistic determinations of orbits for pairs of gravitating bodies and derive in a few simple lines the relativistic advance of 43″ per century in the perihelion of Mercury. Only the two-body problem is considered. In comparison with astronomical observations the effect of NEWTONian perturbations by other planets must also be taken into account. K. P. WILLIAMS in *The Transits of Mercury*, Indiana University Publications Science Series No. 9, 1939, by painstaking reduction and estimate of errors in the data concluded that the non-NEWTONian advance was 42″.93.

To the mathematically illiterate (I use the term not as an insult but as a factual qualification) it is harder to explain relativity than it is to teach the musically illiterate the difference between one canon and another. In music, sound helps; in relativity, sound seems to hinder.

[14] Nothing I state above regarding the Ptolemaic system should suggest that "piling up sufficiently many epicycles" could represent "any conceivable phenomena". I refer only to the phenomena associated with the motions of the centers of the seven great and near heavenly bodies, and I suggest that the approach basic to the Ptolemaic system would not suffice to describe the motions of artificial satellites. I may be wrong in either or both of these opinions. Many common, loose statements about Ptolemaic astronomy are shown to be false in a splendid paper by the late R. C. RIDDELL, "Parameter disposition in pre-NEWTONian planetary theories", *Archive for History of Exact Sciences* **23** (1980): 87–177.

their jobs!" Had machines been available in NEWTON's day, I doubt he would have used them, but I doubt also he would have been impelled to devote years of intense study to the mathematics of the planetary system, and had he done so, I doubt his theory would have been accepted. Had machines been available to the creators of mechanics, I doubt we should have the law of universal gravitation today. To predict the planetary motions, with their obvious near regularities, methods of numerical interpolation can do very well. To get a body out of one orbit and into another is a problem of a different kind entirely. There it is the irregularities that predominate. I doubt that computers of celestial orbits, no matter how large their capacity, could have by any method of mere interpolation, mere fitting of epicycles to data, determined conditions for interplanetary travel. Computers make transplanetation possible today; had they been available 200 years ago, the basis for transplanetation today would never have been discovered.

Computers promote applications of known science; by inhibiting creation of new science, they limit the field of future application. You cannot apply a scientific theory if you do not yet have it.

Here we may return for a moment to FRAUENTHAL's simile of the computer revolution in mathematics to the revolution in physics effected by BOHR and EINSTEIN. To do their work, both BOHR and EINSTEIN used the kind of tools NEWTON had used long before them: their own minds, applied to what physics they knew and aided by what mathematics they knew. I have not perused their writings; in looking over the pages I do not see a single instance where a great computing machine could have helped them. In contrast with the drudges of their day, who sought to determine one more decimal place by measurement or arithmetic, they were content with simple calculations. It is their successors who have made monstrous and inevitable numerics an essential part of physics. Have these successors effected any revolutions? Possibly so, but I must leave it to others to judge whether those revolutions have brought us, in addition to ter-rifying power to destroy human life and works, any clearer under-standing of the nature of matter. There is another difference. The revolutions of BOHR and EINSTEIN were not developed for military aims, promoted by governments, financed by speculative capital, pro-mulged in directives by administrators of industry and bureaucracy, or diffused by floods of popular advertising and armies of salesmen.

The physicists themselves, their intellects already wan and flagging from the ravages of malignant computeritis, may be committing sep-puku by computer. Lest you think I exaggerate, I quote the final sentences of the inaugural lecture of the physicist STEPHEN HAWKING as Lucasian Professor of Mathematics in the University of Cambridge,

29 August 1980:

> At present computers are a useful aid in research but they have to
> be directed by human minds. However, if one extrapolates their
> recent rapid rate of development, it would seem quite possible
> that they will take over altogether in theoretical physics. So maybe
> the end is in sight for theoretical physicists if not for theoretical
> physics.

These words, which in print read like a breathless pronouncement,
Professor HAWKING regards as striking "a slightly alarmist note". Perhaps they were spoken in the witty irony for which the British are
famous, but many a computer addict preaches the same message
in deadly earnest with "theoretical physicists" replaced by "mathematicians".

8. CLASSIC THEORIES USED INDUCTIVE AND DEDUCTIVE MODELS. COMPUTING ENCOURAGES FLOATING MODELS

The old theories, the classic theories of science, provided models of
limited aspects of nature. The example set by the rational mechanics
of EULER and LAGRANGE, based in part upon the discoveries of
HUYGENS, NEWTON, and the BERNOULLIS, illustrates the status of a
"Law" of physics: a clear, precise concept of ideal behavior,
embracing an enormous variety of precisely specifiable cases. The
"Law" when applied to a case restricts but generally does not determine the outcome. Any discrepancy between data of experiment and
such an outcome of theory we attribute first and usually finally to our
own failure to apply the "Law" well, not to the "Law" itself. Only if an
instance can be found for which any direct, not merely *ad hoc* application of the "Law" leads to results contrary to fact, will the "Law" be
questioned. The "Laws" of mechanics have been sharpened and
broadened but never repealed. Some "Laws" in other domains have
indeed been abandoned, but they are few. Lurid journalism of
science gloats over crises and "revolutions", distorts them, expands
them, just as the common press collects floods, earthquakes, volcanic
eruptions, murders, and riots to satisfy the people's thirst for blood.
The predominant character of science is not its crises but its stability.
Of the national systems of government that were in existence when
the laws of rational mechanics were discovered, those laws have out-
lasted all but one, one which is meanwhile become so altered as to be
the same in name only.

The models rational mechanics provides are strictly logical; as POST puts it, they are *deductive models*, articulations of a particular theory. Classic science embraces also *inductive models*, summarizing an organized body of experimental data. Models of both these kinds are systematic. They teach us to find structure in experience, not merely to imitate one or another detail.

I have remarked above in Essay 10 that recent research resorts more and more to *floating models*, which treat phenomena severally, with no subsumption under general theory or organized knowledge gained from experiment. The example developed in some detail there is Applied Catastrophe Theory.

Here I mention another, one that originated in computing and is notorious for the renown of the names associated with it. I return to the attempt of FERMI, PASTA, & ULAM[15], mentioned above in § 3, to find a system such as to show "a gradual, continuous flow of energy from the first mode to the higher modes". Starting, as physicists will, with a simple harmonic assembly, which conserves the energy of each mode forever, they introduced hypothetical "non-linear forces acting between the neighboring points . . .." Thus they arrived at several special members of a class of partial differential equations introduced[16] by EULER (1744, 1766), extended by LAGRANGE (1761, 1781), studied by AIRY (1845), STOKES (1848), EARNSHAW (1861), and many later authors, and familiar to students of mechanics. Apparently knowing nothing of this classical work, FERMI and his collaborators went straight to the biggest computer there then was. It bore the name MANIAC. The results of their long (and no doubt costly) computations, they wrote, showed "features which were, from the beginning, surprising . . .," and they reported them with words of magic about "limits guaranteed by the ergodic theorem" *etc.*, leaving us to guess which ergodic theorem they had in mind. Some of the classical background of the subject was recognized by ZABUSKY[17], who by resort to the familiar hodograph transformation obtained a linear hyperbolic system which may be solved by the method of RIEMANN (1860), still more classic. ZABUSKY thus rediscovered a famous

---

[15] E. FERMI, J. PASTA, & S. ULAM, *Studies of non-linear problems*, Document LA–1940, May 1955 = pages 978–988 of Volume **2** of E. FERMI, *Collected Papers*, Chicago & Rome, University of Chicago Press & Accademia Nazionale dei Lincei, 1965.

[16] *Cf.* C. TRUESDELL, §§ 30 and 55 of *The Rational Mechanics of Flexible or Elastic Bodies, 1638–1788*, LEONHARDI EULERI *Opera omnia* (II) **11₂**, 1960; page CXXI of "Editor's Introduction", LEONHARDI EULERI *Opera omnia* (II) **12**, 1954; and pages LIX–LX and XCVII–IC of "Editor's Introduction", LEONHARDI EULERI *Opera omnia* (II) **13**, 1956.

[17] N. J. ZABUSKY, "Exact solution for the vibrations of a nonlinear continuous model string", *Journal of Mathematical Physics* **3** (1962): 1028–1039.

observation of STOKES and HUGONIOT: After a finite time, the solution ceases to exist, and the outcome is a shock wave or "catastrophe" (*cf.* Essay 8, above). He concluded that "a continuous nonlinear system described by a partial differential equation of second order cannot describe the vibrations of the equivalent discrete system for 'large' times", and "to account for" the results of FERMI, PASTA, & ULAM he proposed "to include terms . . . which involve higher derivatives . . . ." Thus he seemed to imply that the computer's results must have been right despite the floating origin of the discrete problem FERMI and his collaborators had made it solve. In telling the story KRUSKAL[18] decided that the thing to do was replace the problem of the non-linear string by the result of some mysterious manipulations with FERMI, PASTA, & ULAM's numerical system. He thus arrived at a partial-differential equation involving two derivatives of fourth order, which he magically converted to one with a single third derivative: the Korteweg-de Vries equation, which had arisen half a century earlier on a sound basis in hydrodynamics. Thus, apparently, KRUSKAL kicked aside the problem the computer code was designed to solve but could not; he replaced it by one that the computer perhaps did solve. In this way he and ZABUSKY came upon nonlinear waves which pass through each other with no change of form. Such waves, which were named solitons, were found also among the solutions of other nonlinear equations, and an exuberant literature devoted to them resulted and continues[19]. Opinions differ with respect to how much the original work owed to its authors' ignorance of classical hydrodynamics, in which single solitary waves had long been known and studied. There is room for disagreement also on the value of the hints, right and wrong, that the original exploration by computer provided.

If this story seems confusing, that is because it is:

1. The program given to the computer was incorrect for the analytical problem that was to be solved.

2. The correct solution of the analytical problem predicts a catastrophe. (Let the reader reflect on what might have happened, had the problem been one concerning a real nuclear reactor instead of just some physicists' wild guessing, and had the smooth and safe "solution" given by the computer been applied to the real world.)

---

[18] M. D. KRUSKAL, "Asymptotology in numerical calculations: progress and plans on the Fermi-Pasta-Ulam problem", pages 43–62 of *Proceedings of the IBM Scientific Computing Symposium on Large-Scale Problems in Physics* (1963), White Plains (N.Y.), IBM Data Processing Division, 1965.

[19] M. D. KRUSKAL, "The Korteweg-de Vries equation and related evolution equations", pages 61–83 of *Lectures in Applied Mathematics*, Volume **15**, American Mathematical Society, 1974.

3. While the results of the computation were a disappointment at first, the correct solution of the original problem was a still greater one. The physicists threw away the original problem and sought one to which the computer program might apply.

4. This new problem, like the original problem, had a classic foundation, well enough understood that mathematical analysis, making intelligent and directed use of numerical computation when helpful, could develop it further.

Despite the hectic, unprincipled floundering which the story recounts, at least it has a happy ending: hundreds of mathematical papers on a harmless, beautiful topic in classical hydrodynamics, where computing plays a minor or at least directed role. It illustrates an empirical truth called "the BERS principle": GOD watches over applied mathematicians. Let us hope he continues to do so.

9. COMPUTING PROMOTES LOGICAL FRAUD. COMPUTERS PROGRAMMED TO CONFIRM FALSE THEORY CAN DESTROY MANKIND

The collective's war machine of huge computers, always famished for more and more numerical problems and at the same time always insufficient, always needing reinforcement by more and more bigger armaments, not only encourages floating models subject to no laws, it encourages logical fraud. By logical fraud here I mean mathematics that is not rigorous. The old kind of unrigorous mathematics often praised in circles of application was not so dangerous because the "Laws" stood behind us. A computed result which the "Laws" made suspect would be scrutinized at once. But when there are no "Laws", just floating models, there is nothing to check against! To see this, suppose for a moment that a new floating model be a good one, but as usual (in fact *de rigueur*!) so difficult mathematically that nobody can by mathematics assess the general qualities which applications of that model should have. Problems illustrating it in "important" applications are put straight onto the computer, but necessarily of course in some simplified version—further "approximations" they are called, which involve at bottom nothing but finitely many zeroes and ones. Unrigorous mathematics greases the path for wrong answers to slip out of even right assumptions, for something noxious to man to be by hocuspocus with the lingo of formulæ and a bore of computed digits whitewashed into something apparently useful. Here physics provides the worst of examples. In paraphrase of POST I might say that the classic

$$\mathring{\alpha}\epsilon\grave{\iota} \; \mathring{o} \; \theta\epsilon\grave{o}\varsigma \; \gamma\epsilon\text{o}\mu\epsilon\tau\rho\acute{\iota}\zeta\epsilon\iota,$$

the god is always doing mathematics, has degenerated in the minds of modern physicists into

God is a bad mathematician.

Unrigorous mathematics, which is failed mathematics, is fraudulent mathematics. Computerized fraudulent mathematics provides abundant food for research which is aimed at confirming what is known already or what ought to be true even if it is not. This kind of research gets commoner and commoner nowadays. Whatever the proclaimed truth be, the computer can be programmed to support it. Science without "Laws" is fine for fields which, unlike physics, have never had "Laws", only dogmas. A dogma does not apply to cases; it can merely be repeated and rephrased and illustrated; the faithful invoke the dogma as a war cry in whatever they do, and their doings have no purpose but to strengthen the dogma. Of course a revolution can replace one dogma by another, perhaps the very opposite. Computerized floating models can always be adjusted so as to conform with a given dogma, no matter what the inputs. It takes no great genius at computing to make the inputs cancel out.

To reveal what computation can do with a floating model, I return to Applied Catastrophe Theory and quote SUSSMAN & ZAHLER[20] in regard to it:

> The interest it has aroused among the public at large is mostly understandable in terms of the fascination which the mystery of mathematics exerts upon the mathematically uneducated. Mathematics, in the perception of many, is like sorcery. The mathematician performs mysterious passes that others cannot understand, and suddenly a prediction, a theory, emerges. Consider, for instance, the description of how Catastrophe Theory works, as provided to its readers by *Newsweek* magazine (Jan. 19, 1976, pp. 54–55):
>
>> To apply catastrophe theory, a mathematician first selects the variables that are relevant to his problem—these might be 'growth' and 'inflation' in a particular economic environment. He then compiles as much statistical and behavioral data as possible and takes stock of the extraneous

---

[20] H. J. SUSSMAN & R. S. ZAHLER, "Catastrophe theory as applied to the social and biological sciences: a critique", *Synthèse* **13** (1978): 117–216. See page 206.

factors that might influence the economic climate. Finally, using highly complex mathematics and a computer, the mathematician forms a qualitative and quantitative model that, if properly formulated, can make precise forecasts of behavior which Zeeman says are far superior to any that can be achieved with the best statistical techniques known.

The image presented here of the mathematician at work is very much like that of a sorcerer. The statistical data and the computer replace the wand and the flowing robes, but the actual nature of the mathematician's activity is equally mysterious. As in the case of sorcerers who were supposed to have all kinds of powers, yet seldom were able to perform a specific, reproducible feat, the Catastrophist is supposed to be able to make "forecasts" that are "far superior to any that can be achieved with the best statistical techniques available", although not a single example of such a forecast exists.

Their final sentence refers to the status of "the surrealist world of catastrophe theory" in 1978. It prepares us to imagine how computerized research on floating models, particularly in the social sciences, will in the future provide projects ideal for support by the Ministry of Love.

Nothing is easier to apply to human betterment than failed mathematics substantiated by experiments programmed to confirm it. Such mathematics and such computing cannot take men to the moon, but it can destroy all the men on earth.

10. SUMMARY: COMPUTERS ARE HERE TO STAY. THEY ENDANGER THOUGHT, LANGUAGE, SCIENCE, AND THE SURVIVAL OF MAN. LIKE ANY OTHER DANGEROUS TOOL, THEY SHOULD BE PUT UNDER STRICT CONTROLS

A computer, like a knife or a gun or a television network or a nuclear bomb, is an object. An object in itself, even an erupting volcano, is neither good nor bad, but it may be dangerous. Man puts objects to use. Nuclear fission, we know, can produce peaceful power; much of applied engineering today finds the computer indispensable, interplanetary flight being but one of myriad instances of what we can do with the aid of the computer and cannot do without it. I have pointed out what else man can make computers do. As for men, it is not my place to pass judgment on them. Maybe most men are good. Maybe we are entering a new golden age of peace and plenty, in which the lion will lie down with the lamb, the whore with the guileless schoolgirl, the assassin with the prey he has been suborned to shatter.

Maybe man, for the first time in his existence, will turn each of his tools and toys, even the most dangerous, to good uses alone.

Do not misunderstand what I have said. I preach no war on computers. Like Don Quixote's windmills, computers are here to stay, as long as man can afford to make and run them, or until he can replace them by something still more dangerous. I plead only that

(1) As a lead article in *The Wall Street Journal* for 29 September 1980 reminds us, an object code "consists solely of ones and zeroes, the only things even the smartest computer can deal with." The reporter failed to mention that the ones and zeroes are *finite in number*.

(2) Whenever a problem is demonstrably amenable to finite arithmetic, a computer can be used and in most cases should be. Examples: accounting, some aspects of engineering, preliminary exploration of some mathematical problems, *etc.*

(3) Numerics cannot bring understanding of the structure of a mathematical problem unless an informed human being has already conjectured possible structures or inferred them from instances.

(4) Computation is dangerous except in providing details concerning problems whose structure is already understood mathematically.

(5) The importance of numerics to science has been brazenly exaggerated by pressure groups which profit or hope to profit from manufacture, sale, and tending of computers and by addicts who preach computing as the first and last command of Allah.

(6) Preponderance of computation discourages critical analysis, creative thought, and the training of thinkers.

(7) Critical analysis and creative logical thinking are as important today as they were 100 or 2000 years ago—in view of the multitudinous applications of science to the human condition, perhaps even more important.

(8) Mathematics done by human minds should be cherished and fostered.

But it is not only mathematics that the computer vilifies and stifles. It poisons speech. WAN-LEE YIN in a private communication writes:

> Men of all past ages have reserved their better speech for their Gods and for posterity. It was not for the purpose of communicating with their fellow mortals that they invented writing and perfected language. Even in our century, Eliot could write
>
>> Since our concern was speech, and speech impelled us
>> To purify the dialect of the tribe
>> And urge the mind to aftersight and foresight . . ..

I still believe that science ennobles men. But for science to ennoble men, science must speak the language of men and not of the machine. Anyone who has had the misfortune to write his first computer program remembers the humiliation in conversing with a servant or master that insists on a language unworthy of the dullest of intelligences and the lowest of men. For of all human capacities language is traditionally considered the noblest, and hence the impoverishment and adulteration of language is the debasement of the dignity of man. Because freedom consists in an ever-present choice of defying the tyranny of necessity, and because language in its broadest sense as the total medium of expression is the sole means and avenue for that defiance, the abridgement of the structure and form of language by instituting arbitrary yet totally inflexible rules constitutes the most threatening violation of freedom. For what is at stake is not a matter of censoring certain categories of thoughts and ideas; it is rather the suppression of all spontaneous modes of expression and the deprivation of all human elements in speech and gesture for the mere sake of necessity and utility which the machine dictates whenever an individual makes a call and so long as the exchange lasts. The tyranny is total not because the power of the agent is overwhelming, but because the avenue of power is so strictly private and closed to spectators and because the agent himself is so utterly destitute of feeling and understanding (destitute even of sadistic pleasure which, though beastly, is akin to human) that the suffering and debasement of his subject can bear no witness nor meaning—for what is the use of defiance in face of an oppressor who understands not defiance? For such reasons, the tyranny of artificial intelligence represents ontologically a totally wasteful kind of domination in the structure and dynamics of power relations, a kind of domination compared to which even the Hell of Satan or the infamous union of torture chamber and pleasure harem in Marquis de Sade's imagination appears infinitely reasonable and surpassingly humane.

In a communion with machine's intelligence, man's consciousness voluntarily subjects itself to captivity in a barren cell enclosed by stubborn blocks of elementary logic and, like Eliot's spider, suspends its natural operations. It is like the return of the prisoner from the world of ideas to the chains and darkness of the Platonic cave, where he encounters not even the shadows of reality that were granted to him in his earlier captivity, but merely grotesque images, distorted reflections and drawn out echos of his deprived and depraved self. I always hold a low opinion of those obsessed with certain contrived games in which the artistic and communicative elements are totally absent—gadgets like Rubik's cube and video games which are lately in vogue—and I believe any individual so professionally well-disciplined as to entertain a lasting enjoyment in the companionship of machine intelligence has already sucked the Vampire's blood and is condemned to moonlighting as a disciple of the Satan of Bits and Bytes. The future Planet Earth may be ruled by such experts of machine intelligence, but the experts themselves would have to have been so impeccably schooled in the manners and speech of the lowest of slaves as to leave it quite uncertain whether there would be real masters. For once the medium becomes the message, those messages which were previously medium will devour genuine messages until the Vampire's blood runs in the veins of all messages.

In regard to the flamboyant, appalling failures of a computer which, had they not been corrected by human steadiness and human action, would have precipitated a monstrous nuclear war, ART BUCHWALD wrote in the *International Herald-Tribune* for 14/15 June 1980, ". . . war is too serious a business to be left to computers." The dangers potential in application of such sciences as high-energy physics and genetic alteration make them, also, businesses which if not too serious to be permitted at all are at least too serious to be left to computers. Indeed, I think, to renounce critical and creative use of human language and human reason is the greatest of the many present threats to the survival of mankind. Here the computer for "science" is but one of the Satanic instruments bent on the destruction of mind and man. As Mr. YIN puts it,

> The cult of artificial intelligence and agnostic science is not the source but merely a symptom or a catalyst of that larger process of disintegration and demise in which all living men are actors and spectators.

ACKNOWLEDGMENT

Ever since 1946, when for a time I had to take reluctant command of a battery of computing machines and its officers and crew, I have profited from discussions with friends and others regarding the use and misuse of computation in research on pure science. For their critical reading of parts or all of various draughts of the text and for their helpful suggestions I thank Messrs. CAPRIZ, LIEB, CHI-SING MAN, NUNZIATO, PODIO-GUIDUGLI, VILLAGGIO, and WAN-LEE YIN; of course I do not imply that any of them shares any of the views presented in the foregoing essay beyond those which are quoted from their writings.

Note

The first three sections of this lecture and a few sentences in other sections are taken *verbatim* from my lecture of 10 November 1979 in the Biozentrum at Basel: "The Role of Mathematics in Science as exemplified by the work of the Bernoullis and Euler", published in the *Verhandlungen der Naturforschen-den Gesellschaft in Basel* **91** (1981): 5–22. Most of the duplication has been excised in the condensed and revised version of that lecture which is reprinted above as Essay 10 in this volume.

The text printed here is based upon a lecture of the same title read on 7 February 1980 to the international conference "Scientific culture in the contemporary world", organized by *Scientia*, Milano. A version intermediate between that lecture and the final text has been published in translation, "Il calcolatore: rovina della scienza e minaccia per il genere umano", pages 37–65 of *La Nuova Ragione Scienza e Cultura nella Società Contemporanea*, Bologna, Scientia/Il Mulino, 1981.

42. OF ALL AND OF NONE (1964)

After the twelve engineers and three architects had received their diplomas of honorary doctorate from the hand of the President of Italy, Professor Clifford Truesdell on behalf of all those honored delivered the following brief words:

Praeses honoratissime, Rector magnifice, collegae docti, magistratus, clerici, et cives mediolanenses.

Cum mihi detur nomine omnium hoc die a vobis ornatorum, nomine nostrum omnium omnibus ex partibus terrae dicere, mihi oportet et opus est loqui lingua aut omnium aut nullius. Si quis contra dicat, plurimos esse qui familiares litterarum latinarum sint, illi respondeo, verba Ciceronis ab ore trans undas Romanis non cognitas nato et praecepto dicta, sonis non iam in terra olim Romana auditis, Romae vix intelligi. Talis est lingua latina, qualis est scientia. Scientia quoque, cuius servi aut fautores aut pontifices aut haruspices aut poetae sumus, res omnium et nullius est.

Tibi, Rector Magnifice, et Scholae magnae Polytechnicae in civitate mediolanensi maxime et humiliter gratias agimus pro honore humaniter donato. Tibi et Scholae et Italiae hoc centenario gratulamur. Maneat scientia, quae manibus in nostris et vestris stat, semper nostra et vestra. Res vestra et nostra nequit— aut forsitan necesse est in locum verbi «nequit» substituere «nequeat»—fieri hostis hominum.

Note

The foregoing is extracted, partly in English translation, from page 40 of *Cerimonie Celebrative del Centenario del Politecnico, 2–3–4 Aprile 1964*, Milano, Politecnico di Milano, 1964.

INDEX OF NAMES MENTIONED